NONLINEAR OCEAN DYNAMICS

NONLINEAR OCEAN DYNAMICS

Synthetic Aperture Radar

Maged Marghany
*Department of Informatics,
Faculty of Mathematics and Natural Sciences,
Universitas Syiah Kuala,
Darussalam, Banda
Aceh, Indonesia*

ELSEVIER

Elsevier
Radarweg 29, PO Box 211, 1000 AE Amsterdam, Netherlands
The Boulevard, Langford Lane, Kidlington, Oxford OX5 1GB, United Kingdom
50 Hampshire Street, 5th Floor, Cambridge, MA 02139, United States

Library of Congress Cataloging-in-Publication Data
A catalog record for this book is available from the Library of Congress

British Library Cataloguing-in-Publication Data
A catalogue record for this book is available from the British Library

ISBN: 978-0-12-820785-7

Publisher: Candice Janco
Acquisitions Editor: Louisa Munro
Editorial Project Manager: Emily Thomson
Production Project Manager: Sruthi Satheesh
Cover Designer: Mark Rogers

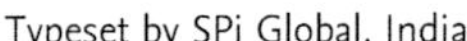
Typeset by SPi Global, India

Dedication

Dedicated to

my mother Faridah

and

COVID-19 Victims

and

Richard Feynman and Michio Kaku who taught me that a poor society is not one which does not have enough wealth, but one which does not respect science and scientists at its academic institutions and organizations.

Contents

9. Wavelet transform and particle swarm optimization algorithms for automatic detection of internal wave from synthetic aperture radar

10. Modeling wave pattern cycles using advanced interferometry altimeter satellite data

11. Multiobjective genetic algorithm for modeling Rossby wave and potential velocity patterns from altimeter satellite data

12. Nonlinear sea surface current mathematical and retrieving models in synthetic aperture radar

Preface

This book was completed when I ranked among the top two per cent of scientists in a global list compiled by the prestigious Stanford University. To date, various oceanography organizations and institutions have been unable to bridge the gap between modern physics, and physical and dynamic oceanography. Conventional physical oceanography data collections, even ocean dynamic model developments, cannot be made to comprehend countless dynamical ocean features, for instance, turbulent flow owing to wave-current interaction or front zone occurrence in coastal waters. There are many ambiguities in ocean dynamic feature studies such as ocean current boundary, internal waves, vorticity, and Rossby waves. Modern physics, as grounded in quantum mechanics and relativity theories, is needed in ocean dynamic studies to comprehend entirely the nonlinearity of ocean dynamic flows. In this view, the *synthetic aperture radar (SAR) imaging mechanism for nonlinear ocean dynamics* describes the analytical tool required to grasp the modern technology in radar imaging of sea surface, particularly microwave radar, as a foremost technology to understand analysis and applications in the field of ocean dynamics. Filling the gap between modern physics, quantum speculation, and depletions of radar imaging of sea surface dynamic features, this book includes technical details allied with the potentiality of synthetic aperture radar (SAR) and the key techniques exploited to extract the value-added information necessary, including ocean wave spectra, ocean surface flow, and wave-current interaction, from the varieties of SAR measurements.

In spite of the extensive development of microwave radar sensors, most institutions still focus on using conventional image processing algorithms or classic edge detection tools, and have isolated or ignored the modern physics behind the measurements.

The understanding of nonlinearity in ocean motion is introduced and overviewed in Chapter 1. It is fundamental to comprehension of the mechanism of synthetic aperture radar (SAR) for imaging ocean dynamics. Conversely, most efforts carried out using SAR imaging of the ocean are dedicated to ocean wave spectra, ship detection, internal wave, and ocean current, without recognizing the physical and dynamic nonlinearity of these phenomena.

It is impossible to avoid quantum mechanics when it comes to ocean dynamics. The science of ocean dynamics is predominantly proven based on quantum mechanics. In this regard, Chapter 2 demonstrates the quantization of ocean dynamics, and Chapter 3 introduces specific quantization of the synthetic aperture radar theories. Chapter 4 then implements the quantum mechanical theories to investigate the nonlinearity of sea surface backscatters. Consequently, radar signal and ocean surface dynamics cannot be tackled separately from quantum mechanics and relativity. Chapter 5 provides a comprehensive platform to understand the relativistic quantum mechanics for the nonlinearity of ocean dynamic features in radar images. This chapter also demonstrates the novel understanding of the radar imaging technique for relativistic quantum mechanics. In Chapter 6, novel formulas are introduced to aid understanding of the radar ocean wave imagined from the point of view of relativity theories. In this chapter, modulation transfer functions and formulas are modified by involving the relativity theories on understanding the radar image. In this sense, Chapter 7 implements the quantum nonlinearity for retrieving ocean wave spectra in the radar image. Consequently, Chapter 8 introduces a new technique for modeling wave refraction in polarimetric radar images.

Ocean dynamic features such as internal waves require accurate detection tools in radar images. In this regard, Chapter 9 delivers a novel technique based on wavelet transforms and the particle swarm optimization algorithm for automatic detection of internal waves in radar images. Chapters 10 and 11 implement radar altimetery sensors for modeling significant wave heights in large-scale areas based on interferometry altimeter satellite data, Rossby wave, and vorticity.

Chapter 12 demonstrates a technique based on the Doppler frequency shift to retrieve sea surface current from SAR images. In addition, it delivers a novel explanation of the Ekman spiral based on quantum mechanics theory. Subsequently, Chapter 13 delivers a novel technique to reconstruct a three-dimensional front zone using relativistic quantum mechanics. Quantum image processing as based on object detection is then implemented in the SAR image for automatic detection of turbulent flows. This is demonstrated in detail in Chapter 14. Lastly, Chapter 15 delivers a novel technique to investigate wave-current interaction using quantum interferometry. This technique delivers a framework for wave-current interactions in four dimensions (4D).

I wish to convey my appreciation to editorial project managers Emily Thomson, Ruby Smith, and Sruthi Satheesh and senior acquisitions editors Louisa Munro and Katie Hammon, who afforded the opportunity to publish this book. Without their intense commitment, this book would not have become such a precious piece of novel knowledge.

Maged Marghany

Department of Informatics, Faculty of Mathematics and Natural Sciences, Universitas Syiah Kuala, Darussalam, Banda Aceh, Indonesia

CHAPTER

1

Nonlinear ocean motion equations: Introduction and overview

1.1 Introduction

We have barely begun to fully comprehend the nonlinear ocean dynamic systems. In developing countries, the concept of nonlinear ocean dynamic systems is totally absent. It is restricted to the partial description of physical ocean parameters such as temperature, salinity, and currents, which are collected by conventional equipment such as current meters and CTD (conductivity, temperature, depth) instruments or "sondes." In spite of using sophisticated advanced devices of oceanography, the complex system of ocean dynamics is still not understood. In developing countries, marine biology, chemistry, or civil engineering graduates can be physical oceanography scientists. This state of affairs delivers such impoverished knowledge in understanding the ocean dynamics that there is no one, for instance, who can tell how the dynamic system of the Indian Ocean affect the trajectory movement of MH370 debris. It might be that a full understanding of dynamic ocean systems can solve the mystery of the "vanishing" of Flight MH370.

This chapter is devoted to a comprehensive introduction of the nonlinear dynamic ocean. In fact, it is key to understanding the mechanism of synthetic aperture radar (SAR) for imaging ocean dynamics. However, the majority of work carried out using SAR imaging of the ocean focuses on ocean wave spectra, ship detection, internal wave, and ocean current, without understanding the physical and dynamic nonlinearity of these phenomena.

1.2 What is meant by ocean dynamics?

Ocean dynamics delineate and describe the mobile flow of water within the oceans. In this regard, the wave-current interaction is the keystone of ocean dynamic features [1]. In fact, ocean dynamics describes and models the interaction between surface gravity waves and

https://doi.org/10.1016/B978-0-12-820785-7.00013-7

mean flow. Consequently, the surface gravity waves are the cornerstone to the synthetic aperture radar (SAR) ocean surface imaging mechanism, as will be discussed in this book. In fact, wave-current interaction implies an exchange of energy, so after the start of the interaction, both the waves and the mean flow are affected, which also impacts the mechanism of SAR imaging of the ocean surface.

Moreover, wave-current interaction is a vital mechanism for the occurrence of rogue waves, as proved by the Agulhas Current. Briefly, when a wave group runs into an opposing current, the waves in the group may heap up on top of each other, which promulgates into a rogue wave [2, 3].

This scenario is just a tiny part of the complicated ocean dynamic system. Through the water column, reaching from the surface to the ocean floor, there are fluctuation changes due to the variations of ocean temperature and motion fields. This dynamic fluctuation can be separated into three distinct layers: mixed (surface) layer, upper ocean (above the thermocline), and deep ocean. The mixed layer borders the surface and can fluctuate in thickness from 10 to 500 m. The temperature, salinity, density, and dissolved oxygen are the main physical properties of the mixed layer. In fact, temperature and salinity are the memory of the ocean in which reflecting a history of dynamic turbulence. Indeed, turbulence is extremely high in the mixed layer [4]. On the other hand, it turns out to be zero at the base of the mixed layer. Turbulence grows once more below the base of the mixed layer owing to shear instabilities. For instance, at extratropical latitudes, the mixed layer is deepest in late winter on account of surface cooling and winter storms, whereas it becomes extremely shallow in summer. Consequently, mixed layer fluctuations are governed by turbulent mixing along with Ekman pumping, trades with the superimposing atmosphere, and horizontal advection [3–5].

Generally, the mixed layer dynamic forces are extremely complex; nevertheless, in particular, zones more or less of its indications are possible. The wind is the main force, which can derive horizontal transport in the mixed layer; which is mainly pronounced by Ekman Layer. In this regard, vertical diffusion of momentum balances the Coriolis effect and wind stress [5]. In fact, the Ekman transport is expanded on geostrophic flow that is mixed with horizontal gradients of density [1, 4]. Therefore, horizontal convergences and divergences within the mixed layer owing, for instance, to Ekman transport convergence enforce a constraint that the ocean beneath the mixed layer must transfer fluid particles vertically. Nevertheless, one of the requirements of the geostrophic association is that the amount of horizontal motion must greatly surpass the magnitude of vertical motion. As a result, the weak vertical velocities amalgamated with Ekman transport convergence (measured in meters per day) create the horizontal motion with speeds of $10\,\mathrm{cm\,s^{-1}}$ or more. The mathematical relationship between the vertical and horizontal velocities can be derived by articulating the notion of conservation of angular momentum for fluid on a rotating sphere, as will be discussed in the following sections [1, 4, 5].

Although some regions of the deep ocean are known to have significant recirculations, the deep ocean is both cold and dark with commonly little motion. The deep ocean is supplied with water from the upper ocean in certain small geographical zones, for instance, the subpolar North Atlantic Ocean and numerous sinking zones close to the Antarctic. Due to the weak flow of water to the deep ocean, the regular residence time of water in the deep ocean is measured in hundreds of years. In general, this layer is dominated by a quite weak mixing due to the fixed correlation between hydrostatic and geostrophic motions.

1.3 What is meant by nonlinear?

In general terms, "nonlinear" denotes a state that has inconsistent cause and effects. The cause represents the force that induces the dynamic motions, i.e., drifts. For instance, the wind is the cause of both wave propagation and current movements. In this sense, complicated nonlinear phenomena, for instance, can result owing to ocean wave-current interactions. Consequently there are many nonlinear effects required to understand this complicated nonlinear ocean dynamic system. In this regard, the nonlinear ocean system differs from nonlinear ocean dynamics [4]. The nonlinear ocean system represents the ocean system in which the output is not proportional to changes in the input. Wave-current interactions, for instance, are not proportional to the input wind stress. Indeed, subsequent to, the beginning of the interaction, both the waves and the mean flow are affected due to the interaction, implying an exchange of energy. The changing of the ocean parameters and variables over time is known as an ocean dynamic system, in which they may appear chaotic, unpredictable, counterintuitive, or contrasting, for instance, with much simpler linear wave propagation descriptions [5].

As a rule, the behavior of a nonlinear ocean dynamic is described in mathematics by a nonlinear system of equations, which is a set of synchronized computations in which the unknowns (or the unknown functions in the case of differential ocean dynamic equations) appear as variables of a polynomial of degree higher than one or in the contention of a function, which is not a polynomial of degree one. In other words, in a nonlinear ocean dynamic of equations, the equation(s) to be solved cannot be written as a linear combination of the unknown ocean parameters or functions that appear in them. Ocean dynamics can be defined as nonlinear, regardless of whether known linear functions appear in the motion equations. In particular, a differential motion equation is linear, if it is linear in terms of the unknown function and its derivatives, even if it is nonlinear in terms of the other variables appearing in it. Conversely, nonlinear ocean dynamical equations are difficult to solve. Therefore, nonlinear ocean dynamics are commonly approximated by linear equations. These nonlinear works emanate to approximate truth and roughly span for the input parameters, but some interesting phenomena such as solitons, chaos, and singularities are hidden by linearization [4, 5].

It follows that specific phases of the dynamic behavior of a nonlinear ocean can appear to be counterintuitive, unpredictable, or even chaotic. However, although such chaotic ocean behavior may be seen as a random behavior, it is in fact not random. For instance, particular features of the wave-current interactions appear to be chaotic, where nominal changes in one part of the wave-current interaction produce complex effects throughout. This nonlinearity is one of the reasons why accurate long-term wave pattern forecasts are impossible even with advanced technology and computer models.

1.4 Classification of ocean dynamic flows

Let us assume that the flow velocity u is a vector field, which is expressed as [6]:

$$u = u(x, t) \tag{1.1}$$

Eq. (1.1) demonstrates the velocity of the fluid parcel at position x and time. Therefore, the length of the flow velocity vector q can be given by:

$$q = \|u\| \tag{1.2}$$

Conversely, Eq. (1.2) shows a scalar field. In this view, the flow velocity of a fluid efficiently describes everything about the motion of the fluid. Various physical properties of a fluid, therefore, can be articulated mathematically in terms of the flow velocity. The fluid dynamic motions can be categorized into steady flow, incompressible flow, irrotational flow, and vorticity [7–9].

1.4.1 Steady flow

In steady flow, the flow of a fluid is said to be *steady* if u does not vary with time. That is if:

$$\frac{du}{dt} = 0 \tag{1.3}$$

The simple Eq. (1.3) refers to the condition where the fluid properties at a point in the system do not change over time. In other words, a flow is not a function of time. However, the time-dependent (t) flow is known as unsteady or transient, which is mathematically described by [4, 9, 10]:

$$u = u(x, t) \tag{1.4}$$

Therefore, both steady and unsteady flows can rely on a particular frame of reference. For instance, in a frame of reference that is stationary with respect to a background flow, the flow is unsteady [6, 11]. On this understanding, the turbulent flows are unsteady by definition. The laminar flow over a sphere, however, is steady in the frame of reference that is stationary with respect to the sphere. Steady flows, consequently, are often more tractable than otherwise similar unsteady flows [11]. The governing equations of a steady problem have one dimension fewer (time) than the governing equations of the same problem without taking advantage of the steadiness of the flow field [6, 10, 12].

In nature, a steady flow does not seem to occur. In this perspective, it is not a trivial ration of the confronted stages. Nonetheless, by limiting the time over which the path of flow is premeditated, the steady, or quasisteady, flow guesstimate may be supposed for numerous phases for instance laminar flow turns into turbulent flow. This is demonstrated in Fig. 1.1.

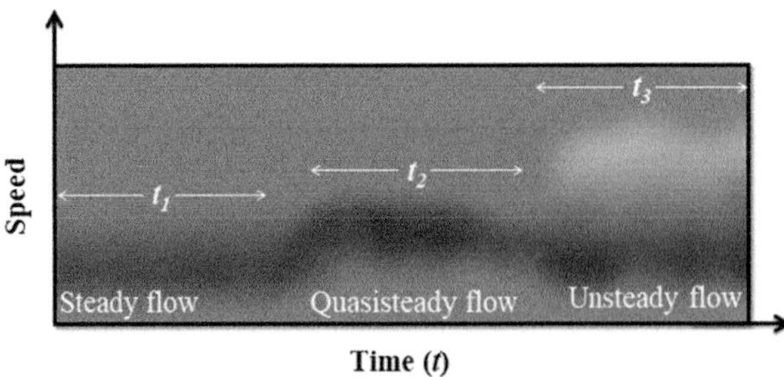

FIG. 1.1 Three fluid states of steady flow, quasisteady flow, and unsteady flow.

In fact, the steady flow is shown during time t_1, the quasisteady flow occurs during time t_2, while unsteady flow develops during time t_3.

During the quasisteady flow, there is a fluctuating, turbulent component superimposed on a mean value equal to that of the steady flow phase. In other words, the mean value is unchanging even though there are fluctuations. Let us assume that u' is turbulent fluctuation velocity, which is generated during quasisteady flow u and mathematically is expressed by [6, 13–19]:

$$u' = \overline{u} - u \tag{1.5}$$

In Eq. (1.5), the turbulent fluctuation is the difference between the mean flow $\overline{u}$ and quasisteady flow. The mean flow $\overline{u}$ is a function of the averaging time τ (Fig. 1.2) and is estimated as:

$$\overline{u} = \tau^{-1} \int_{t-0.5\tau}^{t+0.5\tau} udt \tag{1.6}$$

For oceanic flow the interval time $t-0.5\tau$ to $t+0.5\tau$ could be of the order of hours or perhaps days. However, the time period when experimenting on the consequence of fluid flow on a parcel in a wind tunnel or water basin may be of the order of seconds or even tenths of a second, Unsteady flow is dominated by a tendency for even the mean velocity to change, i.e., $\frac{du}{dt} \neq 0$. In this context, this mechanism leads to turbulent flow, which is unsteady by definition [20]. This turbulent flow has a great effect on radar signal, as will be discussed in the coming chapters.

1.4.2 Incompressible flow

In ocean dynamics or, more generally, scale dynamics, incompressible flow refers to streamflow in which the water density ρ is constant within the fluid parcel that moves with the flow velocity. In other words, the divergence of the incompressible flow (∇) is zero. In this regard, the major constraint for incompressible flow is that the density ρ is constant within a

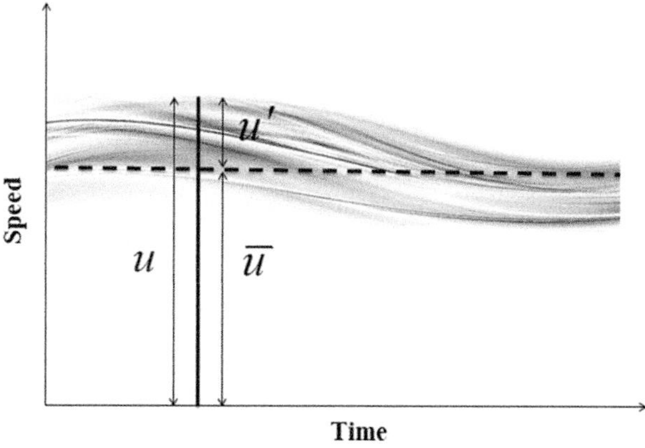

FIG. 1.2 Fluctuating mean speed flow vs time for quasistatic flow.

minor component volume, dV, which moves at the current velocity u. Mathematically, this restraint implies that the physical derivative of the density must vanish to confirm incompressible flow [5, 6, 10, 21]. To implement this constraint, let us assume that u is the mass, which can be computed by a volume integral as follows:

$$m = \iiint_V \rho dV. \tag{1.7}$$

In ocean dynamics, the continuity equation states that the rate at which mass enters a system is equal to the rate at which mass leaves the system plus the accumulation of mass within the system [4, 10, 14, 22]. Substantially, this declaration necessitates that mass is neither formed nor reduced in the constraint volume, and can be decoded into the integral form of the continuity equation:

$$\frac{\partial}{\partial t}\iiint_V \rho dV = \oiint_S \rho \vec{u} \cdot dS \tag{1.8}$$

The left-hand side of Eq. (1.8) represents the rate of growth of mass within the volume and encompasses a triple integral over the constraint volume. On the contrary, the right-hand side of Eq. (1.8) encompasses the integration of the surface of the constraint volume of mass convected into the system. Therefore, the volume mass moves into the system, which is accounted as positive, and since the normal flow vector to the surface, as opposed to the sense of movement into the system the term is invalid.

$$\iiint_V \frac{\partial \rho}{\partial t} dV = \iiint_V \left(\nabla \cdot \vec{J}\right) dV, \tag{1.9}$$

where $\vec{J}$ is the mass flux, which represents $\rho \vec{u}$.

The differential form of the continuity equation can be expressed by the divergence theorem as:

$$\frac{\partial \rho}{\partial t} + \nabla \cdot \left(\rho \vec{u}\right) = 0 \tag{1.10}$$

Eq. (1.10) can be written as:

$$\frac{\partial \rho}{\partial t} = -\nabla \cdot \left(\rho \vec{u}\right) = -\nabla \cdot \vec{J} \tag{1.11}$$

The partial derivative equation of the density ρ as a function of time requisite, which is not neglect to declare incompressible flow. In other words, the partial derivative of the density with respect to time refers to the rate of change within a constrained volume of fixed positions. By permitting the partial time derivative of the density to be nonzero, the flow is not restricted to incompressible fluids. In fact, the density can change as observed from a fixed position as fluid flows through the control volume. This approach maintains generality; and not requiring that the partial time derivative of the density vanish illustrates that compressible fluids can still undergo incompressible flow. What interests us is the change in density of a control volume that moves along with the flow velocity, $\vec{u}$. The flux $\vec{J}$ is related to the flow velocity through the following function:

$$\vec{J} = \rho \vec{u} \tag{1.12}$$

In this regard, the conservation of mass implies that:

$$\frac{\partial \rho}{\partial t} + \nabla \cdot \left(\rho \vec{u}\right) = \frac{\partial \rho}{\partial t} + \nabla \rho \cdot \vec{u} + \rho\left(\nabla \cdot \vec{u}\right) = 0. \tag{1.13}$$

Using the chain rule, the total derivative of density is given by:

$$\frac{d\rho}{\partial t} = \frac{\partial \rho}{\partial t} + \frac{\partial \rho}{\partial x}\frac{dx}{dt} + \frac{\partial \rho}{\partial y}\frac{dy}{dt} + \frac{\partial \rho}{\partial z}\frac{dz}{dt}. \tag{1.14}$$

The control volume in Eq. (1.14) is moving at the same rate of $\frac{dx}{dt}$, $\frac{dy}{dt}$, and $\frac{dz}{dt} = \vec{u}$. On this understanding, the material derivative can be used to simplify this expression as:

$$\frac{D\rho}{Dt} = \frac{d\rho}{\partial t} + \nabla \rho \cdot \vec{u}\,. \tag{1.15}$$

Eq. (1.15) demonstrates that mass flux changes with time as it flows along its trajectory. The term $\frac{d\rho}{\partial t}$ describes how the density of the material element changes with time, which is also known as the unsteady term. Therefore, from the continuity equation, it follows that:

$$\frac{D\rho}{Dt} = -\rho\left(\nabla \cdot \vec{u}\right). \tag{1.16}$$

Eq. (1.16) reveals that the term $\nabla \rho \cdot \vec{u}$ describes the changes in the density as the material element moves from one point to another. This is the advection term (convection term for scalar field). For flow to be incompressible, the sum of these terms should be zero. A change in the density over time would imply that the fluid had either compressed or expanded or that the mass contained in a constant volume, dV, had changed, neither of which are possible. Consequently, the material derivative of the density must have vanished, and equivalently (for nonzero density), so must the divergence of the flow velocity:

$$\nabla \cdot \vec{u} = 0. \tag{1.17}$$

Eq. (1.17) reveals the conservation of mass and the constraint that the density within a moving volume of fluid remainders continuous. It has been shown that an equivalent circumstance prerequisite for incompressible flow is that the divergence of the flow velocity vanishes.

In general, the concept of incompressibility is allied with pressure and specifies that changes in pressure accompanied by the motion of the fluid produce no changes in density. Note that a vertically stratified fluid may be consistent with the condition of incompressibility provided the fluid density changes whenever a parcel moves up or down. In this sense, seawater is not exactly incompressible, but this assumption is a good approximation for many applications because the volume changes are small. For instance, a change of pressure corresponding to a change of depth of 1000 m would change the volume of a sample of average seawater by less than 0.5%; and there are few states in ocean currents where depth (and therefore pressure) changes as large as this occur along the flow path [4].

1.4.3 Irrotational flow

Irrotational flow can be defined as flow with zero vorticity (Fig. 1.3). In other words, it is the net rate of change of angular velocity in all directions, which is zero for the flow. Mathematically, flow is irrotational when the curl of the velocity vector is zero:

$$\nabla \times \vec{u} = 0. \tag{1.18}$$

In this regard, velocity is a vector and if the vector operator is applied, it is called *curl* to the velocity; if the curl is set as equal to zero, then Eq. (1.18) is equivalent to making the flow irrotational. Consequently, irrotational flow is flow in which all the tiny bits of fluid are moving along and translating and going around obstacles, etc., without every rotating about their own infinitesimal centers of gravity. Irrotational flow can only persist if there is no viscosity and all real fluids have viscosity [6].

On this understanding, Eq. (1.18) also demonstrates that the curl of the gradient of anything will always be zero, i.e., the gradient is another vector operator. In this regard, if the curl of $\vec{u}$ is zero and the curl of the gradient of anything is zero, then the velocity must be equal to the gradient of something in the circumstance of irrotational flow. In other words, the velocity $\vec{u}$ equals the gradient of the potential function. In this context, the entire quantity of the fluid dynamics well-known as potential flow. Let us assume the potential flow is ϕ, then the $\vec{u}$ can be mathematically expressed as:

$$\vec{u} = \nabla \phi \tag{1.19}$$

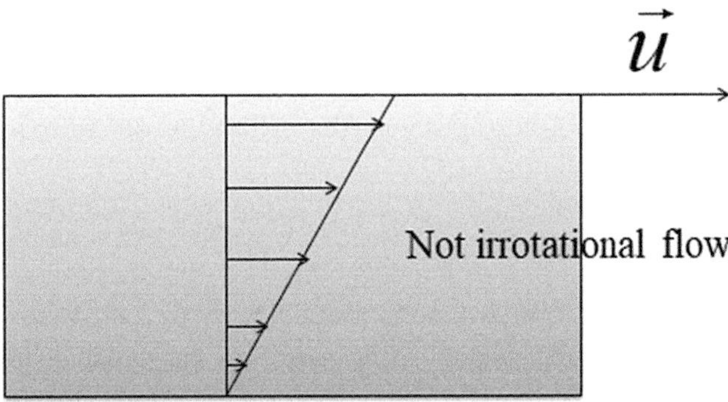

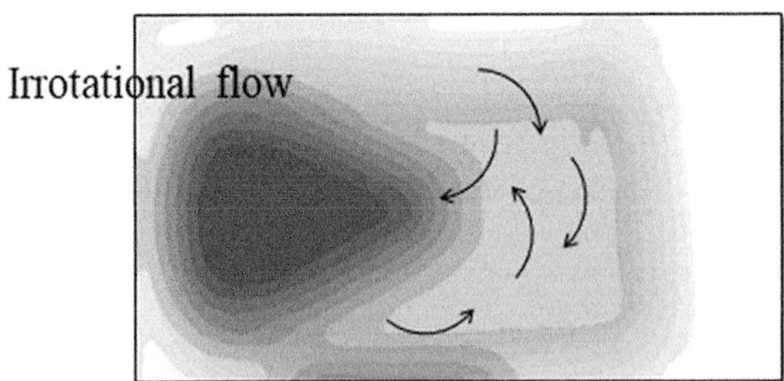

FIG. 1.3 Example of irrotational flow.

This raises the significant question of whether a real fluid can be irrotational. Irrotational flows occur even when they have viscous dissipation. Nonetheless, it is the case that most flows in the vicinity of a wall are rotational because a wall could act as the source or sink of vorticity. For instance, Couette flow between two concentric circular cylinders can be completely irrotational. Therefore, viscous dissipation generates throughout the flow, yet it is irrotational.

Usually, if a flow is irrotational, viscous stresses can occur, but they must satisfy one condition: the divergence of these viscous stresses must vanish. In this regard, the divergence of the viscous stresses evaluated at a point is equivalent to the shear force per unit volume at that point. In this context, the divergence of the velocity is equal to zero if the fluid has a constant density (i.e., is incompressible). The divergence is yet another vector operator courtesy of the mathematicians [7, 22]. If the gradient of the potential function is substituted in place of velocity, the divergence of the gradient of a potential function equals zero for incompressible and irrotational flow. In this circumstance, the divergence of gradient simplifies to the Laplacian and the flow must equal zero as expressed by:

$$\nabla \times (-\nabla \phi) = 0 \tag{1.20}$$

Whenever a velocity potential function exists, then it must be an irrotational flow so they are equivalent to each other. As we are considering ideal fluid, it must follow that the continuity equation is as follows:

$$\frac{\partial u}{\partial x} + \frac{\partial v}{\partial y} + \frac{\partial w}{\partial z} = 0 \tag{1.21}$$

Yielding:

$$\left(\frac{\partial^2 \phi}{\partial x^2}\right) + \left(\frac{\partial^2 \phi}{\partial y^2}\right) + \left(\frac{\partial^2 \phi}{\partial z^2}\right) = 0 \tag{1.22}$$

Eq. (1.22) can be expressed in vector form as:

$$\nabla^2 \phi = 0 \tag{1.23}$$

Eq. (1.23) is known as the Laplace equation. Therefore, any function ϕ that contains the Laplace equation is a possible illustration of irrational flow since there is an infinite quantity of solutions of the Laplace equation, in which each one satisfies specific flow boundaries. However, the main problem is the selection of the appropriate function for a certain flow occurrence. Since ϕ seems to be the first power, which is a linear equation, as a result, the entirety of dual solutions is also a solution [7, 23].

In rectangular Cartesian coordinates, the irrotational flow is calculated as:

$$\frac{\partial w}{\partial y} - \frac{\partial v}{\partial z} = 0 \tag{1.24}$$

$$\frac{\partial u}{\partial z} - \frac{\partial w}{\partial x} = 0 \tag{1.25}$$

$$\frac{\partial v}{\partial x} - \frac{\partial u}{\partial y} = 0 \tag{1.26}$$

Likewise, in cylindrical coordinates, the circumstances for irrotationality are:

$$\frac{1}{r}\frac{\partial w}{\partial \theta} - \frac{\partial v}{\partial z} = 0 \tag{1.27}$$

$$\frac{\partial u}{\partial z} - \frac{\partial w}{\partial r} = 0 \tag{1.28}$$

$$\frac{\partial rv}{\partial r} - \frac{\partial u}{\partial \theta} = 0 \tag{1.29}$$

Therefore, the vanishing of viscous stresses, or of viscosity for that matter, is not a necessary condition for a flow to be irrotational. In most cases of real fluid flow, the regions near walls have viscous dissipation and are not irrotational. Out in the main flow, it is a pretty good approximation that the flow is irrotational [4, 5, 15, 19, 21]. Needless to say, flow without vorticity is called irrotational flow (Fig. 1.4).

1.4.4 Vorticity

The simple definition of what triggers vorticity is a pseudovector field that describes the local spinning motion of an entity near a particular point. In other words, it represents the tendency of a particle to rotate [4, 5, 24], as would be seen by an observer positioned at that point and moving along with the flow. In this regard, one can express the rate of translation vector mathematically by the velocity vector $\vec{U}$ as:

$$\vec{U} = u\vec{i} + v\vec{j} + w\vec{k} \tag{1.30}$$

When it comes to articulate the rate of rotation of a fluid parcel, it becomes a challenging task. In fact, a fluid parcel converts and distorts as it rotates. Let us assume an initially rectangular fluid parcel that commences to rotate [7, 16, 19, 25], while each line of the rectangular has a different angular velocity than the others (Fig. 1.5). In this view, the vorticity vector

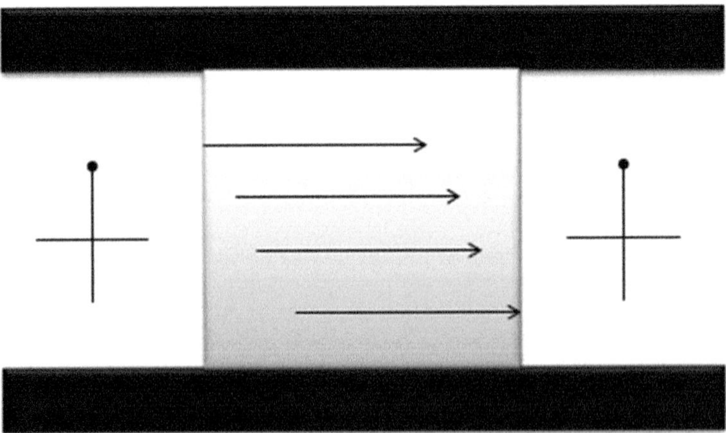

FIG. 1.4 Motion with no vorticity.

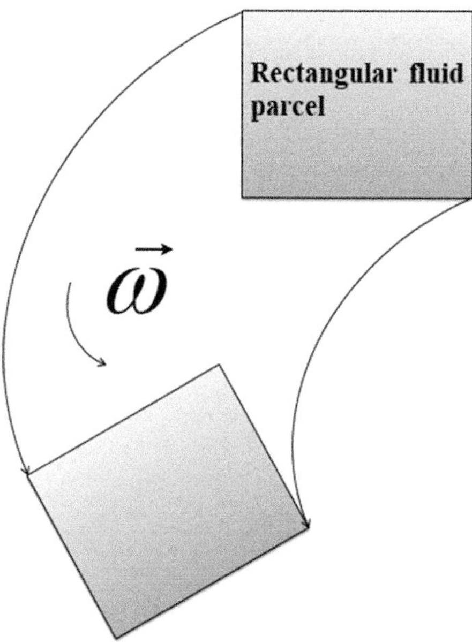

FIG. 1.5 Rotation of rectangular fluid parcel.

would be twice the mean angular velocity vector $\vec{\omega}$ of those parcels, which are relative to their center of mass and oriented according to the right-hand rule. In this regard, the vorticity is a pseudovector field $\vec{\omega}$ defined as the curl of the flow velocity $\vec{u}$, which is expressed mathematically as:

$$\vec{\omega} \equiv \nabla \times \vec{u}, \tag{1.31}$$

In rectangular Cartesian coordinates, the vorticity flow is calculated as:

$$\begin{aligned} \vec{\omega} \equiv \nabla \times \vec{u} &= \left(\frac{\partial}{\partial x} \frac{\partial}{\partial y} \frac{\partial}{\partial z} \right) \times \left(u_x \; u_y \; u_z \right) \\ &= \left(\frac{\partial u_z}{\partial y} - \frac{\partial u_y}{\partial z} \;\; \frac{\partial u_x}{\partial z} - \frac{\partial u_z}{\partial x} \;\; \frac{\partial u_y}{\partial x} - \frac{\partial u_x}{\partial y} \right) \end{aligned} \tag{1.32}$$

Eq. (1.32) indicates how the velocity vector changes when one moves by a tiny distance in a direction perpendicular to vorticity. On the contrary, the vorticity vector must be parallel to the z-axis as long as there is no z component [1, 4, 6, 10, 26]. In this circumstance, the 2-D flow is no longer a function of the z, and therefore can be expressed as a scalar field multiplied by a constant unit vector $\vec{z}$:

$$\vec{\omega} = \nabla \times \vec{u} = \left(\frac{\partial u_y}{\partial x} - \frac{\partial u_x}{\partial y} \right) \vec{z}\,. \tag{1.33}$$

The rotation vector $\vec{\omega}$ can be mathematically expressed as:

$$\vec{\omega} = \frac{1}{2} \left[\left(\frac{\partial w}{\partial y} - \frac{\partial v}{\partial z} \right) \vec{i} + \left(\frac{\partial u}{\partial z} - \frac{\partial w}{\partial x} \right) \vec{j} + \left(\frac{\partial v}{\partial x} - \frac{\partial u}{\partial y} \right) \vec{k} \right] \tag{1.34}$$

Eq. (1.34) reveals that the vorticity vector is twice the angular velocity. Let us get rid of $\frac{1}{2}$ by replacing with ς, then the simplification of Eq. (1.34) is given by:

$$\vec{\varsigma}=\vec{\nabla}\times\vec{U} \tag{1.35}$$

To this end, Eq. (1.35) demonstrates that any fluid parcel in that rotation has nonzero vorticity (Fig. 1.6). In addition, any fluid parcel that subjugates the irrotational has zero vorticity, which means the fluid parcel is not rotating (Fig. 1.7).

Needless to say, circulation and vorticity are the dual major quantities of rotation in a fluid. Consequently, circulation, which is a scalar integral quantity, is a macroscopic measure of rotation for a finite area of the fluid. Vorticity, conversely, is a vector field that provides a microscopic quantity of the rotation at any point in the fluid.

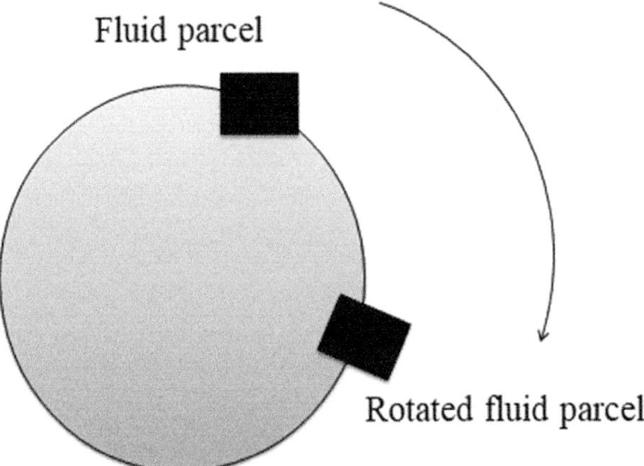

FIG. 1.6 Vorticity due to the rotation of a fluid parcel.

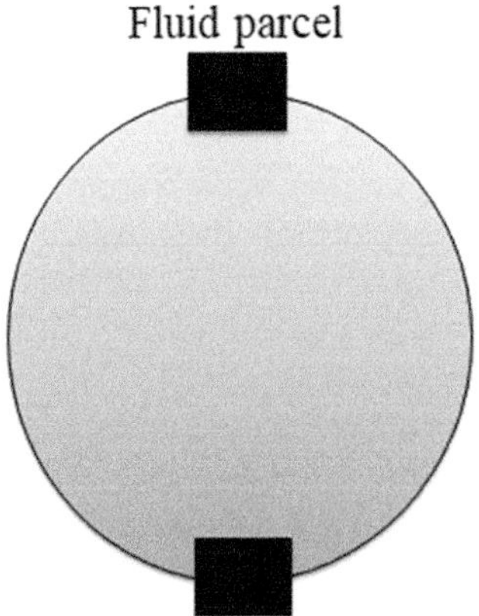

FIG. 1.7 Zero vorticity due to nonrotation of a fluid parcel.

1.5 Ocean dynamic circulation

The circulation around a parcel of fluid ξ bounded by a closed curve may be defined as the line integral of the fluid velocity over the closed curve [5, 6, 26]. Circulation, on the other hand, is a scalar quantity defined as the line integral of the velocity field along a closed contour (Fig. 1.8). Let us assume that $\vec{V}$ is the velocity along a peripheral line segment $d\vec{l}$ (Fig. 1.8) and the circulation $\mathbb{C}$ can be estimated by taking the integral over the closed curve, which is formulated as:

$$\mathbb{C} = \int \vec{V} \cdot d\vec{l} = \int V \cos\theta dl \tag{1.36}$$

In the circumstance of 2-D circulation flow, $\mathbb{C}$ is expressed mathematically as:

$$\vec{V} = u\vec{i} + v\vec{j} \tag{1.37}$$

In this regard, the peripheral line segment $d\vec{l}$ is determined from:

$$d\vec{l} = dx\vec{i} + dy\vec{j} \tag{1.38}$$

The combination of Eqs. (1.37) and (1.38) results in:

$$\vec{V} \cdot d\vec{l} = \left(u\vec{i} + v\vec{j}\right) \cdot \left(dx\vec{i} + dy\vec{j}\right) \tag{1.39}$$

Eq. (1.39) results in:

$$\mathbb{C} = \int \vec{V} \cdot d\vec{l} = \int (udx + vdy) \tag{1.40}$$

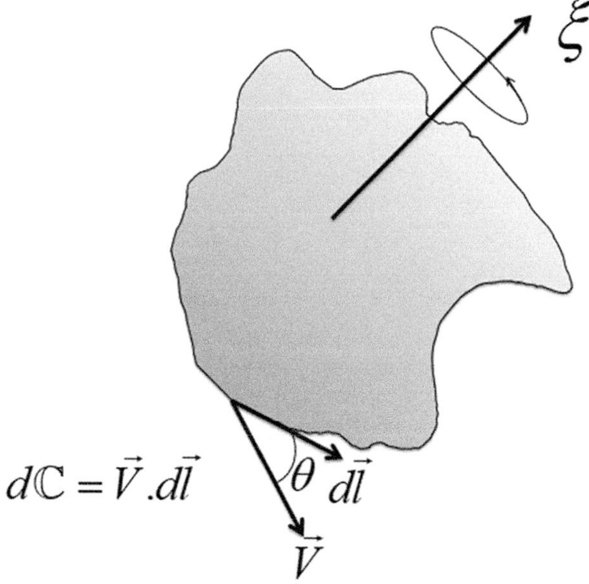

FIG. 1.8 The circulation around a parcel of fluid.

The significant question is: can Eq. (1.40) be used to determine the circulation of small parcel? In order to answer this question, let us consider an elemental rectangular section of fluid moving in the *x*-*y* plane [27], having a cross-section *dx dy* and peripheral velocity, as demonstrated in Fig. 1.9. In this regard, the rate change of velocity in the *x-axis* and *y-axis*, respectively, is presented by Δu and Δv. Based on a truncated Taylor expansion, the velocity changes can be expressed mathematically in terms of the appropriate velocity gradients:

$$\Delta u = \frac{\partial u}{\partial y} dy \tag{1.41}$$

$$\Delta v = \frac{\partial v}{\partial x} dx \tag{1.42}$$

Let us substitute Eqs. (1.41) and (1.42) into Eq. (1.40), which results in [4, 21–25]:

$$\begin{aligned} d\mathbb{C} &= udx + \left[v + \frac{\partial v}{\partial x} dx\right] dy - \left[u + \frac{\partial u}{\partial y} dy\right] dx - vdy \\ &= \frac{\partial v}{\partial x} dxdy - \frac{\partial u}{\partial y} dydx \end{aligned} \tag{1.43}$$

Then Eq. (1.43) can be written as:

$$d\mathbb{C} = \left(\frac{\partial v}{\partial x} - \frac{\partial u}{\partial y}\right) dxdy \tag{1.44}$$

Eq. (1.44) reveals the constitutionally of vorticity in the direction perpendicular to the *x*-*y* plane. This raises a significant question: Can vorticity be thought of as the circulation per unit area? Let us recall Fig. 1.8; the circulation over the closed curve becomes a surface integral along with the changes of *dx* and *dy*, respectively. On this understanding, the scientific explanation of the circulation can mathematically be written as:

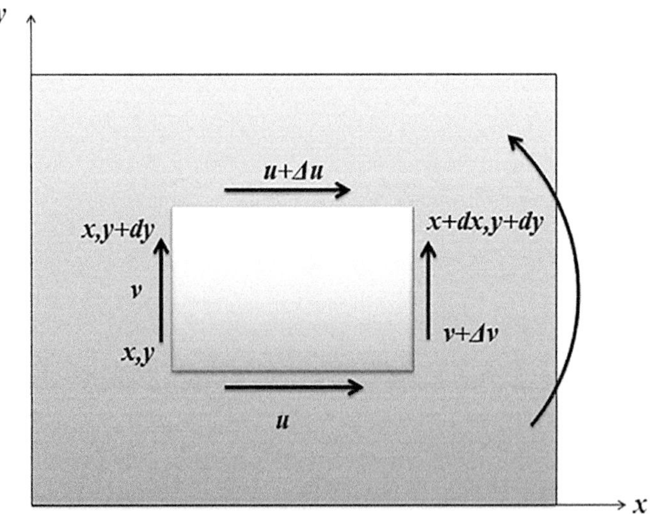

FIG. 1.9 Circulation in an element rectangle.

$$\mathbb{C} = \iint \left(\frac{\partial v}{\partial x} - \frac{\partial u}{\partial y} \right) dxdy \tag{1.45}$$

The magnitude bounded in parentheses is evidently the z-component of the vorticity. What we have revealed is that the line integral of the velocity around an obstructed curve is the surface integral of the vorticity over the encircled area, which is just a form of Stokes theorem [4, 5, 17, 28]. In this regard, the three-dimensional circulation is then expressed as:

$$\mathbb{C} = \int_L \underline{\vec{V}} \cdot d\vec{l} = \iint_S \left(\vec{\nabla} \times \vec{V} \right) \cdot d\vec{s} = \xi_{avg} S \tag{1.46}$$

Eq. (1.46) reveals that the vorticity is averaged over the area S. In this regard, vorticity can be thought of as the circulation per unit area. Needless to say, Eq. (1.46) is the circulation at any point along the vortex boundary, which is delivered by the product of the average vorticity through that unit area and the cross-sectional area [8, 9, 18, 29].

1.6 What is the difference between circulation and vorticity?

Both circulation and *vorticity* are covered when fluid dynamics are studied and are commonly puzzling for many. Both vorticity and circulation are main quantities of rotation in a fluid. In this view, the rotation vector of a fluid equals half of the curl of its velocity vector. Therefore, the vorticity is obtained by multiplying it by two. In point of fact, vorticity is the angular velocity of a particle with respect to another particle. It is a vector quantity, as opposed to circulation, which is a scalar quantity and provides the microscopic idea of rotation at any point of the fluid. Let us assume that a fluid disc of radius a, as shown in Fig. 1.10, is rotating like a rigid body at an angular velocity of Ωs^{-1}. Consequently, the angular velocity of the solid rotating is the same everywhere, which is formulated as:

$$\xi = 2\omega = 2\Omega \tag{1.47}$$

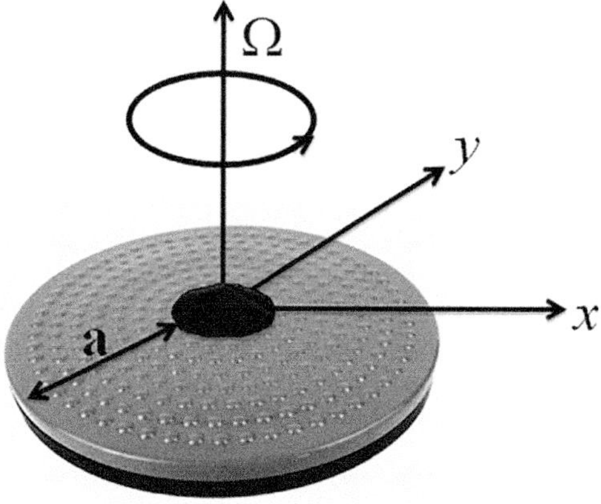

FIG. 1.10 Rotation of a fluid disc.

Using Eq. (1.47) in Eq. (1.36), then:

$$\mathbb{C} = \int \vec{V} \cdot d\vec{l} = \int (\Omega r) dl = (\Omega r)(2\pi r) = 2\Omega(\pi r^2) \tag{1.48}$$

Eq. (1.48) reveals that the maximum value of circulation occurs where $r = a$, i.e., at the rim:

$$\mathbb{C}_{\max} = 2\Omega(\pi a^2) = \xi A \tag{1.49}$$

where A is the disc area.

In this regard, the circulation is once again computed by the product of vorticity and disc area. Generally, the relationship between circulation and vorticity is continuously real for rigid body rotation. These statements illustrate both quantities of vorticity and circulation, as for microscopic and macroscopic, respectively. Both these quantities are basically a gauge of the rotation of the fluid flow [19–30].

1.7 Primitive equation of ocean dynamics

Ocean dynamic is concerned with the forces acting on the ocean waters and with the motions arise. In particular circumstances, the system forces, for instance, winds can induce motions, which are in balance to no resulting force uses. This is the scenario addressed by Newton's First Law of Motion. In other circumstances, there is a resultant force and acceleration ensues. The associations between force and acceleration being governed by Newton's Second Law. Newton's laws of fluids are expressed in the form of the equations of conservation of momentum of a fluid parcel or control volume [4, 8, 14, 19, 21, 31].

1.7.1 Continuity equations

The governing equations can be expressed in both integral and differential forms. The integral form is useful for large-scale control volume analysis, whereas the differential form is useful for relatively small-scale point analysis [1, 6, 32]. Let us express the mass conservation mathematically as:

$$\sum_{S} \rho \vec{U} \cdot A = -\int_{V} \frac{\partial \rho}{\partial t} d\forall \tag{1.50}$$

Mass conservation in Eq. (1.50) reveals the balance between outflow mass across an area and the rate of mass reduction across a volume. The volume change is associated with the rate change of water density ρ over time t. Let us consider a cubic element, oriented so that its sides are perpendicular to the axes x, y, and z, respectively (Fig. 1.11). In this regard, the parcel of fluid is assumed to be infinitesimally small, in which the flow is assumed to be approximately in one dimensional through each face. In this context, the mass flux terms arise on all six faces of the cube, which involves three inlets, and three outlets. The mass flux on x, y, and z faces, respectively, can be given by:

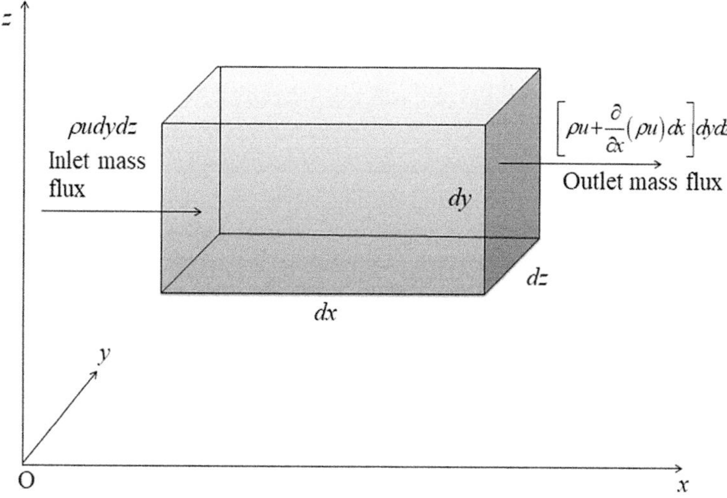

FIG. 1.11 Continuity of volume components of flow in the x, y, and z directions.

$$x_{flux} = \left[\rho u + \frac{\partial}{\partial x}(\rho u)dx\right]dydz|_{outflux} - \rho udydz|_{influx}$$
$$= \frac{\partial}{\partial x}(\rho u)dxdydz = \frac{\partial}{\partial x}(\rho u)d\mathcal{V} \tag{1.51}$$

$$y_{flux} = \frac{\partial}{\partial y}(\rho v)dxdydz = \frac{\partial}{\partial y}(\rho v)d\mathcal{V} \tag{1.52}$$

$$z_{flux} = \frac{\partial}{\partial z}(\rho w)dxdydz = \frac{\partial}{\partial y}(\rho w)d\mathcal{V} \tag{1.53}$$

Combining the above expressions yields the desired result:

$$\left[\frac{\partial \rho}{\partial t} + \frac{\partial}{\partial x}(\rho u) + \frac{\partial}{\partial y}(\rho v) + \frac{\partial}{\partial z}(\rho w)\right]d\mathcal{V} = 0 \tag{1.54}$$

$$\frac{\partial \rho}{\partial t} + \frac{\partial}{\partial x}(\rho u) + \frac{\partial}{\partial y}(\rho v) + \frac{\partial}{\partial z}(\rho w) = 0 \tag{1.55}$$

$$\frac{\partial \rho}{\partial t} + \nabla \cdot (\rho \mathcal{V}) = 0 \tag{1.56}$$

The above equations reveal that the total net mass outflux must balance the rate of decrease of mass within the cross volume change $d\mathcal{V}$, which is satisfied with the condition of $-\frac{\partial \rho}{\partial t}dxdydz = -\frac{\partial \rho}{\partial t}d\mathcal{V}$. In terms of per unit volume $\mathcal{V}$, continuity Eqs. (1.54)–(1.56) can be mathematically expressed as:

$$\frac{D\rho}{Dt} + \rho \nabla \cdot \mathcal{V} = 0 \tag{1.57}$$

Eq. (1.57) represents the nonlinear first-order partial differential equation (PDE) of the volume. Conversely, this equation becomes linear when ρ is constant. If a fluid is incompressible, as may be taken to be the case for seawater in most circumstances, then $\left(\frac{1}{\rho}\right)\left(\frac{d\rho}{dt}\right)=0$, and the equation of continuity becomes:

$$\frac{\partial u}{\partial x}+\frac{\partial v}{\partial y}+\frac{\partial w}{\partial z}=0 \tag{1.58}$$

In order to present the continuity equation in cylindrical polar coordinates (Fig. 1.12), let us assume the velocity $\vec{U}$ at some arbitrary point P can be expressed as:

$$\vec{U}=u_r\hat{\mathbf{e}}_r+v_\theta\hat{\mathbf{e}}_\theta+w_z\hat{\mathbf{e}}_z \tag{1.59}$$

where u_r is radial velocity, v_θ is tangential velocity, and w_z is axial velocity, which are multiplied by unit vectors of at r, θ, and z. The unit vectors of r, and θ can be given by:

$$\hat{\mathbf{e}}_r=\cos\theta\hat{i}+\sin\theta\hat{j} \tag{1.59.1}$$

$$\hat{\mathbf{e}}_\theta=-\sin\theta\hat{i}+\cos\theta\hat{j} \tag{1.59.2}$$

$$\hat{\mathbf{e}}_z=\mathrm{K} \tag{1.59.3}$$

In regard to Eq. (1.59), cylindrical coordinates extend polar coordinates to 3D space. In the cylindrical coordinate system, a point P in 3D space is represented by the ordered triple (r,θ,z). Here, r represents the distance from the origin of the projection of the point P onto the x-y plane, θ is the angle in radians from the x-axis to the projection of the point on the

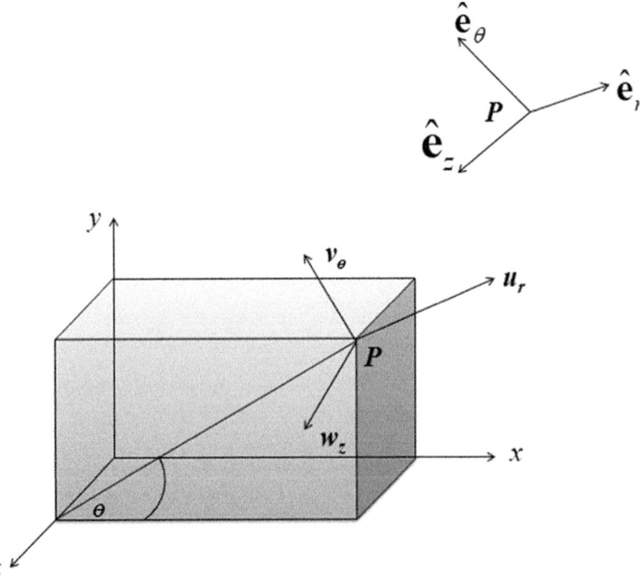

FIG. 1.12 Continuity equation in cylindrical polar coordinates.

x-y plane, and z is the distance from the x-y plane to the point P (Fig. 1.12). In this regard, the continuity equation in cylindrical polar coordinates is given by:

$$\frac{\partial \rho}{\partial t}+\frac{1}{r}\frac{\partial(r\rho u_r)}{\partial r}+\frac{1}{r}\frac{\partial(\rho v_\theta)}{\partial \theta}+\frac{\partial(\rho w_z)}{\partial z}=0 \quad (1.60)$$

Steady, compressible flow can be expressed as:

$$\frac{1}{r}\frac{\partial(r\rho u_r)}{\partial r}+\frac{1}{r}\frac{\partial(\rho v_\theta)}{\partial \theta}+\frac{\partial(\rho w_z)}{\partial z}=0 \quad (1.61)$$

Conversely, incompressible fluids (for steady or unsteady flow) can be represented by:

$$\frac{1}{r}\frac{\partial(r u_r)}{\partial r}+\frac{1}{r}\frac{\partial v_\theta}{\partial \theta}+\frac{\partial w_z}{\partial z}=0 \quad (1.62)$$

where u_r, v_θ, and w_z are the 3-D velocity components at polar coordinates r, θ, and z, respectively.

1.7.2 Stream function based on continuity equations

It was demonstrated early that steady, incompressible, and two-dimensional flows denote the most naive sorts of the significant fluid flow dealing with ocean dynamics. Let us consider two velocity components, u and v, in the x-y plane. For this flow, the continuity equation diminishes to:

$$\frac{\partial u}{\partial x}+\frac{\partial v}{\partial y}=0 \quad (1.63)$$

The main question is how the two velocity components can be related to stream function. Let us consider the stream function to be identified by $\psi(x,y)$, which correlates the velocities as (Fig. 1.13):

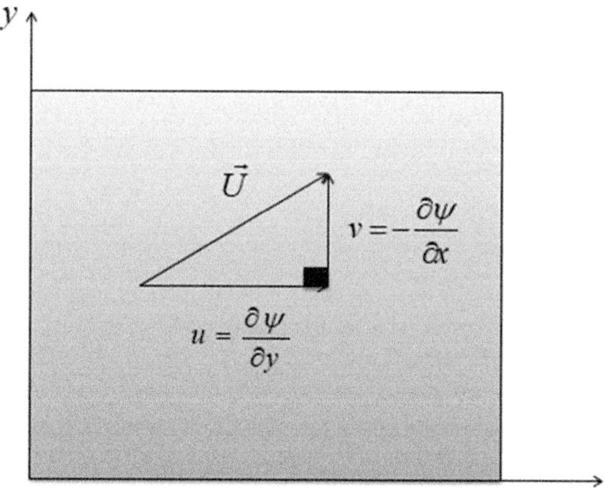

FIG. 1.13 Velocity components of stream function.

$$u = \frac{\partial \psi}{\partial y}, \quad v = -\frac{\partial \psi}{\partial x} \tag{1.64}$$

In this regard, the continuity equation is identically satisfied:

$$\frac{\partial}{\partial x}\left(\frac{\partial \psi}{\partial y}\right) + \frac{\partial}{\partial y}\left(-\frac{\partial \psi}{\partial x}\right) = \frac{\partial^2 \psi}{\partial x \partial y} - \frac{\partial^2 \psi}{\partial x \partial y} = 0 \tag{1.65}$$

Eq. (1.65) shows the velocity components along a streamline (Fig. 1.14). In this regard, the stream function is related to the fact that the lines along which ψ is constant are streamlined. Therefore, there are possible changes of ψ as the flow moves from point (x, y) to the point of $(x + dx, y + dy)$ across points ***A*** and ***B*** (Fig. 1.15). Under this circumstance, the mathematical expression of these changes can be expressed as:

$$d\psi = \frac{d\psi}{dx}dx + \frac{d\psi}{dy}dy = -vdx + udy = 0 \tag{1.66}$$

Eq. (1.66) can be rearranged as follows:

$$\frac{dy}{dx} = \frac{v}{u} \tag{1.67}$$

Eq. (1.67) reveals that the concrete numerical value related to a specific streamline is not of precise significance. On the other hand, the change in the value of ψ is correlated to the volume rate of flow. Let us consider $d\forall$ is the volume rate of flow (per unit width perpendicular to the x-y plane) passing between the two streamlines (Fig. 1.15), which is given by:

$$d\forall = udy - vdx = \frac{\partial \psi}{\partial x}dx + \frac{\partial \psi}{\partial y}dy = d\psi \tag{1.68}$$

In this regard, the volume rate of flow, V, between two streamlines such as ψ_1 and ψ_2, can be determined by integrating to yield:

$$V = \int_{\psi_1}^{\psi_2} d\psi = \psi_2 - \psi_1 \tag{1.69}$$

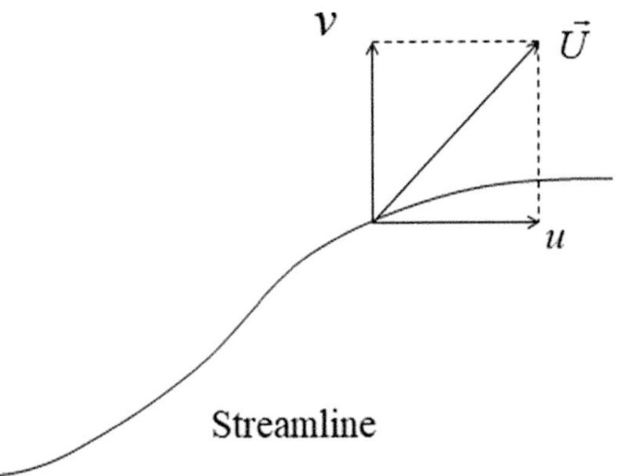

FIG. 1.14 Velocity components across a streamline.

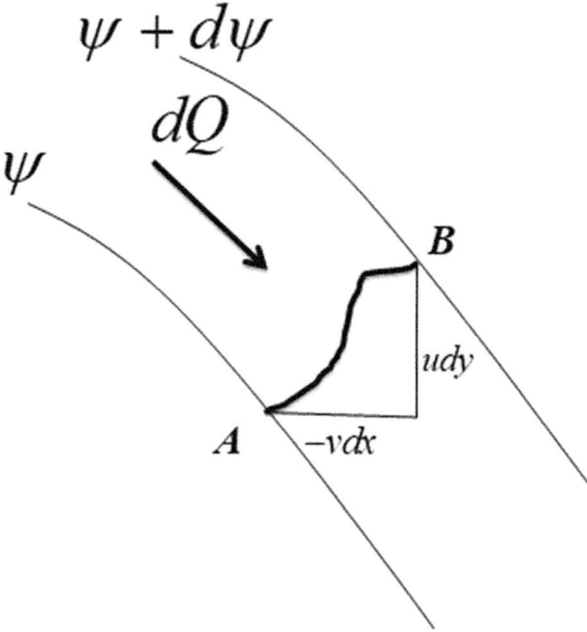

FIG. 1.15 Flow changes along streamlines.

In cylindrical coordinates, the continuity equation for incompressible, plane, two-dimensional flow reduces to:

$$\frac{1}{r}\frac{\partial(ru_r)}{\partial r}+\frac{1}{r}\frac{\partial v_\theta}{\partial \theta}=0 \tag{1.70}$$

The velocity components, u_r and v_θ (Fig. 1.16) can be interrelated to the stream function $\psi(r,\theta)$ through the following equations:

$$u_r=\frac{1}{r}\frac{\partial \psi}{\partial \theta},\quad v_\theta=-\frac{\partial \psi}{\partial r} \tag{1.71}$$

1.7.3 Equation of continuity by the Lagrangian and Eulerian descriptions

Let us assume that S_0 is the region dominated by the portion of fluid density ρ_0 at a certain time $t=0$ and S is the region dominated by the same fluid at any time t. Therefore, the mass of the fluid element at $t=0$ is $\rho\partial x\partial y\partial z=\rho d\text{Ɐ}$. If a, b, and c are assumed as the initial coordinates of fluid-particle P_0 and P (Fig. 1.17) is the subsequent position with density ρ, the total mass inside S_0 must be equal to the total mass inside S, which is given by:

$$\iiint_{S_0}\rho_0\partial a\partial b\partial c=\iiint_{S}\rho_0\partial x\partial y\partial z \tag{1.72}$$

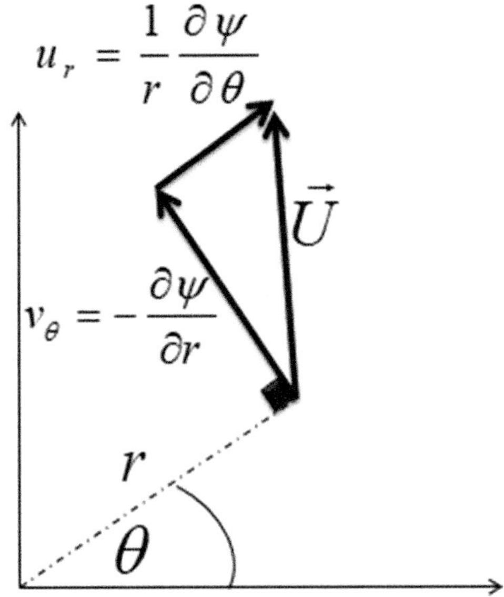

FIG. 1.16 Velocity components of streamlines in cylindrical coordinates.

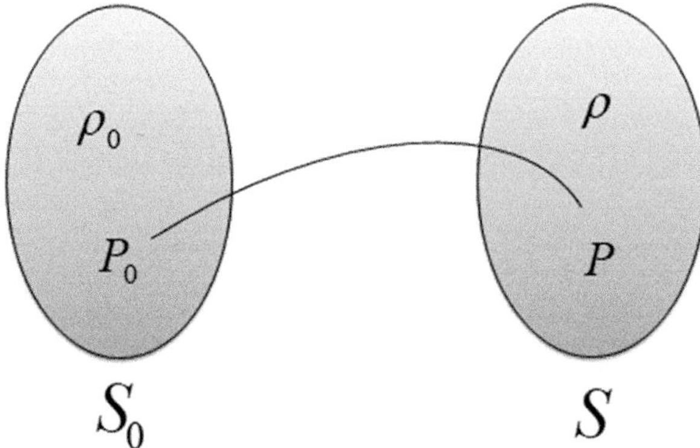

FIG. 1.17 Simplification of Lagrangian flow.

By implementing advanced calculus using the Jacobian determinant, Eq. (1.72) yields:

$$\partial x \partial y \partial z = J \partial a \partial b \partial c \tag{1.73}$$

where:

$$J = \frac{\partial(x,y,z)}{\partial(a,b,c)} = \begin{vmatrix} \frac{\partial x}{\partial a} & \frac{\partial x}{\partial b} & \frac{\partial x}{\partial c} \\ \frac{\partial y}{\partial a} & \frac{\partial y}{\partial b} & \frac{\partial y}{\partial c} \\ \frac{\partial z}{\partial a} & \frac{\partial z}{\partial b} & \frac{\partial z}{\partial c} \end{vmatrix} \tag{1.73.1}$$

Using Eqs. (1.73) and (1.72), we obtain:

$$\iiint_{S_0} (\rho_0 - \rho J)\partial a \partial b \partial c = 0 \tag{1.74}$$

Eq. (1.74) holds for all regions of S_0 if:

$$\rho_0 - \rho J = 0. \tag{1.75}$$

In this regard, Eq. (1.75) gives the equation of the continuity in the Lagrangian form. The Lagrangian description is one in which individual fluid particles are tracked, much like the tracking of billiard balls. In the Lagrangian description of fluid flow, individual fluid particles are "noticeable," and their positions, velocities, etc. are pronounced as a function of time. For instance, the particle transport in the flow field, their positions, and velocities change with time. The physical laws—for instance, Newton's laws and conservation of mass and energy—apply precisely to each particle. If there are only a few particles to consider, as in the case of billiard balls, the Lagrangian description can be an appropriate solution. Nonetheless, fluid flow is a continuum phenomenon at the tiniest particle, such as the molecular, scale. It is difficult to trace each "particle" in a complex flow field. Consequently, the Lagrangian description is of use in fluid dynamics. Awkwardly, the calculation of the Lagrangian description is hard; but it is frequently valuable to deliberate the particle's life history so as to achieve an understanding of flow dynamics. Lagrangian records can be acquired in the atmosphere from balloon flights, or in the Gulf Stream from buoyant devices.

Let us consider fluid particles $x=(a,b,c,t)$, $y=(a,b,c,t)$, and $z=(a,b,c,t)$ where their accelerations can be mathematically expressed as:

$$\frac{\partial u}{\partial a} = \frac{\partial}{\partial a}\left(\frac{dx}{dt}\right) = \frac{d}{dt}\left(\frac{dx}{da}\right) \tag{1.76}$$

$$\frac{\partial v}{\partial b} = \frac{\partial}{\partial b}\left(\frac{dy}{dt}\right) = \frac{d}{dt}\left(\frac{dy}{db}\right) \tag{1.77}$$

$$\frac{\partial w}{\partial c} = \frac{\partial}{\partial c}\left(\frac{dz}{dt}\right) = \frac{d}{dt}\left(\frac{dz}{dc}\right) \tag{1.78}$$

The equation of the continuity in the Eulerian form is given by:

$$\frac{\partial \rho}{\partial t} + \rho\left(\frac{\partial u}{\partial x} + \frac{\partial v}{\partial y} + \frac{\partial w}{\partial z}\right) = 0 \tag{1.79}$$

Differentiating both sides of Eq. (1.73.1) using Eqs. (1.76)–(1.78), we obtain:

$$\frac{dJ}{dt} = \begin{vmatrix} \frac{\partial u}{\partial a} & \frac{\partial u}{\partial b} & \frac{\partial u}{\partial c} \\ \frac{\partial y}{\partial a} & \frac{\partial y}{\partial b} & \frac{\partial y}{\partial c} \\ \frac{\partial z}{\partial a} & \frac{\partial z}{\partial b} & \frac{\partial z}{\partial c} \end{vmatrix} + \begin{vmatrix} \frac{\partial x}{\partial a} & \frac{\partial x}{\partial b} & \frac{\partial x}{\partial c} \\ \frac{\partial v}{\partial a} & \frac{\partial v}{\partial b} & \frac{\partial v}{\partial c} \\ \frac{\partial z}{\partial a} & \frac{\partial z}{\partial b} & \frac{\partial z}{\partial c} \end{vmatrix} + \begin{vmatrix} \frac{\partial x}{\partial a} & \frac{\partial x}{\partial b} & \frac{\partial x}{\partial c} \\ \frac{\partial y}{\partial a} & \frac{\partial y}{\partial b} & \frac{\partial y}{\partial c} \\ \frac{\partial w}{\partial a} & \frac{\partial w}{\partial b} & \frac{\partial w}{\partial c} \end{vmatrix} \tag{1.80}$$

or:

$$\frac{dJ}{dt} = J_1 + J_2 + J_3, \tag{1.81}$$

where:

$$J_1 = \frac{\partial u}{\partial x} J, \tag{1.81.1}$$

$$J_2 = \frac{\partial v}{\partial y} J, \tag{1.81.2}$$

$$J_3 = \frac{\partial w}{\partial z} J, \tag{1.81.3}$$

Then Eq. (1.81) becomes:

$$\frac{dJ}{dt} = J\left(\frac{\partial u}{\partial x} + \frac{\partial v}{\partial y} + \frac{\partial w}{\partial z}\right) \tag{1.82}$$

Eq. (1.82) can be given in the Eulerian forms as:

$$\frac{d\rho}{dt} J + \rho J\left(\frac{\partial u}{\partial x} + \frac{\partial v}{\partial y} + \frac{\partial w}{\partial z}\right) = 0 \tag{1.83}$$

Therefore, Eq. (1.79) can be derived from Eq. (1.83) if the following condition is achieved by:

$$\frac{d\rho}{dt} J + \rho \frac{dJ}{dt} = 0 \tag{1.84}$$

In the Eulerian description of fluid flow, individual fluid particles are not recognized. Instead, a control volume is circumscribed, as shown in Fig. 1.18. In this view, the control volume involves pressure, velocity, acceleration, and all other flow properties. In other words, each property is articulated as a function of space and time, as shown in the velocity field in Fig. 1.18. In the Eulerian description of fluid flow, one is not concerned about the location or velocity of any particular particle. On the other hand, the velocity, and acceleration of

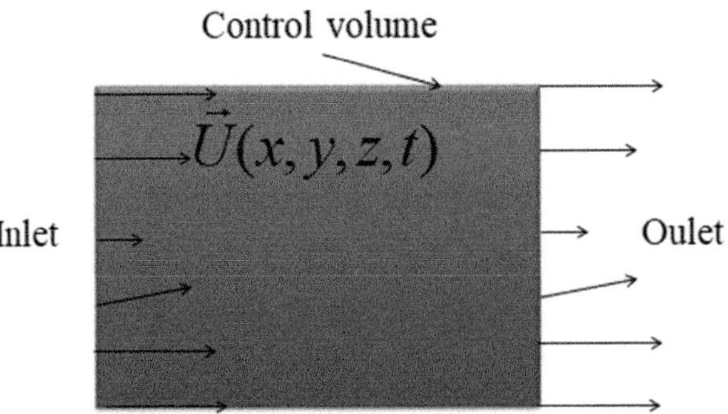

FIG. 1.18 Simplification of Eulerian description of fluid flow.

whatever particle almost occurs in a specific position and in a certain time. Subsequently, the Eulerian description is usually preferred in fluid dynamics. In fact, fluid flow is a variety phenomenon, at least down to the molecular level. Conversely, the physical laws, for instance, Newton's laws and the laws of conservation of mass and energy utilize directly to particles in a Lagrangian description in which monitor their physical characteristics through the fluid motion.

The *material derivative*, which is also called the *total derivative* or *substantial derivative*, is useful as a bridge between Lagrangian and Eulerian descriptions. The material derivative of some quantity is simply defined as the rate of change of that quantity following a fluid particle. It is derived from an arbitrary fluid property Q as follows:

$$\begin{aligned}\frac{DQ}{Dt}&=\frac{dQ}{dt}=\frac{dQ}{dt}\frac{dt}{dt}+\frac{dQ}{dx}\frac{dx}{dt}+\frac{dQ}{dy}\frac{dy}{dt}+\frac{dQ}{dz}\frac{dz}{dt}\\&=\frac{dQ}{dt}+u\frac{dQ}{dx}+v\frac{dQ}{dy}+w\frac{dQ}{dz}\end{aligned} \tag{1.85}$$

Similarly, material acceleration can be derived as follows:

$$\vec{a}=\frac{d\vec{U}}{dt}=\frac{du}{dt}\hat{i}+\frac{dv}{dt}\hat{j}+\frac{dw}{dt}\hat{k}=a_x\hat{i}+a_y\hat{j}+a_z\hat{k} \tag{1.86}$$

$$a_x=\frac{du}{dt}=\frac{\partial u}{\partial t}+\frac{\partial u}{\partial x}\frac{\partial x}{\partial t}+\frac{\partial u}{\partial y}\frac{\partial y}{\partial t}+\frac{\partial u}{\partial z}\frac{\partial z}{\partial t}=\frac{\partial u}{\partial t}+u\frac{\partial u}{\partial x}+v\frac{\partial u}{\partial y}+w\frac{\partial u}{\partial z}=\frac{Du}{Dt} \tag{1.87}$$

where $\frac{Du}{Dt}$ represents the substantial derivative in the *x-axis*.

Similarly, substantial derivatives in the y-axis and z-axis, respectively, can be given by:

$$a_y=\frac{Dv}{Dt}=\frac{\partial v}{\partial t}+u\frac{\partial v}{\partial x}+v\frac{\partial v}{\partial y}+w\frac{\partial v}{\partial z} \tag{1.88}$$

$$a_z=\frac{Dw}{Dt}=\frac{\partial w}{\partial t}+u\frac{\partial w}{\partial x}+v\frac{\partial w}{\partial y}+w\frac{\partial w}{\partial z} \tag{1.89}$$

In vector notation, this can be written briefly as:

$$\frac{D\vec{U}}{Dt}=\frac{\partial\vec{U}}{\partial t}+\vec{U}\cdot\nabla\vec{U} \tag{1.90}$$

where:

$$\nabla=\frac{\partial}{\partial x}\hat{i}+\frac{\partial}{\partial y}\hat{j}+\frac{\partial}{\partial z}\hat{k} \tag{1.90.1}$$

where ∇ is the gradient operator.

Therefore, $\frac{\partial\vec{U}}{\partial t}$ is called local or temporal acceleration, and results from velocity changes with respect to time at a given position. Local acceleration results when the flow is unsteady.

Therefore, $\vec{U}\cdot\nabla\vec{U}$ is called convective acceleration because it is associated with spatial gradients of velocity in the flow field. Convective acceleration results when the flow is nonuniform, that is, if the velocity changes along a streamline. In this view, the convective acceleration terms are nonlinear, which causes mathematical difficulties in flow analysis; also, even in a steady flow, the convective acceleration can be large if spatial gradients of velocity are large.

1.8 Navier-Stokes equations

The ocean dynamic flow in the physical domain is driven by various properties. For the purpose of bringing the behavior of ocean dynamic flow to light and developing a mathematical model, those properties have to be defined precisely as to provide the transition between the physical and the numerical domain. Velocity, pressure, temperature, salinity, density, and viscosity are the main properties that should be considered simultaneously when conducting an ocean flow examination. In accordance with physical mechanisms such as combustion, multiphase flow, turbulent, mass transport, etc., those properties can change substantially, which can be categorized into kinematic, transport, thermodynamic, and other miscellaneous properties.

1.8.1 What are the Navier-Stokes equations?

Thermo-fluid incidents directed by governing equations are based on the laws of conservation. The Navier-Stokes equations are the broadly applied mathematical model to examine changes on those properties during dynamic and/or thermal interactions. The equations are adaptable concerning the nature of the problem and are articulated based on the codes of conservation of mass, momentum, and energy; in other words, they involve the continuity equation, the momentum equation of Newton's second law, and the first law of thermodynamics or energy equation.

Even though some sources identify the articulation of Navier-Stokes equations purely for the conservation of momentum, some of them correspondingly exploit all equations of conservation of ocean physical characteristics. Concerning the ocean dynamic flow circumstances, the Navier-Stokes equations are reordered to deliver confirmatory explanations to diminish the complexity of the nonlinear ocean dynamic flows. For instance, having a numerical case of turbulence along with the precalculated Reynolds number, a proper turbulence model has to be operated to acquire reliable findings.

1.8.2 The Navier-Stokes equations

Let us consider F is the net force applied to any mass particle m that leads to its acceleration. If the particle is of a fluid, it is convenient to divide the equation for the volume of the particle to generate a derivation in terms of density as follows:

$$\sum \vec{F} = \left[\frac{\partial}{\partial x}\left(\rho u \vec{U}\right) + \frac{\partial}{\partial y}\left(\rho v \vec{U}\right) + \frac{\partial}{\partial z}\left(\rho w \vec{U}\right)\right] dxdydz[0, 1] \tag{1.91}$$

Combining and making use of the continuity equation explained previously yields:

$$\sum \vec{F} = \rho \frac{D\vec{U}}{Dt} dxdydz \tag{1.92}$$

where:

$$\sum \vec{F} = \sum \vec{F}_{body} + \sum \vec{F}_{surface} \tag{1.93}$$

$\vec{F}_{body}$ is the applied force on the whole mass of fluid particles as below:

$$\sum \vec{F}_{body} = d\vec{F}_{grav} = \rho g dxdydz \tag{1.93.1}$$

where ρ is the density of the fluid and g is the gravitational acceleration.

External forces deployed through the surface of fluid particles $\vec{F}_{surface}$ are expressed by pressure and viscous forces as shown below:

$$\vec{F}_{surface} = \nabla \cdot \tau_{ij} = \frac{\partial \tau_{ij}}{\partial x_i} = f_{pressure} + f_{viscous} \tag{1.93.2}$$

where τ_{ij} is expressed as stress tensor.

According to the general deformation law of Newtonian viscous fluid given by Stokes, this is calculated as:

$$\tau_{ij} = -p\partial_{ij} + \mu\left(\frac{\partial u_i}{\partial x_i} + \frac{\partial u_j}{\partial x_j}\right) + \partial_{ij}\lambda\nabla \cdot U \tag{1.93.3}$$

Hence, Newton's equation of motion can be specified as follows:

$$\rho \frac{DU}{Dt} = \rho \cdot g + \nabla \cdot \tau_{ij} \tag{1.94}$$

Substitution of Eq. (1.93.3) into Eq. (1.94) results in the Navier-Stokes equations of a Newtonian viscous fluid in one equation:

$$\rho \frac{DU}{Dt} = \rho \cdot g - \nabla p + \frac{\partial}{\partial x_i}\left[\mu\left(\frac{\partial u_i}{\partial x_i} + \frac{\partial u_j}{\partial x_j}\right) + \partial_{ij}\lambda\nabla \cdot U\right] \tag{1.95}$$

where $\rho \frac{DU}{Dt}$ represents the momentum convection, $\rho \cdot g$ is a mass force, ∇p is the surface force, and the last term $\frac{\partial}{\partial x_i}\left[\mu\left(\frac{\partial u_i}{\partial x_i} + \frac{\partial u_j}{\partial x_j}\right) + \partial_{ij}\lambda\nabla \cdot U\right]$ represents viscous force, where μ is viscosity coefficient.

Eq. (1.95) is expedient for ocean flow fields, both transient and compressible. Conversely, $\frac{D}{Dt}$ specifies the substantial derivative in this manner:

$$\begin{aligned} \frac{D()}{Dt} &= \frac{\partial()}{\partial t} + u\frac{\partial()}{\partial x} + v\frac{\partial()}{\partial y} + w\frac{\partial()}{\partial z} \\ &= \frac{\partial()}{\partial t} + U \cdot \nabla() \end{aligned} \tag{1.96}$$

In Eq. (1.95), the viscosity coefficient μ is assumed to be constant and $\nabla \cdot \vec{U} = 0$ as well as the water density ρ being constant. Consequently, the Navier-Stokes equations for an incompressible three-dimensional flow can be formulated as:

$$\rho \frac{D\vec{U}}{Dt} = \rho g - \mu \nabla^2 \vec{U} \tag{1.97}$$

For each dimension when the velocity $U(u,v,w)$:

$$\rho \left[\frac{\partial u}{\partial t} + u\frac{\partial u}{\partial x} + v\frac{\partial u}{\partial y} + w\frac{\partial u}{\partial z} \right] = \rho g x - \frac{\partial p}{\partial x} + \mu \left[\frac{\partial^2 u}{\partial x^2} + \frac{\partial^2 u}{\partial y^2} + \frac{\partial^2 u}{\partial z^2} \right] \tag{1.98}$$

$$\rho \left[\frac{\partial v}{\partial t} + u\frac{\partial v}{\partial x} + v\frac{\partial v}{\partial y} + w\frac{\partial v}{\partial z} \right] = \rho g y - \frac{\partial p}{\partial y} + \mu \left[\frac{\partial^2 v}{\partial x^2} + \frac{\partial^2 v}{\partial y^2} + \frac{\partial^2 v}{\partial z^2} \right] \tag{1.99}$$

$$\rho \left[\frac{\partial w}{\partial t} + u\frac{\partial w}{\partial x} + v\frac{\partial w}{\partial y} + w\frac{\partial w}{\partial z} \right] = \rho g z - \frac{\partial p}{\partial z} + \mu \left[\frac{\partial^2 w}{\partial x^2} + \frac{\partial^2 w}{\partial y^2} + \frac{\partial^2 w}{\partial z^2} \right] \tag{1.100}$$

In Eqs. (1.98)–(1.100), u, v, and w are unknowns where a solution is required by solicitation of both continuity equation and boundary conditions. In addition, the energy equation has to be determined if any thermal interaction is available in the problem.

1.8.3 Conservation of energy

The conservation of energy is an initial law of thermodynamics, which affirms that the sum of the work (W) and heat combined within the system will cause the growth of the energy of the system. Mathematically, this is defined as:

$$dH = dW + dE \tag{1.101}$$

where dH is a heat, which is added to the ocean system, and dE is the growth of the energy of the ocean system.

One of the conventional mathematical expressions of the energy equation is:

$$\rho \left[\frac{\partial H}{\partial t} + \nabla \cdot \left(H\vec{U} \right) \right] = -\frac{\partial p}{\partial t} + \nabla \cdot (k \nabla T) + \phi \tag{1.102}$$

Eq. (1.102) demonstrates the local change with time as represented by $\frac{\partial H}{\partial t}$. Therefore, the convective term is indicated by $\nabla \cdot \left(H\vec{U} \right)$. Consequently, the change of the work overtime is presented by the change of the pressure $\frac{\partial p}{\partial t}$. $\nabla \cdot (k \nabla T)$, subsequently, is a result of heat flux as a function of water temperature T and k is the thermal conductivity. Finally, ϕ is a heat dissipation term.

1.8.4 Time domain

The investigation of ocean dynamic flow can be directed in either steady (time-independent) or unsteady (time-dependent) conditions, depending on the physical circumstances. In the case that ocean flow is steady, it means the motion of the ocean and its parameters do not depend on the change in time. In this regard, the term $\frac{\partial()}{\partial t}$ in Eq. (1.96) equals zero. In this circumstance, the continuity and momentum equations are re-derived as follows:

$$\frac{\partial(\rho u)}{\partial x}+\frac{\partial(\rho v)}{\partial y}+\frac{\partial(\rho w)}{\partial z}=0 \tag{1.103}$$

The Navier-Stokes equations in x, y, and z directions are shown in Eqs. (1.98)–(1.100). In this regard, the steady flow assumption negates the consequence of some nonlinear terms and provides a convenient solution, variation of density is a hurdle that keeps the equation in a complex formation [1, 5, 24, 32, 33].

1.8.5 Compressibility

Due to the flexible nature of ocean flow, the compressibility of particles is a noteworthy question. Although the fact that all sorts of fluid flows are compressible in various ranges concerning molecular construction, most of them can be presumed to be incompressible, in that the density changes are negligible. On this understanding, the incompressible flow assumption delivers rational equations, as the application of steady flow assumption allows us to ignore nonlinear terms where $\frac{\partial()}{\partial t}=0$. Furthermore, the density of the fluid in high speed is excluded as incompressible flow, in which the density changes are notable. In this sense, the Mach number, Ma, is a dimensionless number that is useful in considering ocean flow, whether incompressible or compressible [4, 14, 22, 34]:

$$Ma=Us^{-1}\leq 0.3 \tag{1.104}$$

Eq. (1.104) proves that the ocean flow U is incompressible as Ma is lower than 0.3 under the speed of sound s at 340.29 m s^{-1} at sea level. On the contrary, the change in density cannot be negligible, in that density should be considered as a significant parameter. In fact, as well as velocity, the effect of thermal properties on the density changes has to be considered in geophysical flows [15, 32].

1.8.6 Low and high Reynolds numbers

The Reynolds number (Re) is the ratio of inertial and viscous effects, which is correspondingly applicable on Navier-Stokes equations to abbreviate the mathematical model. While Reynolds number turns into infinite, i.e., $Re\longrightarrow\infty$, the vicious influences are supposed insignificant where viscous terms in Navier-Stokes equations are terrified away. The simplified form of the Navier-Stokes equations, known as the Euler equations, can be defined as [35]:

$$\rho\left[\frac{\partial u}{\partial t}+u\frac{\partial u}{\partial x}+v\frac{\partial u}{\partial y}+w\frac{\partial u}{\partial z}\right]=\rho gx-\frac{\partial p}{\partial x} \tag{1.105}$$

$$\rho\left[\frac{\partial v}{\partial t}+u\frac{\partial v}{\partial x}+v\frac{\partial v}{\partial y}+w\frac{\partial v}{\partial z}\right]=\rho gy-\frac{\partial p}{\partial y} \tag{1.106}$$

$$\rho\left[\frac{\partial w}{\partial t}+u\frac{\partial w}{\partial x}+v\frac{\partial w}{\partial y}+w\frac{\partial w}{\partial z}\right]=\rho gz-\frac{\partial p}{\partial z} \tag{1.107}$$

Even though viscous effects are relatively important for fluids, the inviscid flow model partially delivers a consistent mathematical model to envisage the real development for some precise circumstances. For instance, high-speed external flow over bodies is a broadly used approximation where the inviscid approach fits reasonably well. While $Re \ll 1$, the inertial effects are assumed negligible where related terms in Navier-Stokes equations drop out. These simplified forms of Navier-Stokes equations are called either creeping flow or Stokes flow [32, 35]:

$$\rho g x - \frac{\partial p}{\partial x} + \mu \left[\frac{\partial^2 u}{\partial x^2} + \frac{\partial^2 u}{\partial y^2} + \frac{\partial^2 u}{\partial z^2} \right] = 0 \tag{1.108}$$

$$\rho g y - \frac{\partial p}{\partial y} + \mu \left[\frac{\partial^2 v}{\partial x^2} + \frac{\partial^2 v}{\partial y^2} + \frac{\partial^2 v}{\partial z^2} \right] = 0 \tag{1.109}$$

$$\rho g z - \frac{\partial p}{\partial z} + \mu \left[\frac{\partial^2 w}{\partial x^2} + \frac{\partial^2 w}{\partial y^2} + \frac{\partial^2 w}{\partial z^2} \right] = 0 \tag{1.110}$$

1.9 Turbulence

The behavior of the ocean dynamic conditions is a challenging matter that is characterized by both laminar and turbulent flows. In this regard, the laminar flow is systematic, by which ocean motion can be predicted precisely. On the contrary, the turbulent ocean flow has various difficulties that produce what looks like chaotic behavior and consequently, it is awkward to envisage the ocean flow, In these regards, the Reynolds number—the ratio of inertial forces to viscous forces—predicts the behavior of ocean flow whether laminar or turbulent regarding several properties such as velocity, length, viscosity, and also type of flow. Although the flow is turbulent, an appropriate mathematical model is required to obtain precise numerical solutions. Ocean turbulent flow can be subjected to the Navier-Stokes equations in order to manage accurate solutions to ocean flow chaotic behavior. As well as the laminar flow, transport quantities of the turbulent flow are determined by direct quantities. Direct numerical simulation (DNS) is a method for resolving the Navier-Stokes equation with instantaneous quantities. Having dominion fluctuations differs in an extensive range, DNS requests massive efforts and exclusive computational abilities. To escape those obstacles, the instantaneous quantities are elicited by the sum of their mean and changeable parts as follows [4, 32, 33]:

$$u = \overline{u} + u' \tag{1.111}$$

$$v = \overline{v} + v' \tag{1.112}$$

$$w = \overline{w} + w' \tag{1.113}$$

where $\overline{u}$, $\overline{v}$, and $\overline{w}$ represent the mean flow in x, y, and z directions, while u', v', and w' are the fluctuation flows. In other words, Eqs. (1.111)–(1.113) represent the instantaneous flows, which are the total of mean flows and the fluctuation flows.

In preference to instantaneous flows, which cause nonlinearity, carrying out a numerical solution with mean flows provides an applicable mathematical model, which is known as the Reynolds-averaged Navier-Stokes (RANS) equation. The instabilities of RANS model can be negligible for most engineering cases which cause a complex mathematical model. Thus, RANS turbulence model is a procedure to complete the simulation of mean flow equations. The general form of The Reynolds-averaged Navier-Stokes (RANS) equation can be specified as follows:

$$\rho \overline{U}_j \overline{U}i, j = -\rho P_i + \mu \overline{U}i, jj - \left(u_i' u_j'\right)_j \tag{1.114}$$

where μ is the turbulent viscosity, which is regularly calculated through turbulence models.

Eq. (1.114) contains the Reynolds stress $(u_i' u_j')_j$, which signifies the impact of the small-scale turbulence on the average flow $\overline{U}$ as u' is turbulent flow. In addition, P is the pressure of the fluid as the material parameters are considered uniform. The main idea is to calculate the motion as the sum of a mean flow and a turbulent/unsteady flow. In other words, the steady mean velocity can be computed as the average of the global velocity $\overline{U}_i$:

$$\overline{U}_i = \lim_{t \to \infty} T^{-1} \int_0^T u dt \tag{1.114.1}$$

where T is the averaging time-scale, which must be small enough to have an excellent approximation of the problem, but correspondingly appropriately higher than the turbulence time-scale.

The RANS equations have no exclusive explanation since they are not in a close formula. In fact the unknowns are more than the equations. Accordingly, supplementary equations are required to end the problem. The ultimate conjoined approach, as implemented in Computing Fluid Dynamics (CFD), is to relate the Reynolds stress to the shear rate by the Boussinesq relationship as follows:

$$u_i' u_j' = 2\mu\rho^{-1} \times 0.5\left(\overline{U}_{i,j} + \overline{U}_{j,i}\right) \tag{1.114.2}$$

where $\overline{U}$ is the mean flow in i and j directions, respectively. Consequently, mean flows can be determined from Navier-Stokes as:

$$\rho\left[\frac{\partial \overline{u}}{\partial t} + \overline{u}\frac{\partial \overline{u}}{\partial x} + \overline{v}\frac{\partial \overline{u}}{\partial y} + \overline{w}\frac{\partial \overline{u}}{\partial z}\right] = \rho g x - \frac{\partial \overline{p}}{\partial x} + \mu\left[\frac{\partial^2 u}{\partial x^2} + \frac{\partial^2 u}{\partial y^2} + \frac{\partial^2 u}{\partial z^2}\right] \tag{1.115}$$

$$\rho\left[\frac{\partial \overline{v}}{\partial t} + \overline{u}\frac{\partial \overline{v}}{\partial x} + \overline{v}\frac{\partial \overline{v}}{\partial y} + \overline{w}\frac{\partial \overline{v}}{\partial z}\right] = \rho g y - \frac{\partial \overline{p}}{\partial y} + \mu\left[\frac{\partial^2 v}{\partial x^2} + \frac{\partial^2 v}{\partial y^2} + \frac{\partial^2 v}{\partial z^2}\right] \tag{1.116}$$

$$\rho\left[\frac{\partial \overline{w}}{\partial t} + \overline{u}\frac{\partial \overline{w}}{\partial x} + \overline{v}\frac{\partial \overline{w}}{\partial y} + \overline{w}\frac{\partial \overline{w}}{\partial z}\right] = \rho g z - \frac{\partial \overline{p}}{\partial z} + \mu\left[\frac{\partial^2 w}{\partial x^2} + \frac{\partial^2 w}{\partial y^2} + \frac{\partial^2 w}{\partial z^2}\right] \tag{1.117}$$

The turbulence model of RANS can also differ concerning approaches, for instance, k-omega, k-epsilon, k-omega-SST, and Spalart-Allmaras, which have been exploited to obtain a solution for different sorts of ocean turbulent flows. Correspondingly, large eddy simulation (LES) is another mathematical approach for ocean turbulent flow, which is also systematically utilized for numerous circumstances. The complicated LES guarantees more precise results than RANS but necessitates considerable additional time and computer memory. For instance, both DNS, and LES are believed to resolve the instantaneous Navier-Stokes equations in time and three-dimensional space (Eqs. 1.115–1.117) [1, 4, 9, 14, 23, 30, 33].

1.10 Equations of motion in a rotating frame

The main question is how the continuity equation is modified in a rotating frame. To answer this question, first the concept of solid body rotation must be considered. In this regard, solid body rotation is termed as a fluid or an object that revolves at a constant angular velocity, which means it moves as a solid. On this understanding, the general theory of solid body rotation is a function of the tangential velocity of a parcel (Fig. 1.19) [4, 5, 32].

In this regard, the velocity of the parcel is described as:

$$\vec{v} = \lim_{\Delta t \to 0} \frac{\Delta \vec{r}}{\Delta t} \tag{1.118}$$

An overview of Fig. 1.19 can be useful to correlate $\Delta \vec{r}$ to Ω and Δt. Let us assume that the arc length of the parcel for a given time Δt is $\left|\Delta \vec{r}\right| = r\Omega\Delta t$ and the normalized direction of the parcel is deduced from the right-hand rule as $\frac{\vec{\Omega} \times \vec{r}}{\Omega r}$. In this regard, Eq. (1.118) turns out to be:

$$\vec{v} = \lim_{\Delta t \to 0} \frac{\Delta \vec{r}}{\Delta t} = \lim_{\Delta t \to 0} \frac{\mathrm{r}\Omega\Delta t}{\Delta t} \frac{\vec{\Omega} \times \vec{r}}{\Omega r} = \vec{\Omega} \times \vec{r} \tag{1.119}$$

In this sense, the equation for solid-body rotation velocity is formulated as:

$$\vec{v}_{solid} = \vec{\Omega} \times \vec{r} \tag{1.120}$$

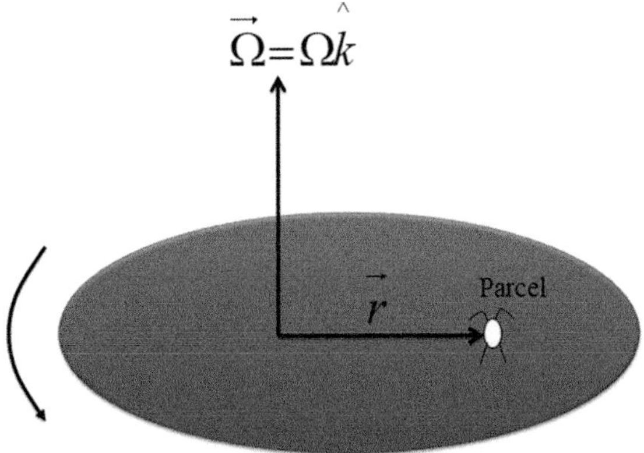

FIG. 1.19 Parcel located at the position $\vec{r}$ moving at constant angular velocity $\vec{\Omega} = \Omega\hat{k}$.

Eq. (1.120) is the cornerstone for deriving how the equation of motion is modified for a parcel of fluid in a rotating coordinate system. In this regard, let us consider an orthogonal coordinate system, which is rotating about an arbitrary axis with constant angular velocity $\vec{\Omega}$ (Fig. 1.20).

Let us assume that $\vec{A}=a_i\hat{e}_i$ is a parcel moving relative to the rotating coordinate system. This motion has two contributions across an external or an inertial frame of reference. These involve the movement of the vector relative to its defined coordinate system and the movement of the coordinate system itself. The scientific explanation of these movements can mathematically be written as [6, 13, 18, 21, 28, 35]:

$$\left(\frac{d\vec{A}}{dt}\right)_I=\frac{da_1}{dt}\hat{e}_1+\frac{da_2}{dt}\hat{e}_2+\frac{da_3}{dt}\hat{e}_3+a_1\frac{d\hat{e}_1}{dt}+a_2\frac{d\hat{e}_2}{dt}+a_3\frac{d\hat{e}_3}{dt} \tag{1.121}$$

The first three terms in Eq. (1.121) are the standard motion of the vector in its relative frame and it can be defined as a relative rate of change of the vector $\vec{A}$.

$$\left(\frac{d\vec{A}}{dt}\right)_R=\frac{da_1}{dt}\hat{e}_1+\frac{da_2}{dt}\hat{e}_2+\frac{da_3}{dt}\hat{e}_3 \tag{1.122}$$

The second three terms require some additional physical insight to express them in a more tractable form. Fig. 1.21 demonstrates that each basis vector of the coordinate system maps out a circular trajectory in space and the velocity of rotation of each of these vectors is simply a solid body rotation with associated velocity for the ith basis vector:

$$\vec{V}\hat{i}=\frac{d\hat{e}_i}{dt}=\vec{\Omega}\times\hat{e}_i \tag{1.123}$$

Exchanging Eq. (1.123) into Eq. (1.122), the rate of change of the vector $\vec{A}$ in the inertial frame of reference is mathematically expressed as:

$$\left(\frac{d\vec{A}}{dt}\right)_I=\left(\frac{d\vec{A}}{dt}\right)_R+\vec{\Omega}\times\vec{A} \tag{1.124}$$

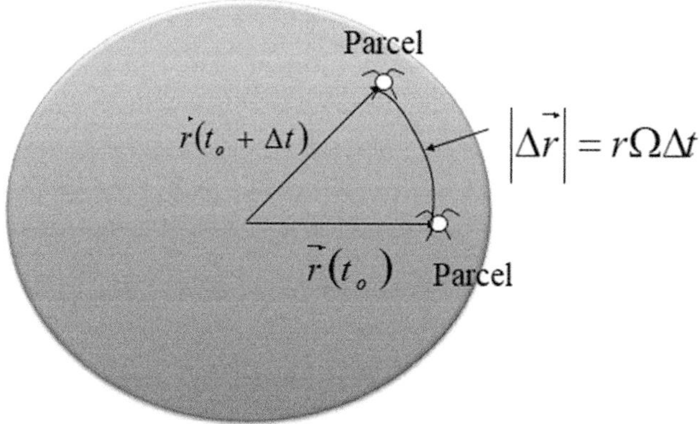

FIG. 1.20 Parcel rotation by angular velocity $\vec{\Omega}$.

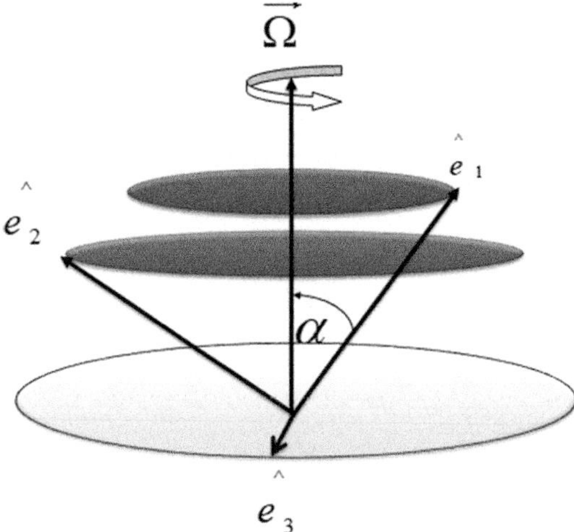

FIG. 1.21 Circular trajectory and rotation velocity along different vectors.

In this regard, there is no physical significance or properties specified to the vector $\vec{A}$ besides that its motion was obtained relative to a moving coordinate system, and so the result of Eq. (1.124) is accurate for any vector that is identified relative to the rotating of the coordinate system. Due to this generality, we can express Eq. (1.120) in an *operator* format as:

$$\left(\frac{d}{dt}\right)_I = \left(\frac{d}{dt}\right)_R + \vec{\Omega}\times \tag{1.125}$$

In this regard, accelerations can be obtained by the second derivative of a position vector with respect to time. The general operator in Eq. (1.125) can be squared to obtain a second-order time derivative of:

$$\begin{aligned}\left(\frac{d}{dt}\right)_I^2 &= \left(\frac{d}{dt}\right)_I\left(\frac{d}{dt}\right)_I = \left(\frac{d^2}{dt^2}\right)_I \\ &= \left(\left(\frac{d}{dt}\right)_R + \vec{\Omega}\times\right)\left(\left(\frac{d}{dt}\right)_R + \vec{\Omega}\times\right) = \left(\frac{d^2}{dt^2}\right)_R + 2\vec{\Omega}\times\left(\frac{d}{dt}\right)_R + \vec{\Omega}\times\vec{\Omega}\times\end{aligned} \tag{1.126}$$

Applying Eq. (1.126) on a position vector $\vec{\ell}$, which is defined in the rotating frame of reference, the apparent acceleration can be formulated mathematically as:

$$\left(\frac{d^2\vec{\ell}}{dt^2}\right)_I = \left(\frac{d^2\vec{\ell}}{dt^2}\right)_R + 2\vec{\Omega}\times\left(\frac{d\vec{\ell}}{dt}\right)_R + \vec{\Omega}\times\vec{\Omega}\times\vec{\ell} = \vec{a}_R + 2\vec{\Omega}\times\vec{u}_R + \vec{\Omega}\times\left(\vec{\Omega}\times\vec{\ell}\right) \tag{1.127}$$

Eq. (1.127) involves three significant terms:

(i) The first term $\left(\frac{d\ell}{dt}\right)_R^2 = \vec{a}_R = \frac{D\vec{u}_R}{Dt}$ signifies the acceleration of a parcel in the rotating frame of reference, which is known as the inertial term.

(ii) The second term $2\vec{\Omega}\times\vec{u}_R$ is called the Coriolis acceleration and measures the deflection of the fluid parcel as observed in the rotating coordinate system. In this regard, its effect is proportional to the magnitude of rotation and the speed of the object. It is this deflection that leads to most of the unique phenomena that are observed in the atmosphere as well as effects in the ocean such as Ekman transport.

(iii) The third term $\vec{\Omega} \times \left(\vec{\Omega} \times \vec{\ell}\right)$ can be represented as two components by using the triple cross product as $\vec{\Omega} \times \left(\vec{\Omega} \times \vec{\ell}\right) = \vec{\Omega}\left(\vec{\Omega} \cdot \vec{\ell}\right) - \Omega^2 \vec{\ell}$. In this regard, let us assume that $\vec{\ell} = x\hat{e}_1 + y\hat{e}_2 + z\hat{e}_3$ and that the rotation axis is in the $\hat{e}_3$ direction so $\vec{\Omega} = \Omega\hat{e}_3$. The two terms of the triple cross product can then be expanded out as:

$$\vec{\Omega}\left(\vec{\Omega} \cdot \vec{\ell}\right) - \Omega^2 \vec{\ell} = -\Omega^2\left(x\hat{i} + y\hat{j}\right) = -\Omega^2 \vec{\ell}_h \tag{1.128}$$

where $\vec{\ell}_h = x\hat{i} + y\hat{j}$ is the horizontal vector perpendicular to the rotation axis to the point of observation.

The resulting term $\Omega^2 \vec{\ell}_h$ represents the centripetal acceleration. It is always directed inward toward the axis of rotation and is at maximum at the equator and zero at the poles. Subsequently, the net force allied with the centripetal acceleration is conservative. In this context, it is a frequent convention to combine this term with any gravitation forces in the problem of the fluid motion [6, 10, 12, 32, 36, 37]. In this sense, it defines the sum of both the gravitational and centripetal force as the *effective or apparent gravitational force*. On this understanding, the equation of motion can mathematically be formulated as:

$$\frac{D\vec{u}}{Dt} + 2\vec{\Omega} \times \vec{u} - \Omega^2 \vec{\ell}_h = -\frac{1}{\rho}\nabla p - g\hat{k} \tag{1.129}$$

Eq. (1.129) can be expressed in component form in a simplified local frame (*f-plane approximation*) as the following [33]:

$$\frac{\partial u}{\partial t} + u\frac{\partial u}{\partial x} + v\frac{\partial u}{\partial y} + w\frac{\partial u}{\partial z} - fv = -\frac{1}{\rho}\frac{\partial p}{\partial x} \tag{1.130}$$

$$\frac{\partial v}{\partial t} + u\frac{\partial v}{\partial x} + v\frac{\partial v}{\partial y} + w\frac{\partial v}{\partial z} + fu = -\frac{1}{\rho}\frac{\partial p}{\partial y} \tag{1.131}$$

$$\frac{\partial w}{\partial t} + u\frac{\partial w}{\partial x} + v\frac{\partial w}{\partial y} + w\frac{\partial w}{\partial z} = -\frac{1}{\rho}\frac{\partial p}{\partial z} - g \tag{1.132}$$

From Eqs. (1.130)–(1.132), the term of $u\frac{\partial u}{\partial x} + v\frac{\partial u}{\partial y} + w\frac{\partial u}{\partial z}$ represents the advective terms, which are called nonlinear. In fact, the velocities occur as squares, i.e., $u\left(\frac{\partial u}{\partial x}\right) = 0.5\left(\frac{\partial\left(u^2\right)}{\partial x}\right)$ or as products between velocity mechanisms and derivatives of other mechanisms, i.e., $v\left(\frac{\partial u}{\partial y}\right)$. Because of these nonlinear terms, a minor perturbation (variation) may develop into a great fluctuation. In this regard, these terms can instigate unsteadiness and leave behind turbulence, which arises whenever they are adequately large compared with the frictional terms, which tend to remove velocity differences [4].

The nonlinear advective terms can be used to derive the *Reynolds number (Re)* for a fluid flow. In this regard, $\frac{\left(\frac{u\partial u}{\partial x}\right)}{\left(\frac{v\partial^2 u}{\partial x^2}\right)}$ represents one of the nonlinear terms to one of the molecular friction terms. To this end, a typical velocity magnitude U is used to replace both u and ∂u. Similarly, L is considered as a typical distance over which the velocity varies by U. In this understanding, the Reynolds number is defined as:

$$\mathrm{Re} = \frac{\left(\frac{U^2}{L}\right)}{\left(\frac{vU}{L^2}\right)} \tag{1.133}$$

Eq. (1.133) measures the ratio of the nonlinear—also known as inertial—terms to frictional terms in the equation of motion. It is clearly stated that the nonlinear effects are very strong compared with the molecular friction effects. In this regard, molecular friction can be ignored in the open sea; it only becomes important very close to solid boundaries and in removing energy from turbulent flow at small scales to prevent it from growing without limit, i.e., molecular friction is important only for low values of Re, which occur at low values of U and/or of L [4].

Though molecular friction may be ignored in most characteristics of the dynamics of ocean motions, it must not be thought that there are no forces with contrasting motions or contributing growth to the reorganization of energy and other characteristics. The turbulent movement contains many inconsistent components as well as an average flow. In this regard, the nonlinear terms provide intensification to terms in the equations of motion, which have the physical character of friction. Consequently, these nonlinear terms can develop a rapid growth of momentum distribution. This leads to the reordering of both forces per unit area and transports per unit area, i.e., fluxes, which can be called Reynolds stresses. In this sense, it appears in the equations for the mean or average motion of a turbulent fluid (Eqs. 1.115–1.117).

Briefly, the acceleration equals the Coriolis frequency times the normal velocity. In this regard, f is used to symbolize the Coriolis frequency, which can be written as:

$$\frac{\partial u}{\partial u} = fv \tag{1.134}$$

$$\frac{\partial v}{\partial t} = -fu \tag{1.135}$$

The magnitude of the Coriolis parameter (which has the units of frequency) changes with latitude. The Coriolis frequency equals $2\Omega\sin(\theta)$ where θ is the latitude in degrees and $\Omega = \frac{2\pi}{T}$ where T is the number of seconds in a sidereal day (86,164 s). At the North Pole $f = 1.45 \times 10^{-4}$, and the inertial period—the time for a particle governed by the momentum balance, described by Eqs. (1.134) and (1.135), to complete one circle—is 11.96 h. For instance, in the tropics, at 18 degrees north $f = 4.537 \times 10^{-5}$, and the inertial period is 36 h. At the Equator, $f = 0$ and the inertial period is infinity. At 40 degrees south $f = -9.37 \times 10^{-5}$ and the Coriolis effect accelerates a moving particle to the left.

In the northern hemisphere, the flow turns to the right. Therefore, northward velocity v is positive and accelerates the flow to the east with positive u. In contrast, an eastward velocity with positive u accelerates the flow to the south with negative v. In this regard, f is positive in the northern hemisphere—so Eqs. (1.134) and (1.135) designate circular motions in a clockwise direction. On this understanding, the flow is always turning to the right and thus flows clockwise. This motion is also called cum *sole*, meaning moving in the same direction that the sun moves across the sky, or anticyclonic, as it is in the opposite direction of the flow around a cyclone.

Finally, inertial currents are commonplace currents in the ocean. They have been perceived at all depths in the ocean and at all latitudes. The motions are transient and decay in a few days. Oscillations at dissimilar depths or at dissimilar adjacent sites are regularly incoherent.

1.11 Conservation equation of ocean waves

It is well-known that ocean waves are created by the wind. The faster the wind, the longer the wind blows, and the bigger the area over which the wind blows, the bigger the waves. The spectrum is an important concept for describing ocean-surface waves quantitatively.

Let us assume that the height of the water column is:

$$h = \bar{h} + \eta \tag{1.136}$$

where $\bar{h}$ is the mean water height and η is the difference from the mean height.

Assume that η is small compared to h and that the wavelength is small compared to the wave amplitude, so that $\eta \ll h$ and $\frac{\partial \eta}{\partial x}$ is small. It can also be shown that the terms $u\frac{\partial u}{\partial x}$ and $u\frac{\partial \eta}{\partial x}$ may be neglected under these conditions. Then the mass and momentum conservation equations can be approximated by:

$$\frac{\partial u}{\partial t} + g\frac{\partial \eta}{\partial x} = 0 \quad \text{Conservation of Momentum (approximate)} \tag{1.137}$$

$$\frac{\partial \eta}{\partial t} + \bar{h}\frac{\partial u}{\partial x} = 0 \quad \text{Conservation of Mass (approximate)} \tag{1.138}$$

Differentiation of Eq. (1.137) with x and the second with t allows Eqs. (1.137) and (1.138) to be combined to deliver:

$$\frac{\partial^2 \eta}{\partial x^2} - \frac{1}{g\bar{h}}\frac{\partial^2 \eta}{\partial x^2} = 0 \quad \text{Wave equation with velocity } \sqrt{g\bar{h}} \tag{1.139}$$

Following Holthuijsen et al. [34], the conservation equation can be used to obtain information about the spectral wave action density. In this context, the conservation equation is parameterized in the frequency domain ω by acquaintance with the zeroth and first moment of the action spectrum as dependent variables, while retaining a fully spectral depiction in the direction domain θ. This precedes to binary combined partial differential equations for the zeroth and the first moments of the action spectrum $A(\omega,\theta)$. Consequently, this can mathematically be expressed as:

$$\frac{\partial(c_{gx}m_0)}{\partial x} + \frac{\partial(c_{gy}m_0)}{\partial y} + \frac{\partial(c_\theta m_0)}{\partial \theta} = S_0 \tag{1.140}$$

$$\frac{\partial(c_{gx}m_1)}{\partial x} + \frac{\partial(c_{gy}m_1)}{\partial y} + \frac{\partial(c_\theta m_1)}{\partial \theta} = S_1 \tag{1.141}$$

where c_g is the wave group velocity and the spectral moments $m_n(\theta)$ are defined by:

$$m_n(\theta) = \int_0^\infty \omega^n A(\omega, \theta) d\omega \tag{1.142}$$

where ω is the absolute frequency and A is the spectral wave action density.

The left-hand side of Eqs. (1.140) and (1.141) comprises the impact of refraction and shoaling. The source terms S_0 and S_1 comprise the special effects of energy input from the wind, bottom dissipation, and wave breaking. A Eulerian finite-difference technique can be used to compute the spatial discretization of the basic partial differential equations. The zeroth and first moments of the action spectrum are calculated on a rectangular grid for a number of discrete directions θ. Linear forward-differencing is applied in the x-direction, while in both the y- and θ-directions it is possible to choose between a number of solution techniques. Therefore, the source terms owing to local wind generation are initiated obviously, while the source terms owing to wave breaking and bottom dissipation are proposed covertly. Henceforth, a nonlinear iteration is performed at each grid point. In fact, the nonlinear equation is resulting from the spatial discretization, which is solved using a once-through marching procedure in the x-direction (the predominant direction of wave propagation). In this regard, it restricts the angle between the direction of wave propagation and the x-axis to be less than about 60 degrees owing to numerical stability considerations [33, 34, 36, 37].

1.12 Water level exchange and water flow

The water levels and flows can be resolved on a rectangular grid covering the area of interest when provided with the bathymetry, bed resistance coefficients, wind field, hydrographic boundary conditions, etc. The instantaneous water levels and fluxes are acquired from the clarification of the continuity and momentum equations as [12, 23, 30, 36]:

$$\frac{\partial \zeta}{\partial t} + \frac{\partial p}{\partial x} + \frac{\partial q}{\partial y} = S - e \tag{1.143}$$

$$\begin{aligned} &\frac{\partial p}{\partial t} + \frac{\partial}{\partial x}\left[\frac{p^2}{h}\right] + \frac{\partial}{\partial y}\left[\frac{p \cdot q}{h}\right] + gh\frac{\partial \zeta}{\partial x} + \frac{g\sqrt{\frac{p^2}{h^2} + \frac{q^2}{h^2}}}{C^2} \cdot \frac{p}{h} - f_w V V_x - \frac{h}{\rho} \cdot \frac{\partial p_a}{\partial x} - \Omega q \\ &- \left[\frac{\partial}{\partial x}\left[M_x \cdot h \cdot \frac{\partial u}{\partial x}\right] + \frac{\partial}{\partial y}\left[M_y \cdot h \cdot \frac{\partial u}{\partial y}\right]\right] = S_{ix} \end{aligned} \tag{1.144}$$

$$\begin{aligned} &\frac{\partial q}{\partial t} + \frac{\partial}{\partial y}\left[\frac{q^2}{h}\right] + \frac{\partial}{\partial x}\left[\frac{p \cdot q}{h}\right] + gh\frac{\partial \zeta}{\partial y} + \frac{g\sqrt{\frac{p^2}{h^2} + \frac{q^2}{h^2}}}{c^2} \cdot \frac{q}{h} - f_w V V_y - \frac{h}{\rho} \cdot \frac{\partial p_a}{\partial y} + \Omega p \\ &- \left[\frac{\partial}{\partial x}\left[M_x \cdot h \cdot \frac{\partial v}{\partial x}\right] + \frac{\partial}{\partial y}\left[M_y \cdot h \cdot \frac{\partial v}{\partial y}\right]\right] = S_{iy} \end{aligned} \tag{1.145}$$

where $\zeta(x,y,t)$ is the instantaneous water surface above datum, $p(x,y,t)$ and $q(x,y,t)$ are the flux densities in x- and y-directions, $h(x,y,t)$ is the total water depth, S is a source magnitude per unit horizontal area, S_{ix} and S_{iy} are sources for impulse in x- and y-directions (for instance, gradients in radiation stress field), e is the evaporation rate, g is gravitational acceleration, and C is Chezy's resistance number, which describes the turbulent mean flow velocity $\overline{U}$ as:

$$C=\overline{U}\left(\sqrt{r\partial u}\right)^{-1} \tag{1.145.1}$$

where r is the cross-sectional area of flow divided by the wetted perimeter (for a wide ocean basin this is approximately equal to the water depth), and therefore ∂u is the flow gradient, which for normal depth of flow equals the bottom slope;f_w is wind friction factor; V, V_x, and V_y are wind speed and its components in x- and y-directions; p_a is barometric pressure; ρ is density of water; Ω is Coriolis coefficient; $M(x,y)$ is the momentum exchange coefficient (eddy viscosity); x and y are space coordinates; and t is time.

Eqs. (1.143) and (1.145) are resolved by implicit finite difference approaches with the variables expressed on a space-staggered rectangular grid. A "fractioned-step" approach in conjunction with an alternating direction implicit (ADI) algorithm is used in the solution to circumvent the inevitability for iteration. Second-order accuracy is confirmed through the centering in time and space of entire derivatives and coefficients. The ADI algorithm denotes that at each time step a solution is first made in the x-direction using the continuity and x-momentum equations followed by a similar solution in y-direction [36].

1.12.1 The alternating direction implicit (ADI) method

This section delivers an example of how ADI is implemented in solving the equation of fluid motion. The ADI is a time splitting or fractional step technique. The principle concept is to utilize an implicit discretization in one direction and an explicit discretization in another direction.

In this regard, let us use the finite difference, the value of water flow u at the node (i, j), which, for iteration $(m+1)$, can be simplified as:

$$u_{i,j}^{(m+1)}=\frac{1}{4}\left[u_{i,j+1}^{(m)}+u_{i,j-1}^{(m)}+u_{i+1,j}^{(m)}+u_{i-1,j}^{(m)}\right] \tag{1.146}$$

Let us add and subtract $u_{i,j}^{(m)}$ from Eq. (1.146) to yield:

$$u_{i,j}^{(m+1)}=u_{i,j}^{(m)}+\frac{1}{4}\left[u_{i,j+1}^{(m)}+u_{i,j-1}^{(m)}+u_{i+1,j}^{(m)}+u_{i-1,j}^{(m)}-4u_{i,j}^{(m)}\right] \tag{1.147}$$

or equally:

$$u_{i,j}^{(m+1)}-u_{i,j}^{(m)}=\frac{1}{4}\left\{\left[u_{i,j+1}^{(m)}-2u_{i,j}^{(m)}+u_{i,j-1}^{(m)}\right]+\left[u_{i+1,j}^{(m)}-2u_{i,j}^{(m)}+u_{i-1,j}^{(m)}\right]\right\} \tag{1.148}$$

Therefore, every iteration is deliberated as a binary-step process in which the first step proceeds to the $(m+0.5)$ level, and the second step proceeds to the $(m+1)$ level. Both steps can mathematically be expressed as:

$$u_{i,j}^{(m+1/2)} - u_{i,j}^{(m)} = \frac{1}{4}\left\{\left[u_{i,j+1}^{(m+1/2)} - 2u_{i,j}^{(m+1/2)} + u_{i,j-1}^{(m+1/2)}\right] + \left[u_{i+1,j}^{(m)} - 2u_{i,j}^{(m)} + u_{i-1,j}^{(m)}\right]\right\} \tag{1.149}$$

$$u_{i,j}^{(m+1)} - u_{i,j}^{(m+1/2)} = \frac{1}{4}\left\{\left[u_{i,j+1}^{(m+1/2)} - 2u_{i,j}^{(m+1/2)} + u_{i,j-1}^{(m+1/2)}\right] + \left[u_{i+1,j}^{(m+1)} - 2u_{i,j}^{(m+1)} + u_{i-1,j}^{(m+1)}\right]\right\} \tag{1.150}$$

The ADI technique uses the tridiagonal set of Eqs. (1.147)–(1.150) at the $\frac{m+1}{2}$ level. The equations can be solved along all rows of the grid, one row at a time. As soon as all nodes have been promoted to the $\frac{m+1}{2}$ level, a parallel process for the column of nodes is operated. The two-step iterations are then accomplished when the new results of $u_{i,j}^{(m+1)}$ are computed.

1.13 Dispersion relation for water waves

Ocean waves are dominated by the dispersion phenomena, because two different forces—gravity and surface tension—contribute enlargement to them. At long wavelengths λ or small wavenumbers $k=\frac{2\pi}{\lambda}$, gravity delivers the dominant restoring force. In this limit, the angular frequency $\omega=2\pi f$ of small amplitude waves (of frequency f) is related to the wavenumber by $\omega^2=gk$. This nonlinear relationship, which is valid for water that is much deeper than λ, may be contrasted with the linear relationship $\omega=ck$ for light waves or sound in air. Consequently, the phase velocity $\frac{\omega}{k}$ and the group velocity $\frac{d\omega}{dk}$ are unequal for water waves. On the other hand, for short-wavelength waves where k is large, the surface tension γ (resistance to stretching of the interface when it is curved, dimensions of force per unit length) delivers the principal restoring force leading to wave motion. In this context, gravity can be negligible by comparison. In this case, the application of Newton's laws to a fluid interface leads to the following prediction:

$$\omega^2 = \frac{\gamma k^3}{\rho} \tag{1.151}$$

where ρ is the density of the fluid.

In general, both effects may be operative at the same time at intermediate wavenumbers, so the general dispersion relation, valid at any wavenumber (for deep water waves), is:

$$\omega^2 = gk + \frac{\gamma k^3}{\rho} \tag{1.152}$$

What occurs if the water is not deep (compared to the wavelength)? In general, Eq. (1.152) for ω^2 is reduced by a multiplicative factor tan$h(kh)$, where h is the depth. In this view of Eq. (1.152)), you will test this dispersion relation and measure the surface tension. Consequently, at large amplitudes, the above relations may no longer be accurate, and uniform waves also may not be stable. On this understanding, viscosity can cause the waves to decay exponentially in time (resembling $e^{-\delta t}$) with a decay rate (inverse decay time) that is approximately $\delta=2\nu k^2$. In this case, the kinematic viscosity n (the ordinary viscosity h divided by the density) is 1×10^{-2} cm^2/s for water. As a consequence, the waves can also decay exponentially in space away from the wave source [4, 5, 23, 37].

1.14 Nonlinear water flow and wave propagation

The research of water waves has philosophical and gripping links to numerous technical exploration scientific fields, utilizing the auxiliary improvement that precise frequently the manners of the waves may be evaluated by explicit observation such as wave tank experiment. The complexity and variability of water-wave phenomena necessitates innovative approaches from countless scientific fields, ranging from physical methods to intellectually accurate scientific considerations. Therefore, it appears to be commonly accepted that the most promising direction for forthcoming exploration in water waves is research of waves of large amplitude, where the conventional linear technique compromises our understanding of the ocean wave mechanics.

It is extensively recognized that the incompressible Euler equations with a free boundary are the prevailing equations for water waves and, essentially, they have been the focus of extensive exploration. Most of the theoretical mechanisms on surface water waves accept that the flow is irrotational (zero vorticity). This is mathematically appropriate, since in this equation, the flow velocity is the gradient of a velocity potential and thus the harmonic function theory applies to our view of wave mechanic. Furthermore, if at a certain moment, the water drift is irrotational, it cannot attain vorticity at subsequent times without nonconservative forces acting on it [4, 18, 34]. While the qualitative comprehending of periodic propagation waves in irrotational flow in excess of a flat bottom topography is relatively complex, it provides understanding of the dynamics of waves of large amplitude [37]. However, there are many vital numerical phases that remain to be revealed; for example, weakly nonlinear model equations, the most important being integrable equations. In fact, a practically comprehensive of wave dynamic propagations [4, 6, 23], unable to tackle linear theory's inability to deliver precise computable predictions for waves of moderate and large amplitude. This could be attributed to the fact that wave geometry structures permit, by means of inverse scattering approaches [37]. Consequently, the irrotational traveling waves have the corresponding item for rotational waves with no stagnation points: explicit dispersion relations which are found for large periods of vorticity distributions.

Nonetheless, nonzero vorticity conveys new phenomena, such as flow conversation and Kelvin's cat's-eye streamline patterns near a critical layer. In contrast to irrotational traveling waves, the wave-current interactions with constant nonzero vorticity result in spherical waves with swinging patterns. In this view, the postulation of constant vorticity for wave-current interactions abridges the gap between mathematics substantially and constant vorticity, which is illustrated when the waves are long compared with the water depth. Indeed, it is the actuality of nonzero mean vorticity that is more significant rather than its precise spreading [22–28, 30].

Needless to say, since the governing equations for water waves are exceedingly nonlinear, solutions that designate accurate fluid motions are vague and the expansion of singularities in conventional solutions (in the form of wave breaking) is one of the most incomprehensible unanswered questions. Nevertheless, on the related matter of the long-time presence of solutions of small amplitude, there has lately been noteworthy progress [10, 12, 33, 37].

1.15 Energy equation of fluid flow

One of the essential rules of nature is that energy is neither created nor destroyed, but only changed from one form to another. Consequently, numerous prescriptions for energy exist. With the intention of understanding how these formulas describe the physical world, it is imperative to comprehend what physicists mean when they talk about energy. Energy is a concept rooted in conventional physics as explained by Sir Isaac Newton. The simple mathematical energy formula is given by:

$$e_k = 0.5mv^2 \tag{1.153}$$

where e_k is kinetic energy, m is the mass in kilograms, and v is the object velocity in meters per second.

In other words, a fluid parcel in motion retains its energy of movement, which is equal to the work that would be necessary to stop it moving, i.e., kinetic energy. In this view, kinetic energy is a function of the square of fluid particle velocity as well as one half of its mass (m). On the contrary, a fluid parcel at rest in Earth's gravitational field retains a potential energy by virtue of its water depth; if it were to fall freely through the water column, it would gain kinetic energy equal to this potential energy. Thus, potential energy (e_p) is a function of the water parcel mass, its water depth (z) and the acceleration due to gravity through the water column (g), i.e., $e_p = gz$.

Consequently, the energy equation corresponds to conservation of energy. In fact, conservation of energy can be specified as the period rate of growth in stored energy of the fluid, which equals the net period rate of energy accumulation of heat transfer into the fluid plus the net rate of energy accumulation by work transfer into the fluid. Consistent with Stern et al. [10–13], the energy equation for fluid flow can be mathematically expressed as:

$$\frac{dE}{dt} = \frac{d}{dt}\int \rho e d\forall + \int \rho e \vec{U} \times d\vec{A} \tag{1.154}$$

where e is energy per unit mass, which equals:

$$e = \hat{u} + e_k + e_p \tag{1.154.1}$$

where $\hat{u}$ is mean flow and e_k and e_p are the kinetic and potential energy, respectively.

In this regard, the basic energy equation, as addressed by Stern et al. [10–13], is given by:

$$\dot{Q} - \dot{W}_s - \dot{W}_{fshear} = \frac{d}{dt}\int_{\forall} \rho\left(\frac{U^2}{2} + gz + \hat{u}\right)d\forall + \int_S \rho\left(\frac{V^2}{2} + gz + \hat{u} + \frac{p}{\rho}\right)\vec{U} \cdot dS \tag{1.155}$$

where $\dot{Q}$ is the rate of heat transfer to the fluid system, $\dot{W}_s$ is the rate of work done by the water particles, and $\dot{W}_{fshear}$ is the work due to shear flows.

Therefore, the net work has been done in the surroundings as a result of normal and tangential stresses acting on the close surfaces. Then the steady flow energy can be expressed as:

$$\dot{Q} - \dot{W}_s = \int_S \rho\left(\frac{U^2}{2} + gz + \frac{p}{\rho} + \hat{u}\right)\vec{U} \cdot dS \tag{1.156}$$

Eq. (1.156) demonstrates that although the velocity diverges across the flow sections, the streamlines are expected to be straight and parallel. As a result, there is no acceleration of the streamlines and the pressure is hydrostatically distributed, i.e., $p/\rho + gz$ is constant. The rate of the change of energy per unit volume is described in terms of the divergence of energy flux $\boldsymbol{E}_F$:

$$\boldsymbol{E}_F = \vec{U}\left(0.5\rho U^2 + p + \rho g z\right). \tag{1.157}$$

In this regard, the law of conservation of energy can be expressed as:

$$\frac{dE}{\partial t} = -\operatorname{div} \boldsymbol{E}_F \tag{1.158}$$

Eq. (1.156) can be modified as the amount of energy passing across a unit surface area normal to the velocity field $\vec{U}$, per unit time is given by:

$$-\int_S \vec{U}\left(0.5\rho U^2\right) \cdot \vec{n}\, dS - \int_S \rho \vec{U} \cdot \vec{n}\, dS - \int_S \rho g z \vec{U} \cdot \vec{n}\, dS. \tag{1.159}$$

Eq. (1.159) contains three terms. The first term is the kinetic energy transported across the surface per unit time by the fluid. The second term is the work done by the pressure forces on the fluid within the surface. The third term is the work done by the gravitational force acting on the fluid particles.

The basis of computational ocean dynamics are the vital governing equations of fluid dynamics—the continuity, momentum, and energy equations. Consequently, the governing equations are considered as a coupled scheme of nonlinear partial differential equations, and hence they are extremely challenging to resolve systematically. As yet, there is no common obstructed-form clarification to these equations. The next chapter will attempt a new approach of quantization of ocean dynamics.

References

[1] Phillips OM. The dynamics of the upper ocean. CUP Archive; 1966.

[2] Dysthe K, Krogstad HE, Müller P. Oceanic rogue waves. Annu Rev Fluid Mech 2008;40:287–310.

[3] Bühler O. Waves and mean flows. Cambridge University Press; 2014.

[4] Pond S, Pickard GL. Introductory dynamical oceanography. Elsevier; 2013.

[5] Talley LD. Descriptive physical oceanography: An introduction. Academic Press; 2011.

[6] Premžoe S, Tasdizen T, Bigler J, Lefohn A, Whitaker RT. Particle-based simulation of fluids. In: Computer graphics forum. vol. 22, No. 3. Oxford: Blackwell Publishing, Inc; 2003. p. 401–10.

[7] Young DF, Munson BR, Okiishi TH, Huebsch WW. A brief introduction to fluid mechanics. John Wiley & Sons; 2010.

[8] Elder S, Williams J. Fluid physics for oceanographers and physicists. Oxford: Butterworth-Heinemann; 1996.

[9] Acheson DJ. Elementary fluid dynamics. Oxford; 19913020.

[10] Stern F, Wilson RV, Coleman HW, Paterson EG. Verification and validation of CFD simulations. Iowa City: Iowa Institute of Hydraulic Research; 1999.

[11] Stern F, Wilson R. Closure to "discussion of 'comprehensive approach to verification and validation of CFD simulations—part 1: methodology and procedures'"(2002, ASME J. Fluids Eng., 124, p. 809). J Fluids Eng 2002;124(3):810–1.

[12] Stern F, Muste M, Xing T, Yarbrough D. Hands-on student experience with complementary CFD educational interface and EFD and uncertainty analysis for introductory fluid mechanics, In: ASME 2004 heat transfer/fluids engineering summer conferenceAmerican Society of Mechanical Engineers Digital Collection; 2009. p. 649–54.

[13] Stern F, Yoon H, Yarbrough D, Okcay M, Oztekin U, Roszelle B. Hands-on integrated CFD educational interface and EFD/ePIV/flowcoach laboratories for introductory fluids mechanics, In: 50th AIAA aerospace sciences meeting including the new horizons forum and aerospace exposition 2012 Jan; 2012. p. 908.
[14] Paterson EG, Stern F. Computation of unsteady viscous marine-propulsor blade flows—part 2: parametric study. J Fluids Eng 1999;121(1):139–47.
[15] White FM, Corfield I. Viscous fluid flow. New York: McGraw-Hill; 2006.
[16] Farmer J, Martinelli L, Jameson A. Fast multigrid method for solving incompressible hydrodynamic problems with free surfaces. AIAA J 1994;32(6):1175–82.
[17] Ghasemian M, Ashrafi ZN, Sedaghat A. A review on computational fluid dynamic simulation techniques for Darrieus vertical axis wind turbines. Energy Convers and Manage 2017;149:87–100.
[18] Owen H, Houzeaux G, Samaniego C, Vazquez M, Calmet H. General Purpose Two Fluid Level Set Simulations With the Parallel Finite Element Code Alya. In: 2nd International Conference on Violent Flows. Editions Publibook; 2012. p. 64.
[19] Hirsch C. Numerical computation of internal and external flows: The fundamentals of computational fluid dynamics. Elsevier; 2007.
[20] Versteeg HK, Malalasekera W. An introduction to computational fluid dynamics: The finite volume method. Pearson Education; 2007.
[21] Chung TJ. Computational fluid dynamics. Cambridge University Press; 2010.
[22] Blazek J. Computational fluid dynamics: Principles and applications. Butterworth-Heinemann; 2015.
[23] Wendt JF. Computational fluid dynamics: An introduction. Springer Science & Business Media; 2008.
[24] Abbott MB, Basco DR. Computational fluid dynamics—an introduction for engineers. In: NASA STI/Recon technical report A. 1989. p. 90.
[25] Roache PJ. Fundamentals of computational fluid dynamics. Albuquerque, NM: Hermosa Publishers; 1998.
[26] Shyy W, Udaykumar H, Rao M, Smith R. Computational fluid dynamics with moving boundaries. AIAA J 1998;36(2):303–4.
[27] Blevins RD. Applied fluid dynamics handbook. New York: Van Nostrand Reinhold Co.; 1984 568 pp.
[28] Hoffmann KA, Chiang ST. Computational fluid dynamics. vol. I. Engineering Education System; 2000.
[29] Leal LG. Advanced transport phenomena: Fluid mechanics and convective transport processes. Cambridge University Press; 2007.
[30] Ferziger JH, Perić M. Computational methods for fluid dynamics. Berlin: Springer; 2002.
[31] Denn MM. Process fluid mechanics. Englewood Cliffs, NJ: Prentice-Hall; 1980.
[32] White F. Fluid mechanics. 4th ed. McGraw-Hill Higher Education; 2002. ISBN 0-07-228192-8.
[33] Pedlosky J. Geophysical fluid dynamics. Springer Science & Business Media; 2013.
[34] Holthuijsen LH, Booij N, Herbers TH. A prediction model for stationary, short-crested waves in shallow water with ambient currents. Coast Eng 1989;13(1):23–54.
[35] Bird RB, Stewart WE, Lightfoot EN. Transport phenomena. 2nd ed. John Wiley & Sons; 2001. ISBN 0-471-41077-2.
[36] Svendsen IA, Jonsson I. Hydrodynamics of coastal regions. Lyngby: Den private Ingeniørfond; 1980.
[37] Constantin A. Nonlinear water waves: introduction and overview. Phil Trans R Soc A 2017;376:1–6. https://doi.org/10.1098/rsta.2017.0310 20170310.

CHAPTER

2

Quantization of ocean dynamics

It is impossible to avoid quantum mechanics when it comes to ocean dynamics. The science of ocean dynamics is mainly established based on quantum mechanics. In this regard, quantum mechanics is a keystone to understand the mechanism of ocean dynamic systems. Both chemical characteristics of seawater and ocean dynamics can only be elucidated by quantum mechanics. The chemical compounds in seawater involve molecules made up of hydrogen and oxygen atoms in addition to minerals and dissolved gas. Understanding quantum phases of seawater and constructing innovative ones can create modern physics and dynamics oceanography theories.

This chapter delivers a wide understanding of ocean dynamics from the point of view of quantum mechanics. This chapter starts with the quantum mechanics of seawater molecules and ends up with large-scale ocean wave propagation as described by the Hamiltonian equation.

2.1 Seawater quantum molecules

Seawater is mainly considered as a solution of salt, which is made of a variety of minerals and dissolved gas molecules. It is essentially comprised of a single bulky oxygen atom and two smaller hydrogen atoms. In this sense, hydrogen bonding has a vital influence on the characteristics of seawater and ice. In this regard, the partial positive charge on the hydrogen is extremely concentrated as its atoms are amalgamated in a polar covalent bond with a tiny atom of extraordinarily coarse-grain such as nitrogen (N), iron (F), or oxygen (O). This reaction is achieved due to the small size of a hydrogen atom. Therefore, the dipole-dipole interaction force occurs when the hydrogen is close to oxygen, fluorine, or nitrogen in another molecule, which has about 5%–10% of the strength of a covalent bond (Fig. 2.1). In other words, the hydrogen bond is not a chemical bond, but a molecular bond.

Generally, hydrogen bonds are weaker than covalent bonds, but still strong enough to create extraordinary water surface tension, great solubility of chemical compounds in water, and infrequent thermal properties of water with an unusual density of water [1–3].

When a mineral such as sodium chloride, NaCl, dissolves in water, the ions are discrete and are bounded by water molecules. The singular pair electron orbitals on the oxygen of

https://doi.org/10.1016/B978-0-12-820785-7.00004-6

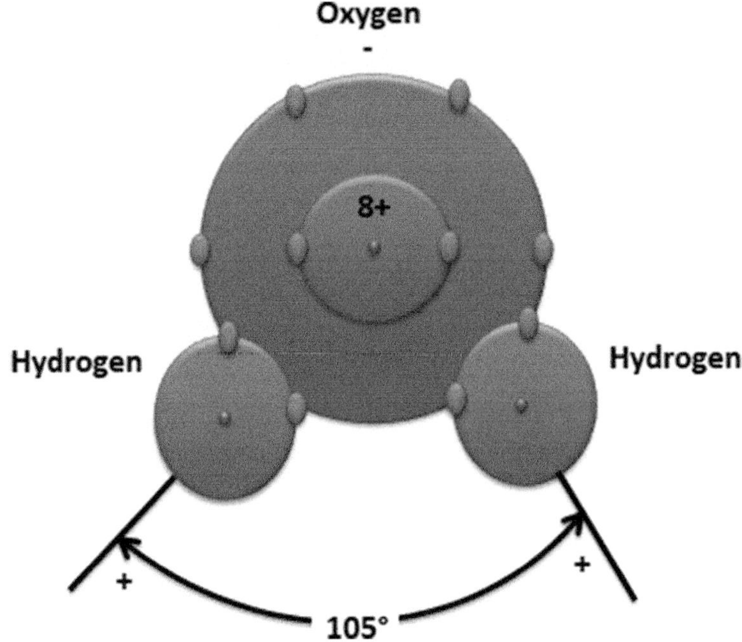

FIG. 2.1 Strong covalent bonds between one hydrogen (H) and two oxygen (O) atoms.

water intermingle with the electron-poor cations, and the anions provide part of their surplus electron density to the electron-poor hydrogen atoms of water. Therefore, other sorts of mineral nutrients (chemical elements) and trace elements including magnesium (Mg), calcium (Ca), and potassium (K) create their structure into crystal salt. Occasionally, this sufficiently disorders the molecular structure of minerals to no longer allow the formation of cube patterns. Every so often in a crystalline structure, other atoms can replace one another. For instance, a potassium (K) or magnesium (Mg) atom can replace a sodium (Na) atom. In this regard, the extra atoms can combine with the crystalline structure, causing substantially greater mineral growth. This is why Dead Sea salt crystals grow so bulky, as such high concentrations of sodium and chlorine atoms are formed. In fact, electrons convey from sodium atoms to nearby chlorine atoms, since the valence electrons in sodium are lightly bound and chlorine has a greater electron attraction. The positively charged sodium ions and negatively charged chloride ions organize into an extended regular array of atoms [3–5].

Conversely, this reaction is mainly ruled by quantum spin. For instance, hydrogen atoms and other substance atoms are arranged based on their quantum spin. In fact, the spin is an attribute of quantum particles that is similar to real balls spinning. However, the quantum spin contains superfluous dimensions and can have just one of the limited discrete momentums. In this regard, atoms can spin in the same direction or in a different direction. In these circumstances, spin generally influences the manner of particles only on the tiny quantum scale. In fact, seawater quantum spin phenomena have a continuing random oscillation even at the lowest temperature of absolute zero [2, 4].

2.2 Ocean dynamics mimic quantum mechanics

Ocean dynamics are precisely analogous with the atmospheric boundary layer characteristics, i.e., temperature, and wind stress; which are tangent to the ocean surface. There is a wealth of complications that are, superficially, so simple that they can be solved. However, most definite problems are too complicated to obtain an exact solution, as was addressed in Chapter 1. A similar technique to the Navier-Stokes equations can be used to acquire approximate solutions for those complex cases.

Ocean dynamics can be described as "chaotic," signifying that there is a constituent of what *looks* like a fundamental probability in many of the problems. An example is turbulence, which is an excellent example of chaotic motion in ocean dynamics. When a ship sails across the sea, it creates a wake. Conversely, there are all kinds of apparently unsystematic ebbs and flows in the wake, which are generally known as turbulence. Consequently, it is believed that the turbulent movements associated with ship wakes are not accurately random. As these turbulent flows fluctuate irregularly as a function of ship physical characteristics such as speed, length, and size. Nonetheless, the turbulent wake autonomous seems random because the waters are therefore complex to all varieties of microscopic and extremely forceful to dynamic rules for causing such turbulent flow.

Quantum mechanics are believed to be a little different. In fact, quantum mechanics are believed to be absolutely probabilistic. With this in mind, the "randomness" in ocean dynamics could eventually be predictable if ocean dynamic scientists reached a low enough level of detail, whereas it is thought that specific parts of quantum mechanics primarily cannot be predicted [6–8].

In this regard, the similarities between ocean dynamics and quantum mechanics can be used to validate and comprehend quantum phenomena. Therefore, it is easier to deal with and comprehend a container of water and some oil than it is to struggle to understand an electron in a ring of ions, which is far too tiny to be seen without using an electron microscope. In this circumstance, it was demonstrated that a fluid dynamics system can be set up that absolutely emulates the effects of one of the "essential problems" in quantum mechanics, for instance, the infinite (circular) well. The critical question is: where is a particle of water existing to be? A particle of water is trapped completely through the water body that it cannot catch out of its existing points or frames.

Since it is believed that approximately perfectly imitate quantum systems can be achieved using an exclusively distinctive set of physics, it might be the case that quantum mechanics is not essentially chaotic. So, if the particles are placed on an approximate variety of an "undulating" medium, such as a droplet of oil placed on the surface of water, this may explain why quantum mechanics seems to be probabilistic [9]. There are a couple of speculations and rules which deviate from that thought; consequently, it is highly questionable, but it is worthy of consideration [10]. By this understanding, the pilot wave theory invented by de Broglie can be proved simply by considering ocean dynamics as the conventional original mechanism of quantum particles' seemingly extraordinary behaviors without resorting to the necessity for an enigmatic and superficially magical understanding of modern quantum theory [6, 9, 11].

2.3 Similarities and differences between quantum field theory and ocean dynamics

All fluids in nature are ultimately described by a microscopic quantum field theory. Conversely, the long-distance, low-frequency limit of a wide class of quantum field theories is expected to be described by fluid dynamics (although there is arguably no rigorous proof of this, merely a great deal of phenomenological evidence). Hence quantum field theoretic techniques to describe classical fluids can prove to be extremely useful.In particular, the great unsolved problem of turbulence in fluid dynamics can in principle be solved using gauge gravity duality, which relates highly nonlinear QFTs (which correspond to highly turbulent fluids can be determined; for instance, in the infrared (IR)) to a weakly coupled gravitational system. On a practical note, the dynamics of fluids can be modeled as a quantum path integral (with density perturbations and velocity acting as "fields" in the quantum theory) [10–12]. This reduces the calculation of correlation functions in weakly nonlinear fluids to the evaluation of Feynman diagrams of weakly coupled QFTs. This approach has various advantages over standard perturbation theory—primarily, renormalization group methods in QFT can be dominant to average out small-scale fluctuations. Further QFT techniques may in principle be used to describe phase transitions in fluids analytically. All of these topics are well worth exploring. The relation between fluids and QFTs is a subject of intense research. One can expect both to have much to contribute to one another.

2.4 Quantum spin of seawater

Seawater, due to the existence of NaCl, has a quantum mechanical character rooted in the Pauli exclusion principle, which is generally known as "exclusion principle repulsion." In other words, seawater particles of half-integer spin must have antisymmetric wave functions, and seawater particles of integer spin must have symmetric wave functions. In this view, when the ions are extensively disjointed, the wave functions of their core electrons do not expressively coincide and they can have matching quantum numbers. For instance, they become closer, the increasing intersection of the wave functions causing particular electrons to be forced into higher energy states. In fact, it is impossible for two electrons to be subjugated at the equivalent state.Consequently, by way of a new set of electron energy circumstances is twisted for the NaCl compound, dual-nucleus system, the lower energy states are occupied and certain electrons are strapped into higher states. This necessitates energy and functions as a repulsion, preventing the ions from coming any closer to each other [1, 10, 12].

Moreover, the quantum correlations of the spins are critical to the description of the seawater components, and they certainly extend out to greater ocean distances than their closest neighbors. Consequently, quantum spin seawater is in a way the opposite of a ferromagnetic or antiferromagnetic material, in which the low seawater temperature state of the orientation of the spins has a high degree of long-range order, all of the spins being parallel in a ferromagnet and all the spins being alternatingly antiparallel in an antiferromagnet. One approach in which spin seawater might result from the ice crystal structure at frozen temperatures is such that the interactions between adjacent electron spins at different lattice sites are antiferromagnetic [3,7], which means that the spins all aim to align in an antiparallel fashion. For instance, at the frozen degree of seawater, quantum spin seawater is a system of electron spins, in a block of ordered crystalline ice, in which the electrons are **strongly localized** to the lattice sites of the crystal, and the electron spin orientations are also **strongly disordered** at low and frozen temperatures [9–14].

It turns out that such materials have very interesting properties when you consider the excitations of the system above the lowest energy state—unlike in a ferromagnet or an antiferromagnet, where the lowest-lying excitations are called spin waves and are uncharged bosons, the excitations in a spin liquid, called spinons can have **fractional** spin. The spinons can also easily move around within the spin liquid by a process of rearranging the quantum state of the electron spins, which costs very little energy since the ground state is, as I implied, very highly degenerate, much like a ferromagnet. This fractionalization has been observed in at least one material quite recently. It is of great interest that such materials apparently exist, because they may be good materials to consider in constructing advanced computer memories and/or quantum computers, with the spinons playing the role of qubits [1, 8, 12].

2.5 Kitaev spin seawaters

Can the crystal structure of seawater atoms be represented by Kitaev candidates? Let us consider that M″ is the transition-metal atom and a NaCl-type of $M''^{2+}O^{2-}$ where all M''^{2+} is octahedrally matched with O^{2-}. Each M″O6 octahedron shares its edges with the neighboring 12 MO6 octahedra. Viewing $M''^{2+}O^{2-}$ along the cubic (111) direction, it can be seen that the structure of $M''^{2+}O^{2-}$ comprises an irregular heap of the triangular M''^{2+} planes and the triangular O^{2-} planes. In this observation, the M″O6 octahedron contains M''^{2+} ions and the two O^{2-} triangles completely overhead and underneath M''^{2+}, respectively (Fig. 2.2).

In the real stage, the seawater is a composite of sodium (Na) and chlorine (Cl) atoms. In this understanding, sodium and chlorine atoms are bosons or fermions, dependent on their spin: Na and Cl particles with half-integer spin (like 1/2, 3/2, and so forth) are fermions, while Na and Cl particles with integer spin are bosons, i.e., (0, 1, 2,…). Consequently, fermions are their own antiparticles and can be described as Majorana fermions. In this regard, the neutral spin-1/2 particles can be described by a real wave equation (the Majorana equation), and would, consequently, be indistinguishable from their antiparticle. Indeed, the wave functions of particle and antiparticle are correlated by complex conjugation. Seawater is considered as a superconducting medium, whereas a Majorana fermion can emerge as a (nonfundamental) quasiparticle. In fact, a quasiparticle in a superconductor is its own antiparticle. Consequently, Majorana fermions become the main concept of a Kitaev quantum spin liquid.

In 2006, the physicist Alexei Kitaev developed a theoretical model of microscopic magnets ("spins") that interact in a fashion that leads to a disordered state called a quantum spin liquid. This "Kitaev quantum spin liquid" supports magnetic excitations equivalent to Majorana fermions—particles that are unusual in that they are their own antiparticles. The presence of Majorana fermions is of great interest because of their potential use as the basis for a qubit, the essential building block of quantum computers. Familiar magnetic materials exhibit magnetic excitations called "spin-waves" that occur in quantized swellings, but in the Kitaev quantum spin liquid, the lumps are split and the Majorana excitations are therefore termed "fractionalized."

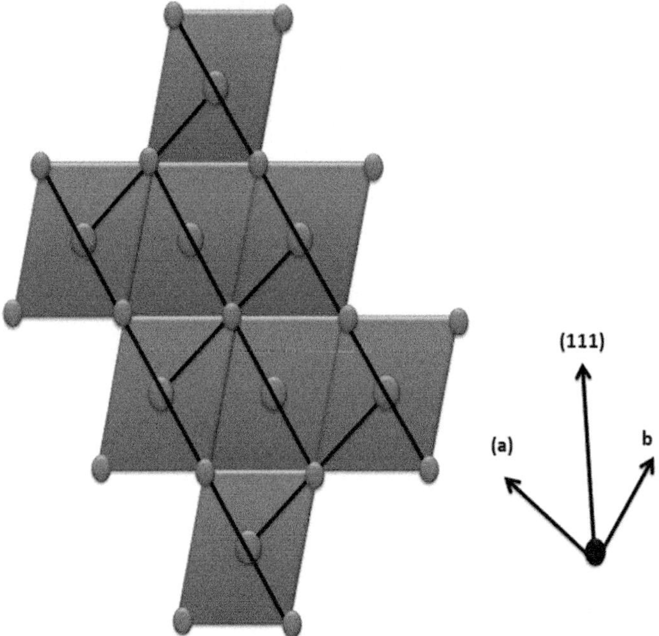

FIG. 2.2 The crystal structure of Kitaev candidate of NaCl.

The form of magnetic excitations created with alpha-ruthenium trichloride was found to be different from spin waves seen in ordinary magnets, but was very well-matched to the spectrum predicted for the Majorana fermions expected in the Kitaev quantum spin liquid. Consequently, the spin-fractionalization in the Kitaev model can be expressed as:

$$H = -\sum_{\langle ij \rangle_\gamma} K_\gamma S_i^\gamma S_j^\gamma \tag{2.1}$$

here $\langle ij \rangle$ is the lattice summation, which is considered between nearest-neighbor spins. The coupling constants K_γ represents a bond index in Eq. (2.1), their value perhaps varying on diverse sorts of bonds. In addition, γ stands for x, y, or z type bonds, and the total is calculated over all honeycomb bonds. In this regard, Eq. (2.1) can be known as Hamiltonian of Kitaev (H) model. S is the quantum spin liquid due to the existing of sodium (Na) and chlorine (Cl) atoms. Eq. (2.1) shows that the seawater quantum spin seawater state can be explained by Hamiltonian of Kitaev (H) model (Fig. 2.2) where the bullets indicate the $S = \frac{1}{2}$ at the corners of a honeycomb structure. Consequently, each spin interacts with three neighboring spins with Ising interactions K^x, K^y, and K^z, respectively (Fig. 2.3). In this regard, Kitaev quantum spin liquid can be formed due to the frustration impacts between the bonds.

Moreover, Eq. (2.1) shows that the fractionalization of quantum spins delivers two sorts of Majorana fermions: itinerant Majorana fermions and localized fluxes. Hence, fundamental excitations in the Kitaev quantum spin liquid (QSL) are not conservative magnons (spin waves), and can reveal an inherent gap in its spin excitation spectrum, relying on the anisotropy of the Ising interactions. Thermodynamic properties of the seawater could show nontrivial behavior, also regarding Eq. (2.1). In this understanding, the nonconservative nature of the z component of the total spin S_z leads to the identical spin liability of the seawater as a function of temperature increment or decrement even though the excitation spectrum of Majorana fermions is integrally separated (Fig. 2.3) [14–17].

Let us assume that the spin of the sweater particles can be considered as an essential dynamic flux. In this view, flux operators F_O can be written as a function of spin S (Fig. 2.4) as:

$$F_{(1-6)} = 2^6 S_1^z S_2^x S_3^y S_4^z S_5^x S_6^y \tag{2.2}$$

Eq. (2.2) demonstrates the possibility of the spin of seawater particles individually around each hexagonal loop as a product of six spin operators S_i^γ (Eq. 2.1), which tolerates every seawater particles eigenstate to be considered by the conserved flux quanta over separate hexagon and conveys the Kitaev Hamiltonian to a block-diagonal form. Consequently, Majorana representation transforms the Kitaev model into the fermionic form, which obviously and usefully imitates the flux-operator $F_{(1-6)}$ as:

$$H = -\frac{1}{4}\sum_{\langle ij\rangle(\gamma)} K_\gamma b_i^\gamma b_j^\gamma \left(a + a^\dagger\right) i\left(a - a^\dagger\right) \tag{2.3}$$

Eq. (2.3) demonstrates that Majorana fermions are constructed from the real or imaginary part of more common complex fermions. In this regard, the dual Majorana fermions can be presented by $(a+a^\dagger)$ and $i(a-a^\dagger)$, respectively, whereas a and $a^\dagger$ are in complex fermion mode. In this understanding, the spin fractionalization of the seawater can be expressed by:

$$S_j^\gamma = \frac{i}{2} b_j^\gamma \left(a + a^\dagger\right) i\left(a - a^\dagger\right) \tag{2.4}$$

Eq. (2.4) reveals that the commutation and conservation rules denote that b_j^γ Majorana particles are constrained to the corresponding γ-type bond connected to the site i and are thus steady (Fig. 2.4). Let us assume that $b_i^\gamma b_j^\gamma$ can be represented by v_{ij}^γ, i.e., $b_i^\gamma b_j^\gamma = v_{ij}^\gamma$. Consequently, the flux-operator $F_{(1-6)}$ conservation law can be given by:

$$F_{(1-6)} = v_{12}^y v_{23}^z v_{34}^x v_{45}^y v_{54}^z v_{67}^x \tag{2.5}$$

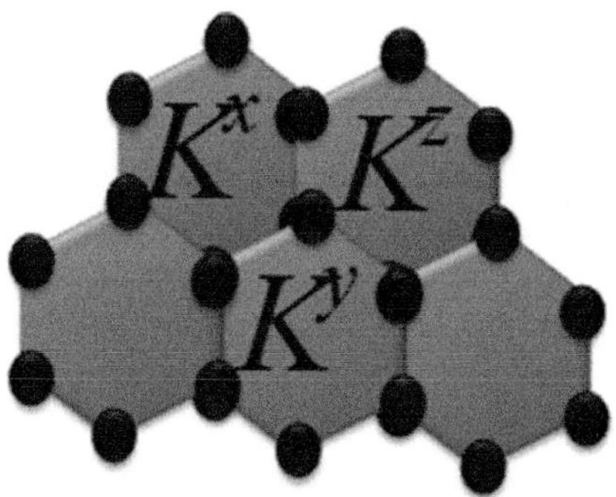

FIG. 2.3 The Kitaev lattice of effective spin.

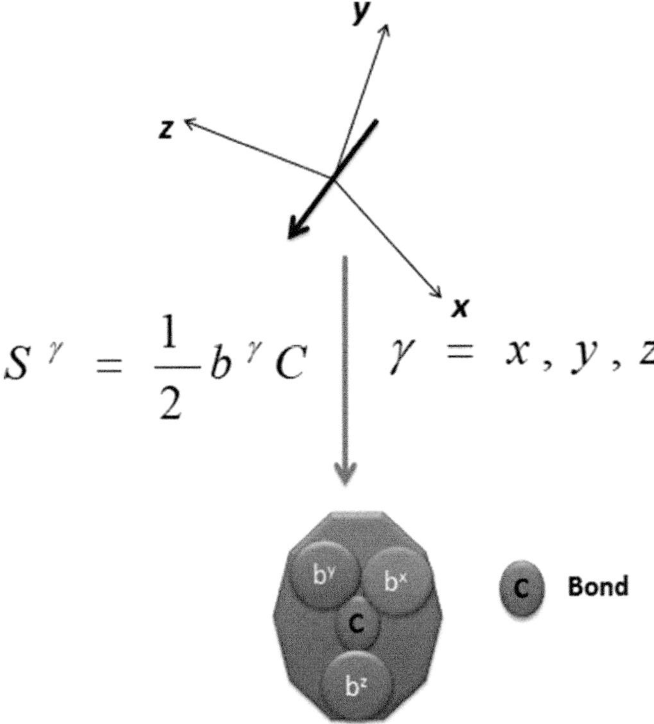

FIG. 2.4 A real spin fractionalizes into two kinds of Majorana fermions.

here bond operators $b_i^{\gamma} b_j^{\gamma} = v_{ij}^{\gamma}$ have eigenvalues $\pm i$ transform with each other and with the Hamiltonian, and their product around a hexagon (Fig. 2.5), which determines the flux $F_{(l-m)} = \pm 1$. Moreover, in each flux sector, the gauge is immobile and the operators v_{ij}^{γ} can be replaced by numbers -1 or $+1$. The equal v_{ij}^{γ} can lead to free flow, which can be a dominant feature of the seawater without wind impacts. Indeed, matter fermions can coherently circulate through the honeycomb lattice of NaCl gaining the maximum kinetic energy [18].

2.6 Hamiltonian mechanics for ocean dynamics

It is agreed that ocean waters are formed of particles, which undergo dynamic motion. The easiest way to describe these particle dynamic motions is through quantum mechanics theory. One of the most important theories to describe the ocean particle dynamics is Hamiltonian. In this understanding, Hamiltonian is a physical function that can be used in computing the motion of particles in conventional dynamics when they are subject to the variability of potentials caused by diverse force fields such as winds, water density fluctuations, and tides. In this understanding, H is implemented to articulate the rate of change in the time of the condition of the ocean dynamic physical system—one regarded as a set of moving particles. In this regard, the Hamiltonian of an ocean dynamic system specifies its total energy—i.e., the quantity of its kinetic energy (that of motion) and its potential energy (that of position)—in

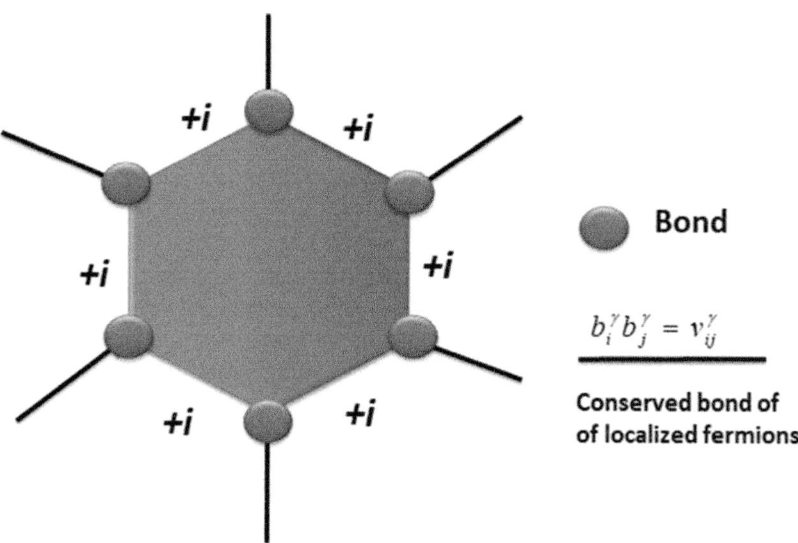

FIG. 2.5 Kitaev quantum-spin-liquid state of flux operator $F_{(1-6)}$.

terms of the Lagrangian function, which have derived in earlier studies of dynamics and of the position and momentum of each of the particles. In other words, $H = E_k + E_p$, where E_k is the kinetic energy of the particle and E_p represents the potential energy function [19–22].

The Hamiltonian function, therefore, is instigated as a comprehensive declaration of the inclination of ocean physical systems to experience fluctuations only by internal and external forces applied to a water body that either minimize or maximize the abstract quantity called the ocean dynamic motion. In this view, the Hamilton operator governs how a quantum ocean system develops with time. Let us assume the quantum state of the ocean system is Φ, and the time evolution of the ocean state is described by:

$$\dot{\Phi} = \frac{i}{\hbar}\widehat{H}\Phi. \tag{2.6}$$

Eq. (2.6) is just the form of the Schrödinger equation, where i is the imaginary unit, and $\hbar$ is the reduced Plank constant $\hbar$ ($6.62607015 \times 10^{-34}$), i.e., $\hbar = \frac{h}{2\pi}$. The expectation value of the Hamiltonian operator, therefore, is the overall energy of the ocean dynamic system; thus, the Hamiltonian is known as the energy operator $\widehat{H}$ [19, 21, 23].

The following critical question can be raised: what are the differences between Newtonian, Lagrangian, Hamiltonian, and quantum mechanics? In this regard, Newtonian mechanics stated three laws of motion:

(i) Objects in motion are inclined to remain in equal motion, unless acted on by a "force."
(ii) The acceleration of an object, which is the rate of change of its velocity, is proportional to a function of its position (and occasionally, its velocity). This function is called the "force" and the proportionality constant is called the "mass."
(iii) If any object A instigates a force on any other object B, then object B triggers a similar force on object A; nonetheless, it must be in the opposite direction.

However, the Newton second law is often hard to solve, especially in more complicated systems such as the ocean, with a large number of interacting forces. Consequently, physicists developed Lagrangian mechanics to obtain a reasonable solution to such problems as ocean dynamic systems. Even though Newtonian mechanics is established on Cartesian coordinates, Lagrangian mechanics are liberated of any particular coordinate system (though the dual interpretations are precisely identical) [20, 22, 24, 25]. This tolerates considerable tractability in performing computing. Let us assume that the Lagrangian function $\ell(v_i, \dot{v}_i; t)$ and the dynamic path can be given by:

$$L[\{v_i(t)\}] = \int_{t_1}^{t_2} dt \ell(v_i, \dot{v}_i; t) \tag{2.7}$$

where v_i are the equivalent generalized velocities.

The generalized coordinates can be any set of coordinates that you like, as long as they completely specify the state of the system. They can be distances, angles or whatever else quantifies the state. But the reason that theoretical physicists went bonkers with the Lagrangian formulation is the elegant physical principle on which the dynamics are based on Hamilton's principle. In this view, the physical path that a dynamic ocean system takes from the time t_1 to t_2 is considered as a stationary movement L. Consequently, ocean dynamic systems have either minimum or maximum fluctuations, which turns out to extremize the dynamic action. Therefore, the following ensues in the Euler-Lagrange equations of motion:

$$\frac{\partial \ell}{\partial v_i} = \frac{\partial}{\partial t} \frac{\partial \ell}{\partial v_i} \tag{2.8}$$

Eq. (2.8) is just a comprehensive form of Newton's second law, where the right-hand side is mass times acceleration, while the left-hand side is the generalized force. The main strength in Eq. (2.8) is that the coordinates v_i can be chosen to facilitate its solution. For instance, spherical coordinates can be used with central forces such as gravity, in which ocean dynamic system can be separated into angular and radial parts. On this understanding, the radial and angular motion of the ocean dynamic system can be described separately by solving the corresponding Euler-Lagrange equations. In this regard, the wider sort of ocean dynamic problems can be solved more easily. Additionally, it expedited the classification of steady-state ocean dynamical systems. However, ocean dynamic parameters fluctuate over time. In this regard, **Hamiltonian mechanics** derived from the Lagrangian formulation can be implemented to describe the steady-state of ocean dynamic systems [20, 22, 24]. The Hamiltonian function H is, therefore, deliberated from the Lagrangian through a modification of variables termed as the Legendre transform:

$$H\left(v_i, \frac{\partial \ell}{\partial \dot{v}_i}; t\right) = \sum_i \dot{v}_i \frac{\partial \ell}{\partial \dot{v}_i} - \ell\left(v_i, \frac{\partial \ell}{\partial \dot{v}_i}; t\right) \tag{2.9}$$

The momentum corresponds to the coordinate $\frac{\partial \ell}{\partial \dot{v}_i}$. Let us assume that $M = \frac{\partial \ell}{\partial \dot{v}_i}$ is the momentum coordinate, which is replaced by the velocity one. In this view, at any time, the dynamic or physical state of the ocean can be described by the set of coordinates $\{v_i, M_i\}$. Consequently, the ocean dynamic motion can be described by using Hamilton's equation of motion, which is derived by merging Eqs. (2.7) and (2.8) together as:

$$\dot{v}_i = \frac{\partial H}{\partial M_i} \text{ and } \dot{M}_i = -\frac{\partial H}{\partial v_i} \tag{2.10}$$

Accordingly, the steady-state motion can be given by Hamilton's equation as:

$$\begin{aligned} \frac{dH}{dt} &= \sum_i \frac{\partial H}{\partial v_i}\frac{\partial v_i}{\partial t} + \frac{\partial H}{\partial M_i}\frac{\partial M_i}{\partial t} \\ &= \sum_i \frac{\partial H}{\partial v_i}\frac{\partial H}{\partial M_i} - \frac{\partial H}{\partial M_i}\frac{\partial H}{\partial v_i} = 0 \end{aligned} \tag{2.11}$$

Furthermore, if the Hamiltonian does not rely explicitly on the coordinate v_i, then subsequently:

$$\frac{dM_i}{dt} = \frac{\partial H}{\partial v_i} = 0 \tag{2.12}$$

Eq. (2.12) demonstrates the steady-state dynamic momentum. In this regard, the time dependence of any the ocean dynamic variable S can be given by:

$$\begin{aligned} \frac{dS}{dt} &= \sum_i \frac{\partial S}{\partial v_i}\frac{\partial v_i}{\partial t} + \frac{\partial S}{\partial M_i}\frac{\partial M_i}{\partial t} \\ &= \sum_i \frac{\partial S}{\partial v_i}\frac{\partial H}{\partial M_i} + \frac{\partial S}{\partial M_i}\frac{\partial H}{\partial v_i} \end{aligned} \tag{2.13}$$

Or more efficiently, Eq. (2.13) can be described as:

$$\frac{dS}{dt} = \{S, H\} \tag{2.14}$$

The Poisson bracket involves S, and H at the right-hand side of Eq. (2.14). Therefore, the steady-state ocean motion occurs when the Poisson bracket of Hamiltonian tends to be zero. In other words, the ocean dynamic system becomes invariant over time [19, 22, 25].

In this understanding, both Lagrangian and Hamiltonian formulations describe similar ocean physical processes. In this regard, both are correspondent to Newtonian mechanics. On the other hand, they have dissimilar compensations reliant on the requested computation. The simple description of ocean motion can be satisfied by using Newtonian mechanics. In contrast, complicated ocean motion can be simplified and standardized by using Lagrangian mechanics. Consequently, homogenous and steady-state motion can be implemented with Hamiltonian mechanics. On the other hand, the combination of these three techniques leads to classical mechanics [22–25].

However, **quantum mechanics** is a fully dissimilar concept of mechanics. In the early stage, it was framed as Hamiltonian mechanics. Feynman then delivered a different interpretation in terms of Lagrangian mechanics.

The ocean dynamics variables such as momentum and positions in point of view of the Hamiltonian version of quantum mechanics are not numbered, and can be considered instead as *operators* Φ. In this context, an operator Φ is simply an ocean dynamic variable that acts on another variable. For instance, ocean current measuring delivers some observable parameters such as velocity and direction, which can be termed as *eigenvalues*, associated with the corresponding operator. In this regard, these eigenvalues are discrete, or *quantized* (hence the name "quantum"). Additionally, the (eigen)values that are obtained from ocean current measurements are *random*. In this regard, the predicted ocean dynamic motion tends to be complicated as only probabilities can be calculated. Subsequently, rather than recognizing the state of an ocean dynamic system with the randomized observables (position, momentum, etc.), it can be identified with the probability variation of observed (eigen)values, or rather its representation—the probability amplitude or *wave function* Ψ. On this understanding, the probability density is the magnitude squared of the wave function. If the wave function, for instance, is well-defined in position space $\Psi = \Psi(x)$, formerly the probability density is $|\Psi(x)|^2$. In other words, the probability of measuring a position in a small neighborhood dx approximately at the point x is $|\Psi(x)|^2 dx$ [20–24]. Consequently, the time dependence of the wave function can be delivered by the Schrödinger equation as:

$$\partial\Psi(i\hbar) = \hat{H}\Psi\partial t \tag{2.15}$$

Eq. (2.15) reveals that the Hamiltonian is the operator $\hat{H}$, which operates on the wave function. In order to obtain this operator, the coordinate of classical Hamiltonian function is replaced by the coordinates and momenta with the corresponding operators under the circumstance of:

$$\hat{v}_i\hat{M}_i - \hat{M}_i\hat{v}_i = i\hbar \tag{2.16}$$

Eq. (2.16) is known as the *canonical commutation relation*. In this regard, any two operators do not transform. An example is a rotation about the z-axis monitored by a rotation about the y-axis, which is dissimilar from the inverse operation. A consequence of this is that arbitrarily precise measurements of both position and momentum simultaneously cannot be obtained [20, 23, 25]. This declaration is frequently abridged as the well-known Heisenberg uncertainty theory:

$$\Delta v \Delta M \geq 0.5\hbar \tag{2.17}$$

where Δv is the uncertainty in the corresponding velocity coordinate v, and ΔM is the uncertainty in the corresponding momentum [19–23].

However, in Newtonian, Lagrangian, and Hamiltonian mechanics the coordinates can be solved as functions of time for ocean dynamic variables such as current movement; here it can be solved for the wave function, which is a function of all these coordinates *and* a function of time. This develops an extra layer of complexity, which creates complicated quantum mechanics problems, which are generally more challenging to unravel than their conventional similarities [19–23].

Consequently, the Feynman's path integral formulation of quantum mechanics is mathematically equivalent to this version of quantum mechanics, nonetheless, rather than implementing the Schrödinger equation for solving the ocean dynamic equation, and annoying to clarify it, the solution can be given as:

$$\Psi = \sum\nolimits_{\{v_i(t)\}} e^{\left[-\hbar^{-1} L[v_i(t)]\right]} \tag{2.18}$$

In Eq. (2.18), the wave function is considered as the sum of all possible ocean motion paths $\{v_i(t)\}$ of this exponential function of the motion over time. In this regard, a particle or ocean dynamic system can consider *any* possible motion path through space and the probability amplitudes are summed up for each path to obtain the total wave function. Nonetheless, this calculation can produce a complicated problem: how can a sum be defined over paths? How do all the motion paths even count? This issue quickly complicates things, and even for simple systems, the sum becomes quite messy. The above equation based on extending Hamiltonian mechanics is generally more uncomplicated [23–25].

In classical mechanics, the Lagrangian is used to derive the equations in order to work with nn second-order differential equations (for a system with nn degrees of freedom), and the Hamiltonian in order to work with 2n2n first-order differential equations. So in that sense, the advantage of the Hamiltonian formalism is that the equations of motion can often be easier to compute than those of the Lagrangian. However, it should be noted that either way, there will be equations of motion describing the same trajectory and the Lagrangian is really "more fundamental" in the sense that the Hamiltonian is defined as a Legendre transform of the Lagrangian, while the Lagrangian comes directly from the principle of least action.

2.7 Incompressible flow with Schrödinger equations

An incompressible, practically inviscid fluid is computed as a continuum controlled by the incompressible Euler equations as [26, 27]:

$$\begin{cases} \frac{\partial}{\partial t}\vec{u} + \left(\vec{u}\cdot\nabla\right)\vec{u} = -\nabla p \\ \text{div } \vec{u} = 0, \end{cases} \tag{2.19}$$

The incompressible Euler equations reveal that the velocity vector $\vec{u}$ with absent forces is completely conveyed by the velocity itself: $\frac{\partial}{\partial t}\vec{u} + \left(\vec{u}\cdot\nabla\right)\vec{u} = 0$. Moreover, a potential force ∇p due to pressure p is merely the prevailing force that acts in the fluid without friction, which leads to the incompressibility constraint. Lastly, under the circumstance div $\vec{u} = 0$, the fluid is incompressible [27].

Ocean flows are considered as the macroscopic phenomenon, which is derived through the inertia and pressure due to colliding molecules and various fluid flowing. In this regard, the fluid is computed by many-body Newtonian mechanics or kinetic models [26–29].

In quantum mechanics, a solitary particle at the atomic size is demonstrated as a complex-valued wave function Ψ, which fulfills the Schrödinger equation:

$$i\hbar\frac{\partial}{\partial t}\Psi + \frac{\hbar^2}{2}\Delta\Psi = p\Psi \tag{2.20}$$

In Eq. (2.20), the dispersive wave equation is presented as the linear formula $i\hbar\frac{\partial}{\partial t}\Psi = -\frac{\hbar^2}{2}\Delta\Psi$. In this context, the wave travels at different speeds with different wavelengths. In this understanding, the Schrödinger equation reveals the wave group pattern as a function of wavenumber variations, i.e., $v = \hbar k$. In this regard, the wavenumber is $k = \nabla \arg \Psi$, where $\hbar \nabla \arg \Psi$ signifies the particle velocity field, while the probability density of the location of the water particle is presented by the amplitude squared of the wave function $\xi = |\Psi|^2$. Group velocity can lead to an inertial motion similar to a fluid, and carries along both the velocity field (wave frequency) and the density (amplitude). Consequently, the potential field $-\nabla p$ in Eq. (2.19) is created by the potential energy $p\Psi$ in Eq. (2.20) [27, 28].

Consequently, a nonlinear continuity equation $\frac{\partial}{\partial t}\vec{u} + \left(\vec{u}\cdot\nabla\right)\vec{u} = 0$ can be implemented to identify the inertial motion in the Euler equation, whereas it is equivalent to the linear comparison $i\hbar\frac{\partial}{\partial t}\Psi - \frac{\hbar}{2}\Delta\Psi = 0$ in the Schrödinger equation. In this understating, the simulation of the equation of motion can be achieved using a Schrödinger equation. In doing so, let us assume that $\Psi = re^{i\theta}$, $q = r^2$ represents the fluid density, and $\vec{u} = \hbar\nabla\theta$ symbolizes the velocity, which is the absence of vorticity flow. In terms of spherical polar coordinate r and θ, the Ψ is formulated as [27]:

$$\frac{\partial}{\partial t}\Psi - \frac{\partial r}{\partial t}e^{i\theta} = ir\frac{\partial \theta}{\partial t}e^{i\theta} \tag{2.21}$$

and:

$$\Delta\Psi = \nabla\cdot(\nabla\cdot\Psi) = \Delta re^{i\theta} + 2i\Delta r\cdot\nabla\theta e^{i\theta} + ir\Delta\theta e^{i\theta} - r|\Delta\theta|^2 e^{i\theta} \tag{2.22}$$

If the potential of the fluid motion to be scrutinized is spherically symmetric, then the Schrodinger equation in spherical polar coordinates can be expended to advantage. In these regards, the Schrödinger equation can be expressed in spherical polar coordinate as [26, 27, 29]:

$$\frac{\partial r}{\partial t} + ir\frac{\partial \theta}{\partial t} = \frac{i\hbar}{2}\Delta r - \hbar\nabla r\cdot\nabla\theta - \frac{\hbar}{2}r\nabla\theta - \frac{i\hbar}{2}r|\nabla\theta|^2 - i\hbar^{-1}pr \tag{2.23}$$

Let us eradicate the common factor $e^{i\theta}$ from Eqs. (2.21) and (2.22), then the real and the imaginary parts of Eq. (2.23) are given by:

$$\frac{\partial}{\partial t}r = -\hbar\nabla r\cdot\nabla\theta - \frac{\hbar}{2}r\Delta\theta \tag{2.24}$$

$$r\frac{\partial}{\partial t}\theta = 0.5\hbar\nabla r - \frac{\hbar}{2}r|\Delta\theta|^2 - \hbar^{-1}pr \tag{2.25}$$

Let us multiply Eq. (2.24) by $2r$ and replace $q = r^2$,and $\vec{u} = \hbar\nabla\theta$ and then implement the mathematical fact of $2r\nabla r = \nabla\left(r^2\right)$,and $2r\frac{\partial}{\partial t}r = \frac{\partial}{\partial t}\left(r^2\right)$ to acquire the continuity equation as:

$$\frac{\partial}{\partial t}q + \nabla\cdot\left(q\,\vec{u}\right) = 0. \tag{2.26}$$

Let us divide Eq. (2.25) by r and take off $\hbar\nabla$ to acquire the momentum equation of water flow as:

$$\frac{\partial}{\partial t}\vec{u} = \nabla\left(-p + 0.5\hbar\frac{\Delta\sqrt{q}}{\sqrt{q}}\right) - \vec{u}\cdot\nabla\vec{u}\,. \tag{2.27}$$

The combination of Eqs. (2.26) and (2.27) produces the quantum Euler equation, which can be expressed as [27]:

$$\begin{cases} \dfrac{\partial}{\partial t}q + \nabla \cdot \left(q\,\vec{u}\right) = 0, \\ \dfrac{\partial}{\partial t}\,\vec{u} + \left(\vec{u}\cdot\nabla\right)\vec{u} = -\nabla\left(p - 0.5\hbar^2\dfrac{\Delta\sqrt{q}}{\sqrt{q}}\right). \end{cases} \tag{2.28}$$

here the term $0.5\hbar^2\frac{\Delta\sqrt{q}}{\sqrt{q}}$ is known as Bohm potential or quantum potential, which depends on the curvature the amplitude of the wave function. Moreover, in the context of the de Broglie-Bohm theory, the quantum potential is an expression within the Schrödinger equation, which acts to escort the travel of quantum particles. On this understanding, the quantum potential is dominated by several characteristics such as the following:

(i) It is derived mathematically from the real part of the Schrödinger equation under polar decomposition of the wave function, which is not derived from other external Hamiltonian sources, and could be said to be involved in the self-organizing process involving a basic underlying field.

(ii) It conveys information about the complete ocean dynamic system in which the particle finds itself.

(iii) It does not change if r is multiplied by a constant, as this term is also present in the denominator, so that $0.5\hbar^2\frac{\Delta\sqrt{q}}{\sqrt{q}}$ is independent of the magnitude of Ψ and thus of field intensity; therefore, the quantum potential fulfills a precondition for nonlocality: it need not fall off as distance increases [30].

2.8 Quantum mechanics of Coriolis force

The significant question that can be raised is: can the Coriolis force be simulated using the Schrödinger equation? In other words, can the Coriolis force add to incompressible Schrödinger flow? In fact, the direct forces for a Schrödinger equation can be appended subjectively to a wave function, which is the delivered forces from the potential term. These potential forces, however, could be excluded through the pressure prediction. Let us consider that Coriolis force is created due to the friction force created from the moving particle in the rotating frame. In this view, a particle is moving at velocity $u \in \mathbb{R}^3$ with the angular momentum of $\Omega \in \mathbb{R}^3$, which creates a Coriolis of $2u \in \Omega$. In this regard, the background motion can be obtained by $curl\ A = 2\Omega = B$. Then the moving particle has experienced the effect of force, which is $u \times curl\ A = u \times B$. Twice the vector field A is assumed to be time-independent. However, the centrifugal force is ignored due to the impact of the pressure projection through the incompressible fluids [27, 31]. Let us consider that F_0 is the initial frame where the water particle locates at the position x, then instigating Schrödinger equation in the frame F_0:

$$i\frac{\partial}{\partial t}\psi(x,t) = \frac{\hat{p}^2}{2m}\psi(x,t) \tag{2.29}$$

where $\hat{p}^2$ is the potential operator and m is the particle mass.

Under the circumstance of the frame rotation from F_0 to F_T due to the impact of the Coriolis force, the Schrödinger equation can be reformulated as:

$$\hat{E}'\psi(u,\partial t) = \frac{\left(\hat{p} - m\vec{V}\right)^2}{2m}\psi(u,\partial t) \tag{2.30}$$

The kinetic energy $\hat{E}'$ is generated due to the water particle mass transfers from F_0 to F_T with resultant velocity $\vec{V}$, and it can be expressed by:

$$\hat{E}' = i\left(\frac{\partial}{\partial t'} + \vec{V}\cdot\nabla'\right) \tag{2.31}$$

Using the nonrelativistic limit of the Lorentz transformation for the energy, one can obtain measured energy E' as:

$$E' = E - \vec{V}\cdot p + 0.5m\vec{V}^2 \tag{2.32}$$

Let us involve $\hat{p}' = \hat{\nabla}' - m\vec{V}$ in Eq. (2.30), which can be modified as:

$$\left(i\frac{\partial}{\partial t'} + 0.5m\vec{V}^2\right)\psi(u,\partial t) = 0.5m^{-1}\left(\hat{p}' - m\vec{V}\right)^2\psi(u,\partial t). \tag{2.33}$$

The nonrelativistic Lorentz boost U expresses the Galilei boost with a velocity $\vec{V}$ as [31, 32]:

$$U = e^{\left(it\vec{V}\cdot\hat{p}' - im\vec{V}\cdot\left(x - \vec{V}t\right)\right)}. \tag{2.34}$$

Since the velocity $\vec{V}$ is given by $\Omega \times x$, the nonrelativistic Lorentz boost U can be reconstructed as:

$$U = e^{\left(it\left(\Omega\times\left(x - \vec{V}t\right)\cdot\hat{p}'\right)\right)} = e^{\left(it\Omega\cdot\hat{L}\right)} \tag{2.35}$$

where $\hat{L}$ is the orbital angular momentum and is given by:

$$\hat{L} = \left(x - \vec{V}t\right)\cdot\hat{p}' \tag{2.35.1}$$

Eq. (2.35) might envisage that the orbital angular momentum is the equivalent in both initial and transfer frames. Those consequences, nonetheless, are evidently incorrect even classically. In fact, nonrelativistic Lorentz boost $U = e^{it\Omega\cdot\hat{L}}$ transforms from the initial frame F_0 to the transform frame F_T, but not to F'_T [31, 33]. On the other hand, a single transformation of the boost cannot be considered as it depends more on successive transformations. In this view, the gauge field A^μ (a special type of quantum field theory) of the rotation frame due to Coriolis force is mathematically expressed as:

$$A^\mu = \left(-0.5(\Omega\times x)^2, \Omega\times x\right). \tag{2.36}$$

In this view, the Hamiltonian for a water particle at relaxation concerning the rotating frame F_T' is given by:

$$H = 0.5m^{-1}(\hat{p} - m\Omega \times x)^2 - 0.5m(\Omega \times x)^2, \tag{2.37}$$

Therefore, the rotation of the water particle due to transferring from the initial frame to frame F_T' under the impact of Coriolis force can make the particle spin $\hat{S}$. On this understanding, Eq. (2.37) requires the spin interaction term, which can be shown as:

$$H = 0.5m^{-1}(\hat{p} - m\Omega \times x)^2 - 0.5m(\Omega \times x)^2 - \Omega \cdot \vec{\hat{S}}. \tag{2.38}$$

In other words, spin is the angular momentum; it is combined with a rotating object, for instance, a spinning golf ball, or the spinning Earth. The angular momentum, therefore, can be estimated by integrating over the contributions to the angular momentum owing to the motion of every one of the tiny masses making up the body. The recognized consequence is that the total angular momentum or spin $\hat{S}$ (Fig. 2.6) is computed by:

$$\vec{\hat{S}} = I\omega \tag{2.38.1}$$

where I is the moment of inertia of the body, and ω is its angular velocity.

In this regard, spin $\hat{S}$ is a vector, which plugs along the axis of rotation in a direction molded by the right hand rule: curl the fingers of the right hand in the direction of rotation and the thumb points in the direction of $\hat{S}$ (Fig. 2.7).

The effect of coupling of the spin, therefore, to the rotation is to act on the initial wave function by the operator:

$$\hat{\Phi}_S = \hat{T}e^{\left[\int dt\vec{S}' \cdot \Omega\right]}, \tag{2.39}$$

where $\hat{T}$ represents the time ordering operator.

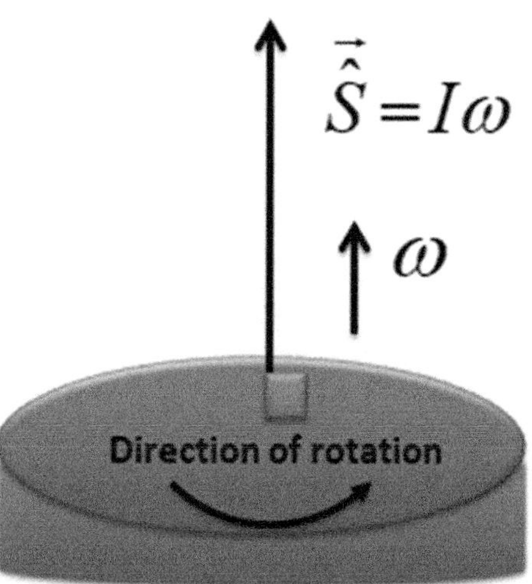

FIG. 2.6 Angular momentum simple concept.

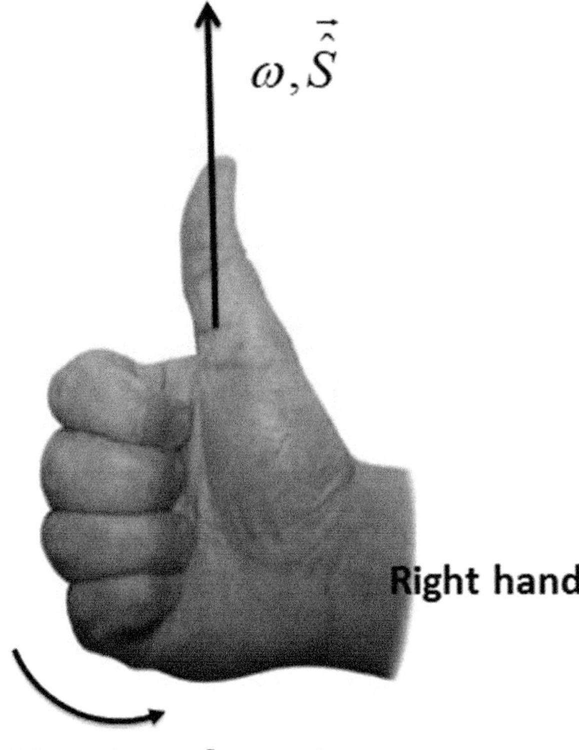

FIG. 2.7 Concept of the right-hand rule.

Eq. (2.39) can be added to the continuity Eq. (2.26) to obtain:

$$\frac{\partial}{\partial t}q + \nabla \cdot \left(q\,\vec{u}\right) + \hat{\Phi}_s = 0. \tag{2.40}$$

Eq. (2.40) reveals that water flows under the impact of the water particle spinning due to the uniform rotation, which is due to the Coriolis force. This reveals that continuity of motion must be governed by water particle spinning. In this view, the Schrödinger equation describes a field of only one complex value [34]. Consequently, the spinning term in Eq. (2.40) is required to be described perfectly as a vector of four complex numbers. In doing so, a relativistic wave Dirac equation is implemented [31, 35]. Therefore, the relativistic phases of the rotating frame in quantum mechanics, exhausting the Dirac equation with a spin connection, are given by:

$$\psi\left(i\gamma^{\mu}\nabla_{\mu}\right) = \psi(m) \tag{2.41}$$

where ∇_{μ} is expressed as:

$$\nabla_{\mu} = 0.5\left[\gamma^{a}, \gamma^{b}\right] \tag{2.41.1}$$

where μ, a, and b run over 0, 1, 2, 3 and the metric in a uniformly rotating frame is given by:

$$G_{00}=1-(\Omega\times x)^2, G_{ii}=-1, G_{0i}=-(\Omega\times x)^i, (i=1,2,3), \tag{2.42}$$

The low energy limit of the Dirac equation can be mathematically expressed as:

$$\left[\gamma^0\left(m+\hat{p}_0-mA_0-0.5A^i\hat{p}_i\right)\right]=-\left[\gamma^i\left(\hat{p}_i-0.5mA_i-\frac{i}{2}\nabla A_0\right)-m\right]\psi, \tag{2.43}$$

Then, by means of the common splitting of the four spinors into upper and lower constituents, the Hamiltonian for the upper constituent and dual spinors in the low energy boundary is formulated as:

$$H=0.5m^{-1}\left(\hat{p}_0-mA-(I\cdot\omega)\times-\nabla A_0\right)^2+mA_0-\Omega\cdot(I\cdot\omega). \tag{2.44}$$

Eq. (2.44) reveals that low energy approximation is in the rotating coordinate system, and not in an inertial coordinate system.

2.9 Quantization of barotropic flow

In ocean dynamics, a barotropic flow is a flow whose density is a function of pressure only. In this understanding, the density of flow is approximately steady (isopycnic) (Fig. 2.8). Hence, the flow densities vary merely indecisively with pressure and temperature. Ocean flow, which fluctuates by only a few percent with temperature and salinity, may be approached as barotropic [36].

The potential barotropic flow and its density fluctuations depend on the phase space and time. In this understanding, their Poisson bracket is represented by:

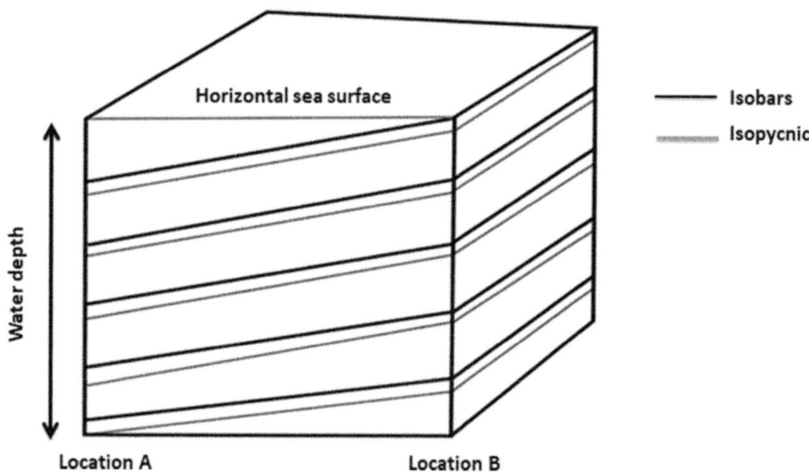

FIG. 2.8 Barotropic flow.

$$\left\{\varphi\left(\vec{x}\right),\rho\left(\vec{y}\right)\right\}=\partial^{d}\left(\vec{x}-\vec{y}\right) \tag{2.45}$$

where φ is the velocity potential and ρ is the density.

In this regard, Eq. (2.45) represents the conjugate field of φ and ρ. For the barotropic flow, the internal energy E_I is the dominant scenario through the water body. Which is a function of pressure variation p and it can be given by:

$$E_I=\rho^{-1}\frac{\partial\rho}{\partial z} \tag{2.46}$$

Eq. (2.46) reveals the barotropic flow with the water density fluctuations $\partial\rho$ through the water depth variation ∂z. In this regard, the equation of motion can be expressed in the Hamiltonian concept as:

$$\frac{\partial\rho}{\partial t}=\frac{\partial H}{\partial\varphi}+\nabla\cdot\left(\rho\cdot\vec{u}\right), \tag{2.47}$$

Eq. (2.47) reveals flow with a vorticity-free and a velocity potential, i.e., $\vec{u}^{\,def}=\partial\varphi$. In other words, vorticity-free expression $\nabla\times\vec{u}=0$.

$$\frac{\partial\rho}{\partial t}+\frac{\partial H}{\partial\rho}+0.5\,\vec{u}\cdot\vec{u}=-\rho^{-1}\frac{\partial\rho}{\partial z}, \tag{2.48}$$

Consequently, Eq. (2.48) leads to the Hamiltonian-Euler equation of the barotropic motion, which is given by [37]:

$$\frac{\partial\vec{u}}{\partial t}+\left(\vec{u}\cdot\nabla\right)\vec{u}+\rho^{-1}\frac{\partial^2\rho}{\partial^2 z}(\nabla\rho)=\rho^{-1}\nabla p \tag{2.49}$$

Eq. (2.49) explains the fluctuation of the barotropic flow over time within the change of the water particle density through the water depth due to the internal energy of the water body. Consequently, the Hamiltonian also has a kinetic-energy part plus an internal-energy part to govern the barotropic motion over a fixed domain D, which can be mathematically described as:

$$H\left[\rho,\,\vec{u}\right]=\int_D\left[0.5\rho\vec{u}^{\,2}+\rho E_{I,K}(\rho)\right]dx \tag{2.50}$$

where x, y, and $z\in D$ and the internal and kinetic energies $E_{I,\,K}$ are a function of the water particle density.

Using Eq. (2.45), the Poisson bracket of barotropic motion can be expressed as:

$$\left\{\varphi\left(\vec{x}\right),\rho\left(\vec{y}\right)\right\}=\int_D\left[\frac{\partial\varphi}{\partial\rho}\frac{\partial}{\partial x}\frac{\partial\rho}{\partial u}-\frac{\partial}{\partial x}\frac{\partial\varphi}{\partial u}\right]dx \tag{2.51}$$

Hamiltonian mechanics can be restated in terms of Poisson brackets. As a result of canonical invariance of the Poisson brackets, the relations so obtained will also be invariant in the form under a canonical transformation [38]. To obtain the total time derivative of some function of the canonical variables and time of $\varphi(q,p,t)$, Hamilton's equations of motion can be implemented as:

$$\frac{\partial\varphi}{\partial t}=\sum_{i=1}^{n}\left(\frac{\partial\varphi}{\partial q_i}\dot{q}_i+\frac{\partial\varphi}{\partial p_i}\dot{p}_i\right)+\frac{\partial\varphi}{\partial t}=\sum_{i=1}^{n}\left(\frac{\partial\varphi}{\partial q_i}\frac{\partial H}{\partial p_i}-\frac{\partial\varphi}{\partial p_i}\frac{\partial H}{\partial q_i}\right)+\frac{\partial\varphi}{\partial t}, \tag{2.52}$$

or:

$$\frac{\partial\varphi}{\partial t}=\{\varphi,H\}+\frac{\partial\varphi}{\partial t}. \tag{2.53}$$

The generalization equation of the barotropic motion for an arbitrary function φ in the Poisson bracket formulation can be revealed by Eq. (2.53). It comprises Hamilton's equations as a singular circumstance when one of the canonical variables is substituted as:

$$\dot{q}_i=\{q_i,H\},\quad \dot{p}_i=\{p_i,H\} \tag{2.54}$$

The steady-state motion can be delivered from Eq. (2.53), if the velocity fluctuation over the time $\frac{\partial\varphi}{\partial t}=0$, which can lead to:

$$\frac{\partial\varphi}{\partial t}-\{\varphi,H\}=0. \tag{2.55}$$

Eq. (2.55) demonstrates that a steady-state motion is conserved throughout the water body motion, presuming, in consequence, a restriction on the motion. In other words, the steady-state barotropic flow is equivalently the Poisson bracket of H, where any constant of the motion must be equal to the partial time derivative of the constant function. In this regard, the steady-state barotropic flow as a function of density flow can be expressed as:

$$\frac{\partial\rho}{\partial t}-\{\rho,H\}=0. \tag{2.56}$$

Eqs. (2.55) and (2.56) prove that the Poisson theorem, in which the Poisson bracket of any two variables of the motion, for instance, φ and ρ, is also a steady-state barotropic motion.

2.10 Quantization of vorticity flow

Topological defects are the keystone of vortex generation. In fact, topological defects occur owing to symmetry breaking of the physical ocean water characteristics, for instance, temperature, density, etc., which lead to phase transitions through the water body [39]. The occurrence of the phase transition can create a vortex. In this regard, one kind of topological defect is quantum vortices, which are exhibited in superfluids such as the ocean. In this understanding, gyres, for example, in the ocean do not occur as we identify them, nevertheless, as an alternative, they are large-scale assemblies of quantized vortices. There is great potential for the fine-scale assembly of these vortices, as a function of temperature and pressure. In this scenario, the superfluid ocean still coincides with density-dependent dynamics, but absences of a collapsing circulation owing to homogenous temperature.

The quantization of vorticity, therefore, is a direct consequence of the existence of a superfluid order parameter as a spatially continuous wave function. Indeed, the quantum vortices describe the circulation of superfluid and it is conjectured that their excitations are a function of superfluid phase transitions. In these regards, a quantum vortex is considered as a cavity with the superfluid circulating the vortex axis (Fig. 2.9); the inside of the vortex can encompass energized particles. Furthermore, quantum vorticity drives the ocean away from equilibrium by increases line length through counterflow [40].

A superfluid has the extraordinary characteristic of the boasting phase, which is determined by the wave function, and its velocity. In this view, the gradient of the phase (in the parabolic mass approximation) is proportional to the velocity fluctuations. The circulation around any closed loop in the vorticity is zero if its locality enclosed is only combined. The superfluid is reckoned as having irrotational flow; however, if the bounded region essentially comprises a lesser region with the nonexistence of vortex—for instance, a rod through a vortex—then the circulation (Fig. 2.10) is mathematically expressed as:

$$\Gamma = \oint_C \vec{V} \cdot \vec{d\,l} = \frac{\hbar}{m} \oint_C \nabla \phi_v \cdot \vec{d\,l} = \frac{\hbar}{m} \Delta^{tot} \phi_v, \tag{2.57}$$

The total phase difference of the water mass m fluctuations around the vortex is identified by $\Delta^{tot}\phi$, which is given by:

$$\Delta^{tot}\phi = 2\pi n \tag{2.57.1}$$

where n is an integer, i.e., 0,1,3,...

Therefore, Eq. (2.57) indicates that the wave function must return to its same value after an integer number of turns around the vortex. Thus, the vortex circulation is quantized as:

$$\Gamma = \oint_C \vec{V} \cdot \vec{d\,l} \equiv \frac{2\pi\hbar}{m} n. \tag{2.58}$$

Moreover, the presence of the $\hbar$ means that quantized vorticity is a consequence of quantum mechanics. In other words, a vortex with such quantized circulation is called a quantized vortex. Any rotational motion of a superfluid is sustained only by quantized vortices. Moreover, the ratio $\frac{\hbar}{m}$ is reasonably macroscopic and the number of vortex lines alters on it.

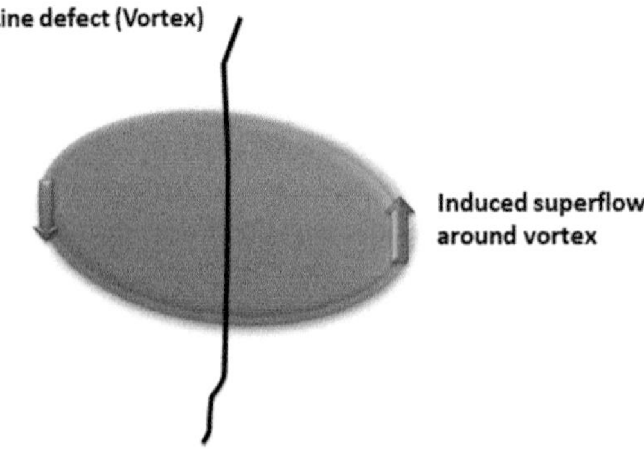

FIG. 2.9 Simplification of inducing a superflow around the line-vortex.

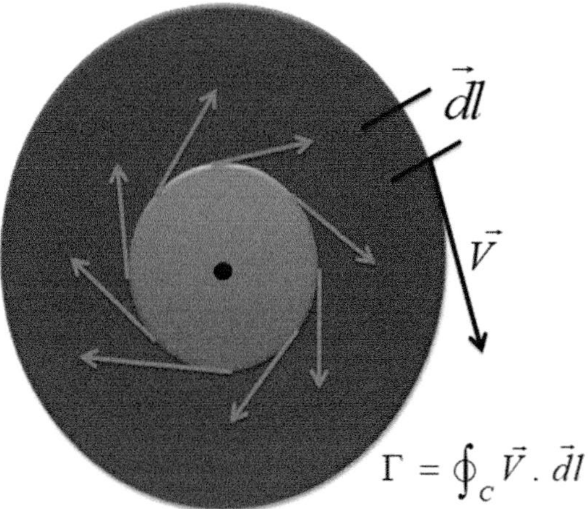

FIG. 2.10 Vortex and circulation.

A quantized vortex, therefore, is a stable topological defect characteristic of a Bose-Einstein condensate (BEC), which exhibits a macroscopic wave function:

$$\psi\left(\vec{r},t\right) = \left|\psi\left(\vec{r},t\right)\right|e^{i\phi(\vec{r},t)} \tag{2.59}$$

Eq. (2.59) validates that the closed-loop circulation in Eq. (2.58) is quantized by $\frac{\hbar}{m}$ since macroscopic wave function $\psi\left(\vec{r},t\right)$ is single-valued for the space coordinate $\vec{r}$. In this regard, a quantized vortex is considered as a vortex of inviscid superflow. In consequence, the quantized vortex cannot be demolished by the viscous diffusion of vorticity owing to its thin core. Vortices, however, are not completely described for a classical viscous fluid. Indeed, they are unsteady, and exist and vanish repetitively. Consequently, the circulation is not sustained and not indistinguishable for every single vortex. On this understanding, the quantized vortex is delineated as a topological excitation in which its potential velocity is given by:

$$\vec{v}_s = \left(\frac{\hbar}{m}\right)\nabla\phi \tag{2.60}$$

As a result, the vorticity $\nabla \times v_s$ can be described from the superfluid velocity, which disappears everywhere in an individually allied zone of the order parameter of the amplitude and phase, i.e., $\psi = \sqrt{\rho}e^{i\phi}$ and all rotational flow is approved merely by quantized vortices. As long as ϕ rotates around the core by 2π, density ρ vanishes at the core. On this understanding, the individual-particle wave function of this rotation can be described as:

$$\psi(r_1, \ldots, r_N) = \prod_{i=1}^{N}\phi(r_i) \tag{2.61}$$

where N is the total number of particles and is computed as:

$$N = \int dr|\psi|^2 \tag{2.61.1}$$

This wave function will be impelled by the kinetic energy, the potential energy, and the interaction energy. In these regards, the interaction energy is mathematically given by the mean-field energy as:

$$\overline{E} = \frac{4\pi\hbar^2 l}{m} \tag{2.62}$$

where l is the growing length of the vorticity.

In order to describe the wave function $\psi(r)$ of the rotation vorticity, the velocity of the particle $V(r)$, the interaction energy $\overline{E}$, and quantized vortices $\frac{\hbar^2}{2m}$ must be included. In this regard, one sort of type of nonlinear Schrödinger equation, which is known as the time-independent Gross-Pitaevskii equation (GPE), can be implemented as:

$$\left(-\frac{\hbar^2}{2m}\nabla^2\phi + \vec{V}(r) + \overline{E}|\psi(r)|^2\right)\psi(r) = n\overline{E}\psi(r), \tag{2.63}$$

The continuity equation of density ρ and superfluid velocity vs can be expressed in the GP equation as:

$$\frac{\partial \rho}{\partial t} + \nabla \cdot \left(\rho \vec{v}_s\right) = 0, \tag{2.64}$$

$$\frac{\partial \vec{v}_s}{\partial t} + \nabla \vec{v}_s^2 = -\frac{\nabla}{m\rho}\left(0.5 g\rho^2\right) + \frac{\hbar^2 \nabla}{2m^2}\left(\frac{\nabla^2\sqrt{\rho}}{\sqrt{\rho}}\right). \tag{2.65}$$

The first term on the right-hand side of Eq. (2.65) represents an effective pressure, i.e., $p = \int_{z_0}^{z} dz\rho(z)g(z)$, which is the variation the hydrostatic pressure through the vortex depth z from the surface z_0. This term arises due to the due to the nonlinearity of the GP equation. Therefore, the second term is known as the quantum pressure, which has no analog in typical fluid mechanics and is significant at minor scales analogous to the healing length such as near vortex cores, where ρ rapidly changes in the scale of ξ [35], which is given by:

$$\xi \approx \sqrt{\frac{\hbar^2}{2m\overline{E}n}} \tag{2.66}$$

Consequently, the condition for irrotationality for a system with a vortex along the z-axis can be generalized to:

$$\vec{\nabla} \times v = z\frac{nh}{m}\partial^2(\rho), \tag{2.67}$$

where n is an integer describing the winding number of the singularity in the wave function, which creates angular momentum through the vortex dynamic rotation.

In this regard, the wave function must encompass a phase factor $e^{i\vartheta}$, where ϑ is the azimuthal angle. Then the azimuthal velocity can be computed using:

$$v_\vartheta = \frac{\hbar}{m}\rho^{-1}\frac{\partial}{\partial\vartheta}\phi \tag{2.68}$$

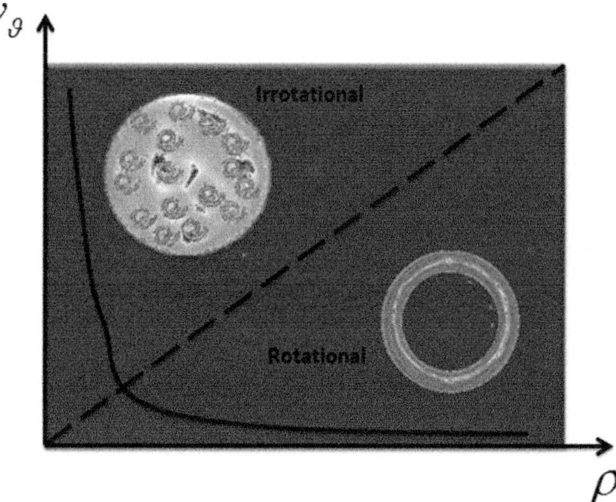

FIG. 2.11 Azimuthal velocity of a vortex.

Eq. (2.68) reveals that the velocity field for a normal rotating fluid behaves like a rigid body, where the azimuthal velocity increases linearly with increasing distance from the symmetry axis (Fig. 2.11). A superfluid shows a nonintuitive velocity pattern, where the azimuthal velocity diverges for decreasing distance from the rotation axis. In order not to let the kinetic energy of the system diverge, the density of the superfluid must go to zero.

Consequently, in a nonlinear quantum fluid, the dynamics and configurations of the vortex cores can be premeditated in terms of effective vortex-vortex couple interactions [41]. The real intervortex potential is expected to distress quantum phase transitions and give rise to diverse few-vortex molecules and countless-body vortex patterns [42].

2.11 Quantum turbulence

Fluid turbulence is an extremely complex and nonlinear chaotic phenomenon. Excessive complications and obstacles are encountered when struggling to compute or simulate fluid flows precisely and robustly, owing to their chaotic, multiscale special effects of turbulence. A significant question is: can ocean turbulence be quantized? Quantum turbulence is the term given to the turbulent flow—the chaotic motion of fluid at high flow rates. In the previous section, the ocean dynamic system was considered as superfluid flows. On this understanding, turbulent superfluid flow involves arbitrarily sized expanses of circulating fluid termed eddies and vortices, which can be orderly, giving rise to large-scale motions, such as tornados or whirlpools, but in general, they are entirely asymmetrical. In this view, quantum flows can describe the free dynamic ocean movement even without the dissipative effect of viscous forces. In addition, the ocean dynamic comprises of vorticities and circulation, their local rotation is discomposed to discrete vortex lines of recognized strength. In contrast, the eddies in ordinary fluids are continuous and can have arbitrary size, shape, and strength. The rough sea surface is a good example of quantum turbulence. In this regard, the wind particles stress the sea surface particles and make them vibrate and spin, which leads to turbulent flows.

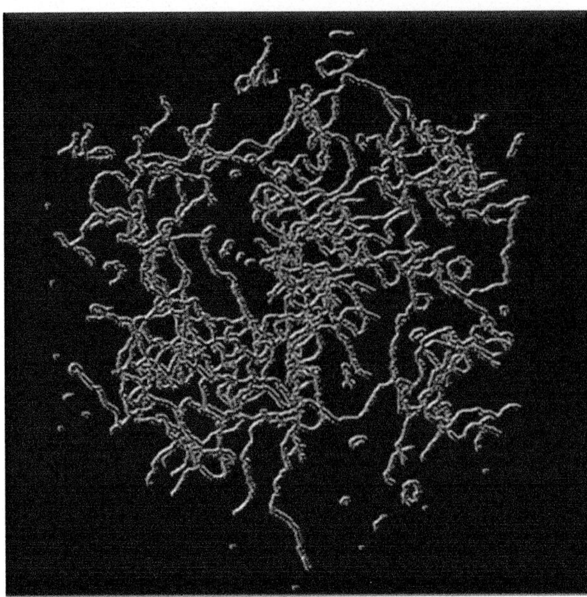

FIG. 2.12 Quantum turbulence flow due to vortex tangle.

Fig. 2.12 demonstrates an example of quantum turbulence flow, which is created by a simulated vortex tangle as presenting the quantized vortices. In this scenario, quantum turbulence is a tangle of these quantized vortices, making it a natural form of turbulence. In this regard, it becomes much easier to explain than conventional turbulence, in which countless conceivable exchanges of the eddies rapidly make the problem too multifaceted to be able to predict what will occur.

In the classical fluid, turbulence is frequently demonstrated merely using virtual vortex filaments, about which there is a definite circulation of the fluid, to acquire an understanding of what is occurring in the fluid. In quantum turbulence, consequently, these vortex lines are real—they can be experimental, and have an identical convinced circulation—and additionally they deliver the sum of the physics of the quantum turbulence flows. Quantum turbulence is considered as hydrodynamic as one is dealing with vortices. This pattern of turbulence—a tangle of quantized vortex lines—can be agitated at minor length scales or vortex rings (Fig. 2.13) or at larger length scales as shown within the accumulation of altocumulus clouds (Fig. 2.14). Moreover, large-scale vortices can be generated, for instance, by the passage of an aircraft wing (Fig. 2.15). In addition, sudden object rotation due to external forces can induce a vortex pattern. For instance, the polar vortex is driven due to the Coriolis force effect (Fig. 2.16).

The possible classification of quantum turbulence, therefore, is also based on the variety of the energy spectrum $E(k)$—the distribution of kinetic energy over the wavenumbers k (inverse length scales). In this regard, a one-dimensional incompressible kinetic energy distribution can be defined as:

$$E_i(k,t) = 0.5m4\pi k^2 \left\langle |\widetilde{w}_i(k,t)|^2 \right\rangle. \tag{2.69}$$

FIG. 2.13 Vortex rings.

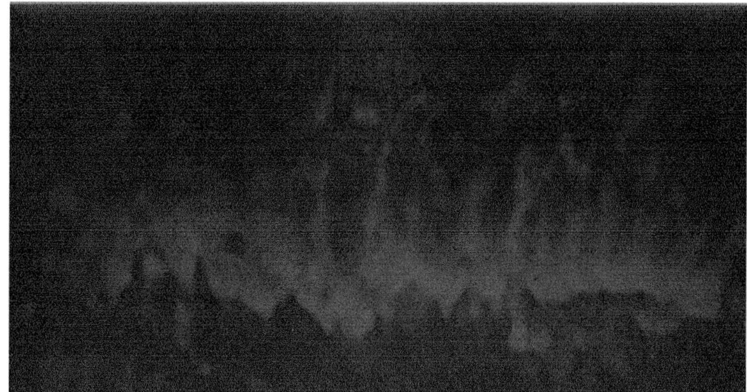

FIG. 2.14 Large-scale vortex due to the accumulation of altocumulus clouds.

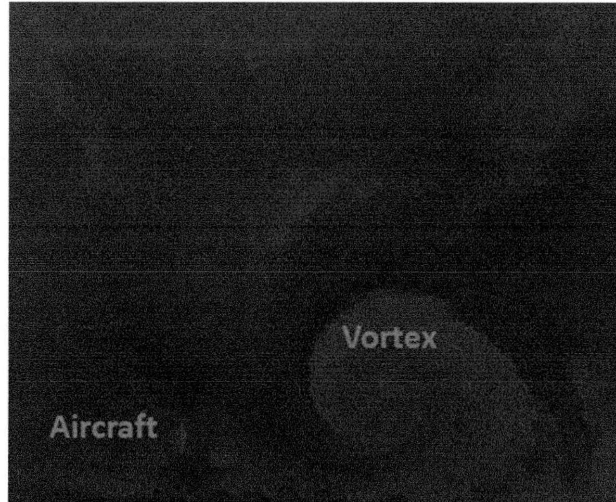

FIG. 2.15 Aircraft wing creates a large-scale vortex.

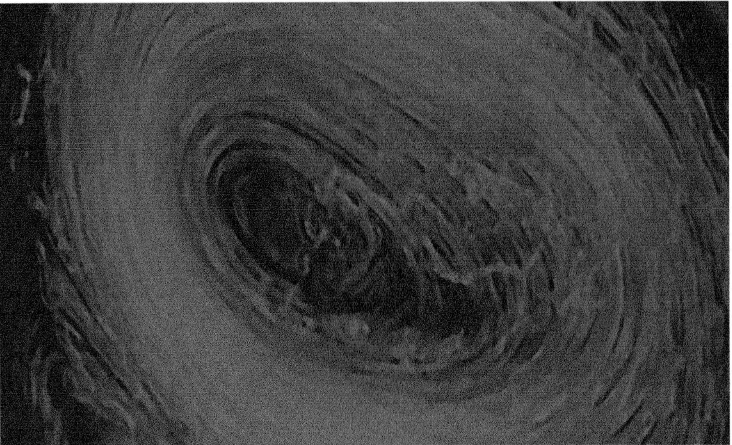

FIG. 2.16 Rotation vortex due to the Coriolis force effect.

Here $\widetilde{w}_i(k,t)$ is isotropically distributed, i.e., exhibits the same properties or behavior in all directions and $\langle \ldots \rangle$ denotes the average over the wavenumber k in the 3D Fourier space. In this regard, for instance, the transform of two sets of nonparallel lines into a 3D set of parallel lines perpendicular to the plane (Fig. 2.17).

Quantum turbulence based on kinetic energy variation can be identified as: (i) Vinen or ultra-quantum turbulence; and (ii) Kolmogorov or semiclassical turbulence [43].

Vinen turbulence is described by arbitrary tangles of vortices without large-scale, energy-encompassing flow configurations. On this understanding, the energy spectrum does not occur for the small wavenumber k region. Subsequently, as the dynamics and interactions of the quantum vortices are extremely nonlinear issues, analytical methods are very restricted. In this circumstance, the numerical simulations are the superlative method to offer the spectral behavior of Eq. (2.69). For instance, the arbitrariness in the alignment of the quantized vortices induces a spectrum that matches a power-law. In this regard, the Vinen turbulence must satisfy [44]:

$$E_i \propto k^{-1}, \text{ for } k < \xi^{-1}, \tag{2.70}$$

$$E_i \propto k^{-3}, \text{ for } k \geq \xi^{-1}. \tag{2.71}$$

In the circumstance of $L \sim t^{-1}$, the Vinen turbulence is decaying as a function of the temporal scaling of the total vortex length within the time increment, as demonstrated in Fig. 2.18. In this regard, the twisted vortex is converted into turbulent flow until decay occurs.

Kolmogorov or semiclassical turbulence is reasonably analogous to conventional turbulence, as the energy spectrum contains an inertial range showing the occurrence of the decaying vortex as $k^{-\frac{5}{3}}$ a power law for $E_i(k)$. This conventional scaling suggests thus some similarity between the turbulent dynamics of $w_i(r,t)$ and the velocity field in an incompressible Navier-Stokes equation, defining the Kolmogorov or quasiclassical quantum turbulence, in which its energy is defined as:

$$E_i \propto k^{-\frac{5}{3}}, \text{ for } k \leq \frac{2\pi}{l}, \tag{2.72}$$

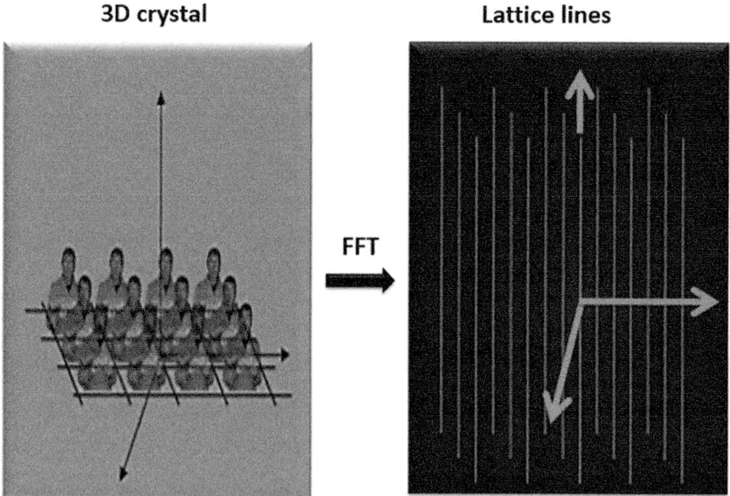

FIG. 2.17 3D Fourier space.

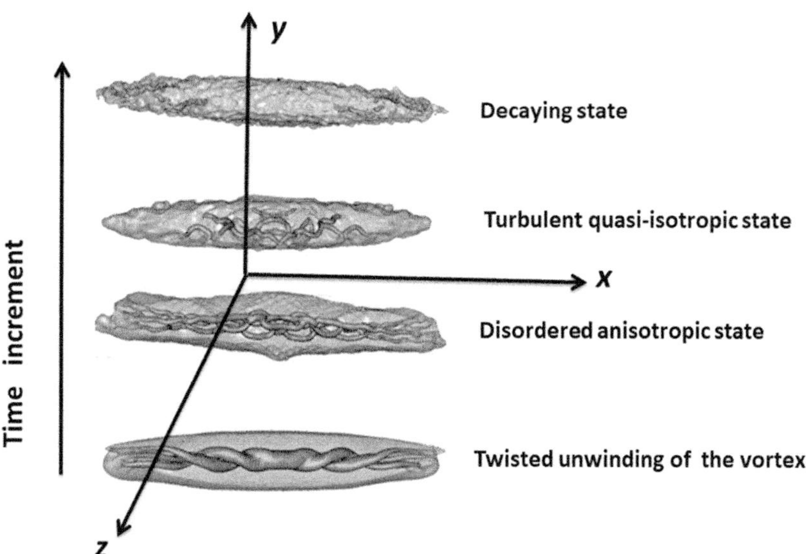

FIG. 2.18 Decaying of the twisted vortex.

Eq. (2.72) indicates the existence of the energy cascade from large to small length scales. It is also determined that if the energy spectrum of the turbulent obeys the Kolmogorov $k^{-\frac{5}{3}}$ scaling, in this scenario the vortex tangle contains transient regions where the vortex lines are oriented in the same direction. The large-scale flow, consequently, is generated as the concentration of the energy in the small-scale zone of wavenumber in the energy spectrum.

$$E_i \propto k^{-3}, \quad \text{for} \quad k \geq \xi^{-1}. \tag{2.73}$$

In addition, Eq. (2.73) is delivered from the universal characteristics of the velocity field profile of the vortex in the superfluid such as ocean dynamic turbulent. Consequently, it can represent the internal structure of the quantized vortex core. Eq. (2.73) also reveals that the spectra for 2D fluid dynamic with quantized vortices behaves likewise for $k \geq \xi^{-1}$, which is independent of dimensionality. Quantum fluid, therefore, can accomplish the 2D limit. In this view, the quantized vorticity with lowering dimensionality implies the characteristics of the vortex line, which are no longer applicable for dynamic ranges. For instance, the 1D vortex is considered as a zero-dimensional vortex in the plane, with appropriately termed positive or negative circulation. In other words, it can be vortex or antivortex. In this understanding of hydrodynamic turbulence, i.e., zero-dimensional vortex, vortices interact with each other forming complicated dynamic patterns [44]. In this scenario, the tangling and vortex reconnections are replaced by the vortex-antivortex annihilation process. In this view, the angle field around a strength +1 vortex at the origin is $\theta^+(x, y) = \tan^{-1}\left(\frac{y}{x}\right)$, and the gradient is $\nabla\theta^+ = \frac{(-y, x)}{x^2 + y^2}$. In this circumstance, the spins may be signified as unit vectors $(\cos\theta^+, \sin\theta^+) = \left(\frac{x}{x^2 + y^2}, \frac{y}{x^2 + y^2}\right)$ (Fig. 2.19). In contrast, antivortex has strength −1 and has spins that rotate by one full turn in the anticlockwise sense (Fig. 2.19). In other words, around a strength −1 antivortex at the origin $\theta^-(x, y) = -\tan^{-1}\left(\frac{y}{x}\right)$, $(\cos\theta^-, \sin\theta^-) = \left(\frac{x}{x^2 + y^2}, \frac{-y}{x^2 + y^2}\right)$, and its gradient is $\nabla\theta^- = \frac{(y, -x)}{x^2 + y^2}$.

The dual vortex cores can merge and annihilate to develop into a solitary wave (Fig. 2.20), but unlike most ordinary particles, they can appear out of nothing or disappear. In other words, they vanish as particles plug into the vacant space (Fig. 2.21).

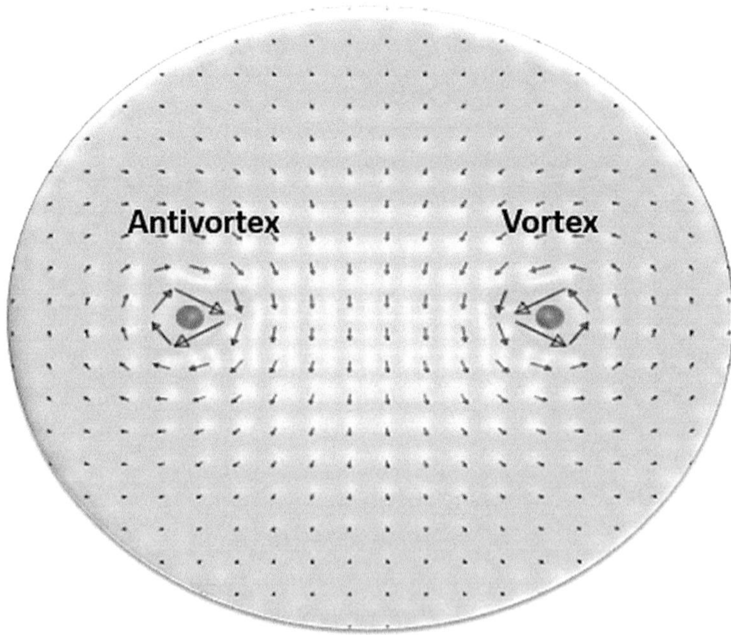

FIG. 2.19 Vortex-antivortex flows.

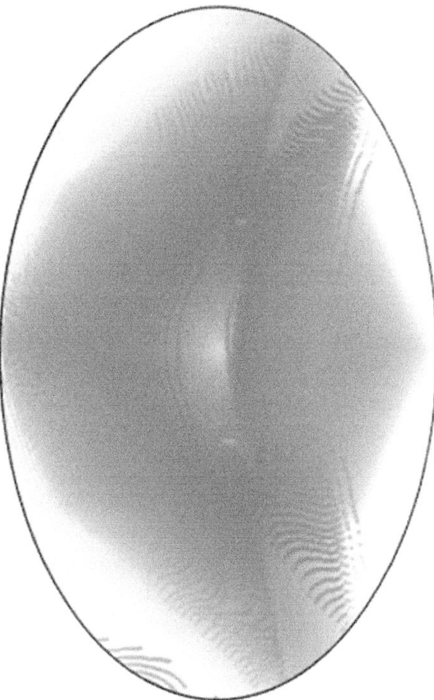

FIG. 2.20 Combination of two vortex.

FIG. 2.21 Disappearing of vortex-antivortex flows.

2.12 Particle theory of ocean waves

The critical question is: can quantum mechanics, which are based on the wave-particle duality, explain ocean wave propagation? In this regard, seawater molecules in sea waves do not transfer vertically and downwards, but rotate in slight circles (Fig. 2.22). Fig. 2.22 reveals that in deep water, the orbital motion of seawater particles decreases rapidly with increasing depth below the surface. Nevertheless, in shallow water, the elliptical movement of a seawater particle flattens with decreasing depth.

From the point view of quantum mechanics, all the information about a particle is encoded in its wave function, a complex-valued feature roughly analogous to the amplitude of a wave at every factor in space. This feature evolves following the Schrödinger equation. For particles with mass, such as seawater, this equation has solutions that comply with the form of the wave equation. The propagation of such waves leads to wavelike phenomena such as interference and diffraction. In these regards, waves are instabilities that promulgate energy through space. Transmission of the energy relies on the interactions between the particles that make up the sea-state across space and time. Particles move as the waves pass through, but there is no net motion of particles. This means, once a wave has passed, the particles return to their original position. Consequently, energy, not matter, is transmitted by waves.

Conversely, the different sorts of waves reveal particular characteristics, which are functions of the orientation of particle motion. On this understanding, the orientation of particle motion, which is relative to the direction of energy propagation, is one approach to categorize the waves. In this context, wave phenomena are classified into: (i) longitudinal waves (Fig. 2.23); (ii) transverse waves (Fig. 2.24); and (iii) surface waves (Fig. 2.25).

Through longitudinal waves, the particles move parallel to the motion of energy. However, across the transverse waves, the movement of the particles is at right angles (perpendicular) to the motion of the energy. Unlike longitudinal waves and transverse waves, through surface waves, particles travel in a circular motion. For instance, these surface waves occur at interfaces, which are seen as waves and ripples in the ocean. One consequence of occurring at an interface is that the motion of the particles diminishes with distance from the interface. The

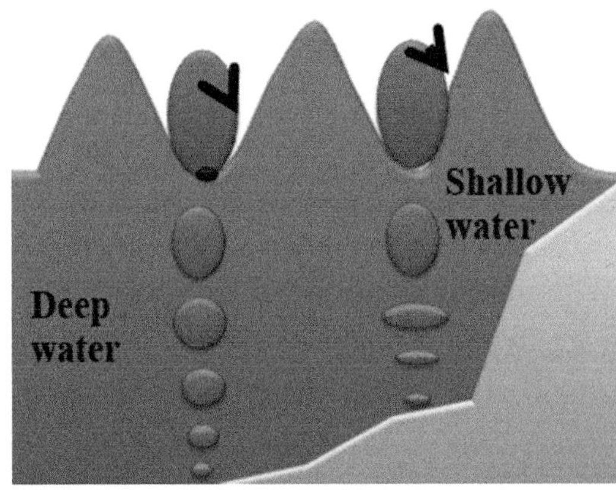

FIG. 2.22 Particle motions of ocean waves.

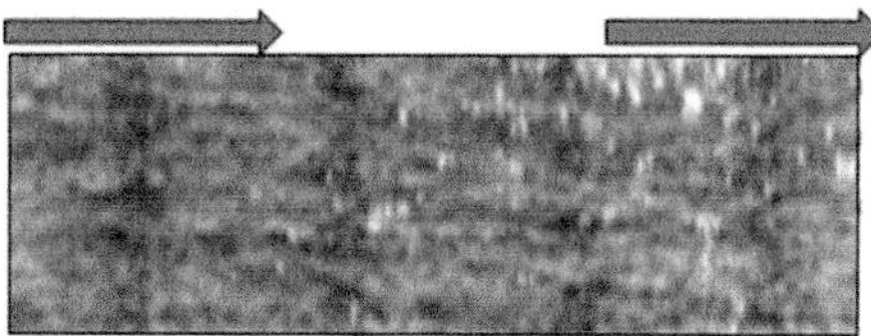

FIG. 2.23 Longitudinal wave.

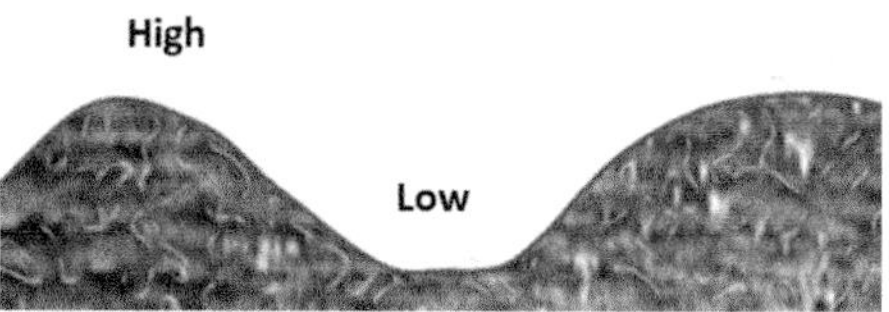

FIG. 2.24 Transverse wave.

FIG. 2.25 Surface wave.

further they are from the interface, the smaller the rotation of the particles until, at some distance from the surface, there is no more movement or energy propagation.

The important question is: how do particles of water in a wave move? In this sense, wind stresses the sea surface and transfers energy to the water, creating waves. Subsequently, waves break on the shore and the water runs down the beach back into the sea. Conversely, waves convey energy, no longer water. When a wave crest rises, the water particles interchange in circular routes.

For irregular waves, the shape of the variance spectrum of the horizontal particle velocity will change with depth, since the wave components of higher frequencies attenuate more rapidly with depth than those of the lower frequencies. The frequency of the depth-averaged horizontal particle velocity turns out to be ω_{20}, which is therefore chosen as the representative frequency for the narrow band process used herein.

$$\omega_{42} = \sqrt{\frac{M_4}{M_2}} \tag{2.74}$$

$$M_J = \int_0^\infty \omega^j S_\eta(\omega) d\omega \tag{2.75}$$

The expected value of particle velocity u in an irregular sea-state relies on the wave amplitude, which is a slowly varying envelope for η, and is, as such, Rayleigh distributed with parameter σ_η. In this context, the mathematical description of the probability density function of sea-state $P(a)$ can be given by:

$$P(a) = a\sigma_\eta^{-2} e\left(-\frac{a^2}{\sigma_\eta^2}\right); \; a \geq 0, \tag{2.76}$$

where a is the amplitude of the ocean wave.

Then the particle velocity of the irregular sea-state is given by:

$$E(a) = \int_0^\infty P(a) E[u|a] da \tag{2.77}$$

where $E[u|a]$ is for points that are submerged or in the splash zone, respectively.

Eq. (2.77) states that, for a given amplitude, the result depends on the expectations of u, summed over all amplitudes, and weighted according to their frequency of occurrence. The ergodicity theorem implies that the calculation of Eq. (2.77) may be done as a time average. Since the wave spectrum is narrow-banded, the components in Eq. (2.77) represent almost harmonic waves occurring sequentially rather than superposition of components that occur simultaneously. Hence the decision whether the point is in the air can be made on a wave-by-wave basis for each value of the amplitude. For a given depth, z, the decision whether the point is submerged or not depends on the amplitude, a. Hence Eq. (2.77) must be split into two integrals, one for $(-z) > a$, which can be evaluated analytically; however, for the points that are always submerged, the mean horizontal velocity can be determined from:

$$\overline{u} = -a\omega k e^{2kz} \tag{2.78}$$

However, the mean value of horizontal particle velocity in the splash zone becomes:

$$\overline{u(0,z)} = \frac{a\omega}{\pi}\left[e^{kz}\cos\theta_0 - ake^{2kz}\left(\frac{\pi}{2} - \theta_0\right)\right]. \tag{2.79}$$

where the phase angle is $\theta_0 = (\omega t)_0$.

The considered contribution represents the value of the integral in the phase angle range $\left(-\frac{\pi}{2} < \theta < \frac{\pi}{2}\right)$, which in view of symmetry yields the average over a full cycle. The above are expressions for regular deep water (Gerstner) waves of amplitude a, period $T = \frac{2\pi}{\omega}$, and wavelength $\lambda = \frac{2\pi}{k}$. The particles move in closed circles, hence the mean velocity when following a particle is zero [45, 46].

2.13 Schrödinger equation for description of nonlinear Sea-state

The nonlinear Schrödinger equation (NLS) is a recognized growth equation, which designates the dynamics of weakly nonlinear wave packets in time and space in a wide range of physical media, such as nonlinear sea surface propagation. Due to its integrability, the NLS delivers relations of precise solutions, portraying the dynamics of localized structures of sea surface state as a function of nonlinear wave propagation. Relying on the coordinate of wave propagation, it is well-recognized that the NLS can be either articulated as a space- or time-development equation. In other words, the theory of weakly nonlinear sea-states has been attained to be precise and valuable for the modeling of sea-state dynamic fluctuations, i.e., ocean waves [45, 47–49]. In shallow and deeper ocean, the nonlinear Schrödinger equation (NLS) is the simplest evolution formula of this sort of nonlinear sea-state that considered dispersion and nonlinearity. Solving it from initial conditions, the NLS delivers precise, systematic solutions, which designate the growth of localized structures of sea-state in time and space, thus permitting successively the revision and identification of the dynamic range of essential localized structures of sea-states [46]. In this regard, the strength and accuracy of the NLS have been experimentally established even in demonstrating of great localizations of sea-state, further than its distinguished asymptotic restrictions [46]. Owing to its interdisciplinary nature equivalences that being proficient to be constructed into other nonlinear dispersive media, such as in wave-wave and wave-current interactions. Moreover, the NLS produces elementary computation models for the depiction of oceanic extreme events identified as breathers [50]. Certainly, the group models of Akhmediev breathers (ABs) [51] and Peregrine breathers [50] are deeply allied with the modulation instability (MI), which is also known as Benjamin-Feir instability [48] of Stokes waves [52].

Indeed, the tentative exploration of precise solutions of the NLS, either mathematically or in water wave facilities, has improved the understanding of nonlinear and unsteady sea-state, besides contributing to the classification of the restraints of uncertainly nonlinear hydrodynamic models of sea-state [46]. Therefore, the selection of the NLS depiction of sea-state growth in either time or space relies on the space or time expansion coordinate of concern.

To discuss the wave packet propagation in terms of space and time, let us assume that $A(x,t)$ and $B(x,t)$ are complex wave amplitudes, which are propagated with phase $\phi = kx - \omega t$, then the mathematical descriptions of two complex wave amplitude propagations are given by:

$$\zeta(x,t) = 0.5\left[A(x,t)e^{i\phi}\right) + C^*C], \tag{2.80}$$

$$\zeta(x,t) = 0.5\left[B(x,t)e^{i\phi}\right) + C^*C], \tag{2.81}$$

where C^*C is a complex conjugate.

The linear dispersion is correlated with both wave frequency ω and wavenumber k, i.e., $\omega = \sqrt{gk}$, where g indicates the gravitational acceleration. Conversely, both space-NLS and time-NLS can be satisfied by weakly nonlinear asymptotic equation as:

$$i\left(A_t + \frac{\partial\omega}{\partial k}A_x\right) - \frac{\omega}{8k^2}A_{xx} - \frac{\omega k^2}{2}|A|^2A = 0 \tag{2.82}$$

$$i\left(B_x + c_g^{-1}B_t\right) - g^{-1}B_{tt} - k^3|B|^2B = 0 \tag{2.83}$$

The boundary conditions of both Eqs. (2.82) and (2.83) are:

$$A(x,0) = A_0(x) \tag{2.84}$$

$$B(0,t) = B_0(t) \tag{2.85}$$

A satisfies space-NLS, and B satisfies time-NLS. Then Eq. (2.82) takes the canonical form, in nondimensional variables as:

$$iQ_T + Q_{xx} + 2|Q|^2 Q = 0 \tag{2.86}$$

In the case of a complex amplitude B, the conical form $Q(X,T)$ is determined from:

$$B = Q^* k^{-1}, \quad X = -\sqrt{2k}(x - c_g t), \quad T = 0.5kx \tag{2.87}$$

The consequence is the canonical NLS equation (2.86), in which the transformations are matching, except that, in case A, and T the transforms are to t, but in case B and T transforms are to x. The solution of Eq. (2.86) can be determined from:

$$A = \alpha Q^*\left(\sqrt{2\alpha k}(x - c_g t), \frac{\alpha^2 \omega t}{4}\right), \tag{2.88}$$

$$B = \alpha Q^*\left(-\sqrt{2\alpha k}(x - c_g t), \frac{\alpha^2 \omega x}{4c_g}\right), \tag{2.89}$$

where α is the wave steepness, and * is conjugate. Consistent with Chabchoub and Grimshaw [48], Eqs. (2.88) and (2.89) are both evidently agreed at the leading order of complex wave amplitudes A and B, respectively, which are being matched in the reliance on the prevailing phase variable $x - c_g t$. However, they differ in the slow t over space x reliance and are in common merely asymptotically correspondent when $x - c_g t = o(\alpha)$, which is achieved as $x = c_g t$ for leading linear nondispersive order $o(\alpha)$. In actual fact, the solutions are cast off for the minor. On the other hand, finite nonzero α and an asymptotic equivalence in the dual sorts of NLS dynamics are predictable [46, 48–50].

The asymptotic equivalence of localized sea-state structures can be observed within the framework of either the space- or the time-NLS. The optimal of one of these two configurations of the NLS relies on whether one is interested in integrating the evolution of wave packets in time, or in space, i.e., as for experimental purposes by tracking the evolution of waves along with a water wave facility.

In practice, generating both complex amplitudes of A and B in a wave packet in respect to the space-NLS allows, for instance, for wave amplitude growing into fixed x position, while a wave packet with respect to time-NLS would investigate the development of several extreme waves with similar envelope amplitude as a function of time. In other words, these envelope waves are developed within different wave phases and different velocity fields at fixed position x. On this understanding, the maximal AB compression can impact on the radar imaging mechanism of oil spill patches due to the scattering fluctuations from the oil spill coverage area and surrounding sea, which are binary coded differently in the radar sensor. These mechanisms will be explained in more depth in the next chapters.

2.14 Hamiltonian formulation for water wave equation

Hamilton's principle is classical mechanics, which is a consequence of quantum mechanics. However, the stationary action method assisted in the expansion of quantum mechanics. In other words, Hamilton's principle exemplifies the principle of stationary action. For quantum mechanics, a Hamiltonian is operationally equivalent to the entire energy of the system in the majority of the circumstances. It is frequently signified by H, also $\check{T}$ or $\hat{H}$. In this sense, its spectrum is the regular of conceivable consequences, when one computes the entire energy of a system.

In these regards, the dynamics study of Hamilton systems has been an imperative issue in mechanics for a long time. Hamilton's principles, therefore, have also the great advantage of certifying that one can establish approximations with optimal "fit" among all the mathematical equations that are proposed to solve ocean wave propagation equations. The principles of Hamilton mechanics established effectively the sequence of problems, which could not be solved by other methods. For instance, Whitham [53] utilized fluid dynamics, Hamilton principles, and variational principles for water waves and correlated difficulties in the theory of nonlinear dispersive waves. Conversely, there are generally two variational formulations for irrotational surface waves that are frequently expended [54, 55], which are used to establish an analytical approximation. Consequently, they are used to determine the velocity potential of the wave particles. In other words, Lagrangian flow is considered as an exact irrotational, which contains a velocity potential, but does not explicitly involve the velocity components. In these regards, Hamilton's principle of incompressible and inviscid fluid is used to derive approximate wave models, which is more effective for handling the highly nonlinear waves to $kh=25$ for dispersion, with accurate velocity profiles up to $kh=25$, where h is the height of the free surface. Conversely, the water elevation and the potential at the free surface are established variables when expressing mathematically the water-waves problem in Hamiltonian formalism. In this regard, Hamilton's principle with Lagrange function is expressed as:

$$L=\int_{0}^{\eta(x,t)}\left\{\varphi_t+0.5|\nabla\varphi|^2+gz\right\}dz \tag{2.90}$$

Eq. (2.90) reveals that an inviscid, irrotational flow of constant density ρ subjected to a gravitational field g acts in the negative z-axis, which is directed vertically downward. In its undisturbed state, the fluid, which is of an infinite horizontal extent, is confined to a region $-\infty<x<\infty,\eta(x,t)$. At the space x and time t, the free sea surface elevation is $\eta(x,t)$ and propagates with the potential velocity of $\varphi(x,t)$. The velocity wave particles can be identified by a Laplacian operator as $u=-\nabla\varphi$. This Lagrangian system can be expressed in Hamiltonian formalism as:

$$H=0.5\int\left(\left(\frac{\partial H}{\partial\phi}\right)^2+\frac{\partial H}{\partial\phi}|\nabla\varphi|^2\right)dx \tag{2.91}$$

Eq. (2.91) explains the irrotational flows where ϕ is a scalar potential and the term $\frac{\partial H}{\partial\phi}$ is equivalent to free surface particle η_t fluctuations over time. In the circumstance of the linear wave, the variation of the η_t leads to the dynamic boundary on the free surface as:

$$\phi_t+0.5(\nabla\varphi)^2=-g\eta(x,t) \tag{2.92}$$

The above equations present the wave-particle potential velocity in Cartesian coordinates. However, in cylindrical coordinates, an inviscid irrotational flow of constant density ρ

subjected to a gravitational field g acts in the negative z-axis, which is directed vertically downward. The fluid with a free surface $z=\eta(r,\theta,t)$ is confined in a region $0<r<\infty, 0<z<\eta$,- $-\pi\leq\theta<\pi$. There exists a velocity potential $\phi=(r,\theta,z)$ lying between $z=0$ and $z=\eta(r,\theta,t)$. Therefore, the fluid velocity is given by:

$$u=-\nabla\phi=-\left(u_r, r^{-1}u_\theta, u_z\right) \tag{2.93}$$

Hamilton's principle of the wave-particle propagations in cylindrical coordinates is cast as:

$$\partial H=\iint_D\left\{\left[\phi_t+0.5(\nabla\phi)^2+gz\right]_{z-\eta}\partial\eta+\int_0^\eta\left(\partial\phi_t+\phi_r\partial\phi_r+r^{-2}\phi_\theta\partial\phi_\theta+\phi_z\partial\phi_z\right)dz\right\}rdrd\theta dt \tag{2.94}$$

Evidently, the Laplace equation, two free-surface conditions, and the bottom boundary condition constitute the nonaxisymmetric water wave equations in cylindrical polar coordinates [56]. This equation of the system has been used to investigate the linearized initial value problem for the generation and propagation of water wave particles [46,57]. The two-dimensional water wave equations have been generalized in Cartesian coordinates and also in cylindrical coordinates. Conversely, a Hamiltonian formulation within a certain flow region for shallow water and Hamilton's canonical equation wave are investigated in the wave-particle propagations.

References

[1] Steiner T, Saenger W. Geometry of carbon-hydrogen. cntdot.. cntdot.. cntdot. oxygen hydrogen bonds in carbohydrate crystal structures. Analysis of neutron diffraction data. J Am Chem Soc 1992;114(26):10146–54.
[2] Frieden E. Non-covalent interactions: key to biological flexibility and specificity. J Chem Educ 1975;52(12):754.
[3] Horsley JA. A molecular orbital study of strong metal-support interaction between platinum and titanium dioxide. J Am Chem Soc 1979;101(11):2870–4.
[4] Jenkins S, Morrison I. The chemical character of the intermolecular bonds of seven phases of ice as revealed by ab initio calculation of electron densities. Chem Phys Lett 2000;317(1–2):97–102.
[5] Carroll MT, Bader RF. An analysis of the hydrogen bond in BASE-HF complexes using the theory of atoms in molecules. Mol Phys 1988;65(3):695–722.
[6] Peres A, Terno DR. Hybrid classical-quantum dynamics. Phys Rev A 2001;63(2):022101.
[7] Fuchs CA, Peres A. Quantum theory needs no 'interpretation'. Phys Today 2000;53(3):70–1.
[8] van Hemmen JL, Sütö A. Tunnelling of quantum spins. Europhys Lett 1986;1(10):481.
[9] Shifren L, Akis R, Ferry DK. Correspondence between quantum and classical motion: comparing Bohmian mechanics with a smoothed effective potential approach. Phys Lett A 2000;274(1–2):75–83.
[10] Schroeck Jr. FE. Quantum mechanics on phase space. Springer Science & Business Media; 2013.
[11] Castagnino M, Laura R. Minimal irreversible quantum mechanics: pure-state formalism. Phys Rev A 1997;56(1):108.
[12] Scully MO, Walther H, Schleich W. Feynman's approach to negative probability in quantum mechanics. Phys Rev A 1994;49(3):1562.
[13] Kitaev A. Anyons in an exactly solved model and beyond. Ann Phys Rehabil Med 2006;321(1):2–111.
[14] Yoshitake J, Nasu J, Motome Y. Fractional spin fluctuations as a precursor of quantum spin liquids: Majorana dynamical mean-field study for the Kitaev model. Phys Rev Lett 2016;117(15):157203.
[15] Yoshitake J, Nasu J, Kato Y, Motome Y. Majorana dynamical mean-field study of spin dynamics at finite temperatures in the honeycomb Kitaev model. Phys Rev B 2017;96(2):024438.
[16] Yoshitake J, Nasu J, Motome Y. Temperature evolution of spin dynamics in two-and three-dimensional Kitaev models: influence of fluctuating Z 2 flux. Phys Rev B 2017;96(6):064433.

[17] Takahashi SK, Wang J, Arsenault A, Imai T, Abramchuk M, Tafti F, Singer PM. Spin excitations of a proximate Kitaev quantum spin liquid realized in Cu2IrO3. Phys Rev X 2019;9(3):031047.
[18] Takagi H, Takayama T, Jackeli G, Khaliullin G, Nagler SE. Kitaev quantum spin liquid-concept and materialization. arXiv; 2019 preprint arXiv:1903.08081.
[19] De León M, Rodrigues PR. Generalized classical mechanics and field theory: a geometrical approach of Lagrangian and Hamiltonian formalisms involving higher order derivatives. Elsevier; 2011.
[20] Mielke A. Hamiltonian and Lagrangian flows on center manifolds: with applications to elliptic variational problems. Springer; 2006.
[21] Batlle C, Gomis J, Pons JM, Roman-Roy N. Equivalence between the Lagrangian and Hamiltonian formalism for constrained systems. J Math Phys 1986;27(12):2953–62.
[22] Weinstein A. Lagrangian submanifolds and Hamiltonian systems. Ann Math 1973;98:377–410.
[23] Giachetta G, Mangiarotti L. New Lagrangian and Hamiltonian methods in field theory. World Scientific; 1997.
[24] Riewe F. Nonconservative Lagrangian and Hamiltonian mechanics. Phys Rev E 1996;53(2):1890.
[25] Pons JM. New relations between Hamiltonian and Lagrangian constraints. J Phys A Math Gen 1988;21(12):2705.
[26] Lin FH, Xin JX. On the incompressible fluid limit and the vortex motion law of the nonlinear Schrödinger equation. Commun Math Phys 1999;200(2):249–74.
[27] Chern AR. Fluid dynamics with incompressible Schrödinger flow. [Doctoral dissertation]California Institute of Technology; 2017.
[28] Gasser I, Lin CK, Markowich PA. A review of dispersive limits of (NON) linear Schrödinger-type equations. Taiwan J Math 2000;4(4):501–29.
[29] Majda AJ, Majda AJ, Bertozzi AL. Vorticity and incompressible flow. Cambridge University Press; 2002.
[30] Fiscaletti D. About the different approaches to Bohm's quantum potential in non-relativistic quantum mechanics. Quantum Matter 2014;3(3):177–99.
[31] Anandan J, Suzuki J. Quantum mechanics in a rotating frame. In: Relativity in rotating frames. Dordrecht: Springer; 2004. p. 361–70.
[32] Obukhov YN, Silenko AJ, Teryaev OV. Manifestations of the rotation and gravity of the Earth in high-energy physics experiments. Phys Rev D 2016;94(4):044019.
[33] Rizzi G, Ruggiero ML. The relativistic Sagnac effect: two derivations. In: Relativity in rotating frames. Dordrecht: Springer; 2004. p. 179–220.
[34] Anandan J. Classical and quantum interaction of the dipole. Phys Rev Lett 2000;85(7):1354.
[35] Mashhoon B. Quantum theory in accelerated frames of reference. In: Special relativity. Berlin, Heidelberg: Springer; 2006. p. 112–32.
[36] Shames IH, Shames IH. Mechanics of fluids. New York: McGraw-Hill; 1982.
[37] Morrison PJ, Francoise JP, Naber GL, Tsou ST. Hamiltonian fluid dynamics. In: Encyclopedia of mathematical physics. Academic Press; 2006. p. 593–600.
[38] Goldstein RE, Petrich DM. Solitons, Euler's equation, and vortex patch dynamics. Phys Rev Lett 1992;69(4):555.
[39] Wells FS, Pan AV, Wang XR, Fedoseev SA, Hilgenkamp H. Analysis of low-field isotropic vortex glass containing vortex groups in YBCO thin films visualized by scanning SQUID microscopy. arXiv; 2018 preprint arXiv:1807.06746.
[40] Tong D. Quantum vortex strings: a review. Ann Phys Rehabil Med 2009;324(1):30–52.
[41] Weiler CN, Neely TW, Scherer DR, Bradley AS, Davis MJ, Anderson BP. Spontaneous vortices in the formation of Bose–Einstein condensates. Nature 2008;455(7215):948–51.
[42] Zhao HJ, Misko VR, Tempere J, Nori F. Pattern formation in vortex matter with pinning and frustrated intervortex interactions. Phys Rev B 2017;95(10):104519.
[43] Barenghi CF, Skrbek L, Sreenivasan KR. Introduction to quantum turbulence. Proc Natl Acad Sci 2014;111(Suppl. 1):4647–52.
[44] Santos AC. Quantum turbulence and multicharged vortices in trapped atomic superfluids. [Doctoral dissertation]Universidade de São Paulo; 2017.
[45] Moe G, Arntsen ØA. Particle velocity distribution in surface waves. In: Coastal engineering 1996. American Society of Civil Engineers; 1997. p. 565–74.
[46] Marghany M. Synthetic aperture radar imaging mechanism for oil spills. Gulf Professional Publishing; 2019.
[47] Kharif C, Pelinovsky E, Slunyaev A. Quasi-linear wave focusing. In: Rogue waves in the ocean. Berlin, Heidelberg: Springer; 2009. p. 63–89.

[48] Chabchoub A, Grimshaw RH. The hydrodynamic nonlinear Schrödinger equation: space and time. Fluids 2016;1(3):23.
[49] Osborne A. Nonlinear ocean waves & the inverse scattering transform. vol. 97. Amsterdam: Academic Press; 2010.
[50] Yuen HC, Lake BM. Nonlinear deep water waves: theory and experiment. Phys Fluids (1958–1988) 1975;18:956–60.
[51] Akhmediev N, Eleonskii VM, Kulagin NE. Generation of periodic trains of picosecond pulses in an optical fiber: exact solutions. Sov Phys JETP 1985;62(5):894–9.
[52] Benjamin TB, Feir JE. The disintegration of wave trains on deep water Part 1. Theory. J Fluid Mech 1967;27(3):417–30.
[53] Whitham GB. A general approach to linear and non-linear dispersive waves using a Lagrangian. J Fluid Mech 1965;22(2):273–83.
[54] Zakharov VE. Stability of periodic waves of finite amplitude on the surface of a deep fluid. J Appl Mech Tech Phys 1968;9(2):190–4.
[55] Luke JC. A variational principle for a fluid with a free surface. J Fluid Mech 1967;27(2):395–7.
[56] Tsubota M, Kobayashi M, Takeuchi H. Quantum hydrodynamics. Phys Rep 2013;522(3):191–238.
[57] Sultana S, Rahman Z. Hamiltonian formulation for water wave equation. Open J Fluid Dyn 2013;3(02):75.

CHAPTER

3

Quantization of synthetic aperture microwave radar

Quantum mechanics is key to understanding the mechanism of synthetic aperture radar (SAR). The principle concept of SAR is based on how to synthesize the antenna. Indeed, the aperture is defined as the area, oriented perpendicular to the direction of an incoming electromagnetic wave, which would intercept the same amount of power from that wave as is produced by the antenna receiving it. The antenna is the main source for generating microwave photons, and for detecting the amount of photon backscattering after interaction with any object in the ground. This means that the basis of SAR is the concept of quantum mechanics.

3.1 Quantize concept of aperture

A quantum of electromagnetic microwaves is known as a photon. Planck's constant, ħ, is the ratio of the energy of a photon to its frequency: $E = f\hbar$, so $\hbar = Ef^{-1}$. Planck's constant, consequently, regulates how much energy is transmitted by a photon of electromagnetic microwaves. The quantum concept articulates that both electromagnetic microwaves and matter consist of tiny particles, with which wavelike properties are associated. Microwaves are composed of particles named photons, and matter is composed of particles called electrons, protons, and neutrons. It is only when the mass of a particle becomes slight sufficient to vibrate on its orbit that its wavelike properties show up. Therefore, from a quantum perspective, the term "aperture" indicates a device that determines the area in which microwave photons or particles with microwave-like properties are collected. Each photon has a wave function, which describes its path from the emitter through the aperture to the screen. In this scenario, the wave function of the path the photon will take is resolved by the physical surroundings such as aperture geometry, screen distance, and initial conditions when the photon is created [1].

An aperture that is wider than a wavelength creates interference upshots in the space downstream of the aperture. These can be explained by assuming that the aperture performs

Nonlinear Ocean Dynamics
https://doi.org/10.1016/B978-0-12-820785-7.00007-1

as though it has a large number of point sources spaced evenly across the width of the aperture.

Let us assume a coherent microwave beam of a single frequency with a similar phase. In this circumstance, the microwave photon intensities at a specified point in the space downstream of the aperture are made up of influences from each of these point sources, and if the relative phases of these contributions vary by 2π or more, the minima and maxima in the diffracted photons can be determined. Such phase differences are caused by differences in the path lengths over which contributing rays reach the point from the aperture. In this regard, the photon path difference is almost $0.5d\sin\theta$ so that the minimum photon energy occurs at an angle $\theta_{\min}$ given by:

$$E = \frac{d\sin\theta_{\min}}{\hbar} \tag{3.1}$$

where d is the width of the aperture and $\theta_{\min}$ is the angle of incidence at which the minimum intensity occurs.

The wave that emerges from a point source has an amplitude ψ at location r that is given by the solution of the frequency domain wave equation for a point source, which can be expressed by the Helmholtz equation:

$$\nabla^2\psi + k^2\psi = \partial(r) \tag{3.2}$$

In the Helmholtz equation, $\partial(r)$ represents the three-dimensional delta function, which has radial dependence only. This $\partial(r)$ can be illustrated in spherical coordinates using the Laplace operator, which is mathematically given by:

$$\nabla^2\psi = r^{-1}\frac{\partial^2}{\partial r^2}(r\psi) \tag{3.3}$$

The solution of Eq. (3.3) by direct substitution is the scalar Green's function, which is shown in the spherical coordinate system [1–3]. Moreover, implementing the physics time convention $e^{-i\omega t}$, the amplitude $\psi(r)$ is given by:

$$\Psi(r) = \frac{e^{ikr}}{4\pi r} \tag{3.4}$$

Consequently, let us assume that electric field $E_{\text{inc}}(x, y)$ is incident on the aperture. The field produced by this aperture distribution can be mathematically expressed as the surface integral:

$$\Psi(r)\alpha \iint_{aperture} E_{\text{inc}}(x', y')\frac{e^{ik|r-r'|}}{4\pi\left|\vec{r} - \vec{r}\,'\right|}dx'dy' \tag{3.5}$$

Eq. (3.5) assumes that the delta function source is located at the origin. If the source is located at an arbitrary source point, denoted by $\vec{r}\,'$ (Fig. 3.1), and the field point is located at the point $\vec{r}$ [2], the expression for the Fraunhofer region field from a planar aperture can be given by [1, 3]:

$$\Psi(r)\alpha\frac{e^{ikr}}{4\pi r} \iint_{aperture} E_{\text{inc}}(x', y')e^{-ik\sin\theta(\cos\phi x' + \sin\phi y')}dx'dy' \tag{3.6}$$

For an aperture of a size much smaller than the wavelength of illuminating radiation, k_z is mainly imaginary value. Hence, the field behind the aperture is mostly composed of nonpropagating near-fields that decay exponentially from the aperture plane [4].

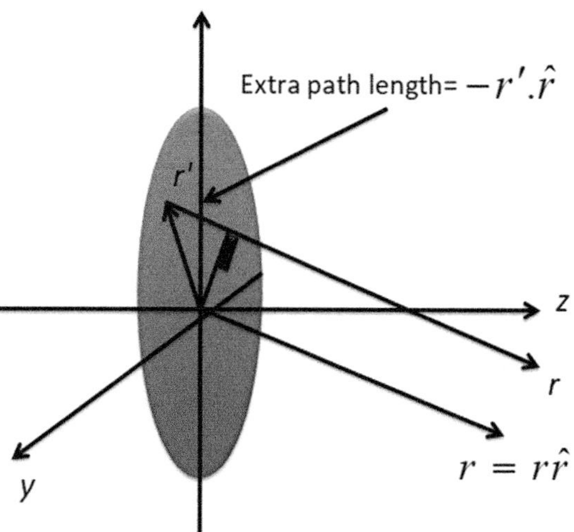

FIG. 3.1 Extra path length through the aperture.

3.2 Aperture antenna

The aperture antenna is an open-ended waveguide. By implementing the field equivalence theory, definite sources can be exchanged with the equivalent electric and magnetic current sources. For the aperture on an infinite flat electric field, only the magnetic current intensity has nonzero values on the aperture. The mathematical computation of the far-field aperture antenna radiation on an infinite electric conducting ground plane can be demonstrated in logical order as follows.

Let us assume a rectangular aperture in an infinite ground plane, which is energized with a photon microwave wave propagating in a uniform waveguide. In this circumstance, the aperture has a vertically polarized wave, its energy E_a can rotate along an axis is given by:

$$E_a = \hat{a}_z E_o \cos\left(\frac{\pi}{a}y\right) \begin{cases} -a/2 < y < a/2 \\ -b/2 < z < b/2 \end{cases} \tag{3.7}$$

here the aperture length, a, for instance, is between $0.5\lambda_o$ and $1.0\lambda_o$, while b is less than a, i.e., $a = 0.55\lambda_o$, $b = 0.25\lambda_o$. The layout of an array of rectangular apertures forms pairwise orthogonal rectangles design. Consequently, the equivalent sources can be computed based on the equivalence principle as:

$$\begin{cases} M_s = -2\hat{n} \times E_a = 2\hat{a}_y E_o \cos\left(\frac{\pi}{a}y\right) \\ J_s = 0 \end{cases} \tag{3.8}$$

Let us assume that L_θ and L_ϕ are correlated with the electric vector potential and are mathematically expressed by:

$$
\begin{aligned}
L_\theta &= \iint_s M_y \cos\theta \sin\phi e^{jkr'\cos\psi} ds' \\
&= \iint_s M_y \cos\theta \sin\phi e^{jk(y\sin\theta\sin\phi + z\cos\theta)} ds' \\
&= 2E_o \cos\theta \sin\phi \int_{-\frac{a}{2}}^{\frac{a}{2}} \cos\left(\frac{\pi}{a}y\right) e^{jky\sin\theta\sin\phi} dy \int_{-\frac{b}{2}}^{\frac{b}{2}} e^{jkz\cos\theta} dz \\
&= 2E_o \cos\theta \sin\phi \cdot \frac{0.5\pi a \cos X_1}{(0.5\pi)^2 - X_1{}^2} \cdot \frac{b \sin Y_1}{Y_1} \\
&= \pi ab E_o \cos\theta \sin\phi \cdot \frac{\cos X_1}{(0.5\pi)^2 - X_1{}^2} \cdot \frac{\sin Y_1}{Y_1}
\end{aligned}
\tag{3.9}
$$

where $X_1 = \frac{k_0 a}{2} \sin\theta \sin\phi, Y_1 = \frac{k_0 b}{2} \cos\theta$:

$$
\begin{aligned}
L_\phi &= \iint_s M_y \cos\phi e^{jk_0 r'\cos\psi} ds' \\
&= \iint_s M_y \cos\phi e^{jk_0(y\sin\theta\sin\phi + z\cos\theta)} ds' \\
&= 2E_o \cos\phi \int_{-\frac{a}{2}}^{\frac{a}{2}} \cos\left(\frac{\pi}{a}y\right) e^{jk_0 y\sin\theta\sin\phi} dy \int_{-\frac{b}{2}}^{\frac{b}{2}} e^{jk_0 z\cos\theta} dz \\
&= 2E_o \cos\phi \cdot \frac{0.5\pi a \cos X_1}{(0.5\pi)^2 - X_1{}^2} \cdot \frac{b \sin Y_1}{Y_1} \\
&= \pi ab E_o \cos\phi \cdot \frac{\cos X_1}{(0.5\pi)^2 - X_1{}^2} \cdot \frac{\sin Y_1}{Y_1}
\end{aligned}
\tag{3.10}
$$

Let us consider that Q_θ and Q_ϕ are related to the magnetic vector potential and both are equal to zero.

Subsequently, as said by Lei [5], the L_θ, L_ϕ, Q_θ, and Q_ϕ are acquired, the far fields of a vertical polarized rectangular aperture, the electrical fields of a vertically polarized aperture E_θ is given by:

$$
\begin{aligned}
E_\theta &\simeq -\frac{jk_0 e^{-jk_0 r}}{4\pi r}\left(L_\phi + \eta Q_\theta\right) \\
&= -\frac{jk_0 e^{-jk_0 r}}{4r} ab E_o \cos\phi \cdot \frac{\cos X_1}{(0.5\pi)^2 - X_1{}^2} \cdot \frac{\sin Y_1}{Y_1}
\end{aligned}
\tag{3.11}
$$

and:

$$\begin{aligned} E_\phi &\simeq \frac{jk_0 e^{-jk_0 r}}{4\pi r}\left(L_\theta - \eta Q_\phi\right) \\ &= \frac{jk_0 e^{-jk_0 r}}{4r} abE_o \cos\theta \sin\phi \cdot \frac{\cos X_1}{(0.5\pi)^2 - {X_1}^2} \cdot \frac{\sin Y_1}{Y_1} \end{aligned} \tag{3.12}$$

Eqs. (3.11) and (3.12) reveal that E_θ (i.e., V, the so-called vertically polarized field) (Fig. 3.2A) is the copolar radiation field, and E_ϕ (i.e., H, is the cross-polar field) (Fig. 3.2B). The co-factor $\frac{jke^{-jkr}}{4r}abE_o$ is used to normalize the electric field radiating from a horizontally polarized aperture. Consequently, whether the H field is the copolar or cross-polar field depends on whether the aperture is illuminated with H or V polarized waves (likewise for the V field).

Following Balanis [6] and Lei [5], the electric field of a horizontally polarized aperture can be calculated by:

$$E_a = \hat{a}_y E_o \cos\left(\frac{\pi}{a} z\right), \text{where } \begin{cases} -b/2 < y < b/2 \\ -a/2 < z < a/2 \end{cases} \tag{3.13}$$

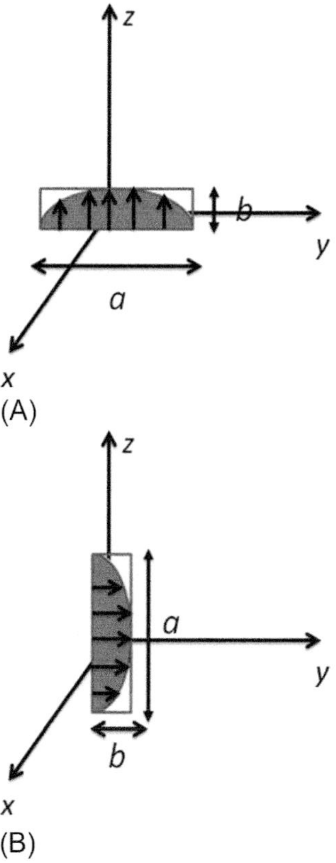

FIG. 3.2 Polarized apertures in (A) vertical and (B) horizontal directions.

As stated by Lei [5], the equivalent magnetic current sources can be expressed by:

$$\begin{cases} M_s = -2\hat{n} \times E_a = 2\hat{a}_z E_o \cos\left(\frac{\pi}{a} z\right), \\ J_s = 0. \end{cases} \tag{3.14}$$

Then, L_θ, L_ϕ, Q_θ, and Q_ϕ can be computed as:

$$\begin{aligned} L_\theta &= \iint_s -M_z \sin\theta e^{jk_0 r' \cos\psi} ds' \\ &= \iint_s -M_z \sin\theta e^{jk_0 (y \sin\theta \sin\phi + z\cos\theta)} ds' \\ &= 2E_o \sin\theta \int_{-\frac{b}{2}}^{\frac{b}{2}} \cos\left(\frac{\pi}{a} z\right) e^{jk_0 z \cos\theta} dz \int_{-\frac{a}{2}}^{\frac{a}{2}} e^{jk_0 y \sin\theta \sin\phi} dy \\ &= 2E_o \sin\theta \cdot \frac{0.5\pi a \cos X_2}{(0.5\pi)^2 - X_2{}^2} \cdot \frac{b \sin Y_2}{Y_2} \\ &= \pi ab E_o \sin\theta \cdot \frac{\cos X_2}{(0.5\pi)^2 - X_2{}^2} \cdot \frac{\sin Y_2}{Y_2} \end{aligned} \tag{3.15}$$

where $X_2 = \frac{k_0 a}{2} \cos\theta, Y_2 = \frac{k_0 b}{2} \sin\theta \sin\phi$, and $L_\phi = Q_\theta = Q_\phi = 0$. Consequently, the electric fields of the horizontally polarized aperture are:

$$E_\theta \simeq -\frac{jk_0 e^{-jk_0 r}}{4\pi r}(L_\phi + \eta Q_\theta) = 0 \tag{3.16}$$

$$\begin{aligned} E_\phi &\simeq \frac{jk_0 e^{-jk_0 r}}{4\pi r}(L_\theta - \eta Q_\phi) \\ &= \frac{jk_0 e^{-jk_0 r}}{4r} ab E_o \sin\theta \cdot \frac{\cos X_2}{(0.5\pi)^2 - X_2{}^2} \cdot \frac{\sin Y_2}{Y_2}, \end{aligned} \tag{3.17}$$

where E_θ and E_ϕ are the cross-polar and copolar fields, respectively [5].

The cross-polar pattern of the V polarized aperture is given by Eqs. (3.16) and (3.17), and it shows that the vertically polarized aperture has an electrical field in the local horizontal direction for directions away from the principal planes [5, 7].

3.3 Quantization of electromagnetic wave and Maxwell's equations

It is recognized that the electromagnetic field comprises discrete energy parcels, which are termed photons. In this sense, photons are massless particles of certain energy, certain spin, and certain momentum. Conversely, the quantization of the electromagnetic field is defined as the amount of discrete energy of the photons.

Let us consider that $\vec{K}$ is the wave vector, which is implemented to designate the transmission of electromagnetic waves. In this case, the photons behave as waves and have the transmission frequency f; otherwise, $\vec{K}$ would have a magnitude and direction. In this regard, the magnitude offers either the wavenumber or angular wavenumber of the wave, which is inversely proportional to the wavelength. Alternatively, the wave direction exhibits the direction of wave transmission. The common equation to quantize the electromagnetic field can be mathematically expressed by:

$$m_{photon} = 0 \tag{3.18}$$

$$H\left|\vec{K},\mu\right\rangle = hf\left|\vec{K},\mu\right\rangle \text{ with } f = c\left|\vec{K}\right| \tag{3.19}$$

$$P_{EM}\left|\vec{K},\mu\right\rangle = \hbar\vec{K}\left|\vec{K},\mu\right\rangle \tag{3.20}$$

$$S\left|\vec{K},\mu\right\rangle = \mu\left|\vec{K},\mu\right\rangle \quad \mu = \pm 1 \tag{3.21}$$

Formulas (3.18)–(3.21) reveal that the photon would have zero rest mass and transmit with energy corresponding to $hf = hc\left|\vec{K}\right|$, where $\vec{K}$ represents the wave vector and c is the speed of light. The electromagnetic momentum of the photon $\hbar\vec{K}$ and polarization vectors, which is an eigenvalue of the z-component of the photon spin state [8]. In quantum mechanics and particle physics, spin is complete—described as an essential approximate of angular momentum state. Photon spin is carried by elementary particles, compound particles (hadrons), and atomic nuclei. However, spin represents one of the dual categories of angular momentum in quantum mechanics, the other being orbital angular momentum. Incidentally, orbital angular momentum is considered to be quantum-mechanical, which is equivalent to the conventional angular momentum of orbital revolution. In addition, it occurs as soon as there is a periodic configuration to its wave function because of the angle dissimilarities. Let us consider that the orthonormal Cartesian vectors are e_x and e_y; as a result, the spin operators can be expressed precisely as:

$$S_z \equiv i\hbar\left(e_x \otimes e_y - e_y \otimes e_x\right) \text{ and cyclically} x \to y \to z \to x. \tag{3.22}$$

The dyadic operator is presented by the twofold operators $\otimes$ between the two orthogonal unit vectors. In this case, the unit vectors would be perpendicular to the transmission trend k (the direction of the z-axis, which presents the spin quantization axis). On the other hand, the angular momentum of the spin S_O is delineated by:

$$[S_x, S_x] = i\hbar S_z \tag{3.23}$$

From the point of view of dyadic operator, Eq. (3.23) can be precisely articulated as:

$$[S_x, S_x] = i\hbar\left[-i\hbar\left(e_x \otimes e_y - e_y \otimes e_x\right)\right] = i\hbar S_z \tag{3.24}$$

By involving Eq. (3.20), Eq. (3.24) can be modified as:

$$-i\hbar\left(e_x \otimes e_y - e_y \otimes e_x\right) \cdot e^{\mu} = \mu\hbar e^{\mu}, \quad \mu = \pm 1, \tag{3.25}$$

where μ identifies the photon spin.

In fact, the photon can be designated a triplet spin with spin quantum number $S=1$. In this case, Eq. (3.21) can be developed to identify the photon spin as:

$$S_z\left|\vec{K},\mu\right\rangle=\mu\hbar\left|\vec{K},\mu\right\rangle \quad \mu=\pm 1. \tag{3.26}$$

Eq. (3.26) demonstrates that the photon has no forward ($\mu=0$) spin component because the vector potential is a transverse field.

In precise description of particle states, spin is like a vector quantity; it has a well-defined magnitude and it also has a direction. Nonetheless, quantization creates a different direction compared to that of an ordinary vector. In other words, entirely elementary particles of a specified type have a similar magnitude of spin angular momentum, which is designated by allocating the particle a spin quantum number. Indeed, a spin quantum number represents a quantum number that parameterizes the inherent angular momentum, or simply the spin of spin angular momentum of a specified particle. Consistent with Marghany [9], the spin quantum (Fig. 3.3) describes the energy, form, and coordination of orbits. In short, each particle has its own Hilbert space, which satisfies the usual angular momentum commutation relations and defined ladder operators.

The quantization of the electromagnetic waves consequently leads to a quantum explanation of electromagnetic waves transmitting in an operating linear media with time-dependent electric permittivity and conductivity. Let us consider that $\vec{u}_T\left(\vec{r}\right)$ is the mode of the vector potential and $\widetilde{\psi}_T(\overline{p},E_k)$ is the amplitude function of every cavity mode. In this case, the combination of the Schrödinger equation and the Hamiltonian is required to quantize the electromagnetic wave propagations. Consistent with Pedrosa [10] and Pedrosa et al. [11], the mathematical portrayal of wave functions for every mode of the electromagnetic field is expressed by:

$$\psi_n(\phi_T,t)=e^{[i\varphi_n(t)]}\left[\frac{1}{\pi^{0.5}\hbar^{0.5}n!2^n\phi_T}\right]^{0.5}\times H_n\left[\left(\frac{1}{\hbar}\right)^{0.5}\frac{\phi_T}{\rho_T}\right]e^{\left[\frac{i\varepsilon_0 e^t}{2\hbar}\left(\frac{\dot{\rho}_T}{\rho_T}+\frac{ie^{-t}}{\varepsilon_0\rho^2{}_T}\right)\phi_T^2\right]} \tag{3.27}$$

where H_n represents the Hermite polynomial of order n and ρ_T is the factor, which is initiated to fulfill the normalization condition.

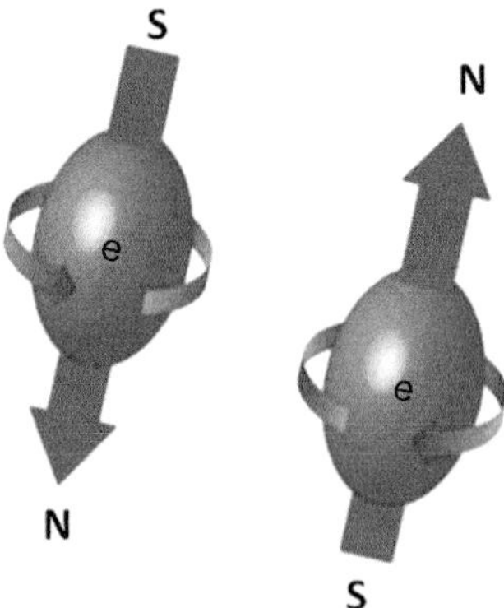

FIG. 3.3 Generalization of spin quantum.

Consequently, ϕ_T is the canonical operator; and consistent with Pedrosa [10], the phase function φ_n is given by:

$$\varphi_n(t) = -(n - 0.5) \int_0^t \frac{e^{-\tau}}{\varepsilon_0 \rho_T^2(\tau)} d\tau. \tag{3.28}$$

where ε_0 presents permittivity and τ is the time delay.

Consequently, Eq. (3.28) validates that a precise and artless quantum description of electromagnetic waves transmitting in time-reliant of conducting and nonconducting media of wave-particle transferring. Likewise, this narrative can be achieved by merging a damped quantum-mechanical oscillator in every mode of the electromagnetic wave spectra. Accordingly, a fusion of the technique was attained to obtain the electromagnetic waves quantum behavior in free space and cavities instructed through a (time-dependent) material medium [10]. It is commonly accomplished, in the former circumstance, by the implication of a regular oscillator through an individual approach of the quantized field. On the contrary, it can be implemented, in the latter one, by uniting a time-dependent harmonic oscillator [10–15].

Marghany [9] stated that Maxwell's theory is considered to be the quantum theory of a single photon, and geometrical optics as the classical mechanics of a photon. Since the photons do not correlate to identical specific approximation for frequencies lower than $m_e c^2/h$ (m_e = electron mass), the theory for one photon corresponds reasonably well to the theory for an infinite number of them. In this understanding, Bose-Einstein symmetry is concerned when the photons are bunched as a function of their energy variations.

Moreover, Feynman's work on quantum electrodynamics revealed that all types of electromagnetic beams that required a wave theory could be explained employing a particle model. In this regard, quantum particles planned out by Feynman diagrams are not free to follow any spatial path, but can correspondingly be bendable in their path through time. A positron, for instance, can be embodied as an electron moving backward in time, which is revealed as the central straight line in the Feynman diagram (Fig. 3.4).

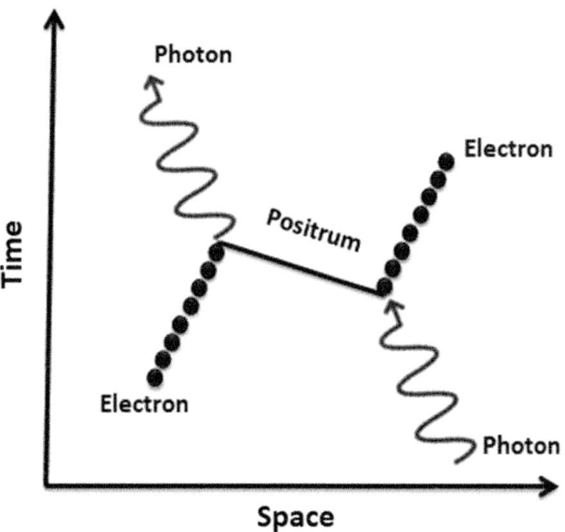

FIG. 3.4 Overview of Feynman diagrams.

Feynman rises in physical postulations that are extremely obstructed to verify Maxwell's equations distinctively. Let us consider a particle at the location x^μ, with the momentum p^μ, which is a function of $F(x^\mu)$. The precise nature of the forces acting on particles can scientifically be expressed as:

$$\frac{dp^\mu}{d\tau} = F_1^\mu(x^\mu) + F_2^{\mu\nu}(x^\mu)p\nu + \cdots \tag{3.29}$$

where F_i is tensor represents the field F, although F_i is requested more physical probabilities to be recognized.

These expectations $\ell(x^\mu, \dot{x}^\mu, t)$ involve the Lagrangian, which is quadratic in velocity. The calculated differentiation of Eq. (3.29) suggests that the force would be at most linear in momentum. Consequently, this linear momentum is expressed as:

$$\frac{dp^\mu}{d\tau} = F_1^\mu + F_2^{\mu\nu}p\nu. \tag{3.30}$$

Eq. (3.30) is derived from conventional Hamiltonian mechanics as follows:

$$[x_i, v_i] = i\frac{\hbar}{m}\partial_{ij}. \tag{3.31}$$

To utilize quantum mechanics in Eq. (3.31), the Dirac prescription of replacing Poisson brackets with commutators is used. This yields the canonical commutation relations $[x_i, p_j] = i\hbar\partial_{ij}$, where x_i and p_j are characteristically canonically conjugate. The momentum can be formulated based on the Lagrangian, and is determined from [9]:

$$p \equiv \frac{\partial \ell}{\partial \dot{x}} \tag{3.32}$$

Eq. (3.32) demonstrates that the Lagrangian is quadratic in velocity; consequently, the force is at utmost linearity in velocity. In the circumstances of $F_1 = 0$ and $F_2 = F$, Eq. (3.30) can be recovered into the Lorentz force law by:

$$\frac{dp^\mu}{d\tau} = F^{\mu v}pv. \tag{3.33}$$

It is a difficult task to acquire Maxwell's equations because there is no information about the field's dynamics. However, to overcome this gap, the simplicity of the Hamiltonian is considered; therefore, the mathematical description of the quadratic momentum is defined as:

$$\mathrm{H} = p^2(2m)^{-1} + \vec{\mathrm{A}}_1 \cdot \vec{p} + A_2 \tag{3.34}$$

The Hamiltonian equation of the momentum under circumstances of gathering $\vec{\mathrm{A}}_1$ and A_2 into a four-vector A^μ is formulated as:

$$\frac{dp^\mu}{d\tau} = (dA)^{\mu v}p\nu \tag{3.35}$$

where d is the exterior derivative. That is, the appliance of the Hamiltonian forces the field F to be designated in expressions of potential, $F = dA$. Subsequently $d^2 = 0$, thus $dF = 0$, which comprises two of Maxwell's equations, specifically Gauss's law for magnetism and Faraday's law.

The significant conclusion is that Feynman's derivation of electromagnetic wave state is numerous, but is not entirely secretive. In particular, it does not combine conventional and quantum mechanics at all—the quantum equivalences that Feynman uses are consistent with conventional ones resulting from the Hamiltonian equation. Feynman cast off the Dirac quantization method. Consequently, Feynman was also responsible for another time-bending method of investigating quantum incidence. On the other hand, Maxwell's equations for electromagnetism had two explanations depicting twofold dissimilar categories of an electromagnetic wave: retarded waves, which are the common ones that can be understood; and advanced waves, which move backward in time from the object to the antenna (Fig. 3.5). The advanced wave description had been disbelieved, as it appeared to have no relationship with reality. Nonetheless, Feynman claimed that the advanced waves did exist, as this overcame a potential difficulty of electrons recoiling after emitting virtual photons by the conservation of momentum (Fig. 3.6). In this regard, when two photons are warped, the advanced photon propagates backward in time from the destination of a retarded photon to the electron that instigated the recoil, removing the feedback.

It is not surprising that electromagnetism bursts out practically as wave-particle "for free" since the space of promising concepts truly is reasonably reserved. Generally, in the framework of the quantum field principle, Maxwell's equations can be obtained by supposing locality, parity, symmetry, and Lorentz invariance, and there occurs a long-range force arbitrated by a spin 1 particle. This has significance for classical physics, since the only classical physics, those quantum fields can be observed in which they have a practical classical constraint. No one, therefore, has ever quantified an electric field—only to establish how it is accomplished on particles [9].

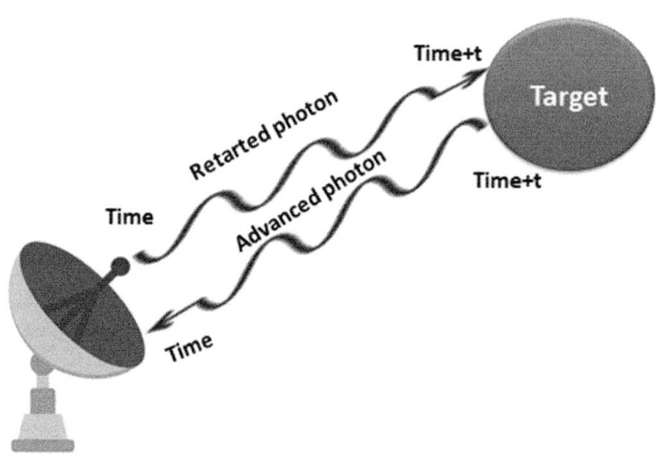

FIG. 3.5 Retarded photon and advanced photon.

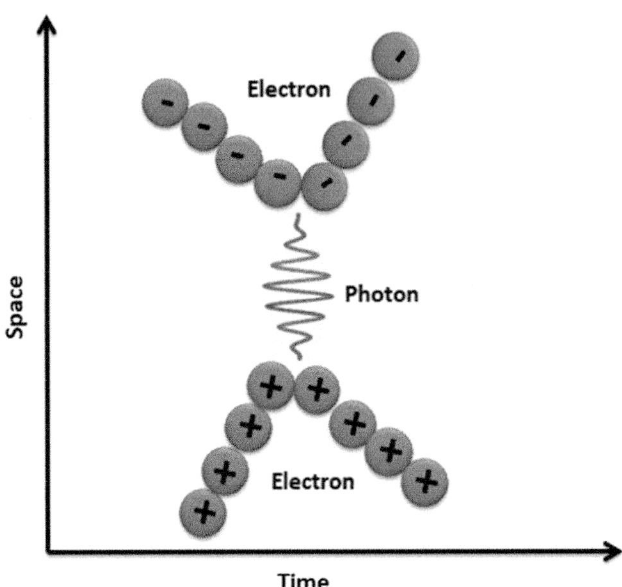

FIG. 3.6 Concept of electron recoil after emitting virtual photons.

3.4 Microwave radar photons

Radar is an acronym for RAdio Detection And Ranging. Radar was initially developed to replace technology for sensing targets and revealing their locations using echolocation, and this remains the primary function of modern radar systems.

Radio waves are the lowest-frequency waves in the electromagnetic (EM) spectrum, and can be used to carry other signals to receivers that subsequently decode these signals into functioning information [9]. The microwave beam is electromagnetic radiation with wavelengths greater than those of visible violet light, the thermal spectrum, and ultraviolet light. This means that microwaves comprise the range of the electromagnetic spectrum between radio and infrared. Microwave photons have frequencies from about 3 GHz up to about 30 trillion hertz, or 30 terahertz (THz), and wavelengths of about 10 mm (0.4 in.) to 100 micrometers (μm), or 0.004 in. The term "radio" is used because the first radar used long wavelengths of radiation (1–10 m) that fell within the radio band of the electromagnetic spectrum.

Microwave is the part of the electromagnetic spectrum that ranges from the wavelength λ of 1 mm to 1 m, corresponding to signal frequencies f equals 300 GHz and 300 MHz ($\lambda f = c$, with the speed of light c), respectively. Unlike the visible spectrum, this wavelength is extremely large and intermingles with quite dissimilar targets, in contrast to passive spectra. In this way, the microwave signal has much lower energy than passive spectra to induce molecular resonance. In contrast, passive spectra of visible and near infrared, for instance, expose their signal patterns into the atmosphere and soil chemical structures due to their high energy. The microwave domain, conversely, still has adequate high energy to model resonant spin of definite dipole molecules consistent with the frequency which is a function of the changing of a signal electric field [16].

In this sense, the object appears dark in radar images because it is smaller than the radar wavelength, which does not imitate abundant energy. On the contrary, short-wavelength radar can discriminate more minor variants of irregularity than long-wavelength radar. In other words, object roughness fluctuates with wavelength. Therefore, because of the diffuse reflection, the incidence angle does not have a significant role in rough surface [9]. Active microwave wavelengths tend to be divided into typical wavelength regions or bands (Table 3.1) (Fig. 3.7).

In microwave sensors, therefore, frequency (in Hertz) can also be used to describe a band range. It is worth mentioning that satellite active microwave sensors do not acquire multispectral microwave images—they attain data for only a single band of wavelength/frequency. Nevertheless, recent airborne sensors can acquire multifrequency bands, for instance, NASA's Jet Propulsion Laboratory AIRSAR system: C-, L-, and P-band.

A theoretical quantum radar operating in the 9 GHz regime could improve object detection under the impact of atmospheric attenuation. In this regard, the photons in the 9 GHz frequency (Table 3.1) correspond to the radar-microwave spectrum. In this sense, the X-band in radar corresponds to the 8–12 GHz region and is extensively used for missile guidance, marine radar, weather, ground surveillance, and airport traffic control [17–19].

The term "radar" is commonly used for all active microwave systems [9]. Radar is defined as a system for detecting the presence, direction, distance, and speed of aircraft, ships, and other objects, by sending out pulses of radio waves that reflect off the object back to the source. A radar device commonly operates in the ultra-high-frequency (UHF) or microwave phase of the radio frequency (RF) spectrum and is deployed to detect the locus and/or movement of targets.

TABLE 3.1 Microwave bands and their physical characteristics [9].

Bands	Wavelength (cm)	Frequency (GHz)
P-band	30–100	1.0–0.3
L-band	15–30	2.0–1.0
S-band	7.50–15	4.0–2.0
C-band	3.9–7.50	4.0–8.0
X-band	2.40–3.75	8.0–12.5
K-band	0.75–2.40	12.5–40 (rarely employed)

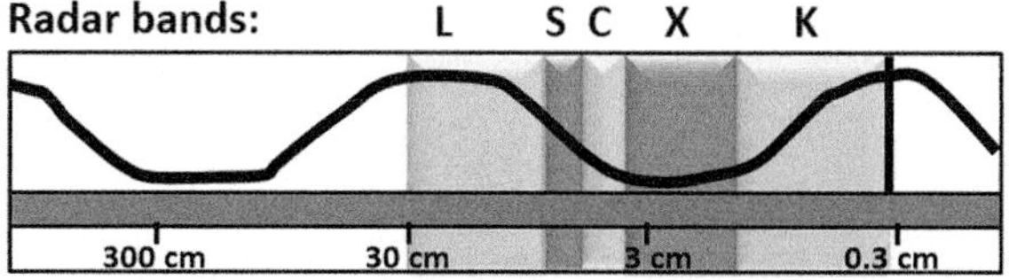

FIG. 3.7 Microwave electromagnetic waves.

Microwaves are used for high-bandwidth communications, radar, and as a source of heat in microwave ovens and industrial applications [17]. Microwaves can overcome obstacles that interfere with radio waves, such as clouds, smoke, and rain. Microwaves can convey radar, landline phone calls, and computer data transmissions, as well as cook your dinner, owing to their higher frequency. Microwave residues of the "Big Bang" radiate from all directions throughout space (Fig. 3.8).

3.5 Microwave cavity main concept

Magnetism is a quantum phenomenon. Nevertheless, most magnets can be defined by conventional physics since they involve large numbers of spins. For instance, the coupling between a magnet and an electromagnetic wave can, in many circumstances, be entirely expressed by Maxwell's equations. The quantum nature of a magnet can, conversely, be revealed if it is precise strongly coupled to a microwave field, for instance, that in a cavity. The starting point is to formulate the electromagnetic field in the cavity in a quantum state, such that the quantity of photons in the field constitute a reasonable quantum number. "Switching on" robust coupling between the cavity and the magnet involves aligning the resonance frequency of the magnet with that of the cavity (Fig. 3.9). In these circumstances, energy in the cavity—and quantum information—can be transferred to the magnet. When the coupling is turned off, the quantum information is "gathered" in the magnet. Switching on intense coupling again relocates the situation of quantum state back into the microwave cavity, where it can be detected with microwaves. The keystone prerequisite of this technique is that the spread of energy is faster than its decay in either the microwave resonator or the magnet [9,20].

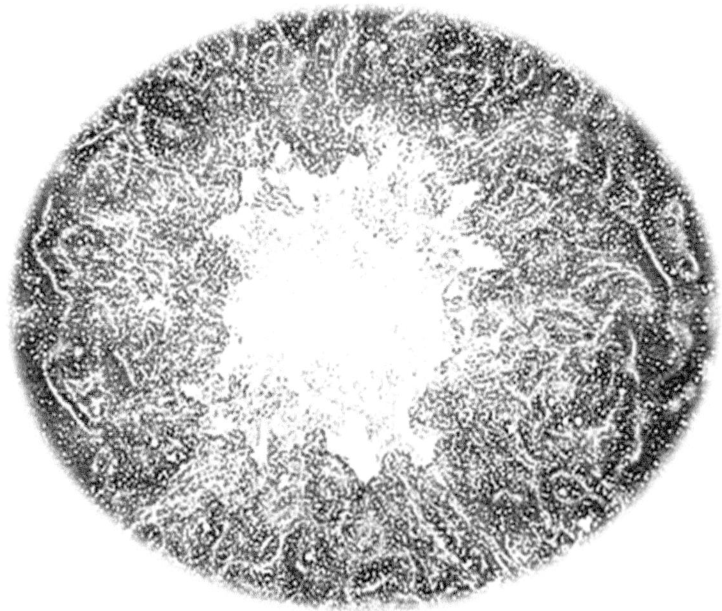

FIG. 3.8 Microwaves emitted from the Big Bang.

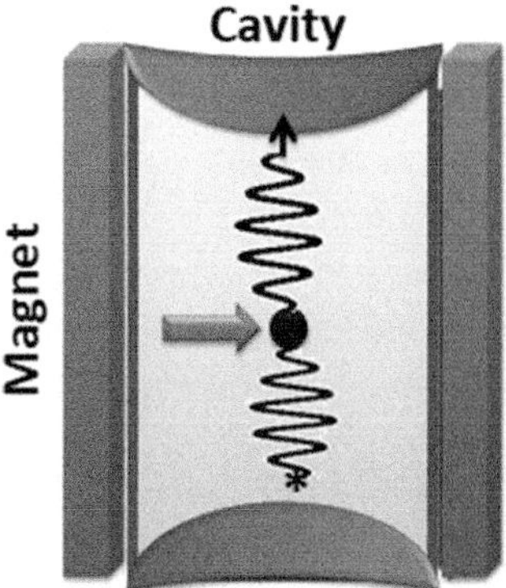

FIG. 3.9 Concept of a microwave cavity.

Consistent with Bai et al. [21], a strong interference between microwave photons and a magnet can create a fusion energy system intensities opposing from both the photons and the magnet on their owns states. Through this "robust combination" system, quantum information can be moved from the photon to the magnet and vice versa. In this view, a robust coupling can be perceived precisely from a magnet, which is derived in a microwave cavity (Fig. 3.9) [21]. On the other hand, a continuous electrical signal can be assimilated continuously from the magnet as opposed to revealing the circumstances of the magnet circuitously with microwaves [22]. This signal is twisted by the declared spin inflating effect, which occurs at the crossing plug between a magnet and a metal [22]. In the circumstance of the magnet is energized by the cavity, it contracts concerning equilibrium by fluctuating a spin-polarized current into the adjoining metal surface. Consistent with Mosendz [23], this spin current, consequently, is carried by a diagonal charge current, i.e., a result of the inverse spin hall influence, which the pumping signal can be detected by conventional electronics. Consequently, the oscillating magnetic field of the microwaves exploits a torque on the systematic spins in the magnet and captivating their exactness. Subsequently, the magnet's resonant frequency is tweaked to that the cavity by interleaving the magnet assembly into an aluminum microwave cavity and utilizing an external magnetic field. Thus, the microwave spectrum is created and dignified from the cavity. Additionally, the spectrum of the electrical spin continues pumping signal from the magnet (Fig. 3.10) [9].

Since the magnet and the cavity are detached, both the electrical signal and the cavity have precise resonances. Nevertheless, if the two constituents are strongly combined, these resonances accelerate and split into two. This phenomenon is known as an "avoided level crossing" that evolves when dual systems with similar energies correlate. Ambiguous of the peaks in the microwave spectrum is formed up, nonetheless, the peaks in the electrical spectrum produce to be lopsided in the exhaustive-coupling radar system. This dissimilarity is consequently expected, since the spin pumping is sensitive to magnetic excitation. Accordingly, it delivers evidence on the robust-coupling routine only from the perception of the magnetic system. From the point of view of physics, the spin-pumping signal regulates the amount of energy by which the magnet is energized in the comprehensive coupling stability [9,24].

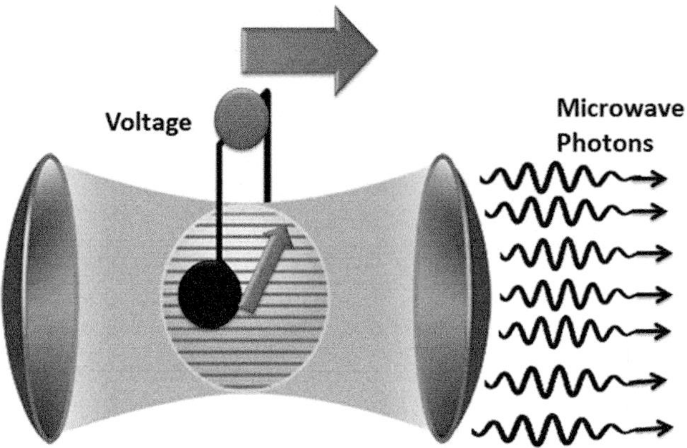

FIG. 3.10 Generating microwave photons by the magnet-electrical device.

3.6 Microwave photon generation by Josephson junctions

The **Josephson** effect **is defined as the** phenomenon of supercurrent, a current that flows for an indefinite period without any voltage applied, across a device called a **Josephson** junction (JJ) (Fig. 3.11). It consists of two or more superconductors coupled by a weak link. The **Josephson** effect is an example of a macroscopic quantum phenomenon.

Marghany [9] stated that the principle theory of superconductivity is required to understand the distinctive and significant attributes of Josephson junctions. Moreover, if numerous metals are cooled and composites to actual low temperatures, i.e., within 20°C or less of absolute zero, a phase transition arises. Marghany [9] concluded that at "critical temperature," within 20°C or less of absolute zero, the metal twists from what is recognized as the standard state, where it has electrical resistance, to the superconducting state, where there is no

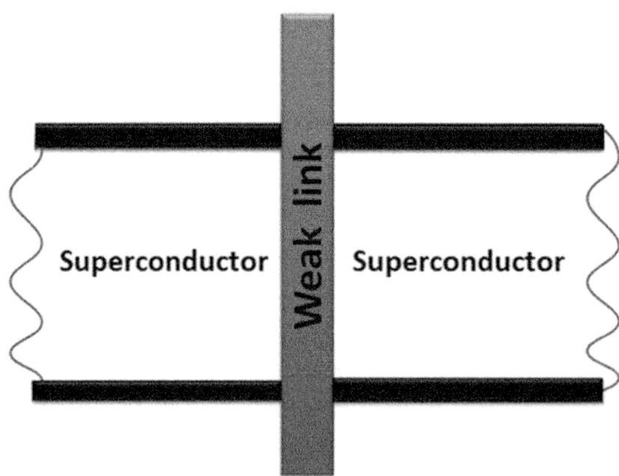

FIG. 3.11 Simplification of Josephson junction.

resistance to the flow of direct electrical current. Contemporary high-temperature superconductors, which are made from ceramic materials, show an analogous performance, although at higher temperatures. In these circumstances, the electrons in the metal turn into an identical state. On the other hand, owing to the high significant temperature, the gain collaboration between twofold electrons is disgusting. Conversely, at the lower significant temperature, though, the general interaction between twofold electrons grows incredibly somewhat attractive, a consequence of the electrons' crossing point with the ionic lattice of the metal. This incredibly trivial attraction permits electrons to charge into an inferior energy level, instigating an energy "gap." In this context, electrons can transmit without being scattered by the ions of the lattice, and current can therefore flow.

Fig. 3.12 shows an insulator separating dual superconductors at a Josephson junction, i.e., a nonsuperconducting obstacle, which must be extraordinarily thin. The insulator has to be in the order of 30 Å thick or less. If the obstacle is nonsuperconducting, it can be as much as a few microns thick. In anticipation of a critical current is captured, a supercurrent can radiate through the conductive metal obstacles. In this way, electron pairs can pass through the obstacle without any resistance. Nonetheless, when the critical current is exceeded, a voltage can be produced at the junction. That voltage depends on time—that is, it is an AC voltage. The frequency of this AC voltage is approximately 500 gigahertz (GHz) per millivolt through the junction. This initiates a lowering of the junction's critical current, creating regular uniform current and a larger AC voltage. Consequently, when the current across the junction falls below the critical current, the voltage becomes zero. As soon as the current exceeds the critical current, the voltage rises above zero and oscillates in time. Sensing and determining the modification from one state to the other is at the cornerstone of the many radar applications for Josephson junctions.

Electronic circuits used in antenna can be made from Josephson junctions, predominantly for digital logic circuitry. Many scientists, for instance, are developing ultrafast computers utilizing Josephson logic, which can be twisted into circuits known as SQUIDs—an acronym for Superconducting QUantum Interference Devices (Fig. 3.13).

Consistent with Marghany [9] and Tarasov et al. [25], a SQUID comprises a loop with dual Josephson junctions interrupting the loop. A SQUID is tremendously sensitive to the total

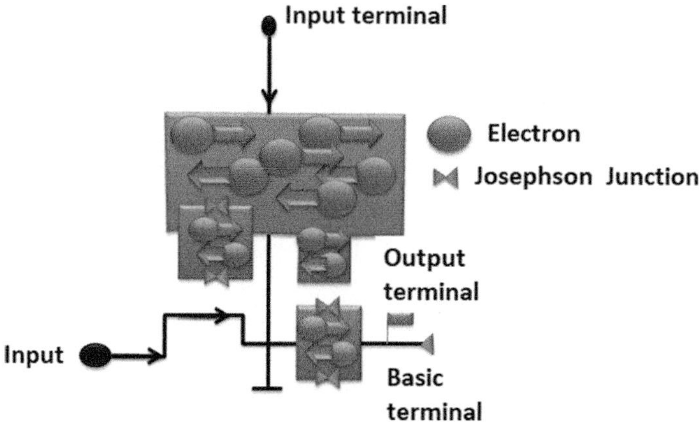

FIG. 3.12 Electronic current flow through a Josephson junction.

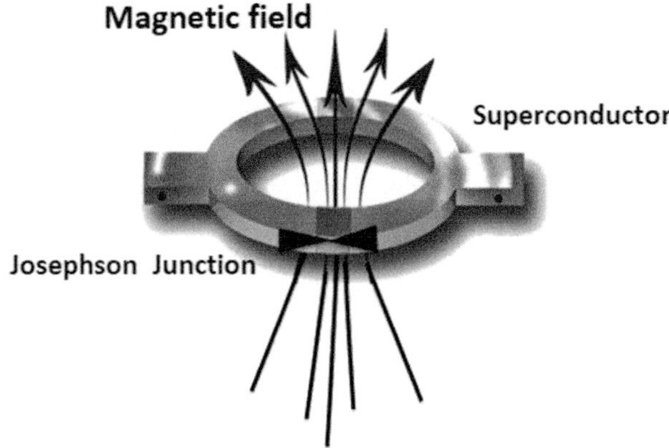

FIG. 3.13 Superconducting quantum interference device (SQUID).

amount of magnetic field that penetrates the area of the loop. In this context, the voltage is generated through the device that is especially powerfully interconnected to the whole magnetic field all over the place of the loop [26]. Arrays of Josephson junctions (JJA) can be utilized to boost the sensitivity of detectors, and improve the output power and diminish the line width of Josephson oscillators [25]. Fig. 3.14 reveals that dual small superconducting electrodes are coupled to each other and to leads by small tunnel junctions. The junctions have equal Josephson coupling and capacitance. The voltage on capacitively connected gates controls the gate-induced charge. The bias current is input via macroscopic leads. As a result, magnetic flux can be applied to the loops.

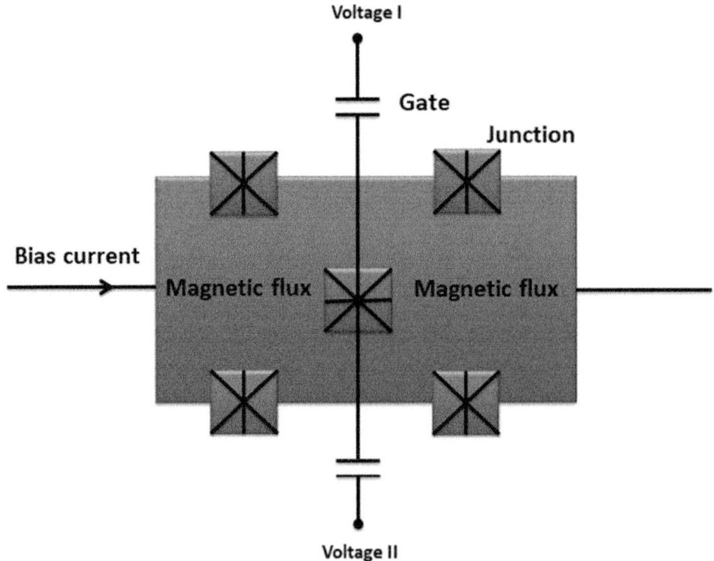

FIG. 3.14 Arrays of Josephson junctions.

As noted by Tarasov et al. [25], arrays of Josephson junctions are accomplished not only a sense of the strengthening in impedance of the sequences array, nonetheless, mainly owing to synchronization of Josephson oscillations in the junctions. In this sense, the Hamiltonian of the Josephson element of a superconducting tunnel junction for a microwave oscillator is expressed as:

$$H=\left(\frac{\hbar}{2e}\right)^2/L_j\cos\varphi+\frac{Q^2}{2(C_J+C_{ext})}-\frac{\hbar}{2e}\varphi\cdot I+H_{env} \tag{3.36}$$

The Josephson energy is expressed in terms of $\left(\frac{\hbar}{2e}\right)^2/L_j$, the total capacitance in parallel with the Josephson element is denoted by $2(C_J+C_{ext})$, and ϕ is the gauge-invariant phase difference across the junction. Moreover, $\mathbf{Q}$ is the charge conjugate to the phase $[\phi,\mathbf{Q}]=2ei$, H_{env} is the Hamiltonian of the transmission line, which counts the pump received signal through this channel, and $\boldsymbol{I}$ is the current operator corresponding to the quantities of freedom of the transmission line. The amplifier functions $\langle\varphi\rangle$ having deviations considerably less than $\frac{\pi}{2}$ and the cosine function in the Hamiltonian can be expanded to fourth order only, with the φ^4 term, which is considered as a perturbation [9]. Thus, the ladder operators of the single mode of the circuit are formulated as:

$$\varphi=\varphi_{ZPE}\left(\vec{\psi}+\vec{\psi}^{\dagger}\right) \tag{3.37}$$

where:

$$\varphi_{ZPE}=\sqrt{\frac{2e^2}{\hbar}}\sqrt[4]{\frac{L_j}{2(C_J+C_{ext})}} \tag{3.37.1}$$

Eq.3.36 can be mathematically formulated in the form of the quantum Langevin equation as:

$$\frac{d}{dt}\vec{\psi}=-i\left(\varpi+k\vec{\psi}^{\dagger}\vec{\psi}\right)\vec{\psi}-\frac{k}{2}\vec{\psi}+\sum k\vec{\psi}^{in}(t) \tag{3.38}$$

Eq. (3.38) can be formulated into Hamiltonian for the degenerated parametric amplifier arising from the pumping of the Josephson junction as:

$$\frac{H}{\hbar}=\left(\varpi_a+2k|\psi|^2\right)\partial\vec{\psi}^{\dagger}\partial\vec{\psi}+\left[\text{h.c.}+\frac{\mu_r\omega_0}{4}e^{i(\Omega_{aa}t+\theta)}\left(\partial\vec{\psi}\right)^2\right] \tag{3.39}$$

where α is defined as:

$$\psi=\frac{i\sqrt{k}\psi^{in}}{(\varpi_a)+\frac{ik_c}{2}-k|\psi|^2} \tag{3.39.1}$$

where c is a complex number, ψ is the classical semisignal amplitude, μ_r is relative amplitude, and ω_0 is the resonant frequency.

Moreover, Eq. (3.39) demonstrates that the center frequency of the amplifier band shifts with the growth in the pump amplitude. According to this understanding, the pump tone needs to be at the center of the band for optimal amplification. Furthermore, $\Omega_{aa}=\Omega_1+\Omega_2$

is the coupling of two pump frequencies, which facilities the use of the amplifier parameter. It is worth noting that for this device, the pump tone and the signal tone must be received by the circuit on the same port, which makes it difficult to deliver the widely diverse amplitude intensities of these dual waves.

3.7 Radar systems

The radar configuration consists of antenna, transmitter, and receiver. The transmitter generates the photon energies to deliver a radar beam and transmits it to the antenna as a function of Josephson junction and, therefore, to the target. The transmitted signal is a short burst, rapidly repeated, to give a pulsed signal. These pulsed signals, traveling at 300,000 km/s, strike the targets in view, where some of the energy is absorbed, some is reflected away, and some refracted or backscattered to the receiver [16]. The receiver, which is usually integrated with the transmitter, receives the pulsed returned signal. It determines the signal strength and relates the signal to the transmission. To estimate target distance, these data are processed into a form suitable for recording. The return signal, consequently, is much lower than the transmitted signal, since not all the signal is backscattered. This signal provides the grayscale tone of the image [16].

This sort of microwave imaging is termed active because it transmits its own microwave energy (pulses) at a particular wavelength (single frequency) for a particular duration of time, known as pulsed coherent radar. Fig. 3.15 shows the common concept of a radar system. The following sections are concerned with the fundamental concepts of SAR image data [9,18].

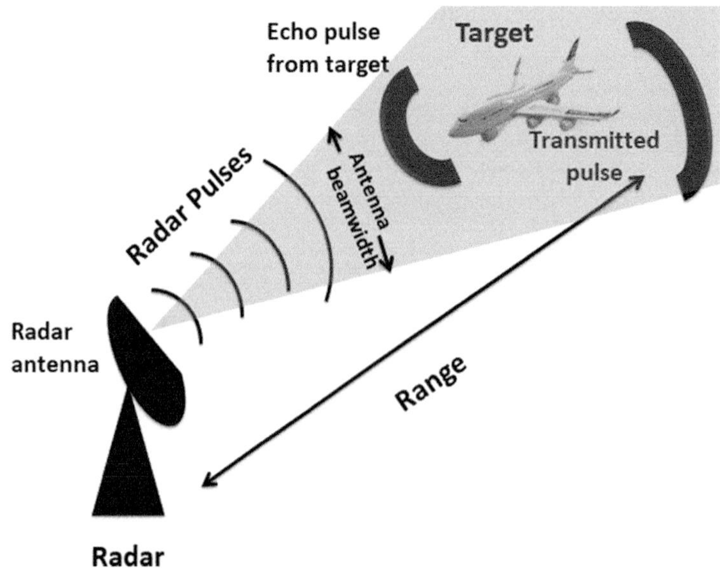

FIG. 3.15 Concept of radar (radio detection and ranging).

3.8 What is meant by echolocation detecting and ranging?

This significant question allows us to understand how radar works. Active radar systems achieve echolocation by conveying an electromagnetic wave and quantifying the reflected field as a time-varying voltage in the radar receiver. To comprehend the basic idea of echolocation (Fig. 3.16), let us consider a radar array that launches, at period $t=0$, a short pulse that propagates at the speed of light c, returns from targets at a range R, and is received by the radar. If the incoming pulse can be identified at a time τ (Fig. 3.17), then because $c\tau=2R$ (speed × time = round-trip distance), it can be shown that the range R must be equal to $R=c\tau/2$. This concept is the fundamental basis for radar systems.

High-range-resolution (HRR) imaging is thus vital. Most targets of interest need to be independently detected. A superposition of reflections, for example, is detected when the short pulse strikes an airplane. This response is called a high-range-resolution (HRR) shape, which can be considered as a one-dimensional "image" of the target (Fig. 3.18).

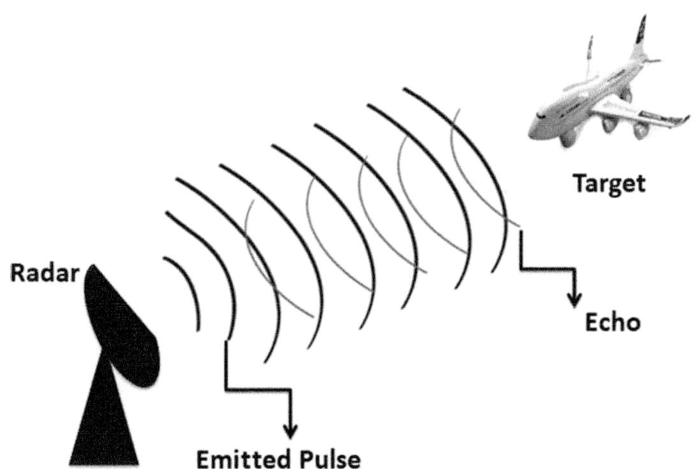

FIG. 3.16 Concept of echolocation.

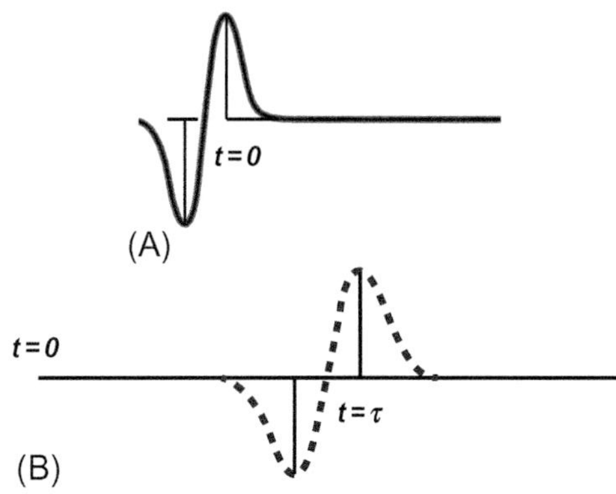

FIG. 3.17 A pulse is transmitted at (A) time zero and (B) received on time τ.

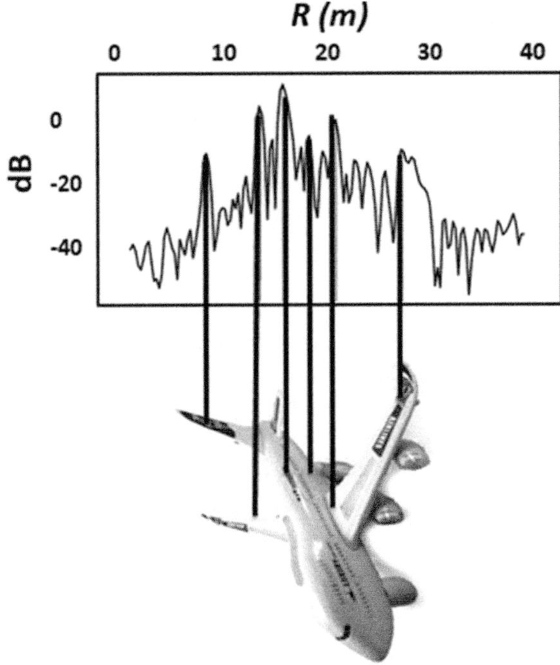

FIG. 3.18 Radar reflections from a complex target.

3.9 Why quantum synthetic aperture radar is necessary

It is certain that quantum questions need to be answered to improve the performance of a broad diversity of conventional information processing synthetic aperture radar (SAR) systems. It is generally accepted that quantum communication and computer devices are required to develop information infrastructure systems. Further, the interface between quantum information science and quantum sensing is significant; for instance, quantum SAR can be designated precisely as noisy quantum modes. Furthermore, quantum computation performance, established from the perspective of quantum control, will be invaluable to exploit quantum sensing hardware.

It is imperative to state that although quantum sensing is not established yet as quantum computation, it delivers humbler manufacturing difficulties of quantum sensing hardware. As quantum computation necessitates a huge quantity of qubits on unfamiliar electron superpositions with adequate coherence times to achieve convoluted computations exploiting a great diversity of accesses. In contrast, quantum sensors necessitate a small number of qubits in a precise, entangled format. As a result, it only needs a minority of quantum processes. In consequence, it seems that the expansion of quantum sensors delivers the possibility for the realistic use of the quantum information technologies; which are demanded the comprehension of a speculative quantum computer processor [9].

By developing entanglement, quantum SAR provides the expectation of improved target recognition proficiencies. In this regard, quantum radar relies on quantum disarrays of electromagnetic waves, i.e., microwave photons, which are maintained on a twisted superposition. These waves are directed toward the object and the backscattered signals detected by the receiver. In this regard, signal detection is improved by utilizing matches between the radiated microwave photons bounced back from the object and the ones coded into the radar. Autonomously, quantum radar provides the option of the automatic physical characteristic of the target identification. Consistent with Lloyd [27], a quadratic resolution enhancement can improve on nonentangled photons by using entangled photons through quantum radar. In this way, the visibility of target detection can be increased using quantum radar. Indeed, a quantum sidelobe structure provides a new mode for the revealing of RF for hiding or extremely tiny objects cannot be detected by classical radar.

3.10 What is meant by quantum SAR?

In general terms, a quantum SAR can be described as an impasse detection sensor that exploits microwave photons and uses some formulas of quantum phenomena to improve its proficiencies to distinguish, recognize, and determine an interesting object. For the purposes of this chapter, quantum SAR is defined as the quantum phenomena required to comprehend the functionality of the SAR mechanism.

In a quantum radar system, the targets are expected to retain around backscatter compare to the classical radar, which is far-off from the sensor. Furthermore, the SAR-target system is assumed to be engrossed in a noisy and lossy environment. Precisely, SAR signal may properly be reduced by absorption or scattering expansions, and the overall operation of the SAR system is exaggerated by the presence of noise.

3.11 What are the classifications of quantum SAR?

The main examples of quantum SAR sensors are single-photon quantum SAR and entangled photon quantum SAR. Three sorts of quantum SAR, therefore, can be categorized as follows:

(i) The quantum SAR conveys unentangled quantum states (momentum, angular momentum, energy, spin, and time) of microwave photons.

(ii) The quantum SAR antenna emits conventional statuses (energy, wavelength, and frequency) of microwave photons, nonetheless, exploits quantum photo-SAR to enhance its operation.

(iii) The quantum SAR emits quantum signal states of microwave photons that are entwined with quantum states of a microwave photon generated at the SAR transmitter.

The basic concept of a single-photon quantum radar is that a single photon is emitted toward a target, and consequently, the photon is scattered back to the receiver (Fig. 3.19). By contrast, in entangled-photon quantum radar, an entangled pair of microwave photons is produced. In this case, one photon is emitted toward the target and the other is stored in the radar sensor. The outward photon is then reflected by the object and consequently

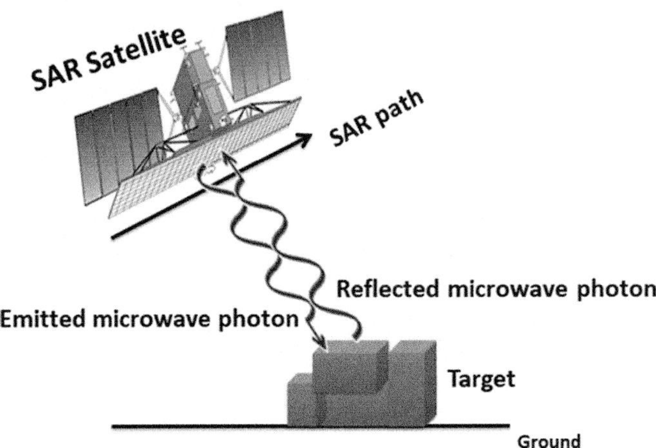

FIG. 3.19 Single-photon quantum radar.

received by the radar system. In this view, the connections between transmitted and received radar signals are established in the entangled condition, which are oppressed to enhance target detection in quantum SAR system (Fig. 3.20) [28].

One of the quantum radar mode is, for instance, LADAR, which stands for LAser Detection And Ranging, and is a similar technology as LiDAR (Light Detection And Ranging) (Fig. 3.21). LADAR operates in the visible and near-visible photon spectra [79]. However, it cannot be counted as a form of quantum radar technology. It cannot penetrate fog or clouds as radars do, and as a limitation, the operational range of a LADAR is frequently constrained within 100 km [9].

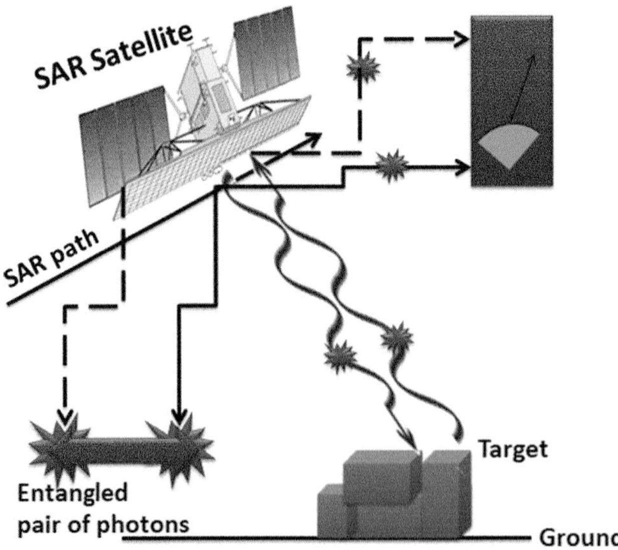

FIG. 3.20 Entangled quantum radar.

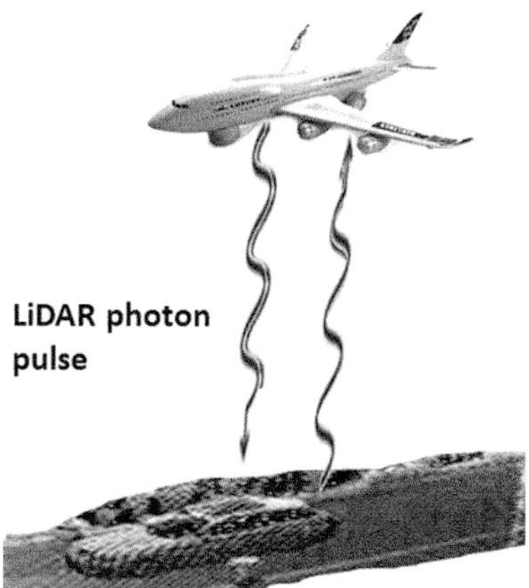

FIG. 3.21 LiDAR photon pulse.

3.12 Classical and quantum radar equations

The radar range equation is the simplest mathematical description of the radar principle. Although it is one of the most effective equations, paradoxically, it is an equation that few radar analysts comprehend and plenty of radar analysts mishandle. The problem lies not with the equation itself, but with the numerous terms that form the equation. A deep understanding of the radar range equation delivers a completely solid basis for the radar principle. The radar equation is the termination of several simpler formulas [9].

The transmitted energy density E_T can be explained as energy per unit area and occurs in a range R. The scientific explanation of the power density at a distance of the emitted signal can be written mathematically as:

$$E_T = \frac{P_T \cdot \tau \cdot G_T}{4\pi R^2} \tag{3.40}$$

In Eq. (3.40), $4\pi R^2$ associates the power transmitted by the radar to an isotropic sphere. In this sense, the electromagnetic energy transmits similarly in all directions (Fig. 3.22).

Further, G_T is the antenna and the focus of the antenna signal is a function of the ratio of $\frac{G_T}{4\pi R^2}$. Consequently, P_T is called the peak transmit power and is the average power when the radar is transmitting a signal. P_T can be specified on the output of the transmitter with a duration time τ, or at some other point like the output of the antenna feed. It is measured in watts.

Eq. (3.40) can be developed as a function of the radar cross-section (RCS) σ, which is a result of the backscatter size of the target on which the radar signal is focused (Fig. 3.23). Therefore, the radiated energy of an object is estimated via:

$$E_\sigma = \frac{P_T \cdot \tau \cdot G_T \cdot \sigma}{4\pi R^2} \tag{3.41}$$

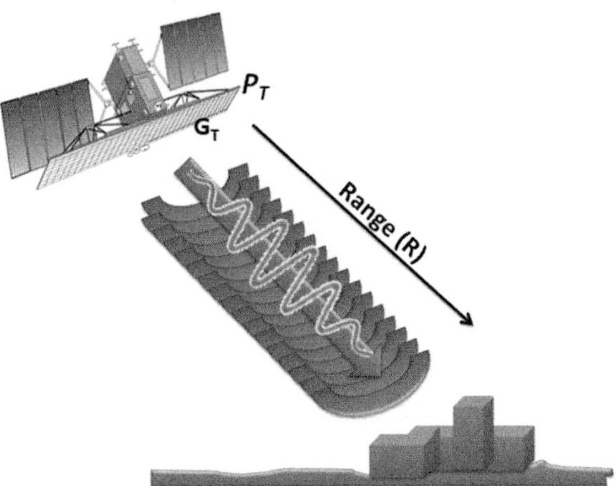

FIG. 3.22 Transmitted pulse from a radar antenna.

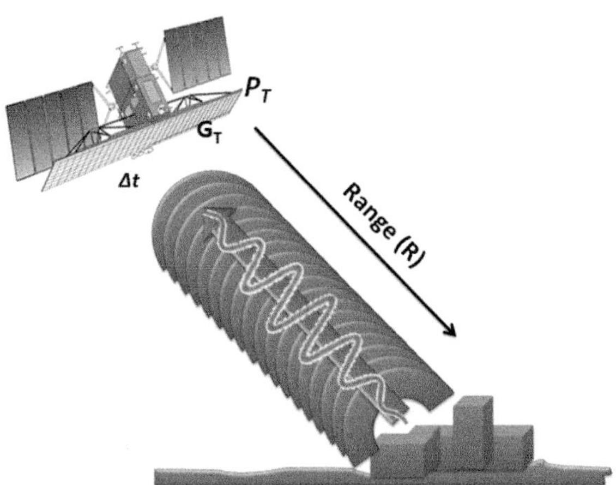

FIG. 3.23 Received reflected pulse from the target.

Eq. (3.41) expresses the backscatter power density as a function of the radar cross-section (RCS). Therefore, RCS relies on the object's unique scattering characteristics. The target radar cross-section or RCS is measured in square meters.

The energy of the received antenna signal S is a function of the backscatter and is located at the same transmit antenna. This is known as monostatic radar. In this regard, the term of $4\pi R^2$ in the previous equation turns into $(4\pi)^2 R^4$ with the additional parameter of A_R in the numerator. The term A_R, conversely, is the operational area of the receiving antenna and is the numerator in a ratio concerning the second $4\pi R^2$. This represents the isotropic radiation owing to a target's radar cross-section [9,16–19]. Under this circumstance, Eq. (3.41) can be modified as:

$$S = \frac{P_T \cdot \tau \cdot G_T \cdot \sigma \cdot A_R}{(4\pi)^2 R^4} \tag{3.42}$$

Eq. (3.42) articulates the quantity of signal that reaches the receiving antenna. Noise, therefore, can be generated in the receiver of the radar system. This noise is known as the signal-to-noise ratio (*SNR*) and is measured in units of watts/watt, or w/w. The radar system is also dominated by additional factors such as thermal noise temperature T_0 and Boltzmann's constant *K*, and is equal to 1.38×10^{23} w/(Hz K). The receiver bandwidth, *B*, and *L* represent losses within the system itself. One last element that needs to be developed for Eq. (3.42) is a relation between antenna gain, its effective area, and signal wavelength λ.

$$A_R = \frac{G_R \lambda^2}{4\pi} \tag{3.42.1}$$

Eq. (3.42.1) expresses the correlation between antenna area and gain, i.e., gain and effective aperture relation. The radar final formula results from substituting A_R and the noise constants into the previously developed formulas. Consequently, the scientific explanation of radar range can be written mathematically as:

$$SNR = \frac{P_S}{P_N} = \frac{P_T G_T G_R \lambda^2 \sigma}{(4\pi)^3 R^4 k T_0 B F_n L} \tag{3.43}$$

where F_n is the radar noise figure and is dimensionless, or has units of w/w, and *L* is a term involving all losses that must be considered when using the radar range equation (*L* is measured in units of w/w). Eq. (3.43) is the final version of the radar equation. Additional terms can be added or substituted for diverse schemes. For instance, occasionally, the transmission interval denotes the combination of countless signals that occur over the total time *t*. The main concept to consider from the radar equation is the interdependence of various diverse aspects of radar [9,16].

The quantum SAR range equation is a function of radar cross-section. In this sense, how do we define a quantum radar cross-section? Let us consider that σ_Q is the quantum radar cross-section, which is mathematically defined as [28]:

$$\sigma_Q = \lim_{R \to \infty} 4\pi R^2 \frac{\langle P_S \rangle}{\langle P_N \rangle} \tag{3.44}$$

The transmitted power P_S is formulated as:

$$P_S \approx \frac{4\pi \varepsilon_0^2 \sigma_Q}{(4\pi)^2 R^4} \tag{3.45}$$

Eq. (3.45) shows the classical radar equation. However, the quantum radar equation, which involves the transmitted P_T^Q and reflected power P_r^Q, respectively, can be formulated as:

$$P_T^Q = 4\pi \varepsilon_0^2 \tag{3.46}$$

$$P_r^Q = \langle P_S \rangle A_R \tag{3.47}$$

The quantum radar equation can be formulated using Eqs. (3.46) and (3.47) as:

$$P_r^Q = \frac{P_T^Q A_R \sigma_Q}{(4\pi)^2 R^4} \tag{3.48}$$

σ_Q has the following significant properties:

Strong dependencies: It can be proved that σ_Q strongly depends on the properties of the target. That is, it depends on the target's geometry (absolute and relative size, shape, and orientation), as well as its composition (material properties).

Weak dependencies: It can be proved that σ_Q is approximately independent of the properties of the radar system. That is, it depends very weakly on the strength and the position of the radar system. In particular, σ_Q is independent of R, the range to the target. As a consequence, it is clear that σ_Q is a property that (approximately) characterizes a specific target, and not the radar system and/or its interaction with the target. In this case, the simulation of σ_Q for the proposed design of a vehicle will provide a good estimate of its "radar invisibility." In addition, σ_Q is also important to characterize the operational performance and capabilities of radar systems—that is, given a radar system, what the minimum σ_Q of a target is that it can detect [9,28].

According to the above perspective, σ_Q relies on the rate of the ratio of the specific size of the target O and the wavelength of the SAR sensor λ. Three scattering regimes characterize the operation of radar [9,16–19,25,27,28].

Rayleigh scattering occurs by the particles that are very small in relation to the wavelength of the light, and in which the intensity of the scattered light varies inversely with the fourth power of the wavelength. In this circumstance, a low frequency has existed, and the incident microwave photons indicate a slight phase variation over the target physical characteristics. That is, at separate moments in time, individually portions of the target are distressed by approximately the similar intensity of the microwave photon emitted energy through the antenna. In this view, at every point in time, the reflected microwave photon energy is approximately determined as a function of the target physical characteristics. The incident microwave photon then induces dipole moments, which merely rely on the size and physical orientation of the target.

The second scattering regime is the **resonant regime, which is achieved if** $O \approx \lambda$. In this understanding, the phase of the incident microwave photon varies along the dimension of the target. In this regard, that part of the incident microwave photon energy is "attached" to the target's surface, generating a surface stream of microwave photons that propagate over the target's surface.

The third scattering regime presents the optical regime which happens if $O \gg \lambda$. In the circumstance of high frequency, combined interactions are minimal and the target can be considered as being made of a gathering of autonomous scattering centers. In this view, the total photons scattered field is the superposition of all the individual scattered fields. As a result, the target shape has a prevailing role in the organization of the photon scattered fields [9,18,19,27,28].

3.13 Quantum SAR illumination

In modern physics, quantum entanglement is the physical phenomenon that arises when a couple or group of particles are created, for instance, by an electronic device such as an antenna. The emitted microwave photon particles and waves are ruled by the quantum state. In other words, the quantum state of every particle of the pair or group of emitted or reflected photons which cannot be described independently of the state of the others, even when these photon particles are separated by a large distance.

Quantum microwave photon entanglement, on interaction with a target, creates reflected microwave photons that are known as quantum photon illumination; this can preserve the information of any objects although the original entanglement is completely destroyed by a lossy and noisy environment [29]. In other words, the interaction between the signal photons and the target can be modeled as a beam splitter with small reflectivity. In this sense, noise is vaccinated into the radar system with an average photon number N_P into each frequency and polarization. For instance, thermal added radiation from the electric circuit in the antenna can be a source of noise, i.e., thermal noise [30].

Let us assume the bandwidth, $\Delta\omega$, and D_T, a temporal detection window; in practice the sensor can distinguish between different radiation modes M as:

$$M \approx \Delta\omega \times D_T \tag{3.49}$$

The photodetector can observe one noise photon per detection event under the circumstance of:

$$M \times N_P \ll 1 \tag{3.50}$$

where $N_P \ll 1$ corresponds to the state where the thermal emission is significantly beneath the signal photon energies. What are the impacts of unentangled and entangled signal microwave photons signals on imagining an ocean dynamic system? The answer to this question will be addressed in depth in the following chapters.

According to Marghany [9], the essential assembly of quantum illumination is target detection. In this context, the SAR antenna, for instance, manages dual entangled systems, so-called signal, and idler. The signal is radiated to probe the existence of a low-reflectivity object in a region with bright background noise while the idler is recollected. The reflection from the target is then associated with the recollected idler system in a joint quantum measurement, which delivers dualistic conceivable consequences: target obtainable or target vague. More specifically, the exploratory progression is repeated frequently with the intention of obtaining countless sets of signal-idler systems that are comprised of the receiver for dual quantum detection.

3.14 Quantum theory of SAR system

3.14.1 Transmitter

The generation of entangled photons is a key element in the SAR transmitter. To this end, a nonlinear crystal is used to generate entangled photons. In other words, it splits an incoming photon into two entangled photons. According to this view, semiconductor nanostructures are implemented to create entangled photons. In particular, the deterioration of bi-excitonic situations in quantum dots contain interband transitions between valence and conduction bands. In this circumstance, entangled photons can be generated with frequencies in the microwave spectra. Consequently, microwave photons are created from impulsive descending conversions between single-particle levels in a quantum dot [9,18,28].

Creating entangled photons can be achieved by developing an array of quantum dots, which are coupled to dual electron reservoirs and inserted inside a cylindrical microwave resonator. To simplify this concept, let us assume four quantum dots (Fig. 3.24), which are labeled as d_1^Q, d_2^Q, d_3^Q, and d_4^Q. In this view, d_2^Q, and d_3^Q are used to deliver distinctive initial and final states for an electron in the conduction band. d_1^Q and d_4^Q afford dual decay paths [29].

At the quantum dot d_2^Q, the electron begins, and tunnels into excited states of quantum dots d_1^Q and d_4^Q, respectively. In this circumstance, the symmetric quantum state occurs and is formulated as:

$$|\Psi\rangle = \frac{1}{\sqrt{2}}\left(\left|d_1^Q\right\rangle + \left|d_4^Q\right\rangle\right) \tag{3.51}$$

Specifically, the electron can be in the distracted state of quantum dots d_1^Q or d_4^Q, with identical probability. Subsequently, the electron position deteriorates into the pulverized of d_1^Q or d_4^Q and in the procedure discharges dual photons. If the electron deteriorations from d_1^Q, then both photons have spiral circular polarization states. Alternatively, if the electron deteriorates from d_4^Q, then both photons have clockwise circular polarization states. The quantum state of this phase can then be defined as:

$$|\Psi\rangle = \frac{1}{\sqrt{2}}\left(\left|d_1^Q\right\rangle \otimes |++\rangle + \left|d_4^Q\right\rangle \otimes |--\rangle\right) \tag{3.52}$$

The electron states can be determined from:

$$|\Psi_\pm\rangle = \frac{1}{\sqrt{2}}\left(\left|d_1^Q\right\rangle \pm \left|d_4^Q\right\rangle\right) \tag{3.53}$$

Conversely, the photonic states can be determined from:

$$|\Phi_\pm\rangle = \frac{1}{\sqrt{2}}(|++\rangle \pm |--\rangle) \tag{3.54}$$

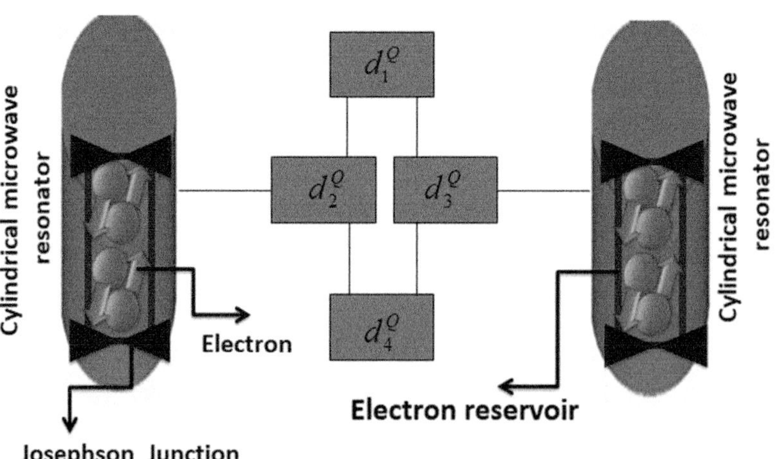

FIG. 3.24 Organization of four quantum dots.

Combining Eqs. (3.53) and (3.54) delivers the generation of quantum photons and electrons, which can be described as:

$$|\Psi\rangle = \frac{1}{\sqrt{2}}(|\psi_+\rangle|\Phi_+\rangle + |\psi_-\rangle|\Phi_-\rangle) \tag{3.55}$$

The coupling with the dot d_3^Q acts to separate the photons from the electron conditions. Moreover, the antisymmetric state $|\Phi_+\rangle$ is restrained owing to critical photon interferences. Subsequently, the state of the photons is located by acquiring the photon state's ledge $|\Psi\rangle$ with $|\psi_+\rangle$. In this circumstance, the radiated or transmitted photon is formulated as:

$$\langle\psi_+|\Psi\rangle\alpha|\Phi_+\rangle\rangle = \frac{1}{\sqrt{2}}(|++\rangle + |--\rangle) \tag{3.56}$$

Eq. (3.56) demonstrates that the radiated photons occur with utmost entanglement.

3.14.2 Receiver

The receiver is made up novel metamaterial which can absorb the reflected microwave photons and can be used to perform single-shot photodetection. A novel metamaterial is critical engineering materials with specific physical characteristics [30]. For instance, the metamaterial lens, located in metamaterial antenna systems, which is depleted as an effective coupler to the exterior, radiates microwave photons. In this regard, it is focusing on radiation alone or from a microstrip transmission line into transmitting and receiving components. Subsequently, it can be exploited as a response device [9,17,19,28].

Conversely, the receiver is designed to acquire or gain a microwave photon from the initial state $|0\rangle$ and transmit it to the new state $|1\rangle$ (Fig. 3.25). In this view, a small operational bandwidth can be correlated to the attenuation rate Γ as:

$$\Delta\omega < \Gamma^{-1} \tag{3.57}$$

Eq. (3.57) reveals the circumstance in which bandwidth $\Delta\omega$ is small in comparison to the time it takes to receive a reflected microwave photon. In this circumstance, the long wavepackets may be affected due to the occurrence of decay at a few frequencies [8].

Consequently, this mechanism deteriorates into an enduring steady ground state $|g\rangle$ with an attenuation rate Γ, which can be linked to the initial state through quantum entanglement (Fig. 3.25). In this regard, the microwave photon growth is identified as the quantum state then decay down to a consideration of attenuation impact in the ground quantum state, i.e., lowest-energy state (Fig. 3.25). Consequently, this sensor acts as a photographic film in the sense that once a photon has been absorbed by the metamaterial, the detector is placed in a steady and mesoscopically distinct state.

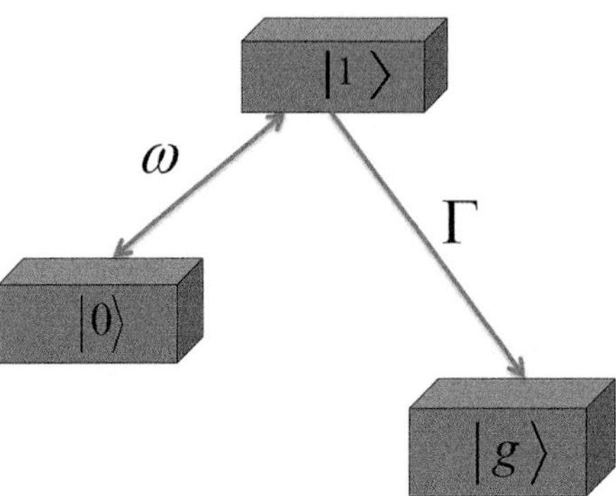

FIG. 3.25 Concept of radar receiver from the point of view of quantum mechanics.

References

[1] Walls DF, Milburn GJ. Quantum optics. Springer Science & Business Media; 2007.
[2] Gerry C, Knight P, Knight PL. Introductory quantum optics. Cambridge University Press; 2005.
[3] Meystre P, Sargent M. Elements of quantum optics. Springer Science & Business Media; 2007.
[4] Luszcz K, Bonvin E, Novotny L. Quantized optical near-field interactions measured with a superconducting nanowire detector. arXiv; 2018 preprint arXiv:1805.02621.
[5] Lei L. Theoretical analysis of and bias correction for planar and cylindrical polarimetric phased array weather radar. [Ph.D. theses]Norman, OK: University of Oklahoma, Graduate College; 2014.
[6] Balanis CA. Fundamental parameters of antennas. In: Antenna theory, analysis and design. John Wiley & Sons; 1997. p. 28–102.
[7] Doviak RJ, Bringi V, Ryzhkov A, Zahrai A, Zrnić D. Considerations for polarimetric upgrades to operational WSR-88D radars. J Atmos Oceanic Tech 2000;17(3):257–78.
[8] Doviak W, Whitmore DL, inventors; Padcom Inc, assignee. Apparatus and method for transparent wireless communication between a remote device and a host system. United States Patent US 5,717,737. 1998.
[9] Marghany M. Automatic detection algorithms of oil spill in radar images. CRC Press; 2019.
[10] Pedrosa IA. Quantum description of electromagnetic waves in time-dependent linear media. J Phys Conf Ser 2011;306(1):012074 IOP Publishing.
[11] Pedrosa IA, Rosas A, Guedes I. Exact quantum motion of a particle trapped by oscillating fields. J Phys A Math Gen 2005;38(35):7757.
[12] de Lima AL, Rosas A, Pedrosa IA. On the quantization of the electromagnetic field in conducting media. J Mod Opt 2009;56(1):41–7.
[13] Pedrosa IA. Exact wave functions of a harmonic oscillator with time-dependent mass and frequency. Phys Rev A 1997;55(4):3219.
[14] Alicki R, Kryszewski S. Completely positive Bloch-Boltzmann equations. Phys Rev A 2003;68(1):013809.
[15] Ünal N. Quasi-coherent states for a photon in time varying dielectric media. Ann Phys Rehabil Med 2012;327 (9):2177–83.
[16] Ahern FJ. Fundamental concepts of imaging radar: Basic level. unpublished manualOttawa, ON: Canada Centre for Remote Sensing; 1995 87 pp.
[17] Lacomme P. Air and spaceborne radar systems. William Andrew; 2001.
[18] Merrill IS. Introduction to radar systems. McGrow-Hill; 2001607–9.
[19] Skolnik ML. Radar handbook. 3rd ed.: McGraw Hill; 2008.
[20] Hofheinz M, Wang H, Ansmann M, Bialczak RC, Lucero E, Neeley M, O'connell AD, Sank D, Wenner J, Martinis JM, Cleland AN. Synthesizing arbitrary quantum states in a superconducting resonator. Nature 2009;459(7246):546.
[21] Bai L, Harder M, Chen YP, Fan X, Xiao JQ, Hu CM. Spin pumping in electrodynamically coupled magnon-photon systems. Phys Rev Lett 2015;114(22):227201.

[22] Tserkovnyak Y, Brataas A, Bauer GE. Enhanced Gilbert damping in thin ferromagnetic films. Phys Rev Lett 2002;88(11):117601.

[23] Mosendz O, Pearson JE, Fradin FY, Bauer GE, Bader SD, Hoffmann A. Quantifying spin Hall angles from spin pumping: experiments and theory. Phys Rev Lett 2010;104(4):046601.

[24] Huebl H, Goennenwein ST. Electrical signal picks up a magnet's heartbeat. Physics 2015;8:51.

[25] Tarasov M, Stepantsov E, Lindström T, Kalabukhov A, Ivanov Z, Claeson T. Antenna-coupled planar arrays of Josephson junctions. Phys C Supercond 2002;372:355–9.

[26] Newrock R. What are Josephson junctions? How do they work?, Available from:https://www.scientificamerican.com/article/what-are-josephson-juncti/.

[27] Lloyd S. Enhanced sensitivity of photodetection via quantum illumination. Science 2008;321(5895):1463–5.

[28] Lanzagorta M. Quantum radar. Synth Lect Quantum Comput 2011;3(1):1–39.

[29] Smith JF. Quantum entangled radar theory and a correction method for the effects of the atmosphere on entanglement. In: Quantum information and computation VII. vol. 7342. International Society for Optics and Photonics; 2009. p. 73420A.

[30] Yao XW, Wang H, Liao Z, Chen MC, Pan J, Li J, Zhang K, Lin X, Wang Z, Luo Z, Zheng W. Quantum image processing and its application to edge detection: theory and experiment. Phys Rev X 2017;7(3):031041.

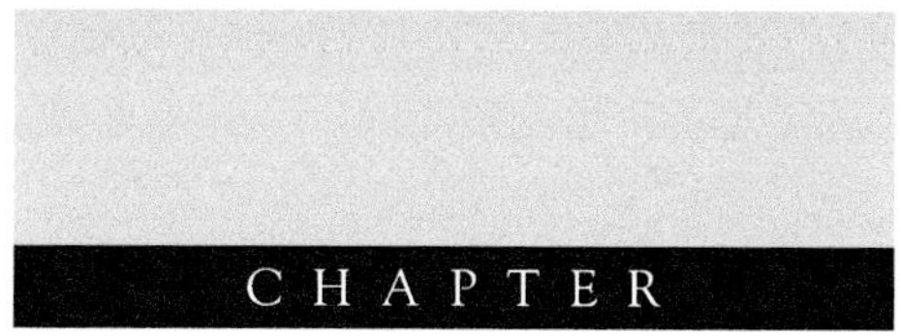

CHAPTER

4

Quantum mechanism of nonlinear ocean surface backscattering

4.1 What is meant by scattering?

In previous chapters, it was agreed that a photon is the quantum of electromagnetic radiation. The term quantum represents the smallest elemental unit of a quantity or the smallest discrete amount of something. Thus, one quantum of electromagnetic energy is called a photon. The plural of quantum is quanta.

The concept of photons and quanta comes from quantum mechanics and quantum theory. Quantum mechanics is a mathematical model that describes the behavior of particles on an atomic and subatomic scale; it demonstrates that matter and energy are quantized—or come in small, discrete parcels—on the smallest scales imaginable. A photon propagates at the speed of light.

In physics, it is rare for microwave photon energy to travel in a completely straight line after interacting with rough surfaces such as the ocean surface. Numerous processes can arise that make the photons depart from its path. This chapter will cover one of those processes: scattering or backscattering.

The significant question is: how can we define photon scattering? When microwave energy photons are caused to deviate from a straight line owing to deficiencies in the medium, it is called scattering. Scattering, therefore, is unique in that the microwave photon energy is commonly deflected in numerous directions that are difficult to envisage or compute. The basic scattering concept, for instance, is to think of how the sun shines on a target through a thin cover of clouds. Instead of hitting the target directly, the sun's light is defused. This is because as the sun shines through the clouds, its light is scattered and only some of it ends up striking the target [1].

It is a great challenge to determine the position of these scattered photons using a conventional approach. In fact, photon particles propagate and interact with a target in a highly localized manner. Therefore, the quantum electrodynamics (QED) theory can be used to predict the probabilities of locating these photons at any given point in space-time. The main question that arises is how the QED can describe microwave photon scattering. Let us

https://doi.org/10.1016/B978-0-12-820785-7.00001-0

consider a Feynman diagram where positron 1 absorbs B and converts to positron 4, which radiates C and turns into positron 3. Likewise, electron 2 and positron 3 annihilate and produce D (Fig. 4.1). Consequently, Fig. 4.1 reveals that vertex 1 could be turned in either direction. In this view, four possible directions could be described as follows:

(i) B splits into electron 1 and positron 4;
(ii) Electron 1 absorbs A and becomes electron 2;
(iii) Electron 2 emits D and becomes electron 3; and
(iv) Electron 3 and positron 4 annihilate, producing C.

Consistent with the above perspective, the photon scattering comprises two states: (i) elastic scattering (Fig. 4.2), where the kinetic energy of the scattered particles is preserved; and (ii) inelastic scattering (Fig. 4.3), where kinetic energy is not conserved, e.g., due to a photon taking off with part of the energy [2, 3].

Figs. 4.2 and 4.3 reveal the differences between elastic and inelastic scattering concepts. During elastic scattering, the state of the emitted photon changes, and the direction of motion of the electron accounts for the momentum of exchange. Nonetheless, the bound state of the electron does not change. On the contrary, there is an additional modification of the bound state of the electron from the elastic state to state for inelastic (Raman) scattering [1, 4].

4.2 Comparison between coherent and incoherent multiple scattering

The most significant question is what the differences are between coherent and incoherent scattering. Indeed, coherence influences in the electromagnetic waves in disordered systems are at the cornerstone of many phenomena in countless areas of technology. Coherent

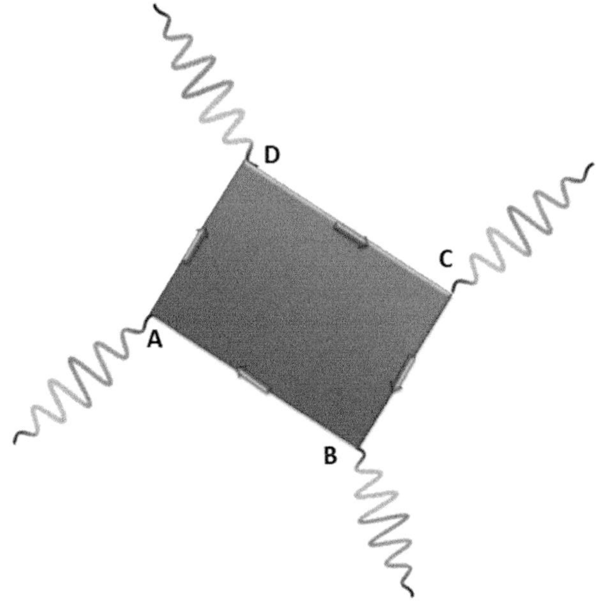

FIG. 4.1 Four possible electron-photon interaction in the Feynman diagram.

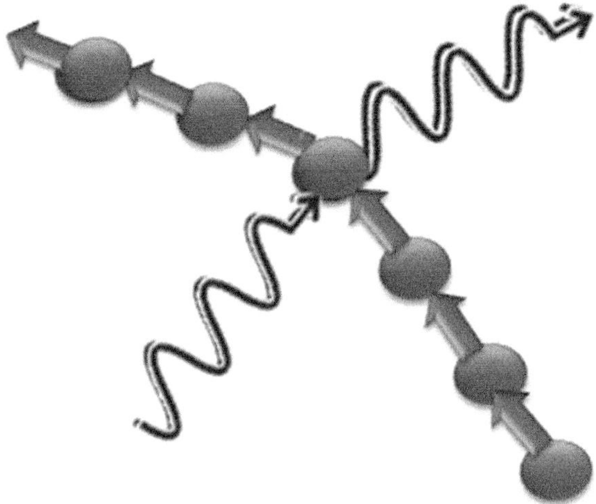

FIG. 4.2 Feynman diagram for elastic scattering.

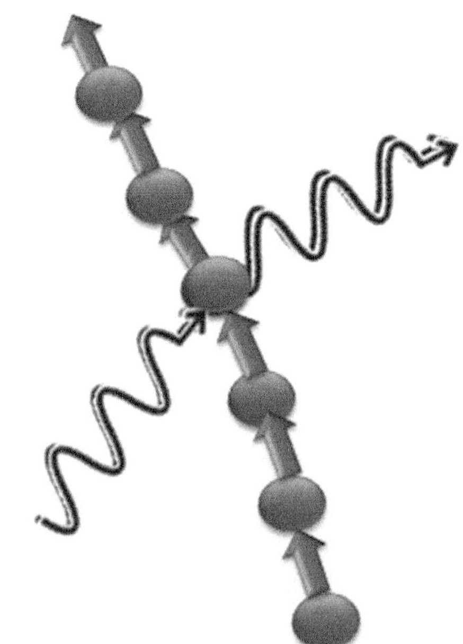

FIG. 4.3 Feynman diagram for inelastic scattering.

scattering is also known as unmodified, Rayleigh, conventional, or elastic scattering, and is one of three configurations of photon interaction. Mostly, coherent scattering occurs when the energy of, for instance, an X-ray or gamma photon is smaller than the ionization energy of an atom. Consequently, it happens with low energy radiation. In this understanding, when a very low energy X-ray photon interacts with the electrons in an atom, it triggers the electrons to vibrate at a similar frequency as the incident photon. The main source of the coherent scattering is due to similar emitted photons that have the same frequency or wavelength, similar amplitude, and a constant phase relationship between them.

Let us consider a plane wave allied with a photon, which is scattered by narrowly spaced objects. Along with or very near the wavefront A-B (Fig. 4.4), dual neighboring objects that are spaced apart such that their scattered waves in a direction θ are 0.5λ out of phase and cancel each other by destructive interference. In this regard, the spacing d can be given as [1, 3]:

$$d = \lambda(2\sin\theta)^{-1} \tag{4.1}$$

Eq. (4.1) reveals that for a given scattering angle θ different from 0 and π, one can always realize two consistent scattering constituencies with spacing such that their scattered photon has a phase difference of half a wavelength (λ). This results in destructive interference for photons scattered in all but the forward ($\theta = 0$) and backward $\theta = \pi$ directions.

On the contrary, incoherent scattering is a sort of scattering phenomenon in physics. Mostly it is used when discussing the scattering of an electromagnetic wave (usually light or radio frequency) by accidental variations, which are most regularly electrons. In other words, incoherent scatter echo comes from a very large number of electrons. These are not stationary, but rather are in random thermal motion. Thus, the echo will not be at a single frequency but instead will contain a range, or spectrum, of frequencies near the transmitter frequency. Incoherent scatter radar theory is the most well recognized applied technique for studying the Earth's ionosphere, which was first proposed by Professor William E. Gordon in 1958. In this sense, a radar beam scattering off electrons in the ionospheric plasma produces an incoherent backscatter. To this end, the scattering function of the ionospheric electrons is

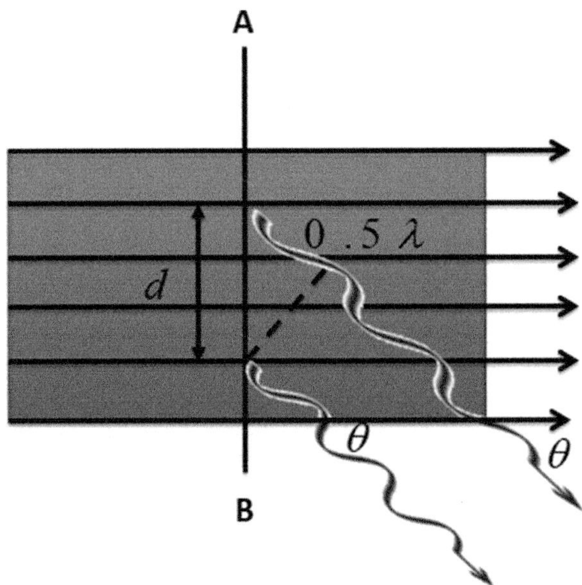

FIG. 4.4 Simplification of coherent scattering.

altered by the extremely slow and massive positive ions. In this circumstance, electron density variations correlate with ion temperature, mass distribution, and motion. Consequently, the incoherent scatter signal permits discovery of electron density, electron temperature, ion temperature, ion composition, and plasma velocity [2, 3].

Let us assume an ensemble of randomly positioned identical atoms separated regularly from one another by a distance considerably larger than the wavelength of the incident microwave photons. In this circumstance, the quantity of interference between the photons caused or scattered by each interaction would be diminished. In this regard, photons' joint interactions are neglected as they are far apart from each other. The general consequence could be approximated by a totaling of discrete interactions. In that regard, photon intensities can be summed and joint interference abandoned. In this understanding, the mathematical descriptions of the ensemble of scatterers whose discrete scattering patterns are axially symmetrical, for instance, spheres, or for an ensemble of randomly oriented scattering particles are expressed as [1]:

$$\frac{dI}{dz} = I_0 e^{-b(v)z} 2\pi \int_0^{\pi} \beta(v,\theta) \sin\theta d\theta \tag{4.2}$$

Eq. (4.2) demonstrates that function $\beta(v,\theta)$ signifies the angular distribution of scattered photons intensity I_0 in units of inverse distance times inverse solid angle. Moreover, it is normalized by dividing with the scattering coefficient and the result is termed as the phase function $p(v,\theta)$. Therefore, The volume scattering function, $\beta(v,\theta)$, is arguably the most important data required to estimate the photon field in scattering media. In this regard, $-b(v)$ represents the volume scattering function at a specified frequency v in units of inverse distance d^{-1}. In other words, Eq. (4.2) shows that an average photon intensity has been performed over the azimuth angle, measured from an arbitrary plane (say, the scattering plane, containing the incident and scattered directions) about the incident direction.

In general, coherent scattering is a form of electromagnetic wave in which photons share the same frequency and wavelengths are in phase with one another. On the contrary, incoherent scattering does not contain photons with the same frequency and does not have wavelengths that are in phase with one another.

4.3 What is the role of spin in understanding scattering?

A photon is an electrically neutral, massless, spin-1 vector boson, whereas an electron is an electrically charged, massive, spin-1/2 fermion. This concept can explain why electrons with spin parallel to the magnetization (henceforth, to the mainstream-spin direction of the electrons) scatter less than those with spin antiparallel to the magnetization (thus, parallel to the marginal-spin).

The chance of scattering depends on the number of prevailing quantum occurrences for the electron to scatter into and that depends on the comparative path of the electron's spin and the magnetic field inside the ferromagnet. The electron also revolves or spins around its axis. The spinning of the electron creates a magnetic dipole. If the majority of electrons in the atom spin in the same direction, a strong magnetic field is created. The direction of the electron's spin regulates the direction of the magnetic field. Consequently, the magnetic fields created by the discrete atoms repel each other. When a significant amount of ferromagnetic is positioned into an external magnetic field, two states occur. The spins in each domain are shifted with the intention of the magnetic moments of the electrons that more developed and affiliated with the direction of the magnetic field.

Matthew [5] stated that if the spin and magnetic field are antiparallel, more circumstances are available for electron scattering. In this circumstance, the electrical resistance is greater if the spin and the magnetic field are parallel. This is the basis of spin-dependent scattering. Alternatively, the spin relies on the inelastic scattering cross-section (inverse average free pathway) and the elastic scattering cross-section are premeditated for polarized electrons, which scatter from orientated atoms in the Born-Ockhur approximation, which attempts to understanding spin-dependent scattering in ferromagnets. The Born-Ockhur approximation consists of taking the incident field in place of the total field as the finding field at each point in the scatterer. For instance, the scattered radio waves can be estimated by assuming that each part of the elastic scattered wave is polarized by the similar electrical field that would be present at that point without the column and then computing the scattering as emission integral over that polarization distribution.

Elastic spin dependence seems to be larger than inelastic, and the exchange effects fall off rapidly with increasing energy. In this circumstance, Matthew [5] found that in the medium-to-high-energy range ($\gtrsim 100\,\text{eV}$) the elastic scattering for parallel spins is higher than for antiparallel spins. On the contrary, the inelastic cross-section for parallel spins is less than for antiparallel.

The inconsistency and the complete scattering cross-section could be determined as the as part of the investigation into the scattering. The persistence of the scattering principle is, then, to guess information on the force, or the interaction, responsible for the scattering. It is a critical issue as SAR object imagines, which is exclusively based on the scattering theory. The scattering theory accomplishes this challenge by finding, for instance, an association between the cross-section and the wavefunction of the SAR system.

4.4 Spin of scattering of particles

Let us assume interaction between two spinful particles that leads to corrected scattering as a function of solid angle Ω as $d_\Omega\sigma$. The spin of the electron $e-e$ equal $\frac{1}{2}$. In this understanding, two spins $\frac{1}{2}$ perhaps integrate into a singlet as $(\uparrow\downarrow - \downarrow\uparrow)$ or three triplet$(\uparrow\uparrow, \downarrow\downarrow, \uparrow\downarrow + \downarrow\uparrow)$ circumstances. In this sense, singlet would have a probability of 0.25, while triplet would have a probability 0.75, which are considered when the particles are randomly polarized. In this circumstance, the corrected $d_\Omega\sigma$ is mathematically expressed as:

$$d_\Omega\sigma = \frac{3}{4}|f(\theta)|^2 + |f(\pi-\theta)|^2 + \frac{1}{4}|f(\theta) + f(\pi-\theta)|^2 \tag{4.3}$$

Singlet state is achieved if the particles scattered into $\theta = 0.5\pi$ while if $\theta = 0$, the triplet state occurs. Moreover, if the spins are not mainly arbitrary, but are entirely polarized in the similar θ, the scattering into $\theta = 0.5\pi$ is zero [6–10].

4.5 Scattering of identical particles

For the sake of simplicity, the scattering potential is assumed as spherically symmetric, which implies that the motion of the dual particles is restricted to a stationary plane passing through the origin. In this view, the two-particle states are considered $\psi(1,2)=\psi(2,1)$. In this circumstance, a two-particle state ψ that is symmetric (antisymmetric) r_1 and r_2 would be an even (odd) function of the relative coordinates $r\equiv r_1-r_2$. Nonetheless, in spherical coordinate $r\rightarrow -r$, which designates that $(r,\theta,\phi)\rightarrow(r,\pi-\theta,\phi+\pi)$, the wave function is neither symmetric nor antisymmetric. Subsequently, for identical particles, it must be swapped and given by:

$$\psi(1,2)=e^{i\vec{k}\cdot\vec{r}}\pm e^{-i\vec{k}\cdot\vec{r}}+[f(\theta)\pm f(\pi-\theta)]\frac{e^{i\vec{k}\cdot\vec{r}}}{r} \tag{4.4}$$

Eq. (4.4) reveals that the upper sign is expended for asymmetric wavefunction, and the lower for an antisymmetric wavefunction. In this view, the spherical wave $\frac{e^{i\vec{k}\cdot\vec{r}}}{r}$ deliberates the scattering of particles in entirely contrary directions. In physics, the scattering of both particles $\psi(1,2)=\psi(2,1)$ would occur when $\psi(1,2)=\psi(2,1)$ are matching. In this regard, an abrupt change of the scattering cross-section $d\sigma$ is recognized by the ratio of the particle flux into a solid angle $d\Omega$ and the incident particle stream for one of two plane waves (x,y). This can be formulated mathematically as:

$$\frac{d\sigma}{d\Omega}=|f(\theta)\pm f(\pi-\theta)|^2 \tag{4.5}$$

Eq. (4.5) reveals two possibilities: (i) scattering of particles with spin, and (ii) scattering of spin-0 particles $d_\Omega\sigma=|f(\theta)|^2+|f(\pi-\theta)|^2$, which has no interference term between $f(\theta)$ and $f(\pi-\theta)$.

4.6 Schrödinger equation for scattering particles

Let us consider the Schrödinger equation for scattering a particle 1 with the coordinate $\vec{r}_1$, the mass m_1 and the momentum k_1 on the dual particles bound system, as given by:

$$\left[-\sum_{i=1}^{3}\frac{\hbar}{2m_i}\Delta_i+\sum_{i>j=1}^{3}V_{ij}\left(\vec{r}_i-\vec{r}_j\right)\right]\Psi\left(\vec{r}_1,\vec{r}_2,\vec{r}_3\right)=E\Psi\left(\vec{r}_1,\vec{r}_2,\vec{r}_3\right) \tag{4.6}$$

Eq. (4.6) reveals that the coordinates, masses, and momenta of bound particles are $\vec{r}_2$ and $\vec{r}_3$, m_2, and m_3, k_2, and k_3, respectively. In this regard, the dual-particles interaction potentials $V_{ij}\left(\vec{r}_i-\vec{r}_j\right)$ are the real functions and the probability density is conserved. Using Jacobi coordinates $\vec{r}=\{x,y,z\}=\vec{r}_2-\vec{r}_3$, $\vec{R}=\{X,Y,Z\}=\vec{r}_1-\vec{R}_{23}$, where $\vec{R}_{23}=\frac{m_2\vec{r}_2+m_3\vec{r}_3}{m_2+m_3}$, the Schrödinger equation can be written in the form

$$\left[-\frac{\hbar}{2M}\Delta_{\vec{R}}-\frac{\hbar}{2\mu}\Delta_{\vec{r}}+V_{12}\left(\vec{R}-\frac{\mu}{m_2}\vec{r}\right)+V_{13}\left(\vec{R}+\frac{\mu}{m_3}\vec{r}\right)+V_{23}\left(\vec{r}\right)\right]\Psi\left(\vec{r},\vec{R}\right)=E\Psi\left(\vec{r},\vec{R}\right). \tag{4.7}$$

where μ and M are reduced masses in the systems of particles [11–13], respectively.

Eq. (4.7) relies on dual variables $\vec{r}$ and $\vec{R}$ but it is impossible for them to be discrete in the common situation because of the potentials of V_{12} and V_{13}, respectively. Let us assume that the bound system of particles present $\left|\vec{R}\right| \to \infty$, which is mathematically expressed as:

$$\left[-\frac{\hbar}{2\mu}\Delta_{\vec{r}} + V_{23}\left(\vec{r}\right)\right]\Phi_0\left(\vec{r}\right) = E_0\Phi_0\left(\vec{r}\right) \tag{4.8}$$

where $\Phi_0\left(\vec{r}\right)$ is the ground state normalized solution of this equation at energy E_0.

Hence, we select the wave function $\Psi\left(\vec{r}, \vec{R}\right)$ in the quantum-optics:

$$\psi\left(\vec{R}, \vec{r}\right) = e^{if\left(\vec{R}, \vec{r}\right)}\Phi_0\left(\vec{r}\right), \tag{4.9}$$

where $f\left(\vec{r}, \vec{R}\right)$ is pure real function.

Then substituting Eq. (4.9) in Eq. (4.7), differentiating in $\vec{r}$ we obtain:

$$\left\{-\frac{\hbar}{2M}\Delta_{\vec{R}} - \frac{\hbar}{2\mu}\Delta_{\vec{r}} f\left(\vec{R}, \vec{r}\right) - i\frac{\hbar}{\mu}\vec{\nabla}_{\vec{r}} f\left(\vec{R}, \vec{r}\right) \cdot \vec{\nabla}_{\vec{r}} \ln\Phi_0\left(\vec{r}\right) + V_{12}\left(\vec{R} - \frac{\mu}{m_2}\vec{r}\right)\right.$$
$$\left. + V_{13}\left(\vec{R} + \frac{\mu}{m_3}\vec{r}\right) + \frac{\hbar}{2\mu}\left[\vec{\nabla}_{\vec{r}} f\left(\overleftarrow{R}, \vec{r}\right)\right]^2\right\}\psi\left(\vec{R}, \vec{r}\right) = (E - E_0)\psi\left(\vec{R}, \vec{r}\right) \tag{4.10}$$

Eq. (4.10) is considered a Schrödinger equation. Therefore, the complex potential of a Schrödinger equation can be written as:

$$\left\{-\frac{\hbar}{2M}\Delta_{\vec{R}} + V\left(\vec{R}, \vec{r}\right)\right\}\psi\left(\vec{R}, \vec{r}\right) = (E - E_0)\psi\left(\vec{R}, \vec{r}\right) \tag{4.11}$$

Both Eqs. (4.10) and (4.11) contain information about the system that is only involved in the imaginary part of the potential. In this circumstance, we can accept the solution of Eq. (4.10) on the assumption of scattering problems regarding the coordinate $\vec{R}$.

4.7 How do the Lippmann-Schwinger equation and the scattering amplitude generalize when spin is included?

The Lippmann-Schwinger equation is named after Bernard Lippmann and Julian Schwinger. In quantum mechanics, it is one of the most widely implemented formulas to explain particle collisions—or, more specifically, scattering. It may be exploited in the scattering of molecules, atoms, neutrons, photons, or any other particles, and is foremost in atomic, molecular, and optical physics, nuclear physics, and particle physics. It correlates the scattered wave function with the interaction that creates the scattering (the scattering potential) and therefore tolerates computation of the applicable experimental parameters, for instance, scattering amplitude and cross-sections. The Lippmann-Schwinger equation, therefore, is equivalent to the Schrödinger equation plus the classic boundary circumstances for scattering problems. For scattering problems, the Lippmann-Schwinger equation is often more convenient than the original Schrödinger equation [14]. The Lippmann-Schwinger equation has a similar form, whether the spin is included or not, as given by:

$$\left|\psi^{(\pm)}\right\rangle = |\phi\rangle + \mathrm{X}\left|\psi^{(\pm)}\right\rangle (E - \mathsf{H}_0 \pm i\varepsilon)^{-1} \tag{4.12}$$

The potential energy X designates the interface between the dual colliding systems. H_0 represents Hamiltonian, which designates the situation in which the dual systems are extremely far apart and do not intermingle. $|\phi\rangle$ represents the Eigen functions, E indicates energies, and $i\varepsilon$ represents a mathematical aspect, which is required for computing the integrals required to reveal the equation. It is a consequence of interconnection between electron and photon, verifying that scattered waves comprise merely of photoelectric scattering. In quantum mechanics, which is the perspective of the Lippmann-Schwinger equation, the photon spins come into (anti-)symmetrization of the multiparticle wavefunction under particle replacement. Therefore, the Lippmann-Schwinger equation is obviously assembled for potential scattering and could merely be depleted to describe the scattering amplitude of the photon-electron interaction [15,16].

This intuitive representation is not very precise, because $\psi^{(\pm)}$ is an Eigen function of the Hamiltonian and consequently at various times merely varies by a phase. Consequently, the physical state does not change and so it cannot develop to be noninteracting [17]. This difficulty is avoided by accumulating $\psi^{(\pm)}$ and ϕ into wavepacket $g(E)$ with some spreading of energies E above a characteristic scale ΔE. Plugging the Lippmann-Schwinger equations into the mathematical definitions are given by:

$$\psi_g^{(\pm)}(t) = \int dE e^{-iEt} g(E) \psi^{(\pm)} \tag{4.13}$$

and:

$$\phi_g(t) = \int dE e^{-iEt} g(E) \phi \tag{4.14}$$

Eqs. (4.13) and (4.14) show the alteration between the $\psi_g^{(\pm)}(t)$ and $\phi_g(t)$ wavepackets, which is delivered by an integral over the energy E. This integral could be gauged by describing the wave function over the complex E plane and closing the E contour by means of a semicircle on which the wavefunctions vanish. In this circumstance, the integral over the closed contour could then be appraised with the Cauchy integral theorem as a total of the residues at the infinite poles. It can be argued that the residues of $\psi_g^{(\pm)}(t)$ approach those of $\phi_g(t)$ at the time $t \to \mp\infty$ and so the corresponding wavepackets are equal at temporal infinity [14, 17, 18].

In fact, for every positive time t, the e^{-iEt} factor in a Schrödinger picture state forces one to close the contour on the lower half-plane. The pole in the $(\phi, \mathrm{X}, \psi^{\pm})$ from the Lippmann-Schwinger equation reveals the time-uncertainty of the interaction, while that in the wavepacket weight function reveals the length of the interaction. Both of these variabilities of poles arise at finite imaginary energies and, subsequently, they are suppressed at very great periods. The pole in the energy transformation in the denominator is on the upper half-plane in the case of ψ^{-} and accordingly, it does not locate within the integration contour and does not impact the ψ^{-} integral. The remainder, therefore, is equivalent to the ϕ wavepacket. Hence, at very late times, $\phi = \psi^{-}$, recognizing ψ^{-} as the asymptotic noninteracting out of quantum state [14, 18].

Likewise, one may integrate the wavepacket corresponding to ψ^+ actual negative periods. In this circumstance, the contour needs to be locked over the upper half-plane, which consequently oversees the energy pole of ψ^+, which is in the lower half-plane. This means that the ψ^+ and ϕ wavepackets are corresponding in the asymptotic past, identifying ψ^+ as the asymptotic is noninteracting in this state [15–18].

4.8 Seawater atom-photon scattering

In Chapter 2, Section 2.4, it is agreed that the ocean or seawater consists of approximately 35 g (1.2 oz) of dissolved salts (predominantly sodium (Na^+) and chloride (Cl^-) ions). Therefore, the most abundant dissolved ions in seawater are sodium, chloride, magnesium, sulfate, and calcium, the osmolarity of which is about 1000 mOsm/L [19]. In addition, small quantities of other substances are observed, comprising amino acids at concentrations of up to 2 micrograms of nitrogen atoms per liter [20]. By weight these ions make up about 99% of all sea salts. The amount of these salts in the global volume of seawater varies because of the addition or removal of water locally (e.g., through precipitation and evaporation). The significant question arises: how do the seawater atoms interact with microwave photons?

Let us assume that a seawater atom O_i interacts with a microwave photon γ_i, which directs to the atom O_f and a photon γ_f in various possible quantum states. In this case, the seawater atomic scattering of microwave photon energy can mathematically be articulated as:

$$O_i + \gamma_i \rightarrow O_f + \gamma_f \tag{4.15}$$

Eq. (4.15) reveals that microwave photon and sea surface atom interaction causes microwave photon scattering in various quantum states, which mathematically is defined as:

$$E_i + \hbar\omega_i = E_f + \hbar\omega_f \tag{4.16}$$

The conversion quantum state is revealed in Eq. (4.6), where E designates the seawater atomic energies and $\hbar\omega$ denotes the photon energy. Certainly, the incident microwave photon energy interaction with seawater atomic energy corresponds to the scattering energy of microwave photons owing to diverse seawater atomic energies. Generally, there are four classes of atom-photon scattering methods: (i) Rayleigh scattering; (ii) Raman scattering; (iii) Thomson Scattering; and (iv) Compton Scattering [21]. Consistent with Marghany [22], both Rayleigh scattering and Raman scattering are existing low-energy elastic scattering and inelastic scattering methods, where $\omega_i = \omega_f$ and $\omega_i \neq \omega_f$, respectively. On the other hand, both Thomson scattering; and Compton scattering are high-energy elastic scattering and inelastic scattering methods, where $\omega_i = \omega_f$ and $\omega_i \neq \omega_f$, respectively. The interaction of photon and atom, consequently, triggers the electron to swing from higher orbits to lower orbits or from lower orbits to higher orbits. There is a change of electrons in the orbits, which causes the scattering.

From the point of view of nonrelativistic perturbation speculation, two conceivable Feynman diagrams can clarify the microwave photon scattering by seawater atoms (Fig. 4.5). Nonetheless, in relativistic quantum field theory, both time-ordered diagrams would be denoted by a single Feynman diagram. Subsequently, the microwave photon is permanently a relativistic particle, while the seawater atom would be preferred as in nonrelativistic kinematic phases. In this view, nonrelativistic perturbation theory can be exploited to specify the ocean dynamics as the fluctuations of seawater atoms in time and space.

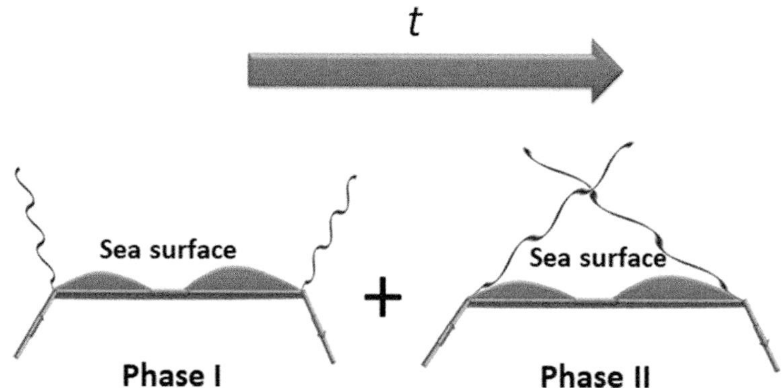

FIG. 4.5 Feynman diagrams of microwave photon scattering from the sea surface.

In phase I, the seawater atom O_i absorbs the photon γ_i, which transforms to the intermediate O_n form; then the seawater atom emits a photon γ_f and transforms to the O_f state. Both quantum states can be mathematically formulated as:

$$O_i + \gamma_i \rightarrow O_n \rightarrow O_f + \gamma_f \tag{4.17}$$

In phase II, however, the seawater atom O_i radiates the photon γ_f and transforms into the intermediate state of O_n; then the seawater atom absorbs the photon γ_i and transforms into the O_f state. The quantum status of phase II is then mathematically expressed as:

$$O_i + \gamma_i \rightarrow O_n + \gamma_f \rightarrow O_f + \gamma_f \tag{4.18}$$

Let us consider that absorption or emission of photons can be symbolized through four quantum states:

(i) Q^a_{ni}: O_i absorbs a photon ω_i and transforms into O_n.
(ii) Q^e_{fn}: O_n emanates a photon ω_f and transforms into O_f.
(iii) Q^e_{ni}: O_i radiates a photon ω_f and changes into O_n.
(iv) Q^a_{fn}: O_n absorbs a photon ω_i and converts into O_f.

In general, the quantum states of seawater atomic and microwave photon interaction can be formulated in four mathematical states as:

$$\begin{aligned}
Q^a_{ni} &= \left\langle O_n \right| \otimes \langle 0 | \hat{Q}^a \left| O_i \right\rangle \otimes |\omega_i\rangle \\
Q^e_{fn} &= \left\langle O_f \right| \otimes \langle \omega_f | \hat{Q}^e \left| O_f \right\rangle \otimes |0\rangle \\
Q^e_{ni} &= \left\langle O_n \right| \otimes \langle \omega_f | \hat{Q}^e \left| O_i \right\rangle \otimes |0\rangle \\
Q^a_{fn} &= \left\langle O_f \right| \otimes \langle 0 | \hat{Q}^a \left| O_n \right\rangle \otimes |\omega_i\rangle
\end{aligned} \tag{4.19}$$

Eq. (4.19) contains dual quantum operators $\hat{Q}^a$ and $\hat{Q}^e$, which denote the seawater atomic absorption and emission of a microwave photon, respectively. Accordingly, nonrelativistic perturbation speculation specifies the transition amplitude for the seawater atom and microwave photon scattering technique as:

$$\tilde{Q}_{fi} = \sum_n \left(\frac{Q^e_{fn} Q^a_{ni}}{(E_i + \hbar\omega_i) - E_n} + \frac{Q^a_{fn} Q^e_{ni}}{E_i - (E_n + \hbar\omega_f)} \right) \tag{4.20}$$

Consistent with Marghany [22], Ryder [23], and Weinberg [24], Eq. (4.20) reveals the energy states E. In addition, Eq. (4.20) demonstrates that in marginal pairing quantum electrodynamics (QED), the relations between a photon of minimum energy and seawater is quantified by an expression of the energy that couples a charged particle field with the quantum electromagnetic field.

4.9 Scattering from roughness surface

Roughness at the seawater border disturbs carrier transport in the succeeding approaches. First, roughness produces a fluctuating oxide thickness, causing variations in the electrostatic potential inside the water body and thus in the electron subband profile. In addition, roughness changes the thickness of the seawater quantum well, causing extra variations in the subband energy and, correspondingly, differences in the wavefunction form. The subband energy variations, caused by both these impacts, lead to fluctuating elements in the diagonal terms of the Hamiltonian operator and behave as a scattering potential. Simultaneously, seawater particle-to-particle fluctuations in the wavefunction form along the ocean induces deformation and coupling components in both diagonals and off-diagonals, known as the Hamiltonian operator, and consequently lower the transmission. This approach is termed as wavefunction deformation scattering.

Following Marghany [22] and Valavanis et al. [8], interface roughness scattering is most noticeable in regulating systems such as ocean dynamic surfaces, in which the energies for charge carriers are regulated by the sites of interfaces. Ocean-atmosphere interactions are considered as a quantum well, which is caused due to fluctuations of different layers above, across, and below the ocean surfaces. These layers act as a semiconductor. Differences in the thickness of these layers, consequently, cause the energy of particles to be reliant on their in-plane location on the ocean surface. Though the roughness $\Delta_z(\vec{r})$ differs in detail on a microscopic scale, it can be reproduced to reveal a Gaussian Fourier transform $\Delta_z(\vec{r})$, which is deliberated by a height Δ and correlation length ℓ. The mathematical expression of the scattering from the roughness surface can be expressed as:

$$\left\langle \Delta_z(\vec{r})\Delta_z(\vec{r}\,') \right\rangle = \Delta^2 e^{\left(-|r-r'|^2 \ell^{-2}\right)} \tag{4.21}$$

The frequently diminished expression for the deriving scattering rate accepts a sudden photon interface with object geometry. This has been precisely fitted for investigation of an abrupt change of object crossing-section structure, but Eq. (4.21) is incompatible with smooth envelope potentials. In this view, The perturbation $\Delta_V(\vec{R})$ due to a position shift

$\partial_z(\vec{r})$ in an arbitrary confining potential $U(z)$ is assumed to be correlated over the length of a single interface. At the point $r = x + y$, assuming isotropy across the x- and y-plane:

$$\Delta_V(\vec{R}) \sim -\partial_z(\vec{r})\frac{dU(z)}{dz} \tag{4.22}$$

Under time-independent perturbation theory, the perturbation must be small, i.e., $\partial_z(\vec{r}) \to 0$, and the perturbing potential simplifies to:

$$\Delta_U(\vec{R}) = U_0(z_I)\partial_z(\vec{r})\partial(z - z_I) \tag{4.23}$$

where the I-th interface in a multilayer structure is centered about the plane $z = z_I$ and extends over the range $(z_{L,\ I}, z_{U,\ I})$.

The scattering matrix S can be formulated as:

$$S = \langle f|U_{0,I}\partial(z - z_I)|i\rangle \tag{4.24}$$

where $|f\rangle$ and $|i\rangle$ are the final and initial wave functions, respectively.

Consequently, surface roughness scattering or interface roughness scattering is the elastic scattering of a charged particle by a defective interface between two different particles. It is a significant consequence in electronic devices, which encompass narrow layers, for instance, field-effect transistors and quantum cascade lasers.

4.10 Mathematical depiction of SAR backscattering cross-section

Let us assume SAR backscattering is an elastic scattering of the dual comparable particles before and after. Meanwhile, the potential velocity of a photon is $U(r_1 - r_2)$ and can be abridged to an operative one-body mathematical problem, as merely the comparative signal of the particle matter, the backscattering problem can be deliberated in the equivalent center-of-mass (CM) frame. Let us consider that a constant fluctuation of microwave photons with density F_{in} is incident on a scattering center S.

In this regard, the scattered photon particles are counted in a solid angle detector, which includes the direction. The mathematical description of the scattering cross-section per solid angle is given by:

$$\frac{d\sigma}{d\Omega} = \gamma_S(d\Omega \cdot F_{in})^{-1} \tag{4.25}$$

Eq. (4.25) shows the ratio quantity of microwave photons backscattered γ_S into a solid angle per unit time to the flux of photons with density F_{in} into a solid angle $d\Omega$. In the circumstance of $F_{in} = |F_{in}|$ and $\frac{d\sigma}{d\Omega}$ has a dimension area, the mathematical description of backscattering cross-section is articulated as [22]:

$$\sigma = \int \frac{d\sigma}{d\Omega} d\Omega = \int_{\phi=0}^{2\pi} \int_{\theta=0}^{\pi} \frac{d\sigma}{d\Omega} \sin\theta d\theta d\phi \tag{4.26}$$

Marghany [22] stated that the width area of σ which resembles the complete target area in which the incident microwave photons are stroked it, and then causes backscattering. The number of microwave photons surpassing σ transversely will be as great as the quantity of microwave photons that ultimately are backscattered in a precise direction and can be gathered in a detector receiver. Let us consider a solid sphere with radius R; its range $\sigma=\pi R^2$, which corresponds to microwave photons scattering on a sphere of radius R. On account of equilibrium, the central potential of photon velocities $U(r)=U(|r|)$ do not rely on the azimuthal angle ϕ but on incident angle variations.

4.11 Wave function of SAR backscattering cross-section

Elastic scattering and inelastic scattering are dissimilar. In inelastic scattering, the particles' interior phase is revolutionized.In this circumstance, elastic scattering conceivably will excite almost the highest number of electrons of a scattering atom. In other words, inelastic scattering is the widespread annihilation of a scattering particle and the formation of exclusively different particles. Elastic scattering assumes that the internal statuses of the scattering particles do not exchange—the particles develop unaffected from the scattering evolution. On this understanding, the scattering obstruction is calculated as time-independent and primarily is amplitude-dependent. On the other hand, the time-independent Schrödinger formula of photon scattering is articulated as [22–24]:

$$\left[U(r)-\hbar^2(2\mu)^{-1}\Delta\right]\Psi\left(\vec{r}\right)=E\Psi\left(\vec{r}\right), \tag{4.27}$$

where:

$$E\equiv E_k=\hbar^2k^2(2\mu)^{-1}, \tag{4.28}$$

where $\Delta\equiv\nabla^2$ is the Laplacian operator, and $\hbar\vec{k}$ is the momentum of incidents particles or photons.

In Eq. (4.27) it is assumed that the potential $U(r)$ is of finite range and it tends to zero faster than r^{-1} as $r\rightarrow\infty$. Eq. (4.28) proves that a homogenous beam of photons with momentum $\hbar\vec{k}$ is bound in the positive z-direction and is specified by:

$$\Psi_{incident}\equiv\Psi_k^i\underset{r\rightarrow\infty}{\sim}e^{\left(i\vec{k}\cdot\vec{r}\right)}=e^{(ikz)} \tag{4.29}$$

Eq. (4.29) indicates that photons distribute in all directions from the scattering center long after scattering. In this understanding, the flux of photons is designated by a spherical wave. In contrast, Eq. (4.29) signifies the flux of photons, which resembles plane waves (Fig. 4.3). Conversely, a plane wave is characterized by a definite linear momentum $\hbar\vec{k}$ but no significant angular momentum. A plane wave, being in principle of infinite extension, matches to effect parameters fluctuating from zero to infinity. Similarly, the angular momenta enclosed in a plane wave (x, y) also vary from zero to infinity. In this regard, it is conceivable to investigate a plane wave into an infinite number of constituents, of which each resembles a certain angular momentum. Each of such constituents is precisely recognized as a partial wave, and the procedure of decomposing a plane wave into the partial waves is known as a partial wave analysis [5, 6, 25–29]. In this regard, $e^{(i\vec{k}\cdot\vec{r})}$ is considered as a solution of the free-particle Schrödinger equation, which defines the linear momentum representing the system as:

$$\hat{H}\Psi\left(\vec{r}\right)=E\Psi\left(\vec{r}\right) \tag{4.30}$$

with:

$$\hat{H}=-\hbar^2\Delta(2\mu)^{-1} \tag{4.30.1}$$

$$\Delta=\left[r^{-2}\frac{\partial}{\partial r}\left(r^2\frac{\partial}{\partial r}\right)-\vec{L}^2(\hbar^2r^2)\right] \tag{4.30.2}$$

where:

$$\left[H,\vec{L}^2\right]=0, \tag{4.31}$$

$$\left[H,\widehat{L}_z\right]=0, \tag{4.32}$$

where $\vec{L}^2$ and $\widehat{L}_z$ are angular momentum operators and are expressed as:

$$\vec{L}^2=-\hbar^2\left[\sin^{-1}\theta\frac{\partial}{\partial\theta}\left(\sin\theta\frac{\partial}{\partial\theta}\right)+\sin^{-2}\theta\frac{\partial^2}{\partial\phi}\right] \tag{4.33}$$

$$\widehat{L}_z=-i\hbar\frac{\partial}{\partial\phi} \tag{4.34}$$

Both Eqs. (4.33) and (4.34) indicate the eigenvectors of $\hat{H}$, which are likewise the eigenvectors of $\vec{L}^2$ and $\widehat{L}_z$. Nonetheless, the eigenvectors of $\vec{L}^2$and $\widehat{L}_z$ are spherical harmonics $\gamma_{lm}(\theta,\phi)$ expressed as:

$$\gamma_{lm}(\theta,\phi)=\left(-1^{l+m}\right)2^{-l}l!\left[\frac{(2l+1)!}{4\pi}\frac{(l-m)!}{(l-m)!}\right]\sin^{|m|}\theta\times\left(\frac{d}{d(\cos\theta)}\right)^{l+m}\sin^{2l}\theta\cdot e^{im\phi} \tag{4.35}$$

where $\gamma_{lm}(\theta,\phi)$ is termed the spherical harmonics of order l, $\gamma_{lm}(\theta,\phi)$ is the solution of the Laplace equation on the unit sphere ($r=1$) (hence the name spherical harmonics), and m represents $-l,-l+1,\ldots,+l$.

In this context, l would be larger than m and would be a positive integer or zero, where $l=0$, 1, 2, 3, …, $+\infty$. On this understanding, $e^{(i\vec{k}\cdot\vec{r})}$ can be articulated in the formula of radiated spherical waves and scattering spherical waves, respectively, as:

$$e^{(i\vec{k}\cdot\vec{r})}=4\pi\sum_{l=0}^{\infty}\sum_{m=-l}^{+l}i^{-l}j_l(kr)\gamma^*{}_{lm}(\theta_k,\phi_k)\gamma_{lm}(\theta_r,\phi_r) \tag{4.36}$$

where (θ_k,ϕ_k) describes the direction of $\vec{k}$ and (θ_r,ϕ_r) that of $\vec{r}$.

$\frac{e^{-(i\vec{k}\cdot\vec{r})}}{r}$ signifies the scattering spherical wave as $\vec{k}$ and $\vec{r}$ are antiparallel, so that $\vec{k}\cdot\vec{r}=kr\cos\pi=-kr$. Nonetheless, $\frac{e^{(i\vec{k}\cdot\vec{r})}}{r}$ denotes that an incident spherical wave through radar antenna $\frac{e^{-(i\vec{k}\cdot\vec{r})}}{r}\vec{k}$ and $\vec{r}$ are parallel. On the other hand, the wave function of the incident beams and the scattering one can be articulated as:

$$\Psi_k\left(\vec{r}\right)\underset{r\to\infty}{\sim}0.5(ikr)^{-1}\sum_{l=0}^{\infty}(2l+1)\left[(-1)^{l+1}e^{(-ikr)}+e^{(2i\partial_l)}e^{(ikr)}\right]P_l(\cos\theta). \tag{4.37}$$

The first term of Eq. (4.37) represents a scattering spherical wave while the second term where P_l is the Legendre polynomial of an order l representing the incident beam as a spherical wave. In these regards, both terms have a similar intensity. Hence, the phase of the incident wave is shifted comparative to the phase of the corresponding wave in Eq. (4.37). Consequently, the spherical wave phase is swung by the amount ∂_l, which is termed as the phase-shift [8–10, 14, 15, 22].

On the contrary, the mathematical explanation of spherical wave scattering (Fig. 4.6) is expressed as:

$$\Psi_{scattering}\equiv\Psi_k^S(r)\underset{r\to\infty}{\sim}f_k(\Omega)e^{(ikr)}r^{-1}. \tag{4.38}$$

where $f_k(\Omega)$ is the scattering amplitude.

On this understanding, the scattering particles would be in the configuration of the spherical wave with similar energy as the incident one. In addition, Eq. (4.38) confirms that for greater distances there is a similar amount of particles spreading through any cross-section of the specified solid angle element (Fig. 4.7), as assumed.

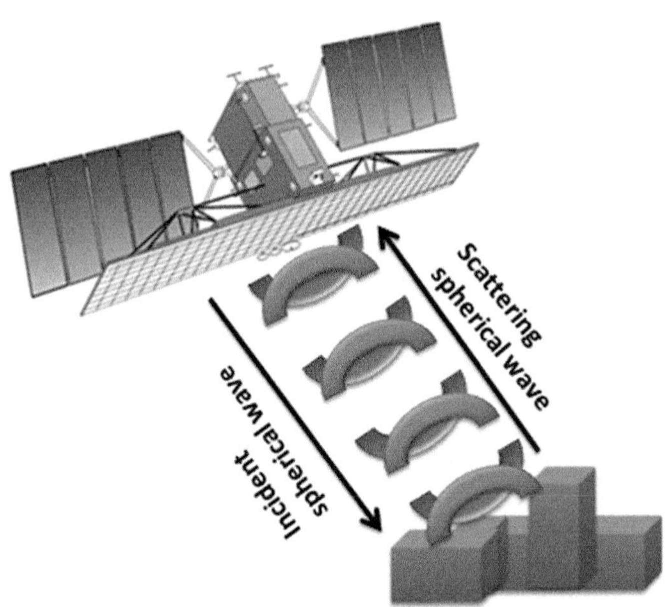

FIG. 4.6 Mechanism of spherical wave scattering.

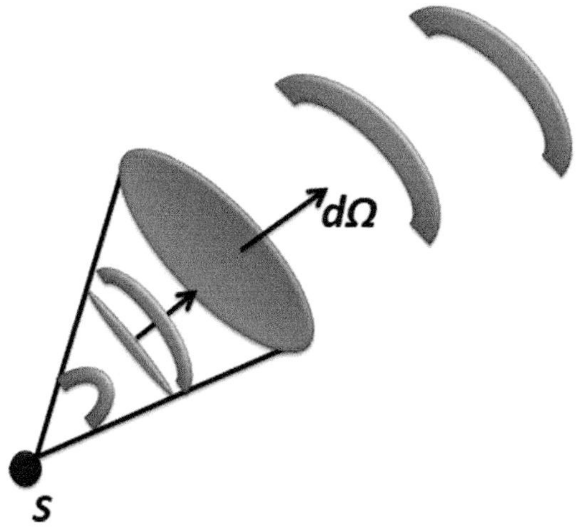

FIG. 4.7 Cross-section of a given solid angle.

Eq. (4.38) can be depleted to acquire the scattering cross-section based on Eqs. (4.25) and (4.26) as:

$$\frac{d\sigma}{d\Omega} = |f_k(\Omega)|^2 \tag{4.39}$$

The scattering amplitude $|f_k(\Omega)|^2$ is independent of the azimuthal angle. In this regard, Eq. (4.39) can be written as a function of Eq. (4.37):

$$\frac{d\sigma}{d\Omega} = (k^{-2}) \sum_{l=0}^{\infty} \sum_{l'=0}^{\infty} (2l+1)(2l'+1) \times e^{[i(\partial_l - \partial_r)]} \sin\partial_l \sin\partial_r P_l(\cos\theta) P_{l'} \tag{4.40}$$

Then the total cross-section is obtained from Eqs. (4.26) and (4.40) as:

$$\sigma = \int \frac{d\sigma}{d\Omega} = \frac{4\pi}{k^2} \sum_{l=0}^{\infty} (2l+1) \sin^2\partial_l = \sum_{l=0}^{\infty} \sigma^l \tag{4.41}$$

In Eq. (4.41), $\sum_{l=0}^{\infty} \sigma^l$ represents the contribution to the total scattering cross-section from the partial wave, which is constrained by:

$$\sigma^l \le \left(\frac{4\pi}{k^2}\right)(2l+1) \tag{4.42}$$

From Eqs. (4.41) and (4.42) we have:

$$\sigma^l = \left(\frac{4\pi}{k}\right) \mathrm{Im}\{f_k(\theta = 0)\} \tag{4.43}$$

where $\mathrm{Im}\{f_k(\theta = 0)\}$ is the imaginary part of scattering amplitude.

Eq. (4.43) shows that the sum cross-section exemplifies the loss of power undergone by the incident microwave photon signal; in other words, some particles have been bounced away from the incident angle. Consequently, this loss of energy is signified by the imaginary part of the scattering amplitude in the forward direction. Consistent with the above perspective, the thickness of the object has to be large enough to initiate abundant scattering power, but small enough to preserve multiple scattering at a minimum [22].

4.12 Quantization of Bragg scattering

Conventional Bragg scattering concept is precisely recognized that the reflection and backscatter from the ocean surface count on the wavelength spreading of the surface waves concerns to the SAR beam wavelength λ. The incident SAR photon beam is scattered from the capillary waves and reflected from long waves. Generally, the wind creates a continuous spectrum of capillary waves, which leads to a resonant wave. In this view, the mean-square wave slope and surface roughness increase with wind speed as well as Bragg scattering. In other words, Bragg scatters also occur from capillary waves riding on the long swell.

In this regard, the wave surfaces can be understood as a gathering of similar isosceles triangles, known as facets (Fig. 4.8), which operate as a specular reflector. This speculation is usable if each facet's length is much greater than λ and if the deviation of the approaching facet from the wave surface is much less than λ. This state exists if the part of the sea surface with the curvature R_c radius satisfies:

$$R_c \gg \lambda \tag{4.44}$$

Generally, the Bragg scattering is given by the large angle backscatter. On the basis of specific incidence angles and frequencies, the backscatter exhibits strong resonance. For the ocean, if the surface wave spectrum involves a wavelength component with a similar relation to the incident beam, Bragg resonance occurs (Fig. 4.9). In other words, Bragg resonance occurs if there exists a surface component with λ_w equal to half the surface projection of the radar wavelength λ, or when:

$$\lambda_w = 0.5\lambda / \sin\theta \tag{4.45}$$

Eq. (4.45) indicates that the intensity backscatter from two contiguous wave crests is in phase. In this circumstance, the SAR beams that are incoherently backscattered from the

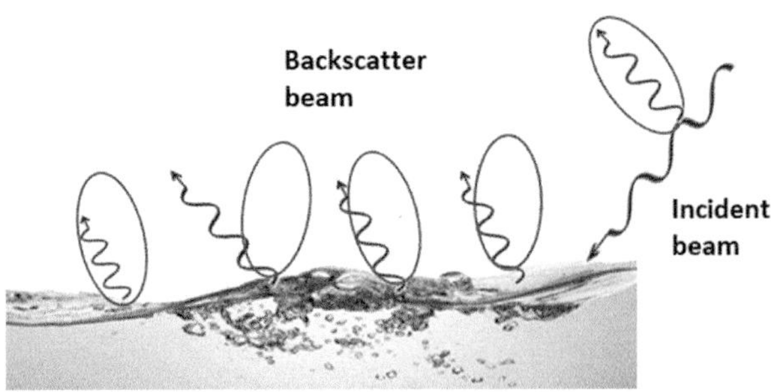

FIG. 4.8 Facet combinations with backscatter beam.

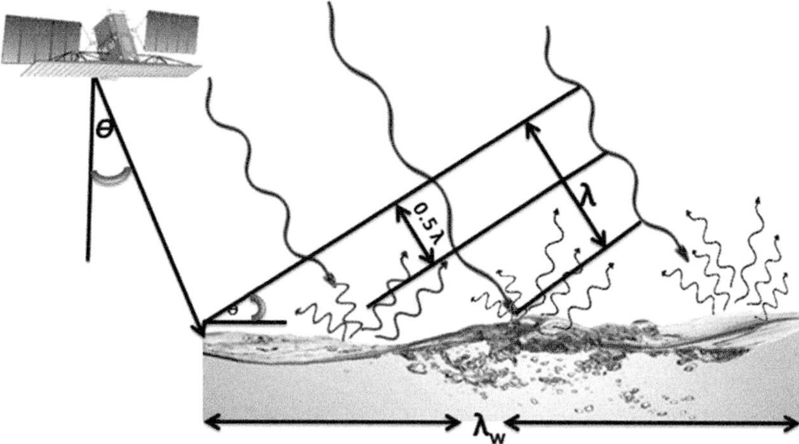

FIG. 4.9 Bragg scatter.

sea surface enlarge coherently at the SAR antenna. In other words, the antenna receives strong backscatter as a function of $\theta > 15$ degrees [30].

Consistent with the Bragg scattering theory, the normalized radar cross-section (NRCS) is proportionate to the spectral energy density of the Bragg waves, i.e., of those surface waves of wavelengths λ_w that satisfy the Bragg resonance circumstance [31]. According to Gade and Alpers [32], the normalized radar cross-section (NRCS) for a sea surface (σ_w) can be formulated as:

$$\sigma_w = |g_{ij}|^2 (16\pi k_r^4 \cos^4\theta) \psi(k_B) \tag{4.46}$$

where $\psi(k_B)$ is the spectral power density of the Bragg waves with wavenumber $k_B = 2\pi\lambda_w^{-1}$, while k_r is the radar wavenumber, and the function g_{ij} relies on radar wavenumber, dielectric constant ε_w of the water, incidence angle, and polarization.

In this regard, *ij* involves HH, VV, HV, and VH; H and V representing horizontal and vertical polarization, respectively. The first letter denotes the polarization of the transmitted radiation and the second letter denotes that of the received radiation. Consistent with the composite sea surface model, the roughness of the sea surface can be seen as small-scale capillary waves superimposed on a large-scale wind wave or swell. Conversely, the backscattered radar signal can be treated as Bragg scattering modulated by the tilted scattering surface created by large-scale gravity waves. The mathematical expression of the total quantity of backscatter from the sea surface is formulated as:

$$\sigma_w = \sigma_s + \sigma_B \tag{4.47}$$

where σ_w is the total backscatter from the entire sea surface, which involves specular reflection σ_s and Bragg scattering σ_B.

Consistent with the above perspective, primary SAR imaging of the ocean surface is dependent on the radar cross-section variation across the composite sea surface. From the point view of quantum mechanics, the radar cross-section involves the photon wave function owing to the interaction between the signal photons and atoms of the sea surface particles. The signal photon wave function is given by:

$$\psi_\gamma(\Delta R, t) = \frac{\varepsilon_0}{\Delta r}\Theta\left(t - \frac{\Delta R}{c}\right)e^{-\left(0.5(i\omega+\Gamma)\left(t-\frac{\Delta R}{c}\right)\right)} \tag{4.48}$$

Eq. (4.48) comprises Θ, a step function that merely specifies that the signals cannot travel faster than the speed of light. In addition:

$$\varepsilon_0 = -\frac{\omega^2\left|\hat{\vec{\mu}}\right|\sin\eta}{4\pi\varepsilon_0 c^2\Delta r} \tag{4.48.1}$$

where ω is the frequency of the radiated signal photon, ε_0 is a polarization basis vector, either linear or circular, and η is the angle between the electric dipole moment of the atom μ. and Δr is the distance between sea surface particles and SAR antenna.

On the other hand, Γ is given by:

$$\Gamma \equiv \tau^{-1} = \frac{1}{4\pi}\frac{4\omega^3\left|\hat{\vec{\mu}}\right|^2}{3\hbar c^3} \tag{4.48.2}$$

Eq. (4.48.2) signifies the inverse backscattered signal photon delay time owing to dynamic oscillation of the sea surface. The quantum of the intensity of signal illumination of the sea surface is formulated as:

$$\left\langle \hat{I}_s(r_s, r_d, t)\right\rangle = N^{-1}\left|\sum_{i=1}^{N}\psi_\gamma(\Delta Rt)\right|^2 \tag{4.49}$$

Eq. (4.49) states that the SAR image shows the oscillations of the photon wave function, which is backscattered owing to its interaction with an infinity of dynamic sea surface particles. From the quantum point view, the radar cross-section of sea surface dynamic has been just a quantum radar cross-section, which determines the rate of microwave photon interaction with ocean wave particles under the circumstance of perturbation theory, i.e., Bragg scattering. In other words, σ_Q is a measure the quantum illumination of sea surface dynamic particle oscillations. In this sense, the quantum radar cross-section of the sea surface roughness is formulated as:

$$\sigma_Q = \lim_{R\to\infty} 4\pi R^2 \frac{\left\langle \hat{I}_s(r_s, r_d, t)\right\rangle}{\left\langle \hat{I}_i(r_s t)\right\rangle} \tag{4.50}$$

where $\left\langle \hat{I}_i(r_s, t)\right\rangle$ and $\left\langle \hat{I}_s(r_s, r_d, t\right\rangle$ are the incident and scattered energy density, respectively, and r_s and r_d are the positions of the transmitter and the receiver.

For quantum radar applications, Rayleigh scattering is the process that characterizes the interaction between the SAR signal photons and sea surface dynamic fluctuation. That is, the final state of the atom is the same as the initial state, the circular frequency of the scattered photon is equal to that of the incident photon, and the photon energy is low relative to the ionization energy of the interacting atom. The interaction between the photon and an atom is described using an essential QED process of absorption and emission of photons [33] and the scattering effect only occurs in the two-order approximation of perturbation theory [34]. Let us assume the sea surface area struck by the microwave photon $A_\perp$ is proportional inversely with the power spectra of Bragg waves $\psi(k_B)$. Consequently, the corresponding analytical expression of the radar cross-section of the sea surface can be obtained:

$$\sigma_Q = 4\pi\left(\frac{\sigma_w}{|g_{ij}|^2\left(16\pi k_r^4 \cos^4\theta\right)}\right) \lim_{R\to\infty} \frac{\left|\sum_{i=1}^{N} e^{\left(j\omega\Delta R_i c^{-2}\right)}\right|^2}{\int_0^{2\pi}\int_0^{0.5\pi}\left|\sum_{i=1}^{N} e^{\left(j\omega\Delta R'_i c^{-2}\right)}\right|^2 \sin\theta_d d\theta_d d\phi_d} \tag{4.51}$$

Eq. (4.51) confirms sea surface dynamic area in a radar system is imagined as an area of existing Bragg wave power, which is the construction of the photon wave function (Eq. 4.51) due to the interaction with capillary wave particles. Indeed, capillary waves can develop to swell across the sea surface in the form of wave-duality. This leads to quantum coherence when a superposition forms between the quantum microwave photon states and the capillary wave. In other words, the SAR mechanism of capillary wave imagine is just considered as quantum coherence state, where the gaining of information about Bragg waves from a from the sea surface into SAR system through the backscattered photon [30]. Consequently, coherence arises when diverse portions of the backscattered photon wave function create entangled with the SAR receiver [33].

References

[1] Jonasz M, Fournier G. Light scattering by particles in water: Theoretical and experimental foundations. Elsevier; 2011.

[2] Dalberg PS, Bo/e A, Strand KA, Sikkeland T. Quasielastic light scattering study of charged polystrene particles in water. J Chem Phys 1978;69(12):5473–8.

[3] Koirtyohann SR, Pickett EE. Light scattering by particles in atomic absorption spectrometry. Anal Chem 1966;38 (8):1087–8.

[4] Babin M, Morel A, Fournier-Sicre V, Fell F, Stramski D. Light scattering properties of marine particles in coastal and open ocean waters as related to the particle mass concentration. Limnol Oceanogr 2003;48(2):843–59.

[5] Matthew JA. Spin dependence of the electron inelastic mean free path and the elastic scattering cross section—a high-energy atomic approximation. Phys Rev B 1982;25(5):3326.

[6] Gilmore R. Elementary quantum mechanics in one dimension. JHU Press; 2004.

[7] Ter Haar D, editor. Problems in quantum mechanics. Courier Corporation; 2014.

[8] Valavanis A, Ikonić Z, Kelsall RW. Intersubband carrier scattering in n-and p−Si/Si Ge quantum wells with diffuse interfaces. Phys Rev B 2008;77(7):075312.

[9] Dirac PA. The Lagrangian in quantum mechanics. In: Feynman's thesis—A new approach to quantum theory. World Scientific Publishing Co Pte Ltd; 2005. p. 111–9.

[10] Chandler D, Wolynes PG. Exploiting the isomorphism between quantum theory and classical statistical mechanics of polyatomic fluids. J Chem Phys 1981;74(7):4078–95.

[11] Jackson DF. Nuclear sizes and optical model. Rep Prog Phys 1974;34:55–146.

[12] Auger JP, Lombard RJ. Proton–nucleus elastic scattering at 1 Gev. in the Glauber model. Ann Phys 1978;115 (2):442–66.
[13] Aguiar CE, Zardi F, Vitturi A. Low–energy extension of the eikonal approximation to heavy–ion scattering. Phys Rev C 1997;56(3):1511–5.
[14] Zhu W, Huang Y, Kouri DJ, Arnold M, Hoffman DK. Time-dependent wave-packet forms of Schrödinger and Lippmann-Schwinger equations. Phys Rev Lett 1994;72(9):1310.
[15] Kadyrov AS, Bray I, Stelbovics AT, Saha B. Direct solution of the three-dimensional Lippmann–Schwinger equation. J Phys B At Mol Opt Phys 2005;38(5):509.
[16] Alt EO, Grassberger P, Sandhas W. Reduction of the three-particle collision problem to multi-channel two-particle Lippmann-Schwinger equations. Nucl Phys B 1967;2(2):167–80.
[17] Heller EJ, Reinhardt WP. Comment on the direct-matrix solution of a singular Lippmann—Schwinger equation. Phys Rev A 1973;7(1):365.
[18] Mongan TR. Note on the numerical solution of the lippmann-schwinger equation. Il Nuovo Cimento B (1965–1970) 1969;63(2):539–48.
[19] Tada K, Tada M, Maita Y. Dissolved free amino acids in coastal seawater using a modified fluorometric method. J Oceanogr 1998;54(4):313–21.
[20] Maeda M, Taga N. Alkalotolerant and alkalophilic bacteria in seawater. J Mar Ecol Prog Ser 1980;2:105–8.
[21] Cohen-Tannoudji C, Dupont-Roc J, Grynberg G. Atom-photon interactions: Basic processes and applications. Wiley-VCH; 1998. ISBN 0-471-29336-9678.
[22] Marghany M. Automatic detection algorithms of oil spill in radar images. CRC Press; 2019.
[23] Ryder LH. Quantum field theory. Cambridge University Press; 1996.
[24] Weinberg S. The quantum theory of fields I. Cambridge University Press; 1995.
[25] Hofheinz M, Wang H, Ansmann M, Bialczak RC, Lucero E, Neeley M, O'connell AD, Sank D, Wenner J, Martinis JM, Cleland AN. Synthesizing arbitrary quantum states in a superconducting resonator. Nature 2009;459(7246):546.
[26] Bai L, Harder M, Chen YP, Fan X, Xiao JQ, Hu CM. Spin pumping in electrodynamically coupled magnon-photon systems. Phys Rev Lett 2015;114(22):227201.
[27] Tserkovnyak Y, Brataas A, Bauer GE. Enhanced Gilbert damping in thin ferromagnetic films. Phys Rev Lett 2002;88(11):117601.
[28] Mosendz O, Pearson JE, Fradin FY, Bauer GE, Bader SD, Hoffmann A. Quantifying spin Hall angles from spin pumping: experiments and theory. Phys Rev Lett 2010;104(4):046601.
[29] Huebl H, Goennenwein ST. Electrical signal picks up a magnet's heartbeat. Phys Ther 2015;8:51.
[30] Marghany M. Synthetic aperture radar imaging mechanism for oil spills. Gulf Professional Publishing; 2019.
[31] Wright JW. A new model for sea clutter. IEEE Trans Antennas Propag 1968;16(2):217–23.
[32] Gade M, Alpers W, Hühnerfuss H, Masuko H, Kobayashi T. Imaging of biogenic and anthropogenic ocean surface films by the multifrequency/multipolarization SIR-C/X-SAR. J Geophys Res Oceans 1998;103 (C9):18851–66.
[33] Liu K, Xiao H, Fan H, Fu Q. Analysis of quantum radar cross section and its influence on target detection performance. IEEE Photon Technol Lett 2014;26(11):1146–9.
[34] Fang C, Chen Y, Xu Y, Hua L. The analysis of change factor of the simulation of the bistatic quantum radar cross section for the typical ship structure, In: 2018 IEEE Asia-Pacific conference on antennas and propagation (APCAP) 2018 Aug 5IEEE; 2018. p. 190–3.

CHAPTER

5

Relativistic quantum mechanics of ocean surface dynamic in synthetic aperture radar

Imagining ocean surface dynamic in synthetic aperture radar requires modern physics speculations to comprehend perfectly how the radar signal can carry large quantities of information regarding ocean surface dynamics. Radar signal and ocean surface dynamic cannot be tackled separately from quantum mechanics and relativity. In this view, we are dealing with two systems: the radar system as the instruments and the dynamic system as ocean surface fluctuations. Both systems can be explained through quantum and relativity theories. However, they cannot be correlated together without understanding quantum mechanics and relativity theories. SAR ocean surface image is considered as the correlation of quantum mechanics and relativity theories. Quantum mechanics speculations have contributed to the creation of radar ocean images, and relativity theory also contributes to shaping the ocean surface dynamic pattern in radar images. This chapter answers the significant question: what is the relationship between radar ocean dynamic images with quantum and relativity theories? Part of this answer was demonstrated in Chapter 4. Continuing from that chapter, the correlation between quantum mechanics and relativity theory is addressed here to aid a full understanding of SAR image mechanism of ocean surface dynamic.

5.1 What is meant by relativity?

Albert Einstein formed two important physical theories in 1905 and 1915: special relativity and general relativity, respectively. He defied numerous expectations underlying formerly physical principles, redescribing in practice the essential theories of space, time, matter, energy and gravity. According to this understanding, both energies of sea surface dynamic and microwave photon propagation are functions of space, time, and gravity. In conjunction with quantum mechanics, relativity is the cornerstone of modern physics. To create a SAR image,

Nonlinear Ocean Dynamics
https://doi.org/10.1016/B978-0-12-820785-7.00010-1

successive pulses of radio waves are transmitted to "illuminate" a ocean dynamic scene, and the echo of each pulse is received and recorded as a function of relativity.

"Special relativity" is restricted to how space and time are linked for objects that are shifting at a steady velocity in a straight line. In this regard, SAR cannot, by purely mechanical experiments, distinguish sea surfaces dynamic features such as wave dynamic motions and current movements; one from the other in the circumstance of identical sea surface owing to the impact of special relativity. Special relativity revealed that the speed of light is a constraint that can be considered, but not reached by any sea surface dynamic features; it is the origin of the most famous equation in science: $E = mc^2$. It has also led to other tantalizing consequences, such as the "twin paradox." In this circumstance, the SAR at orbit has a different frame than the ocean surface dynamic on the Earth's frame. The earthbound sea surface dynamic is at rest in the same inertial frame throughout the procedures of SAR sea surface imaging, while the two-dimensional SAR image of the sea surface dynamic is not: in the simplest version of the thought-experiment, the SAR sea surface image switches at the midpoint of the trip from being at rest in an inertial frame, which moves in one direction (away from the Earth), to be at rest in an inertial frame which moves in the opposite direction (toward the Earth). In this approach, determining which SAR switches frames and which does not is crucial.

Though both real ocean surface dynamic and SAR images can reasonably be considered to be at rest in their frames, only the SAR image experiences an increase of sea surface fluctuation rate compared to that of the real one. In other words, inertial references frames—a SAR moving with constant velocity. However, in noninertial references frames—a dynamic ocean movement with constant speed during SAR overpasses; which perhaps changes after SAR ends its mission.

"General relativity" involves gravity, one of the essential forces in the universe, which comprises electricity and magnetism unified together—for instance, to generate microwave photon radiation from a SAR sensor. Gravity expresses macroscopic behavior, and so general relativity describes large-scale physical phenomena such as planetary dynamics. Up to date, special and general relativity theories in no way been used in investigation of radar sea surface imaging mechanism, but used commonly in different physics applications; for instance, nuclear energy and nuclear weapons industries.

Ocean dynamics is the study of the behavior of ocean movements related to external and internal forces such as wind and density gradient. If we study ocean dynamics in the arena of special or general relativity, it can perhaps be termed as relativistic ocean dynamics.

Special relativity can come into play when the velocities attained by certain portions of the ocean or by the ocean as a whole approach the speed of light. General relativity comes into play when there are sufficiently strong gravitational fields. One example is tide generation, either because the ocean's flow environment features such fields, or because the mass and energy of the ocean flow are sufficient to generate their own strong gravity.

5.2 Relativistic quantum mechanics versus ordinary quantum mechanics

Relativistic quantum mechanics (RQM) represents the growth of quantum mechanics, combined with the perceptions of the special theory of relativity. The relativistic formulation has been far more effective than conventional quantum mechanics in some circumstances, such as in the prediction of antimatter, electron spin, spin magnetic moments of elementary

$-1/2$ fermions, fine structure, and quantum dynamics of charged particles in electromagnetic fields. In contrast to RQM, nonrelativistic quantum mechanics (non-RQM) denotes the mathematical verbalization of quantum mechanics in the framework of Galilean relativity and quantizes the formulations of conventional mechanics by changing dynamical variables by operators. Relativistic effects in a dynamic system such as the ocean can be considered to involve perturbations, or small corrections, to the nonrelativistic theory, which is developed, for instance, from the solutions of the Schrödinger equation of *wave* propagation. Schrödinger was searching for a relativistic wave equation, but he ended up with the new simple "Newtonian" theory. Consequently, the relativistic version of the Newtonian theory is significant, in particular due to the high-speed relativistic effects of electrons in atoms.

However, apart from the relativistic speed (v/c) consequence that needs to be considered properly, (v/c) is to bridge between $\boldsymbol{E} = \boldsymbol{mc^2}$ and the quantum relation $E = hc\lambda^{-1}$.

It is known that smaller scales of λ are dominated by the large energy scale E. Nevertheless, on a specific situation, such as when a microwave beam nears the mass-energy scale of the electron, instead of looking at small things, perturbing electrons begin to be created from the microwave beam. However, this phenomenon has not been clarified in regular quantum mechanics. In this view, for high energy photons, a high quantum probability is presented in which the photon spins into a couple of electron-positrons.

Consequently, relativity involves a nonconserved quantity of particles, which is not the case for quantum mechanics. In different words, a huge particle is relativistic when its whole mass-energy (rest mass + kinetic energy) is at least twice its relaxation mass. This circumstance implies that the particle's velocity is close to the velocity of light. [1].

This singularity is very significant since it asserts that energy exchanges at an essential stage are implicit as particles change into one another, which is considerably further understated perception than the "action-reaction forces" from the Newtonian view, which is immobile depleted in conservative relativity and quantum mechanics." According to this understanding, the formalism of quantum field theory has arguably addressed this [2].

Relativity has also recognized the loss of simultaneity of events for different SAR observers. SAR has unique pulse geometry in a range, and the backscatter intensity of this pulse differs from near-range to far-range (Fig. 5.1). In other words, the radar photon beam has a certain beam width which is known as the "elevation beamwidth." The microwave photon beam illuminates an area on the ground between "near-range" and "far-range." At a specific propagation time, the pulse is transmitted as a wavefront, between the trailing edge of the pulse and the leading edge of the pulse with an incident beam θ_i (Fig. 5.1).

This is a precise and well-identified consequence from conventional relativity and its spacetime structure: for two events of ocean dynamic and SAR photon signal that are "space-like," the first SAR photon can backscatter from the sea surface at zone (A) as a near-range (Fig. 5.2A) before one at zone (B) as far-field. Reciprocally, for other SAR photon signals, a sea surface one at near-range is imaged before one at far-range. In this understanding, both events are the same; the SAR photon particle used in the description of the phenomenon has to be the same one, with the same mass and same characteristics, except for its backscatter pattern, which is opposite. The near-range one (Fig. 5.2A) is dominated by stronger backscatter than the far-range one (Fig. 5.2B).

The antenna as the observer in the above examples plays a significant role in proving the relativistic quantum mechanics. In this regard, the microwave photon wavelength λ and dimension of antenna D deliver the relativistic quantum mechanics in Fig. 5.1. The near-range

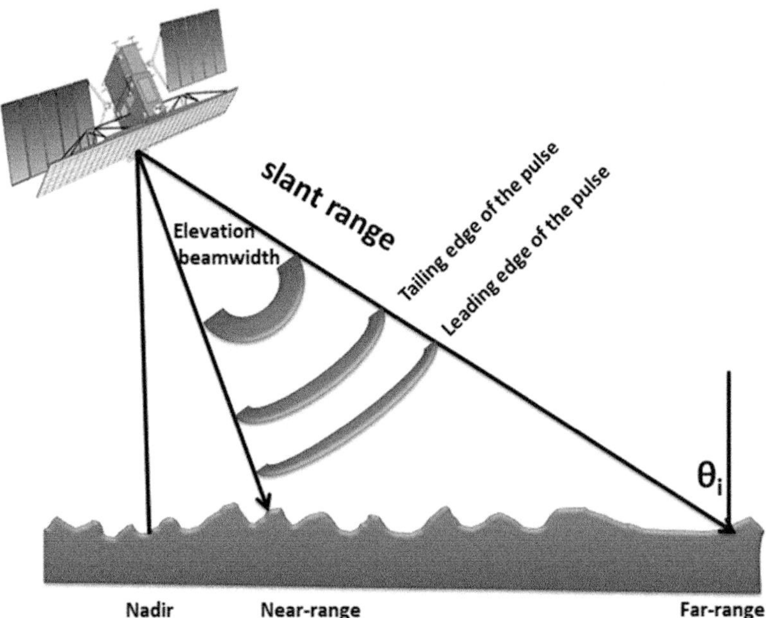

FIG. 5.1 SAR pulse geometry in a range.

presents the Fresnel region while the far-range represents the Fraunhofer region. The near-range radiated many photon particles and the range distribution is dependent on the distance from the antenna. In the near-range area, $r<\lambda$, fields with r^{-2} term will prevail and remain. The near-range is defined by a sphere with a radius $R_1=2D_1\lambda^{-1}$. In the far-range, $r\gg\lambda$ fields with the r^{-1} term will prevail and remain. This region is defined by a sphere with a radius $R_2>R_1$ [3].

5.3 SAR backscatter in relativistic quantum mechanics

SAR backscatter demonstrates the fact that quantum mechanics does not envisage the consequence of a discrete reflection, but rather the statistical distribution of all conceivable consequences. SAR backscattering is significant in quantum mechanics and is closely associated with certain state problems. For instance, much of the information about the interaction between photons and ocean wave particles is derivable from scattering experiments. Essentially, in nonrelativistic quantum mechanics, scattering theory is concerned with the solution of the Schrödinger wave equation for different potentials with appropriate boundary conditions. In general, the primary connection of relativistic quantum field theory to SAR is through scattering theory, i.e., the theory of the collision of photons due to the rough sea surface. In circumstances where the kinetic energy of the colliding particles from the sea surface is large in comparison with the interaction energy, the scattering potential may be considered as a perturbation. The mattered wave function at points far from the interaction region may

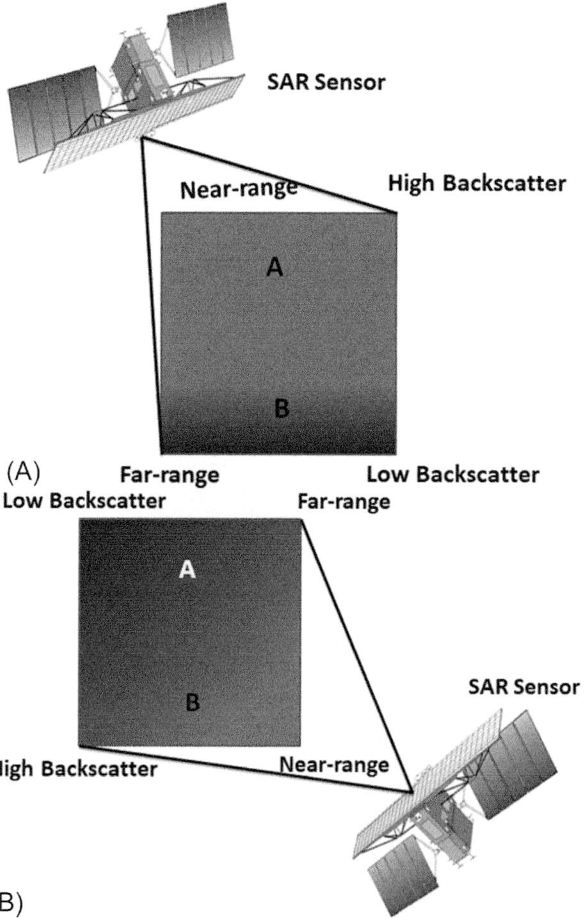

FIG. 5.2 Space-like sea surface backscatter in SAR image (A) near-range and (B) far-range.

then be expressed as plane wave solutions. In this regard, the Born approximation can be implemented to determine SAR scattering in relativistic quantum mechanics as follows:

$$\psi_k\left(\vec{r}\right) \sim e^{i\vec{k}\,\vec{r}} - + \frac{e^{ikr}}{r} f_k(\theta, \phi) \tag{5.1}$$

where:

$$f_k(\theta, \phi) = -(4\pi)^{-1} \int e^{-i\vec{q}.\vec{r}'} V_{eff}^S\left(\vec{r}\,'\right) d^3\vec{r}\,' \tag{5.1.1}$$

where:

$$V_{eff}^S = 2EV - V^2\left(C^2\hbar^2\right)^{-1} \tag{5.1.1.1}$$

Eq. (5.1.1) reveals that the first-order relativistic scattering amplitude is presented by $\vec{q} = \vec{k}' - \vec{k} \cdot f_k(\theta, \phi)$. Consistent with relativistic quantum mechanics, a variety of scattering amplitudes are reliant upon the kind of combination between the scattering amplitude of the effective potential, and matter interaction. Knowing the relativistic scattering amplitude, the differential scattering cross-section may be found using:

$$\frac{d\sigma}{d\Omega} = |f_k(\theta, \phi)|^2. \tag{5.2}$$

Subsequently, as relativistic quantum mechanics essentially concerns high energies, the Born approximation is appropriate to describe relativistic scattering phenomena in the SAR system.

5.4 Duality of wave packages in relativistic quantum mechanics

Let us assume that the ocean surface dynamic and SAR signal backscatter can be correlated together using the Lorentz transformation. This transformation relates the measures of space and time for both physical events, as given by:

$$\begin{aligned}
\Delta t' &= \left(\Delta t - \frac{v}{c^2}\Delta x\right)\frac{1}{\sqrt{1-\frac{v^2}{c^2}}};\\
\Delta x' &= (\Delta x - v\Delta t)\frac{1}{\sqrt{1-\frac{v^2}{c^2}}};\\
\Delta y' &= \Delta y;\\
\Delta z' &= \Delta z
\end{aligned} \tag{5.3}$$

Let us consider that measures of frequency ω and wave vector k of the sea surface fluctuate between real ocean and SAR according to the relationship of:

$$\begin{aligned}
\omega' &= (\omega - vk_x)\frac{1}{\sqrt{1-\frac{v^2}{c^2}}}\frac{1}{\sqrt{1-\frac{v^2}{c^2}}};\\
k_x' &= \left(k_x - \frac{v}{c^2}\omega\right)\frac{1}{\sqrt{1-\frac{v^2}{c^2}}};\\
k_y' &= k_y;\\
k_z' &= k_z
\end{aligned} \tag{5.4}$$

From Eq. (5.4), one can comprehend the Doppler effect, which can also be used for a wave in material media. Therefore, motion is an association between space intervals (Δx, Δy, Δz) and time intervals (Δt) [4–6]. The straightforward associations are allied with definitive vector $\overline{W}$ algebra:

$$(\Delta x, \Delta y, \Delta z) = \overline{V}\Delta t; \\ \Delta t = \overline{W} \cdot (\Delta x, \Delta y, \Delta z) \tag{5.5}$$

Eq. (5.5) shows that W and V have the same transformation rule as in the case of motion. Substituting Eq. (5.4) into Eq. (5.5), the motion of a wave—namely, the movement of a constant phase state—can be given by:

$$\overline{k} \cdot \Delta\overline{r} - \omega\Delta t = 0 \Rightarrow \frac{\overline{k}}{\omega} \cdot \Delta\overline{r} = \Delta t; \\ \overline{W} = \frac{\overline{k}}{\omega} \tag{5.6}$$

Eq. (5.6) reveals that the k/ω vector transforms between inertial frames of ocean and SAR just as the vector W does. Eqs. (5.5) and (5.6) describe the duality of motion for both ocean surface and microwave beams as wave and particle in relativistic quantum mechanics as encoded in SAR signal and then included $\sqrt{1-\frac{v^2}{c^2}}$ of the velocity of particle v corresponding to the speed of light c [5–7].

Let us assume that the dispersions of the linear combination of several plane waves (i.e., a wave package) are characterized by a wavelength Δk and frequency dispersion $\Delta\omega$.

$$\Delta\omega = \overline{V} \cdot (\Delta k_x, \Delta k_y, \Delta k_z); \\ (\Delta k_x, \Delta k_y, \Delta k_z) = \overline{W}\Delta\omega \tag{5.7}$$

The first formula of Eq. (5.7) corresponds to the wave package group velocity V associated with the dispersion. The second formula is also for a wave package, but now the dispersion holds a constant value for W, and the group velocity is equal to phase velocity (i.e., the inverse of W):

$$\frac{\Delta\overline{k}}{\Delta\omega} = \frac{\overline{k}}{\omega} \tag{5.8}$$

Eq. (5.8) shows that the wave package holds its shape while traveling in space and time. In addition, in the wave-particle duality context, Eq. (5.7) shows that any interaction produces a dispersion in the wave package; therefore, regarding the De Broglie formulas, the values of energy and mechanical impulse have an additional margin or amplitude in energy and mechanical impulse. To investigate the SAR backscatter σ_p due to interaction with rough sea surface fluctuation energy ∂E_w, let us assume that particle p exchanges its energy and momentum with a wave w [4–7]. The backscattered energy and mechanical impulse of the system are preserved as:

$$\partial\sigma_p = \partial E_w \Rightarrow \partial\sigma_p = \overline{V} \cdot \overline{W}\partial E_w; \rightarrow \overline{V} \cdot \overline{W} = 1 \tag{5.9}$$

Eq. (5.9) reveals that the variation of the backscatter is a function of the change of the sea surface spectra energy due to the wave-particle fluctuations. In this regard, the spectrum energy in space-time can mathematically be expressed by:

$$E_w\left(\overrightarrow{k}\right) = \left\{ \int_{-\infty}^{\infty} f(z) e^{(-jk_z)dz} \right\} \partial(k_x x) \tag{5.10}$$

where $f(z)$ is the delay function which corresponds to the wavefront, k_x is the wavenumber in x direction, and $E_w\left(\vec{k}\right)$ is obtained by implementing the Fourier transform. The delay function carries information as the scattering elements contributing to the "bulk scattering cross-section"; in which is distributed spatially. For the monostatic backscatter case, $\theta = 180$ degrees and the circumstance for scatter is that $\left|\vec{k}\right| = 4\pi\lambda^{-1}$, which is recognized as the Bragg scattering concept [8].

5.5 Relativities of SAR time pulse range traveling

The main struggle with range measurement and pulsed radars is how to determine unambiguously the range to the target when the target scatters a strong echo. Ambiguous range is a consequence of the transmission of a series of pulses. According to this understanding, pulse repetition interval (PRI) is twisted and is identified as the spaces between emitted pulses. It is also designated as a pulse repetition frequency (PRF) (Fig. 5.3). Accordingly, the delay time is formed due to tiny amounts of time; just a few seconds are taken for transmitted and received pulses. In this scenario, the delay time is caused and triggers an ambiguity along the range direction or vagueness. The unambiguous range R_{amb} as a function of delay time τ_{PRI} is cast as:

$$R_{amb} = \frac{c\tau_{PRI}}{2} \tag{5.11}$$

Under the circumstance of Eq. (5.11), radar range can be determined unambiguously when the target range is smaller than R_{amb}. In contrast, an impossible range of unambiguously can

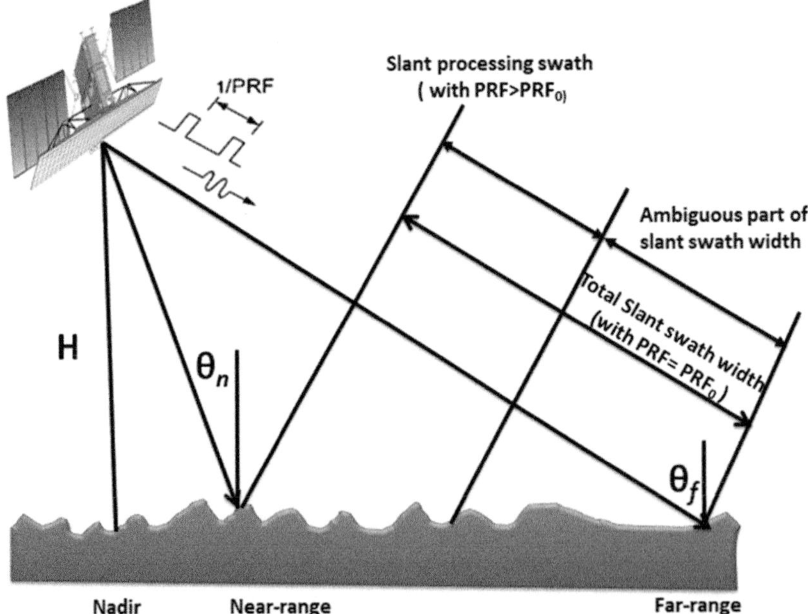

FIG. 5.3 Range ambiguity geometry.

be accounted when the object range is larger than R_{amb}. In these circumstances, the radar must have PRI greater than the range delay frequented with the longest target ranges to circumvent range ambiguities. Subsequently, the SAR image is allocated in dual parts: (i) an unambiguous scene where the positioned reflectors are properly imaged by the SAR system; and (ii) an ambiguous scene within a single range ambiguous reflector.

Let us assume that SAR is transmitting in orbit $O: x^{\mu} = x^{\mu}_{O}(t)$with the proper time. In this circumstance, the SAR parameters can be established and can be verified by the special relativity as: $t^{+}(x) \equiv$, which munificence an appropriate time at which a pulse, i.e., technically, a null geodesic, leaving point x could intercept O. $t^{-}(x) \equiv$ demonstrates a suitable time at which a pulse (null geodesic) leaves O, and approaches a point x. Subsequently, the SAR time pulse transmitting is formulated as:

$$t(x) \equiv 0.5(t^{+}(x) + t^{-}(x)) \tag{5.12}$$

Consequently, the SAR distance R is given by:

$$R(x) \equiv 0.5(t^{+}(x) + t^{-}(x)) \tag{5.13}$$

Therefore, the novel definition of SAR hypersurface of simultaneity $\sum_{t_0}$ at a time t_0 is introduced as (Fig. 5.4):

$$\sum_{t_0} \equiv \{x : t(x) = t_0\} \tag{5.14}$$

Eq. (5.14) simply articulates that SAR can assign a time to a distant event by sending a sequence pulse to the objects and backscatter, and averaging the (proper) times of sending and receiving. Nevertheless, the practice of describing hypersurfaces of simultaneity in terms of "SAR time" has rarely been applied to noninertial satellite orbit or nonflat space-times. This is perhaps due to Bondi's statement [9] that "how a clock reacts to acceleration is reliant on how the clock is created." In line with Dolby [10], it is conjectured that "appropriate times" can operate identically as a function of acceleration or as a function of gravitational fields, which is a simple principle of general relativity, without which proper time would have no physical

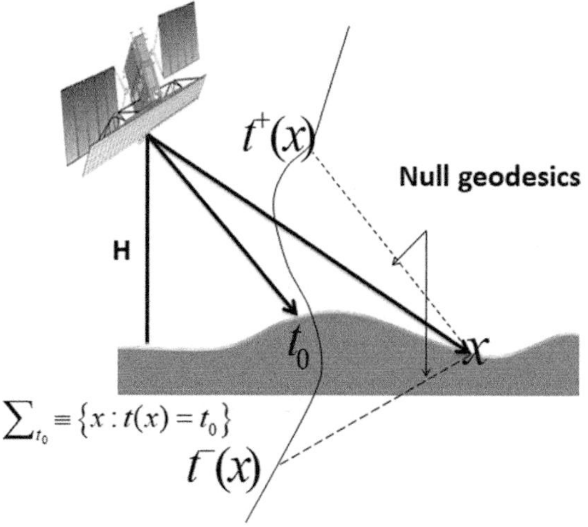

FIG. 5.4 Schematic of the definition of "SAR time."

meaning. There is, therefore, no reason to use "radar time" to speed up SAR or to warp space-time. Subsequently, SAR time can be designated short of slight observation of the noninertial satellite orbit or nonflat space-times it is, by construction, independent of our choice of either noninertial satellite orbit or nonflat space-times frames. In this context, radar time or SAR time is single-valued, resembles appropriate time on the SAR path, and is invariant as a function of "time-reversal"—that is, under reversal of the sign of the SAR's proper time [10]. Consistent with the above perspective, the delay time between the transmitted pulse and the received pulse from the dynamic sea surface can precede to special relativity as causing the range ambiguities. In this understanding, the relativities of the SAR time pulse range produce a copy of the target, which appears offset in range. In this circumstance, simultaneous returns from the desired illuminated region and of a previously or successively transmitted pulse create range ambiguities, as the original SAR image dominates, due to high brightness level (Fig. 5.5), over the original SAR image without range ambiguity. Therefore, nadir ambiguity can be demonstrated as bright sharp linear features that appear at an approximately constant range (Fig. 5.6). In this regard, the signal scatters from nadir are extremely strong due to near-specular reflection of photon particles from targets within a very narrow range of distance which creates a bright tone in SAR images in the form of the sharpest line on the SAR image (Fig. 5.6).

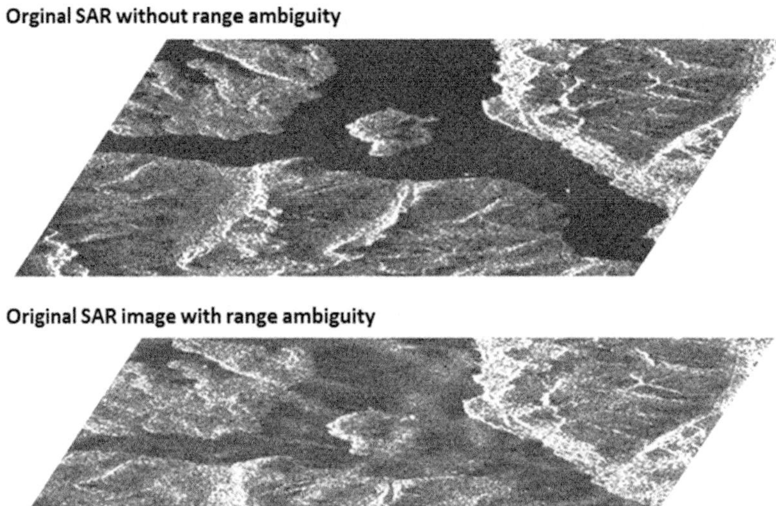

FIG. 5.5 SAR range ambiguity.

FIG. 5.6 SAR nadir ambiguity.

5.6 SAR space-time invariance interval

In general relativity, the space-time interval is an invariant quantity. This means that all SAR observers will estimate space-time to be a similar number, no matter what their reference frame is. The space-time interval can be mathematically expressed as:

$$\Delta s^2 = -c^2\Delta t^2 + \Delta x^2 + \Delta y^2 + \Delta z^2. \tag{5.15}$$

Eq. (5.15) reveals that if Δt^2 does not have a minus sign, then this would be the distance formula in four dimensions. The minus is what makes time dissimilar from space in relativity. Therefore, Eq. (5.15) for the space-time interval is relevant only in special relativity. In general relativity, Eq. (5.15) becomes more convoluted due to what the persuading microwave photon particles in space-time are doing.

Let us consider a four-vector (c,t,r) which is transformed into the four-vector (c,t',r'); we then use the fact that:

$$\Delta s = \left\langle \left(ct, \vec{r}\right), \left(ct, \vec{r}\right)\right\rangle = c^2t^2 - \vec{r} \cdot \vec{r} \tag{5.16}$$

where Δs is the "interval," which is formulated as:

$$c^2t^2 - \vec{r} \cdot \vec{r} = c^2t'^2 - \vec{r}\,' \cdot \vec{r}\,' \tag{5.17}$$

Eq. (5.17) demonstrates that the interval is essentially the dot product of the four-vector—although a minus sign exists where one might expect a plus sign for a normal inner product. The space-time interval, therefore, is not reformed by a Lorenz transformation. It is, therefore, liberated of the frame of reference. In addition, in cases where s^2 is positive, the space-time interval is described as "space-like." If s^2 is negative, however, the space-time interval is called "time-like."

Consistent with the above perspective, let us consider the transmitter and receiver platforms move with a similar continuous velocity so that their spots can be considered immobile in a platform-centered reference frame. Additionally, the transmitted and received pulses are assumed to be coordinated within this frame, liberated of the devoted synchronization repetition of signal transmission and receiving (Fig. 5.7). In this scene, let us consider TSX and TDX as signifying the transmitter and receiver satellites, which are separated by a fixed baseline B. In this scenario, as long as both satellites are immobile, the scene target reasonable moves to the specific frame with a velocity v. In this specific reference frame, the time interval between the transmission and reception of a radar pulse is denoted by Δt.

On the word of the special theory of relativity, the speed of light has eternally the same rate, and enlightened of the inertial reference frame one practices to designate a SAR physical system. An instantaneous significance of SAR two-dimensional ocean imagine can be termed as nonsimultaneity of events, i.e., an inertial frame that represents SAR orbital motion and noninertial that represents ocean dynamic movements. According to this understanding, dual spatially separated events, which occur at similar periods in one reference frame, may no be longer simultaneous in another reference frame that transfers relative to the first

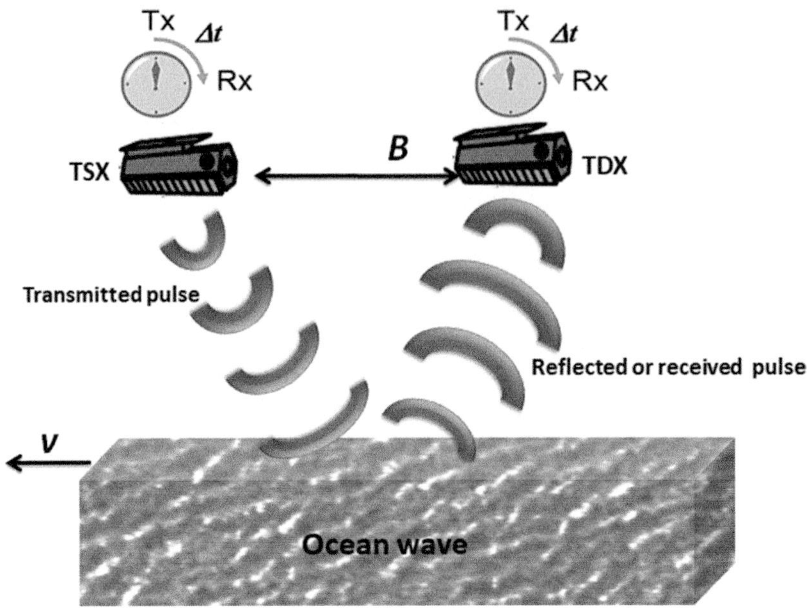

FIG. 5.7 A reference form of bistatic SAR data.

one. Accordingly, radar transmitted and received pulses that are totally synchronous in the platform-centered frame are no longer synchronous in the Earth Centred Earth Fixed (ECEF) reference frame. The scientific explanation of the theory of relativity is the consistency of the space-time interval that can be formulated by:

$$\Delta s = \sqrt{(c \cdot \Delta t^2) - \sum_{i=1}^{3} \Delta x_i^2} \tag{5.18}$$

where Δt is time and Δx position differences between dual occurrences as examined in an assumed reference frame. Furthermore, Δs relics invariant in the Lorentz group of linear space-time transformations between SAR signal transmitting and receiving and ocean dynamic movements. Thus, in both platform reference frames, the space-time interval between the transmit (Tx) and receive (Rx) platforms is calculated using:

$$\Delta s = \sqrt{(c \cdot \Delta t^2) - \left\| \vec{B} \right\|^2} \tag{5.19}$$

Let us considered, along-track baseline between the two satellites changes by $v \cdot (r_{bi} c^{-1})$ from that conveyed in the platform-centered reference frame if one compares the transmit and receive events. Here, v denotes the receiver velocity and r_{bi} shows the bistatic range (Fig. 5.8).

Based on Fig. 5.8, the interval between these events is delivered by:

$$\Delta s = \sqrt{\left(c \cdot \frac{r_{bi}}{c}\right)^2 - \left\| \vec{B} + \vec{v} \cdot \frac{r_{bi}}{c} \right\|^2} \tag{5.20}$$

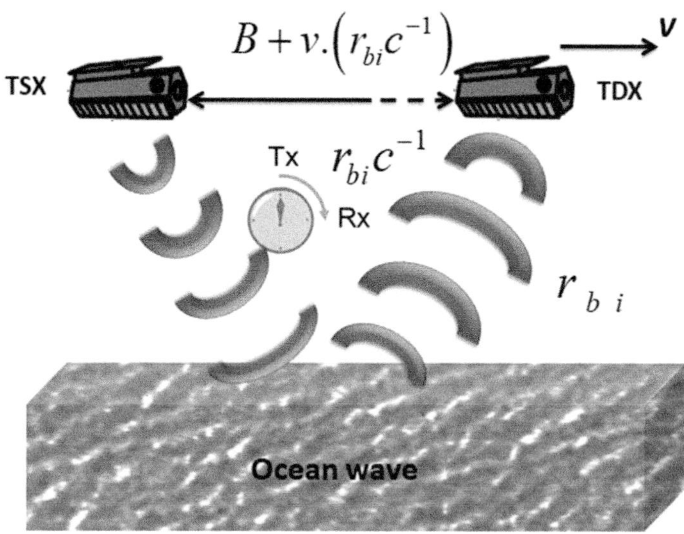

FIG. 5.8 Bistatic SAR data acquisition as seen from the Earth Centred Earth Fixed (ECEF) reference frame.

Both Eqs. (5.19) and (5.20) are equivalent to each other as:

$$\sqrt{(c \cdot \Delta t^2) - \left\|\vec{B}\right\|^2} = \sqrt{\left(c \cdot \frac{r_{bi}}{c}\right)^2 - \left\|\vec{B} + \vec{v} \cdot \frac{r_{bi}}{c}\right\|^2} \tag{5.21}$$

The quantity can be determined from Eq. (5.21) as:

$$r_{bi} \approx c \cdot \Delta t + \vec{B} \cdot \vec{v}\, c^{-1} \tag{5.22}$$

Eq. (5.22) is designated such that the platform velocities are smaller than the speed of light. This equation, therefore, validates that $c \cdot \Delta t$ symbolizes the product of the velocity of light with the time difference between the T_x and R_x occasions as distinguished in the platform frame, where SAR data acquisition and recording are accomplished. The term of $c \cdot \Delta t$ causes such relativistic special effects along the bistatic radar range that apperas as range spin or range ambiguity.

Let us consider that the space-time structure of special relativity $\vec{B} \cdot \vec{v}\, c^{-1}$ arises. On the other hand, $\vec{B} \cdot \vec{v}\, c^{-1}$ is proportional to the scalar product between the platform velocity vector $\vec{v}$ and the baseline vector $\vec{B}$. In other words, it grows with both the along-track baseline between the satellites and the satellite velocity. For instance, a satellite tandem is hovering with a speed of 7.5km/s and an along-track baseline of 1km; the $\vec{B} \cdot \vec{v}\, c^{-1}$amounts to a bistatic range error of 2.5cm. In this regard, this relativistic range offset may announce errors, nonetheless for an interferometric X-band system with a wavelength of 3.1 cm, the phase error is at 290 degrees, owing to the ambiguity level Variations.

Following Krieger and De Zan [11], the procedure of different reference frames for bistatic SAR processing and bistatic radar synchronization is predisposed to the distinguished phase and time errors. Marghany [12], therefore, stated that these inaccuracies are a straightforward concern of the relativity of simultaneity and can be clarified in close approximation within the framework of Einstein's special theory of relativity. Subsequently, expending the invariance of the space-time interval, an analytic countenance results, which illustrates that the time and phase errors increase in proportion with the along-track distance between both satellites (Figs. 5.7 and 5.8). Generally, in special relativity, a coordinate system plugs all of the space-time—for instance, the coordinates carry it in a conventional line endlessly. In general relativity, therefore, the scenario of object representation in the coordinates is much more complicated because of the curvature of space-time.

5.7 How is quantum entanglement consistent with the time relativity?

It is well-recognized that relativity predicts time moves slowly near massive objects, e.g., time moves slowly for clocks on the sea surface compared to clocks on SAR satellites by about 40,000ns. In this understanding, let us consider dual quantum entangled particles separated by a large distance, but with one of them being near a massive ocean body. In this context, time slows down near massive oceans. Consequently, when using quantum mechanics, in which time is impartially preserved as a parameter, to describe ocean particles imagined in the SAR system, we encounter a problem: which clock's time should we use? A critical question that follows on from that is: how can quantum mechanics be implemented in the context of general relativity and special relativity, considering the curvature of space-time, for instance?

It is common to discover that accomplishing one system instantly distresses an entangled partner. In other words, in a SAR system, the receiving signals from the ocean surface can be entangled owing to a variety of sea surface dynamic fluctuation; for instance, wave-wave interaction, and wave-current interaction. How is this reliable with relativity, which explains that there is no absolute time-frame, and hence there is no consistent approach to conclude that one of two spatially separated measures arose prior to the other?

Let us consider a scenario using the previous example in Figs. 5.7 and 5.8 within two pulses of the transmitted pulse $|\psi\rangle$ and reflected or received pulse $|\psi'\rangle$; these are spatially separated and are entangled or they share an entangled resource. In other words, according to the edge, an entanglement is a bond that forms between ocean dynamic surface particles and microwave photon particles in the form of the radar cross-section backscatter. In this regard, the entangled of both states can mathematically be expressed as:

$$|\psi'\rangle = |\psi\rangle \tag{5.23}$$

Then the nature of entanglement, as a function of the state, should also be invariant under this symmetry:

$$E_{|\psi\rangle}(\alpha) = E_{|\psi\rangle}(\alpha') \tag{5.24}$$

Eq. (5.24) reveals that mode transformation, under a symmetry transformation, is that a mode α is relabeled as mode α', existing on the same basis. It can be seen that the entanglement between modes $(1, \omega, k)$ and $(2, \omega, \mathrm{k})$ is equal to the entanglement between each of them and the rest of the whole system [13]:

$$S_{\omega,\vec{k}} = -\sum\nolimits_n \frac{e^{-4n\omega\pi}}{\sum_n e^{-4n\omega\pi}}(n) \ln \sum\nolimits_n \frac{e^{-4n\omega\pi}}{\sum_n e^{-4n\omega\pi}}(n) \tag{5.25}$$

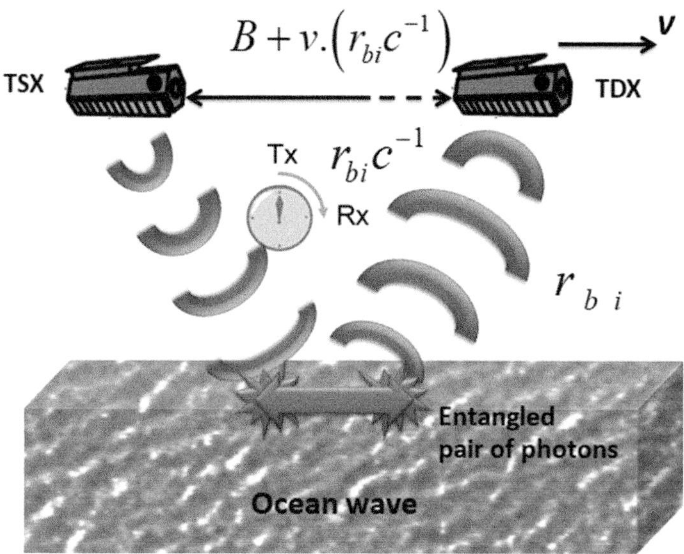

FIG. 5.9 Relativity of entanglement of dual antenna observers.

This is the entanglement between two parts of the space-time. It is crucial to understand that, as long as the transmitted pulse and reflected pulse do not in some way interact or share information, their being entangled is entirely irrelevant. To absolutely impact, providing the transmitted pulse does not in any way interact or share information with a reflected pulse, which can be stated as a predictable mixture of interaction between the transmitted signal and radar backscatter. In other words, the transmitted signal and received signal can be twisted in each other. In this view, no transmitted pulse can work out its physical characteristics, which can develop the fact that there is an entangled party somewhere else in the SAR system, and same for a reflected pulse would be entangled with its physical environment [14].

This means that a transmitted pulse and reflected pulse can somehow exploit their shared entangled resource, which cannot occur faster than light speed, i.e., involving a time delay. In this understanding, the reflected pulse is affected by the interaction behavior of the transmitted pulse with the ocean surface.

To put it differently, to exploit the entanglement, the dual parties have to be connected by a space-time line following the usual rules of general relativity; in particular, it must be less than the speed of light (Fig. 5.9). Prior to entanglement, there must be some deep sense in which it is not meant to say that for instance, the SAR transmitted signal and ocean surface backscatter are entangled in all [15]. In general relativity, a transmitted pulse describes the path taken by a microwave photon emerging from a single point in time and space; along with all paths taken by the pulse it reaches the point as the radar illumination occurs. The antenna shape emerges since all transmitted pulses must reach the same distance from the point at a given time, no matter its direction. The boundary of the antenna arises because the speed of light is the highest conceivable velocity through the entire universe [13–15].

5.8 SAR time dilation

Marghany [12] stated that time dilation is one point of view of relativity theory. In this sense, time dilation is a division in the intervened period restrained through dual SAR sensors, either due to a speed distinction relative to each other or through being otherwise located relative to a gravitational field. According to the description of space-time, a time that is fluctuating relative to a SAR receiver can be measured to impulse slower than a time that is on relaxation in the antenna's frame of reference, where the signal is transmitted. In this understanding, the time of the received signal is slower than the transmitted signal owing to the superior influence of the gravitational field. This is recognized as a delay time and symbolized as τ (s). The scientific explanation of the time interval between two pulses $\Delta\tau$ in that SAR frame can be mathematically formulated as:

$$\Delta\tau = -\Delta t\sqrt{1-\frac{v^2}{c^2}} \tag{5.26}$$

Eq. (5.26) reveals the time delay which is a function of SAR beam velocity v, the speed of light c, and time dilation effect Δt. In this view, the space shuttle and SAR timers are barely slower than reference clocks on Earth, or clocks on GPS and Galileo satellites running very slightly faster [12]. In other words, for GPS satellites, for instance, to operate, they should adjust for analogous bending of space-time to the Earth coordinate systems [16].

5.9 SAR length contraction in polarized data

The SAR sensor transmits a longitudinal electromagnetic wave. It is possible to transmit the longitudinal wave in a single plane (polarization). Commonly, two polarizations are used: horizontal (H) and vertical (V) (Fig. 5.10).

Polarization has tremendous effects on quantum radar cross-section σ_Q in circumstances where atomic dipole moments are randomly oriented in relation to each other. Consequently,

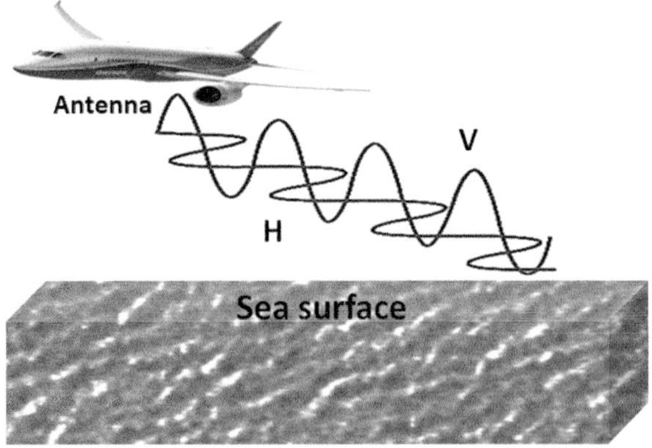

FIG. 5.10 Dual polarization H and V in SAR antenna.

strong Bragg scattering is enveloped in a strong uniform electric field and the atomic dipoles are oriented in the same direction, which leads to the effect of signal photon polarization on σ_Q. For instance, if the polarization vector is parallel to the dipole moment vector, their dot product is equal to 1. In contrast, the dot product equals 0 if the polarization vector is perpendicular to the dipole moment vector, which causes a skewed response. Thus the equation of the polarization vector, in terms of the incidence angle, is as follows:

$$\mathrm{e}_K = \left(\sin\left(\theta_i + \frac{\pi}{2}\right)\cos\phi_i,\ \sin\left(\theta_i + \frac{\pi}{2}\right)\sin\phi_i,\ \cos\left(\theta_i + \frac{\pi}{2}\right)\right) \tag{5.27}$$

The impact of the polarization vector e_K can be involved in the quantum radar cross-section equation as follows:

$$\sigma_Q = 4\pi\left(\frac{\sigma_w}{|g_{ij}|^2\left(16\pi k_r^4\cos^4\theta\right)}\right) \lim_{R\to\infty} \frac{|(\mathrm{d}\cdot\mathrm{e}_K)|^2\left|\sum_{i=1}^{N} e^{\left(j\omega\Delta R_i c^{-2}\right)}\right|^2}{\int_0^{2\pi}\int_0^{0.5\pi} |(\mathrm{d}\cdot\mathrm{e}_K)|^2\left|\sum_{i=1}^{N} e^{\left(j\omega\Delta R_i' c^{-2}\right)}\right|^2 \sin\theta_d d\theta_d d\phi_d} \tag{5.28}$$

The dipole vector d is defined as:

$$\mathrm{d} = d(\sin\varsigma\cos\delta,\ \sin\varsigma\sin\delta,\ \cos\varsigma) \tag{5.29}$$

The angles θ and φ describe the direction of polarization, i.e., incidence angles, of the photon in mode **k**, and the angles ζ and δ delineate the direction of the dipole moment vectors. It is worth mentioning that the direction of the polarization angle changes as well as the incidence angle. In this view, a change of the interaction of polarization vector and the atomic dipole moments occurs. In fact, the correlation between d and e_K is in the form of the dot product, i.e., $|(\mathrm{d}\cdot\mathrm{e}_K)|^2$. The direction of every constituent of the polarization vector can be articulated in terms of being offset from the direction of incidence [16–18].

On the other hand, the polarization of the backscatter pulses can be received by the SAR antenna. Frequently, the highest scatterers reflect the wave in the equivalent polarization (copolarized: HH, VV). Some of the pulses, nonetheless, perhaps backscatter in a disparate plane (cross-polarized: HV, VH). In these observations, HH indicates a horizontal transmitted pulse and horizontal received pulse (Fig. 5.11), while HV stands for horizontal transmit and vertical receive (Fig. 5.12). However, VH involves vertical transmit and horizontal receive (Fig. 5.13), while VV presents vertical transmit and vertical receive (Fig. 5.14).

Both the copolarized signal and cross-polarized signal have diverse physical characteristics. The copolarized signal is usually strong and triggers specular, surface, and volume scattering. The cross-polarized signal, on the other hand, is typically frail, allied with manifold scattering, and has strong associations with the coordination of scatters [19]. Accordingly, the polarized signal commonly depends on incidence angle owing to the strong overtone with SAR geometry. In this view, VV polarization is extensively depleted for investigation of the capillary waves on the sea surface while cross-polarization is valuable when volume

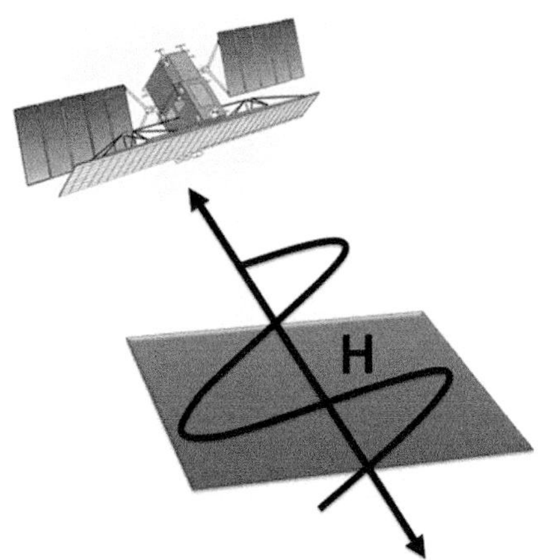

FIG. 5.11 HH polarization.

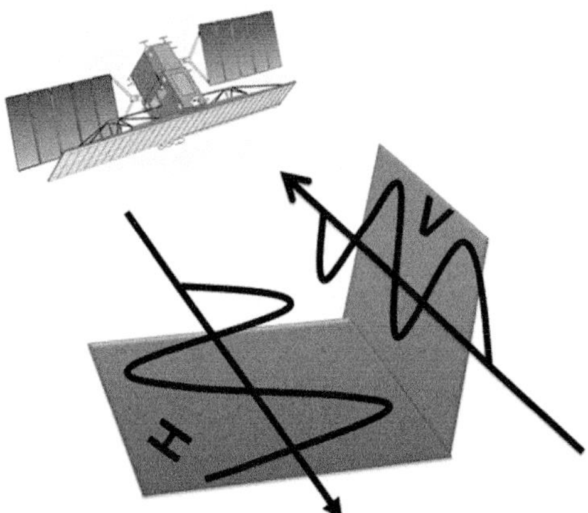

FIG. 5.12 HV polarization.

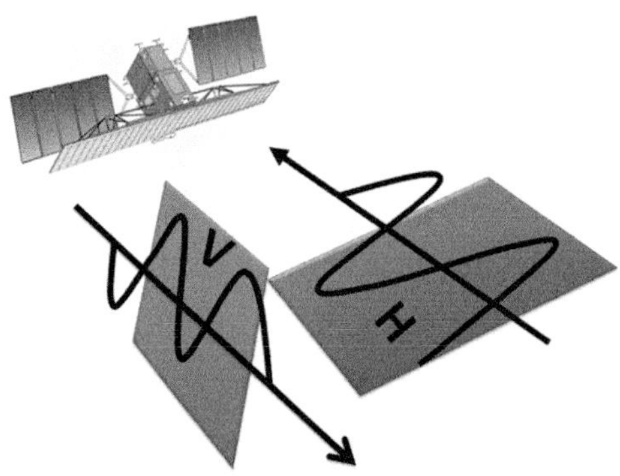

FIG. 5.13 VH polarization.

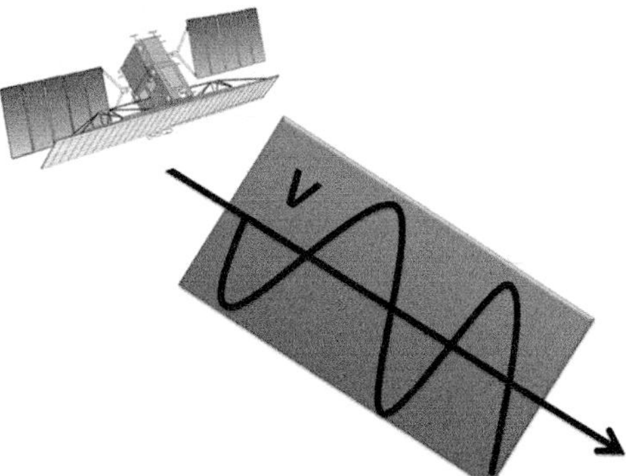

FIG. 5.14 VV polarization.

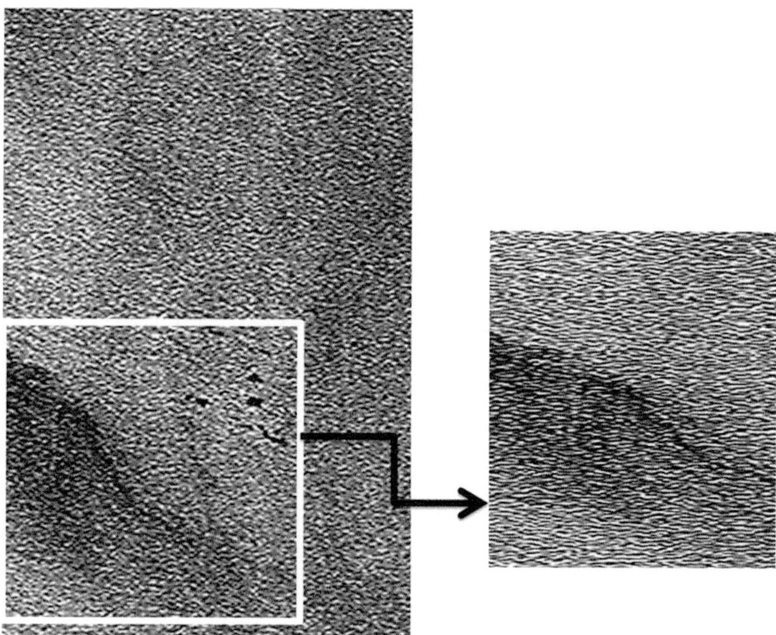

FIG. 5.15 Swell wavelength in VV polarization.

(multiple) scattering occurs. In this circumstance, cross-polarization can be used easily to distinguish between soil and vegetation of forest and nonforest [20].

Consistent with the above perspective, VV polarization provides better results than the other polarizations imaging swell wavelength (ocean wave) than VH, HV, and HH polarizations. From the point of view of length contraction, swell wavelength is imagined with different sorts of polarization. Fig. 5.15 demonstrates that the swell wavelength is clearer in VV polarization than quadrature-polarimetric (quad-pol) (Fig. 5.16) Quad-pol SAR data allow the measurement of the 2×2 complex scattering matrix S, which is usually built up

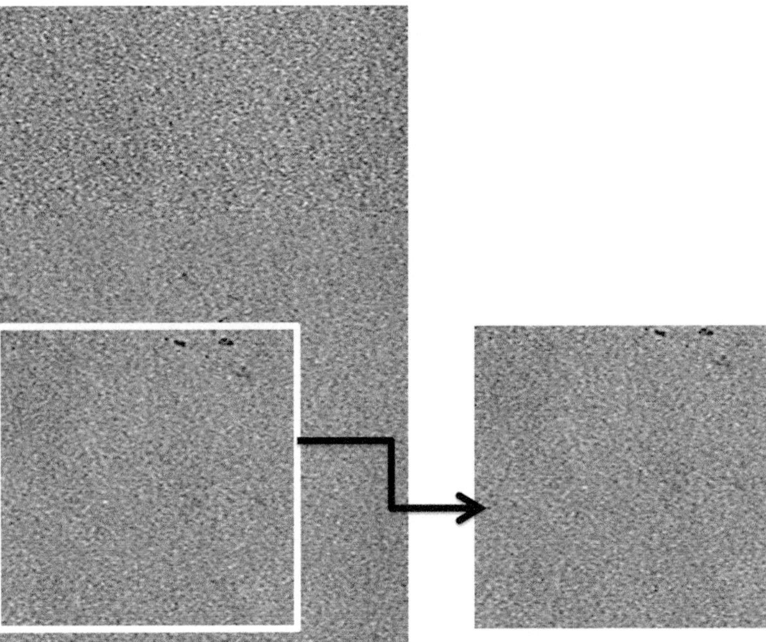

FIG. 5.16 Swell wavelength in quadrature-polarimetric.

by operating the radar with the interleaved transmission of alternate H- and V-polarized pulses and simultaneous reception of both H and V polarizations (conventional quad-pol SAR). When investigating the small-scale roughness of (capillary) waves on the water surface, VV is better than HH or cross-polarized combinations, which indicates suitability for extensive use for surface wind speed extraction. It is clear that length contraction takes place in the quad-pol SAR data (Fig. 5.16) as the swell wavelength is totally constricted.

The length contraction appears obviously in quad-pol SAR data owing to its narrow swath width. This restricts the application of routine retrieving ocean wave spectra, causing further obscurity in retrieving large-scale ocean surface features. According to this understanding, the swell wavelength contraction is considered as a sort of optical illusion relying on the sensor. It is merely intended to accentuate that length contraction seems different in diverse SAR polarization frames.

In other words, the swell wavelength in a moving SAR frame will look foreshortened or contracted in the direction of motion. This contraction (also known as Lorentz contraction or Lorentz-FitzGerald contraction after Hendrik Lorentz and George Francis FitzGerald) is frequently merely obvious at an extensive fraction of the light speed [21–23]. Length contraction, therefore, is merely in the direction in which the swell is traveling: $2\pi k^{-1}$. In general, the swell wavelength contraction only occurs along the line of SAR motion, i.e., the azimuth direction, which is mathematically formulated as:

$$L' = \frac{2\pi k^{-1}}{\gamma(v)} \tag{5.30}$$

where $\gamma(v)$ is the Lorentz factors, which are the factors by which time, length, and relativistic mass change for an object while that object is moving. Additionally, an inertial SAR reference frame in relative motion with the moving frame of the dynamic sea-state measures a period dilated by the γ Lorentz factor. If one wishes to compute the time dilation in the oscillator period of sea surface dynamic as imaged by the SAR moving frame, the γ factor must be considered. Swell propagations $2\pi k^{-1}$ in SAR images appear to undergo the length contraction L' since they are imagined in space-time cross-section. In this view, if the imagined target in SAR data experiences deformation, target form geometry merely the one decoded in SAR signals. In this circumstance, the deformation SAR imagined target actually appears to be rotated and stretched larger by the Lorentz boost in the space-time diagram, and it is merely post a nonperpendicular cross-section is deliberated that this appears from γ down to $\gamma^{-1}$is corrected. The Euclidean distance should be $\gamma\sqrt{1+v^2c^{-2}}$, which directs the target in an azimuth-coordinate to stretch out by a factor of γ [22–25].

References

[1] Bjorken JD, Drell SD. Relativistic quantum mechanics. McGraw-Hill; 1965.

[2] Greiner W. Relativistic quantum mechanics. Berlin: Springer; 1990.

[3] Wang BC. Digital signal processing techniques and applications in radar image processing. John Wiley & Sons; 2008.

[4] Mandelstam LI. Lectures on optics, relativity theory and quantum mechanics. Moscow: Nauka; 1972 440 pp. Under the revision by corresponding member of the Academy of Sciences of USSRSM Rytov.(Russian Title: Lektsii po optike, teorii otnositel'nosti i kvantovoi mehanike). 1971 Dec.

[5] Rindler W. Introduction to special relativity. 2nd ed. 1991.

[6] Baierlein R. Newton to Einstein: The trail of light: An excursion to the wave-particle duality and the special theory of relativity. Cambridge University Press; 1992.

[7] Elbaz C. Wave-particle duality in Einstein-de Broglie programs. J Mod Phys 2014;5(18):2192.

[8] Gjessing DT, Hjelmstad J, Lund T. Directional ocean wave spectra as observed with multifrequency continuous wave radar. Int J Remote Sens 1985;6(7):979–1008.

[9] Bondi H. Assumption and myth in physical theory: The tanner lectures delivered at Cambridge in November 1965. CUP Archive; 1967.

[10] Dolby CE, Gull SF. On radar time and the twin "paradox" Am J Phys 2001;69(12):1257–61.

[11] Krieger G, De Zan F. Relativistic effects in bistatic SAR processing and system synchronization, In: EUSAR 2012; 9th European conference on synthetic aperture radar 2012 Apr 23VDE; 2012. p. 231–4.

[12] Marghany M. Automatic detection algorithms of oil spill in radar images. CRC Press; 2019.

[13] Shi Y. Entanglement in relativistic quantum field theory. Phys Rev D 2004;70(10):105001.

[14] Li H, Du J. Relativistic invariant quantum entanglement between the spins of moving bodies. Phys Rev A 2003;68 (2):022108.

[15] Datta A. Quantum discord between relatively accelerated observers. Phys Rev A 2009;80(5):052304.

[16] Kumar A, Ghatak AK. Polarization of light with applications in optical fibers. SPIE Press; 2011.

[17] Johnson SG, Joannopoulos JD. Block-iterative frequency-domain methods for Maxwell's equations in a planewave basis. Opt Express 2001;8(3):173–90.

[18] Zhang Z, Satpathy S. Electromagnetic wave propagation in periodic structures: Bloch wave solution of Maxwell's equations. Phys Rev Lett 1990;65(21):2650.

[19] Chatziioannou K, Yunes N, Cornish N. Model-independent test of general relativity: an extended post-Einsteinian framework with complete polarization content. Phys Rev D 2012;86(2):022004.

[20] Da-hai YA, De-bao MA. Refined polarimetric SAR speckle lee filtering algorithm based on relativity of polarization vector. J Inf Eng Univ 2010;3(6):20.

[21] Brown HR. The origins of length contraction: I. The FitzGerald–Lorentz deformation hypothesis. Am J Phys 2001;69(10):1044–54.

[22] Sastry GP. Is length contraction really paradoxical? Am J Phys 1987;55(10):943–6.
[23] Goldreich D, Tong J. Prediction, postdiction, and perceptual length contraction: a Bayesian low-speed prior captures the cutaneous rabbit and related illusions. Front Psychol 2013;4:221.
[24] Redžić DV. Towards disentangling the meaning of relativistic length contraction. Eur J Phys 2008;29(2):191.
[25] Ashby N. Relativity in the global positioning system. Living Rev Relativ 2003;6(1):1.

CHAPTER

6

Novel relativistic theories of ocean wave nonlinearity imagine mechanism in synthetic aperture radar

It was demonstrated in Chapter 1 that the nonlinearity is a term exploited in statistics to designate a state where there is not a straight-line or direct relationship between ocean wave parameters and a SAR backscattered signal. In a nonlinear relationship, therefore, the simulation of wave spectra does not change in direct proportion to a change in any of the SAR transmitted and backscattered pulses.

As noted in previous chapters, it is well-known that "radar" stands for RAdio Detection And Ranging. Ranging is the distance from the radar where the microwave photons travel across space and time and interact with the objects. In this regard, space-time is the keystone of relativity theory. The main question that can be raised is whether SAR imagine mechanism of ocean surface wave can be elucidated by relativity theories. This chapter is devoted to finding an accurate correspondence between ocean wave pattern appearance in SAR and relativity theory.

6.1 What is meant by waves and flows?

It is challenging to distinguish between waves and flows in nature. Many significant questions are raised: can the ocean dynamic be described from the wave-duality speculation? This chapter attempts to exploit the wave-duality concept in-depth to contribute to an understanding of the dynamics of ocean waves and ocean flows. Water consists of particles; if they receive external or internal energy forces, they oscillate vertically and horizontally, creating a wave-like or flow-like effect, respectively. In this regard, it is a great challenge to articulate an inclusive description of a wave or flow. In different circumstances, waves arise in the diversity of configurations, such as tidal waves, wind waves, or swell. Subsequently, deep water waves do not comprise of net movement. However, in the swash zone along the beach, a great displacement takes place [1–3].

https://doi.org/10.1016/B978-0-12-820785-7.00009-5

Likewise, tidal waves can instigate bores that transfer up rivers for extensive dimensions. In other words, At certain times of the year, the lower tidal reaches of the trent experience a moderately large bore (up to five feet (1.5 m) high). The Aegir occurs when a high spring tide meets the downstream flow of the river. In this understanding, it is complicated to distinguish between waves and flows.

6.2 Description of ocean waves

For wave-duality, the ocean surface is composed of countless particles that are oscillated randomly in space-time, forming a wave-like pattern. These particles spin and are propagated in random but well-organized arrays which the space-time separates in wavelength configuration.

Waves are spread out—for instance, waves are the large breakers on the open ocean, or ripples in a pond. If at one moment the wave is localized, sometime later it will have spread out over a large region, like the ripples when a pebble is dropped in a pond. The wave carries with it energy related to its motion. Unlike the particle, the energy is distributed over space because the wave is spread out [3].

Colliding particles will bounce off each other, but colliding waves pass through one another and emerge unchanged. However, overlapping waves can interfere—for example, where a trough overlaps a crest, the wave can disappear altogether (Fig. 6.1). The waves spread out in all directions and interfere with each other, leading to regions in space where the wave disappears and regions where it becomes stronger.

Somehow, those parts of the wave distant from the point of interaction "know" that the energy has been lost and disappear instantaneously. If this happens to ocean waves, a hypothetical surfer on the wave would receive all the energy, and at that moment the ocean wave would disappear, all along the length of the beach. One surfer would be shooting along the surface of the water while all the other surfers would be sitting becalmed on the surface.

Ocean surface waves may be categorized by their period, which is the time reserved by successive wave crests to pass a fixed location. Fig. 6.2 shows the organization of ocean waves

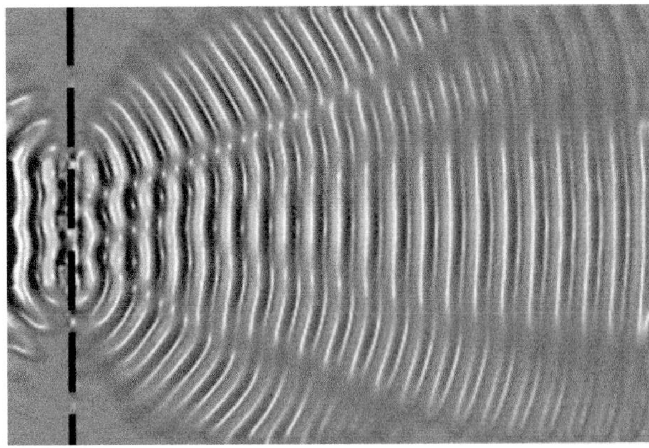

FIG. 6.1 Overlapping wave propagations.

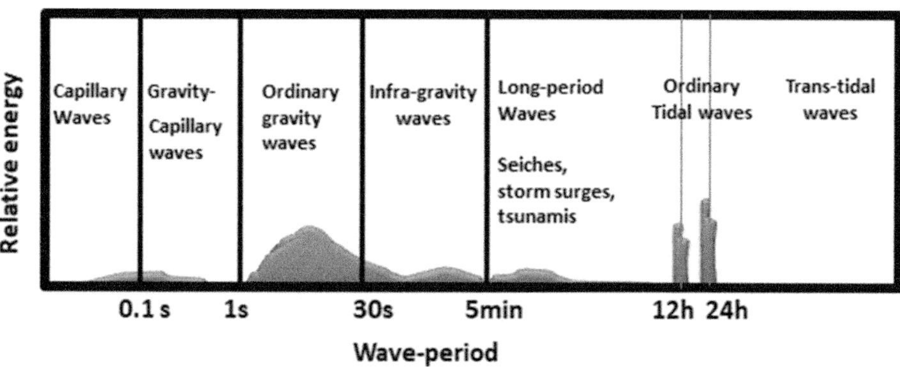

FIG. 6.2 Classification of ocean surface waves.

FIG. 6.3 Wind waves.

by the wave period. This chapter is concerned with remote sensing of ocean surface gravity waves having periods between 3 and 30 s. The gravity waves are well-known in two circumstances: wind waves, when the waves are being generated by the wind (Fig. 6.3), and swell waves, when the waves have traveled away and escaped the influence of the generating wind. In other words, they are the series of waves are generated at a distance; well-sorted according to their wavelengths, heights, and periods (Fig. 6.4).

The ocean surface seems to be composed of random waves with various heights, lengths, and directions. The velocity of idealized traveling waves on the ocean is wavelength λ reliant on wave period and group velocity. Therefore, for shallow adequate depths, it is also contingent upon the depth of the water d. In this regard, the wave speed v can be given by:

$$v = \sqrt{\frac{g\lambda}{2\pi}\tanh\left(2\pi\frac{d}{\lambda}\right)} \tag{6.1}$$

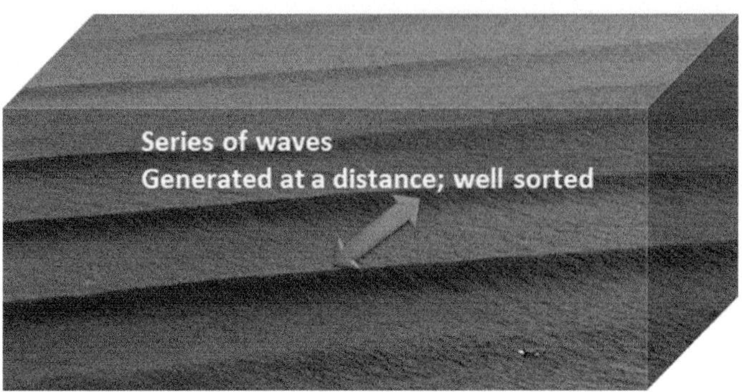

FIG. 6.4 Swell waves.

where g is the acceleration of gravity, which equals $9.81\,\mathrm{m/s^2}$. The celerity presents the velocity of the progressing wave concerning stationary water, meaning that any current motion or other net water velocities would be added to it. Ocean waves obey the basic wave relationship [1, 3, 4]:

$$v_g = f\lambda = T^{-1}\lambda \tag{6.2}$$

Eq. (6.2) reveals the motion relationship of "distance = velocity × time," which is key to the fundamental wave relationship. This equation is a general wave relationship that applies to sound and light waves, other electromagnetic waves, and waves in mechanical media. With the wavelength as distance, this relationship becomes $\lambda = vT$ where the wave frequency equals the inverse of the period, i.e., delivering the standard wave relationship [5, 6]. A novel investigation of wind-wave generation is: ocean wave patterns are considered as the hidden wavy wind energy. In other words, wind blowing is just a form of hidden wave propagation. It is clearly seen when the calm ocean taking a form of a wave. The evidence is when the wind blows on the plain sand, it turns into the shape of wavy sands.

6.3 How sea waves are formed based on Spooky Action at a Distance

Is it possible to be used as a novel theory to understand how the sea wave is formed under the impact of the wind? Let us consider wind as particles fluctuating in space-time; they have angular momentum that allows these wind particles to spin in all directions. As long the spinning particles of wind stress the sea surface, the wind angular momentum acts on the surface water particles causing wave propagation in space-time [7]. In this view, spin is built into wind particles; which involve gaseous mixtures such as oxygen, carbon dioxide, and nitrogen. In this understanding, the novel approach to address the wind flow is the upward and downward spinning of the gaseous mixtures (atoms) in the air are causing horizontal flow due to the differences in atmospheric pressure over large-scale. In this circumstance, the downward wind spanned particles have down stress on the water particles while the upward wind spanned particles have lower upward stress on the water particles. In other words, the wind particles and sea surface particles are entangled, causing them to affect one another instantaneously no matter how far apart they are. In this regard, the wind-wave generation is simply a phenomenon of entanglement which occurs when a pair of particles, such as wind-wave, interact physically.

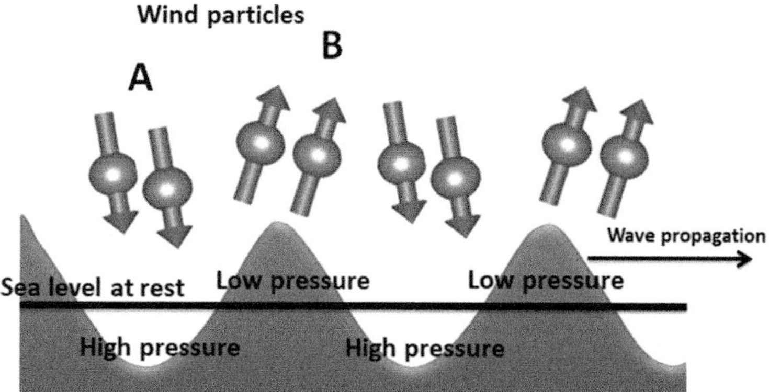

FIG. 6.5 Wind-wave generation based on "Spooky Action at a Distance" theory.

According to this understanding, wind particles blowing across the sea can cause individual wind particles to be split into pairs of entangled wind particles, which are separated by a certain distance [8]. When observed, wind particle A takes on the downspin state, while entangled wind particle B, though now far away, takes a state relative to that of particle A, i.e., it has an upspin state (Fig. 6.5). The oscillation of the sea particles are correlated with upward and downward spinning of the wind particles, creating wave propagation in space-time. This comprehends investigation of how can implement quantum entanglement to address a new concept of wind-wave generation is novel proof of "Spooky Action at a Distance" theory in the real world.

The sea surface applies the gravitational force on the bottom layer of the wind as long as the wind is blowing on the ocean surface. This, in sequence, retains the attraction on the layer above it until it influences the highest layer. In this view, the wind particles travel at a variable rate at each layer with fluctuations in gravitational attraction. According to these mechanics, a circular wind particle movement is formed at the peak layer, which forms a downward pressure at the front and upward pressure at the rear of the surface, causing a wave [7–9]. There are tidal waves that are twisted by the gravitational pull of the sun and moon on the earth. It should be noted that a tidal wave is a shallow water wave, not a tsunami.

6.4 What is doing the waving?

The regular wave composes a contour which is revealed by a traced function. A trochoid (from the Greek word for wheel, *trochos*) is a roulette twisted by a circle rolling along a line. In other words, trochoid shape is the curve drawn out by a point patch up a circle, where the point perhaps is on, inside, or outside the circle as it spins along a straight line. If the point is on the circle, the trochoid is termed a cycloid. If the point is inside the circle, the trochoid is curtate, and if the point is outside the circle, the trochoid is prolate.

The trochoid shape may be traced to the sine curve for small amplitudes, although the shape is different than sinusoid, with a narrowing of the peaks of the trochoid. In this view, the narrowing or steepening of the peak becomes more noticeable as the wave amplitude increases. Fig. 6.6 demonstrates that the trochoid curve is traced out from point A as the outer circle rolls along the underside of line B. In other words, the ocean as shown in Fig. 6.7 reveals that point A traces a sine curve as the moving circle rotates with constant angular velocity

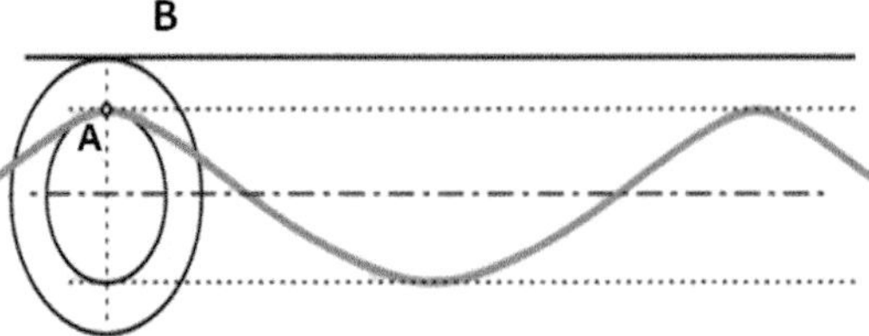

FIG. 6.6 Example of the trochoid curve.

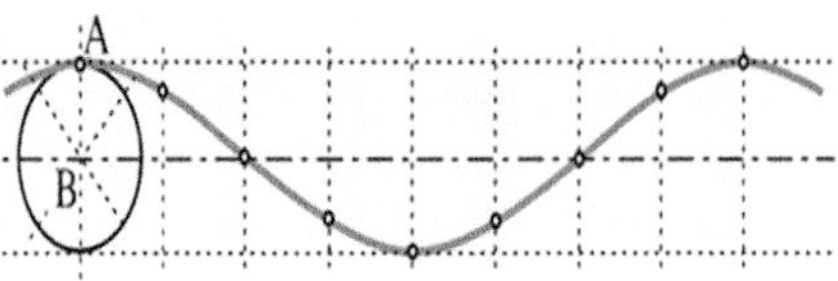

FIG. 6.7 Trochoid curve tracing out a sine wave.

about B. Both figures prove that the spinning of the water particle in trochoid form perhaps transfers through space-time based on the kinetic motion involved in these particles [2–6, 10].

Consistent with the above perspective, the water particles would accomplish a circular motion as a wave passed without significant net advance in their position. The motion of the water is forward as the peak of the wave passes, but backward as the trough of the wave passes, reaching the same position again when the next peak arrives (Fig. 6.8).

The trochoidal wave can be described mathematically by:

$$x = \frac{L_w}{2\pi}\left[\widehat{\alpha} - \pi\frac{H_w}{L_w}\cdot\sin\alpha\right] \tag{6.3}$$

where:

$$\xi = \frac{H_w}{2}\cdot\cos\alpha \tag{6.3.1}$$

Eq. (6.3) reveals the parameters of a trochoidal wave involving the wave steepness $\frac{H_w}{L_w}$ as the ratio of wave height and wavelength and the relative water depth $\frac{d}{L_w}$. For a large water depth from $\frac{d}{L_w}$ between 1 and 0.5, the wave contour by Stokes to the third order is given by:

$$\xi(x,t) = 0.5H_w\cos(kx-\omega t) + \frac{\pi}{4}\cdot\frac{H_w^2}{L_w}\cos(kx-\omega t) + \frac{3\pi^2}{16}\cdot\frac{H_w^3}{L_w^2}\cos(kx-\omega t) \tag{6.4}$$

Eq. (6.4) demonstrates that the second-order constituent upturns the wave crest and raises the level in the wave trough, where ξ is the wave ordinate expressing surface elevation.

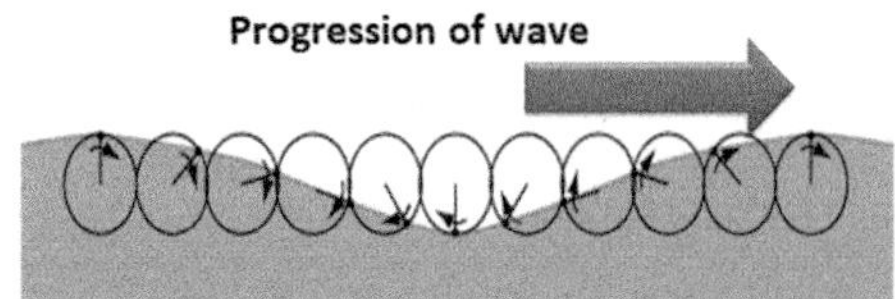

FIG. 6.8 Progression of the wave due to water particle rotations.

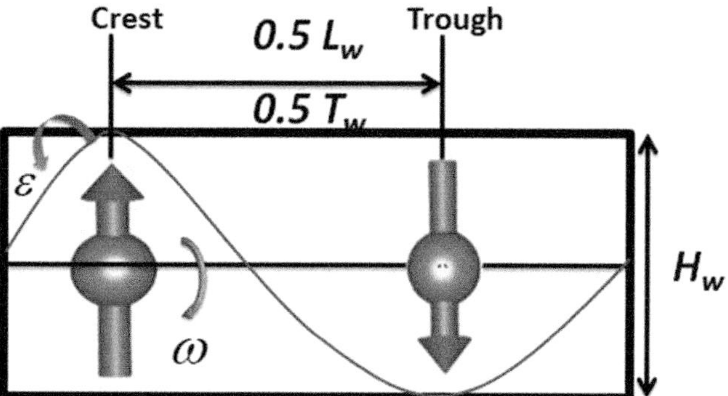

FIG. 6.9 Entanglement between wave particles to generate wave propagation.

The third-order constituent equals 0.5mm. The wave crest shows a sharper peak than the sinusoidal wave, and the wave trough appears wide and flat, which is close to reality. Moreover, the maximum wave steepness of 0.142 or 1/7 occurs in deep water. In other words, wave steepness cannot be higher than 1/7 of the wavelength. Conversely, the wave breaks as long its steepness reaches 1/7 [2, 3, 10, 11].

At the water surface, the spinning motion of the water particles has its maximum. The spinning radius r subsides exponentially as the vertical distance from the surface increases.

$$r(z) = 0.5 \cdot H_w e^{(kz)} \quad \text{for} \quad z \leq 0 \tag{6.5}$$

Eq. (6.5) demonstrates that at the water surface, the radius is equal to half the wave height, $0.5 \cdot H_w$. At $0.5 \cdot \lambda_w$ below the surface, the spinner radius of the wave motion has been reduced to 4% of the radius of the sea surface. Subsequently, the celerity of wave propagation is shown from left to right. In this understanding, wave particles spin clockwise with the negative angular velocity ω. In contrast, the angle of the spinner position ε is counter-clockwise, i.e., it is mathematically positive (Fig. 6.9).

Entanglement occurs when the water particles spin up and down to form the crest and trough, respectively. It can additionally be seen that the entanglement also occurs between ε and ω. Entanglement occurs between wind particles, contributing to wave generation. It also happens between wave particles that are spinning up and down, contributing to wave propagation and generation.

6.5 Hamiltonian formula for nonlinear wave description

The Hamiltonian formula can be used to describe wave propagation across space-time. Let us consider the spectral model of the sea surface—that is, a superposition of uncertainly interrelating major waves, whose directional wave spectrum is decided by a proper radiative transport formula:

$$\frac{\partial N\left(\vec{k}\right)}{\partial t} = S_{in} + S_{nl} + S_{dis} \tag{6.6}$$

Eq. (6.6) involves the source term for wave generation by wind or other causes S_{in}, the source term for nonlinear wave-wave interactions S_{nl}, and dissipation processes S_{dis}. The nonlinearity allied with wave-wave interactions is particularly relevant to SAR, where gravity waves are the main factors that influence the backscattering process.

A linear input source term is required for the consistent spin-up from the initial stage of wind-wave generation to an initial state of the wave growth development. Hence the input term S_{in} is mathematically formulated as:

$$S_{in}(k,\theta)=80\left(\rho_a\rho_w^{-1}\right)^2\frac{1}{g^2k}\max\left[0,u,\cos\left(\theta-\theta_w\right)\right]^4 e^{\left[-\left(\frac{\frac{f}{g}}{28u}\right)^{-4}\right]} \tag{6.6.1}$$

The last term $e^{\left[-\left(\frac{\frac{f}{g}}{28u}\right)^{-4}\right]}$ in Eq. (6.6.1) presents the Tolman filter with filter frequency f which equals the Pierson-Moskowitz frequency (P-M), i.e., $f_{PM}=g\cdot(28u)^{-1}$ [10]. ρ_a and ρ_w are the air and water densities, respectively. In addition, θ is the wave spectra direction, while θ_w is the mean wind direction and u is the wind velocity over the sea surface. The energy spectrum $E(f)$, which is based on Pierson-Moskowitz (P-M), is given by:

$$E(f_{PM})=\frac{8.10\times10^{-3}g^2}{(2\pi)^4f_{PM}^5}\exp\left[-0.032\left(\frac{g/H_s}{(2\pi f_{PM})^2}\right)^2\right] \tag{6.6.1.1}$$

where H_s is the significant wave height. Consequently, the source term for nonlinear wave-wave interactions S_{nl} is mainly calculated based on discrete interaction approximation (DIA). In this regard, let us assume four-wave quadruplets with wavenumber $\vec{k}_1$ through $\vec{k}_4$ as resonant nonlinear interactions occur; in DIA, it is considered that $\vec{k}_1=\vec{k}_2$. In this understanding, S_{nl} can be formulated as:

$$\begin{pmatrix}\partial S_{nl,1}\\ \partial S_{nl,3}\\ \partial S_{nl,4}\end{pmatrix}=D\begin{pmatrix}-2\\ 1\\ 1\end{pmatrix}\frac{C}{g^4}f_{r,1}^{11}\left[F_1^2\left(F_3(1+\mu_{nl})^{-1}+F_4(1-\mu_{nl})^{-1}\right)-2F_1F_3F_4\left(1-\mu_{nl}^2\right)^4\right] \tag{6.6.2}$$

where $F_1=F(f_{r,1},\theta_1)$ presents the discrete frequency and $\partial S_{nl,1}=\partial S_{nl}(f_{r,1},\theta_1)$ etc., C, and μ_{nl} are constants. In this regard, Hasselmann et al. [11] investigated that $0<\mu_{nl}<1$, where C is 2.8×10^7 if μ_{nl} is 0.25. D is the shallow water scaling factor that equals:

$$D=1+5.5\left(\vec{k}d\right)^{-1}\left[1-\frac{5}{6}\vec{k}d\right]e^{-1.25\vec{k}d} \tag{6.6.2.1}$$

Dissipation processes S_{dis} occur due to wave breaking as the breaking of all waves arises in the random field when they exceed a threshold height. In other words, when the wave amplitude reaches the point that the crest of the wave actually overturns, the wave breaking occurs. Therefore, S_{dis} can be given as:

$$S_{dis}(k,\theta) = -0.25 Q_b f_m \frac{H_{\max}^2}{E} F(k,\theta) \tag{6.6.3}$$

The total spectra energy is presented by E, while the mean frequency is f_m. When the wave height reaches its maximum $H_{\max}$, the wave breaks down and the fraction of breaking waves Q_b is expressed as:

$$-Q_b = 1 - \left[H_{rms} (H_{\max})^{-1} (-\ln Q_b) \right] \tag{6.6.3.1}$$

The breaking wave height root mean square H_{rms} is related to the water depth at breaking, h_b, through the following relation:

$$H_{rms} = \gamma h_b \tag{6.6.3.1.1}$$

or:

$$H_{rms} = \sqrt{\frac{\rho_w g H^2}{\rho_w g}} \tag{6.6.3.1.2}$$

here γ is the breaking index, which ranges between 0.35 and 0.5. The fact that wave heights within the surf zone are depth-limited means that wave heights approach a linear function of water depth [12]. In addition, ρ_w is seawater density, and H is the wave height. Therefore, the term $\frac{1}{8}\rho_w g H^2$ represents the wave energy variation. The maximum wave height $H_{\max}$ is then given by:

$$H_{\max} = \overline{H}_{1/3}\sqrt{0.5 \ \ln N} = 1.9\sqrt{2}\sqrt{\frac{\rho_w g H^2}{\rho_w g}} \tag{6.6.3.1.3}$$

The expectation of $H_{\max}$ must be grounded in a realistic duration, e.g., 6 h, in addition to the standard confidence limits of the $H_{1/3}$ expectation. This suggests $N = 2000$–5000 (in 6 h there are approximately 2700 waves if the peak period is 8 s) [2, 4, 10].

The unique consequence of nonlinear interactions is to revise the rapid surface geometry, using the Stokes expansion, for instance, continuous in sharpened wave crests and flattened troughs. This was agreeable to field measurements; for example, the precise quantity of surface slope or the statistical distribution of surface slope delivers an incomplete frame onto this phase of nonlinearity. The gravity wave field as a sum over orders of wave interaction can be described by the Hamiltonian formula $\mathcal{H}$ as:

$$\begin{aligned} \mathcal{H} &= \sum_k \omega_k a_k^+ a_k + \sum_{k,j} V_{k,j} a_k^+ a_j a_{k-j} + \sum_{j,k,l} \Gamma_{j,k,l} a_k^+ a_j a_{k+l-j} + \dots \\ &\equiv \mathcal{H}_2 + \mathcal{H}_3 + \mathcal{H}_4 + \dots, \end{aligned} \tag{6.7}$$

here a_k^+ and a_k are the creation and annihilation operators for the regular approaches $\psi_k = e^{i(kx-\omega_k t)}$, respectively. The state vector $\psi(t)$ is implemented to describe the time evolution of the surface as:

$$\psi(t) = e^{i\mathcal{H}(t-t_0)} \psi(t_0) \tag{6.8}$$

The nonlinearity of sea surface variation is then formulated as:

$$\psi_{Lin}(t) = \sum_k e_k e^{i(kx-\omega_k t)} = \sum_k e_k \psi_k(t) \tag{6.9}$$

Eq. (6.9) is the state consequential from self-governing propagation of the usual sea surface movements, in place of predictability. The accuracy with which the nonlinear dynamics of a system initially in state $\psi(t_0)$ can be measured rest on the capability of the SAR scattering kernel to discriminate between the state space trajectory $\psi(t_0) \xrightarrow{dt} \psi_{obs}(t)$ and the trajectory $\eta_x(x) = \int_{k,\omega} T_{\eta_x}(k,\omega) e^{i(k.x-\omega t)} \sum_k e_k e^{i(kx-\omega_k t)}$ [13].

6.6 SAR image mechanism for ocean wave

Synthetic aperture radars (SARs) deliver two-dimensional records of the radar cross-section $\sigma(x,y)$, where x and y represent the range and the azimuth coordinates, respectively, as demonstrated in Fig. 6.10. The SAR scene is generally designated as a massive gathering of point scatterers. This section briefly describes how and with what resolution the location of a given point scatterer can be retrieved by the radar to create this image. Fig. 6.10 illustrates how the antenna hovers in the positive y-direction, at an altitude h with velocity V. The antenna footprint with azimuth size Y_a is represented by the elliptical shape on the ground. The incidence angle is θ_0 and R signifies the slant-range distance from the antenna to the point scatterer.

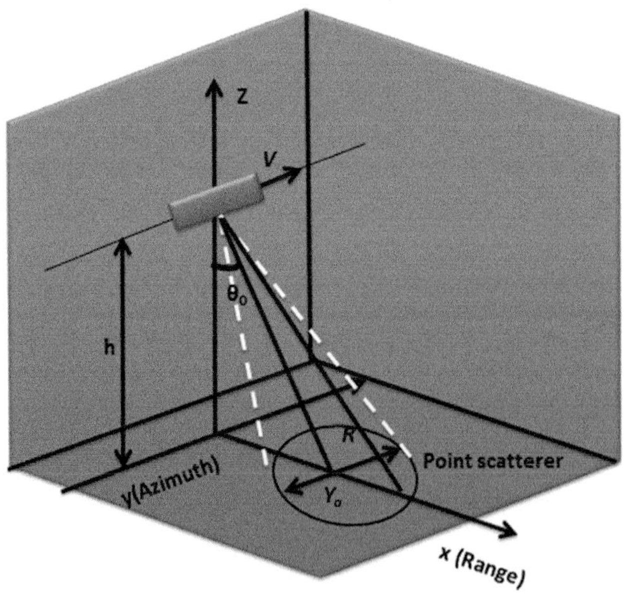

FIG. 6.10 The geometry of SAR.

6.6.1 SAR range processing

In Chapter 5, the relativity of the SAR image as based on the range was discussed. Briefly, the range is generated with a delay τ time, which creates what is known as range relativity. In other words, range ambiguities are considered a feature of SAR relativity (Chapter 5). The range coordinate is regulated by computing the time reserved by the signal to accomplish the round trip between the SAR platform and the ocean surface. The signal conveyed by the radar is a pulse of duration and can be mathematically expressed as:

$$\partial t = \frac{2\partial R}{c} \tag{6.10}$$

Let us consider SAR has a sinusoidal pulse (these are sine or cosine waves); the two echoes must not overlap to resolve the two targets and the pulse length τ has to be such that $\tau < \frac{2\partial R}{c}$ as c is the speed of light. The de Broglie relations state that if a particle has specific values of energy and momentum, which can correlate pulse wavelength (λ), and its momentum (p): $\lambda = hp^{-1}$. Moreover, de Broglie stated that the momentum of a pulse is given by $p = Ec^{-1}$ and the wavelength (in a vacuum) by $\lambda = cv^{-1}$. In this regard, the energy of the SAR pulse can be given by:

$$E_p = P_p \tau \tag{6.11}$$

Eq. (6.11) indicates that the pulse transmitted duration τ can increase the energy of the SAR pulse E_p. However, the instantaneous peak power P_p cannot be amplified on account of hardware restrictions. The minimization of the pulse duration τ and the maximization of the pulse energy are thus mismatched. In this circumstance, implementing the chirp (sending frequency-modulated pulses) can overcome such an issue by using a compression de-ramping algorithm to process the received signal [12]. The slant-range resolution can be given by:

$$\partial R = c(2B)^{-1} \tag{6.12}$$

Eq. (6.12) denotes that the variation in the slant range can inversely correlate with pulse bandwidth, which presents a frequency difference between the transmitter and the receiving of the pulse. The ground slant-range can be computed [13] as:

$$\partial x = \frac{\partial R}{\sin \theta_0} \tag{6.13}$$

Eq. (6.13) signifies that the ground-range resolution is the minimum distance ∂x along the x-axis for dual dissimilar scatterers positioned on the ground to be determined, which is associated with the slant-range resolution through the incidence angle θ_0. The fluctuation in pulse energy ΔE_p leads to energy uncertainty. In other words, if the microwave particles do not remain in the same state indefinitely owing to the change of sea state conditions, then $\Delta t$is finite and nonzero. The energy uncertainty of the unsteady microwave particles has a minimum uncertainty of:

$$\Delta E_p \approx \hbar c(2\Delta tc)^{-1} \tag{6.14}$$

The transmitted pulse would also have energy uncertainty as its peak is known as resonance. The peak has a width that is due to uncertainty in the energy, as indicated in Eq. (6.12).

6.6.2 SAR azimuth distribution processing

The principle of the azimuth imaging for a SAR is based on the change of the Doppler frequency shift from a given scatterer as the platform passes by with a constant velocity. Azimuth compression has two significant constraints: (i) Doppler centroid frequency f_{Dc} and (ii) Doppler rate f_R, which can be expanded from the SAR satellite state vectors. These contain the satellite's location and velocity vectors. Sporadically, these two principles need to be recomputed again specifically, which can be done by improving and managing of the SAR data itself; for instance, shortening the duration of delay pulse. In this perspective, Yigit et al. [14] stated that the Doppler centroid frequency f_{Dc} is exposed to the SAR data by the clutter lock approach. Consequently, the Doppler rate f_R can be modeled by an autofocus method implementing the azimuth subaperture correlation technique. Multiple pulse repetition frequency (PRF) ambiguity of Doppler centroid, which is important for range migration, can be resolved by the range supporter correlation approach. Consistent with Zhu et al. [15], azimuth spectral filtering can be smeared on dual SAR signals, creating an interferometric pair to recompense decorrelation from disparate Doppler centroids of the dual SAR data. The azimuth signal is then compressed using matched filtering.

Let us undertake that the angular carrier frequency ω_c from the range-compressed signal can be eliminated by demodulation. We select the value at $t = t_n$ that compromises maximum $|g_n(t)|$ by particular sampling and representing t as relaxed time s delivers:

$$\hat{g}(s|x_c, R_c) = \exp[-j4\pi R(s)/\lambda]. \tag{6.15}$$

The range function $R(s)$ can be protracted as a Taylor series about $s_c = x_c/V_s$, the relaxed time at which the center of the SAR beam brightness is the target, where V_s is the speed of the SAR platform relative to the target location.

$$R(s) = R_c + \dot{R}_c(s - s_c) + \ddot{R}_c(s - s_c)^2/2 + \cdots. \tag{6.16}$$

The Doppler frequency can be expressed as the time rate of the phase $\varphi(s)$, which is articulated as:

$$\varphi(s) = -j4\pi R(s)/\lambda, \tag{6.17}$$

$$f_D(s) = \dot{\phi}/2\pi = -2\dot{R}(s)/\lambda, \tag{6.18}$$

$$\dot{f}_D(s) = \ddot{\phi}/2\pi = -2\ddot{R}(s)/\lambda. \tag{6.19}$$

The Doppler centroid and Doppler rate can be determined at $s = s_c$ as:

$$f_{Dc} = -2\dot{R}_c/\lambda, \tag{6.20}$$

$$f_R = -2\ddot{R}_c/\lambda, \tag{6.21}$$

Consistent with Eqs. (6.16)–(6.21), one can achieve:

$$R(s) = R_c - (\lambda f_{Dc}/2)(s - s_c) - (\lambda f_R/4)(s - s_c)^2. \tag{6.22}$$

In this sense, the range-compressed signal can be formulated in terminologies of f_{Dc} and f_R as:

$$\hat{g}(s|s_c, R_c) = \exp(-j4\pi R_c/\lambda)\exp\left\{j2\pi\left[f_{Dc}(s - s_c) + f_R(s - s_c)^2/2\right]\right\}, \quad |s - s_c| < S/2, \tag{6.23}$$

where S presents the azimuth integration time, which indicates a linear frequency modulation FM wave with center frequency f_{Dc} and frequency rate f_R. In this circumstance, the azimuth compression is to compute the correlation as:

$$\varsigma(s|s_c, R_c) = \int_{s_c - S/2}^{s_c + S/2} h^{-1}(s' - s|s_c, R_c)\hat{g}(s'|s_c, R_c)ds'. \tag{6.24}$$

Associated with range compression, the azimuth compression can be approximately obtained using a correlator function as:

$$h^{-1}(s|s_c, R_c) = \exp\left[-j2\pi\left(f_{Dc}s + f_R s^2/2\right)\right]. \tag{6.25}$$

Eq. (6.25) can offer the sequence of azimuth compression as:

$$\varsigma(s|\, s_c, R_c) = \exp\left(j2\pi f_{Dc}s\right)\exp\left(-j4\pi R_c/\lambda\right)\cdot \frac{\sin \pi f_R S(s - s_c)}{\pi f_R(s - s_c)} \tag{6.26}$$

Derivation of Eq. (6.26) corresponds to that of range compression. The peak of this pulse occurs at $s = s_c$, the target azimuth locality [14, 15]. Hence, the 3 dB width of this pulse can determine the azimuth time resolution as:

$$\delta s = 1/|f_R|S = 1/B_D \tag{6.27}$$

where $B_D = |f_R|\, S$ is the Doppler bandwidth. The azimuth spatial resolution (or azimuth resolution) is then computed as:

$$\delta x = V_s \delta s = V_s/B_D = V_s/|f_R|S. \tag{6.28}$$

From the simple geometry of a radar antenna, with a physical length L_a along-track, the nominal beam width is $\theta_H = \lambda/L_a$ so that any particular object point at the range R_c is illuminated for a nominal time $S = \lambda R_c/V_s L_a$. Based on a squint angle θ_s (Fig. 6.11), the Doppler parameters can be geometrically clarified as:

$$R^2(s) = R_c^2 + V_s^2(s - s_c)^2 - 2R_c V_s(s - s_c)\sin\theta_s \tag{6.29}$$

$$R(s) \approx R_c + V_s^2(s - s_c)^2/2R_c - V_s(s - s_c)\sin\theta_s \tag{6.30}$$

$$f_{Dc} = (2V_s/\lambda)\sin\theta_s, \tag{6.31}$$

$$f_R = -2V_s^2/\lambda R_c. \tag{6.32}$$

In the context of a real aperture radar, azimuth resolution exchanges in reverse with its antenna physical length as $\delta x = \lambda R_c/L_a$. In this view, the Doppler bandwidth of SAR is hence $B_D = 2V_s/L_a$, and the system azimuth resolution is formulated as:

$$\delta x = L_a/2. \tag{6.33}$$

Consistent with Marghany [16], δx changes in proportion to the physical length of SAR antenna. With intensive signal compression processing, high azimuth resolution can be achieved with a SAR in a significantly smaller physical measurement than that in a real aperture radar.

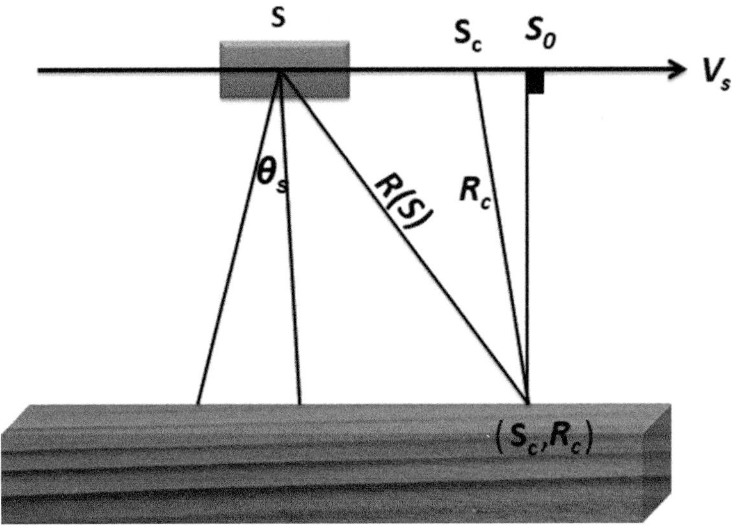

FIG. 6.11 Doppler geometry parameters.

The misplacing of moving scatterers on the image is of primary importance to the synthetic aperture radar technique. Let us assume that u_r presents a moving target with slant-range velocity, taken positively from the ground to the radar. The azimuth position of the latter on account of the modification in Doppler frequency instigated by the motion of the scatterer can be computed as:

$$y' = \frac{R}{V_s} u_r(x, y) + y \tag{6.34}$$

Eq. (6.34) reveals that the azimuth spot of the identical but motionless scatterer is y, R is the slant range distance to the antenna, and V_s is the radar-platform velocity [13]. In this understanding, $\frac{R}{V_s}$ can lead to blurring of the real ocean-field due to the phase of the sea surface radar echoes, locally modified by the triggering the sea surface dynamic fluctuations. In other words, $\frac{R}{V_s}$ leads to a systematic distortion of the imaged spectrum. In most circumstances, the random blurring is consequently robust so that the shortest wave constituents are filtered out.

6.6.3 SAR ocean wave imaging theory

The two-dimensional SAR imaging system can deliver ocean surface images. Real aperture radar (RAR) imagines the ocean wave close to range direction, i.e., a cross-track. The modulation in backscatter is the dominant feature of RAR owing to the long wave-induced varying surface tilts (i.e., local relative orientation) and straining surface tilt (i.e., local roughness orientation). According to this view, 2D SAR (RAR) ocean wave image can describe the waves shorter than the resolution cell of SAR that contribute to the composite Bragg resonant backscatter mechanism (Chapter 4). As a result, these shorter waves travel on the weakly random sea of abundant longer waves. The statistical properties of such shorter waves are used to determine their physical characteristics in the domain of elevation and slope variances, and elevation spectrum amplitude and directionality. These statistical ocean wave parameters are functions of the fluctuation of mean radar cross-section owing to wind spinning stress on the sea surface. In this context, the SAR ocean wave retrieving is extremely complicated as it involves several mechanisms.

To begin with, the local radar backscatter is directed by the sea surface roughness on the centimeter scale. It relies on the wind velocity and the physical properties of the air-sea interface. First, the local backscatter is modulated by waves longer than about twice the SAR resolution cell, which is known as real aperture radar (RAR) modulation. It designates the radar backscatter as seen by a conventional radar without a synthesized antenna. Lastly, the complex motion of the ocean surface disturbs the Doppler shift attendant with the azimuth processing of the image and induces strong distortions, which is well-known as velocity bunching [13].

6.6.4 Differences between linear and nonlinear backscatters

The following significant question is raised: how to distinguish between linear and nonlinear backscatter? To tackle the linearity and nonlinearity of SAR backscatters, the modulation transfer function (MTF) must be involved. In other words, SAR backscatters are termed as backscatter modulation. In this context, the important question is: what is meant by MTF?

The modulation transfer function (MTF) is a technique of demonstration "the level to which a SAR degrades the ocean wave images" created in it due to signal backscatter from the ocean surface. Consequently, MTF is a measure of the ability of the imaging system to handle contrast as a function of spatial frequency; the greater the MTF of a system, at high spatial frequencies, the more adept it is at capturing fine detail. In other words, MTF is equivalent to the modulation in a SAR image at a particular spatial frequency.

MTF is only well-defined for a linear system where the output is linearly proportional to the input. Additionally the ocean wave whose backscatter varies sinusoidally with a frequency of lines/pixels is imaged through the SAR sensor. The modulation of the wave is defined as the ratio between the maximum intensity minus minimum intensity and a total of maximum intensity and minimum intensity, i.e., maximum intensity − minimum intensity/maximum intensity + minimum intensity. Let us assume that the RAR image $I_R(x,t)$ can be expressed mathematically as:

$$I_R(x,t) = \int_{k,\omega} T_l(k,\omega)e^{i(k\cdot x-\omega t)}\sum_k \omega_k a_k^+ a_k + \sum_{k,j} V_{k,j} a_k^+ a_j a_{k-j} + \sum_{j,k,l} \Gamma_{j,k,l} a_k^+ a_j a_{k+l-j} + \cdots \tag{6.35}$$

Eq. (6.35) shows that RAR involves the tilt modulation T_l, which is the summation of tilt modulation T_t and hydrodynamic modulation T_h. The long waves tilt the capillary waves so that the local incident angle is changed (Fig. 6.12), thus modifying the Bragg wavelength and the backscattered energy. The tilt modulation is thus the linear correlation between the slope of the long waves with a proportionality coefficient that is associated with the derivative radar cross-section as a function of incidence angle θ_0. The initial computation of tilt

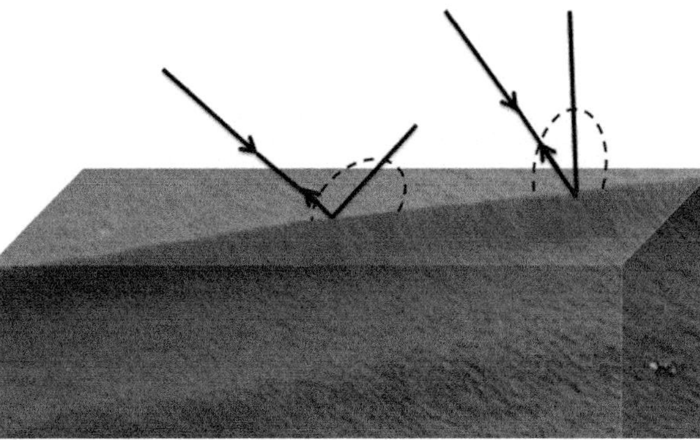

FIG. 6.12 Tilt modulation.

modulation was introduced by Hasselmann and Hasselmann [17] in their seminal work on the nonlinear mapping of the ocean surface wave spectrum by a SAR instrument. A later computation was delivered by Mastenbroek and De Valk [18] in their semiparametric algorithm to retrieve ocean wave spectra from SAR. This technique is used to avoid the uncertainty associated with the use of a pure Bragg scattering model, and certainly offers an efficient alternative for C-band VV polarized SAR data [19]. The tilt modulation is then mathematically formulated as:

$$T_I = 4i\frac{\cot\theta_0}{1+\sin^2\theta_0}k_x \tag{6.35.1}$$

here k_x is wavenumber along the azimuth direction and θ_0 is the incidence SAR beam angle.

The tilted part of the RAR MTF is virtually invisible since for range-traveling waves the equivalent maximum backscatter arises where the wave slope is maximized. The second modulation that involves RAR is known as hydrodynamic modulation. The hydrodynamic interaction between the long waves and the capillary waves causes divergence and convergence, and thus modulates the energy returned. The hydrodynamic modulation process depends on the fact that small-scale (centimeter-decimeter wavelength) ripples control the magnitude of the radar echo through the Bragg mechanism when the radar is pointing obliquely at the surface. These small ripples are themselves modulated by convergent and divergent surface currents allied with long waves. The elongating and compression of the sea surface as ocean waves pass along it create configurations of smoother and rougher texture at Bragg wavelengths, and these configurations are hidden by the phase of long waves (Fig. 6.13). The consequential configurations of bright and dark patterns on the radar image seem to present troughs and crests of longer surface waves [20]. The hydrodynamic modulation T_h is then given by:

$$T_h = 4.5\omega\frac{k_x^2}{|k|}\frac{\omega - i\mu}{\omega^2+\mu^2} \tag{6.35.2}$$

Hasselmann and Hasselmann [17] demonstrated the hydrodynamic relaxation rate μ, chosen equal to $0.5s^{-1}$, which is correlated to the response time of the modulation of the short waves by the longer ones. Eq. (6.35.2) indicates that for deep water, the wave

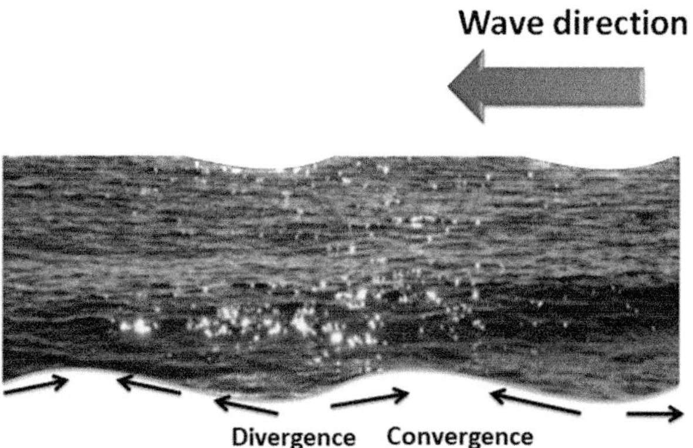

FIG. 6.13 Hydrodynamic modulation.

frequency is attained through the dispersion relation $\omega^2 = \left|\vec{k}\right| g$, where g is the acceleration of the gravity. For short waves, the hydrodynamic part of the RAR MTF is real so that the maximum of the backscatter is in phase with the crests of the waves. In the circumstance of long waves, T_h is a complex number and the maximum of the hydrodynamic part of the backscatter travels toward the side of the wave. Tilt and hydrodynamic modulation are essentially explicated at the scatter level, presuming a frozen sea surface during the SAR coherent integration period. Conversely, for the motion allied with the traveling waves, a frozen sea surface assumption is no longer valid. In other words, the RAR modulation is zero for ocean waves that propagate parallel to the flight direction since the RAR MTF is proportionally k_x. Consequently, RAR cannot image such waves. In the circumstance of SAR, the distortion of the real wave pattern is clear in the SAR data pattern owing to surface motion impacts, as will be explained in Section 6.6.5.

At the initial stage, let the linear backscatter take a Gaussian form with possible negative values. Following Schulz-Stellenfleth et al. [21], the nonlinear model for the normalized backscatter of RAR image I_R is formulated as:

$$\sigma_{NL}\langle\sigma_{NL}\rangle^{-1} = [1+\rho_{II}(0,0)]^{-0.5} e^{\left[\ln\left(\frac{1+\rho_{II}(0,0)}{\rho_{II}(0,0)}\right]^{0.5}(I_R)\right.} \tag{6.36}$$

Eq. (6.36) demonstrates that the nonlinear backscatter σ_{NL} and normalized nonlinear backscatter $\langle\sigma_{NL}\rangle$ are functions of the variance of the RAR image $\rho_{II}(0,0)$. In addition, the steep waves can cause quasispecular reflection, which can easily be implemented in an ocean-to-SAR integral transform.

6.6.5 Velocity bunching modulation

In deep water, discrete water particles take a circular trajectory; consequently, long waves have a periodic orbital motion, which causes a deceptive growth and diminution in the strength of scatters. Let us assume the orbital velocity component u_r, which has a positive value as it moves toward the radar. In addition, facet scattering due to a the sea surface radial velocity u_r causes an azimuth shift η_x. The simple mathematical formula can be expressed as:

$$\eta_x = \frac{R}{V_s} u_r \tag{6.37}$$

Eq. (6.37) reveals that a moving scatterer with slant-range velocity u_r can be imaged at $y' = y + \eta(x)$. In addition, R represents the slant-range distance to the antenna and V_s the radar-platform velocity. This shift produces the azimuthally traveling waves detectable in a SAR image; conversely, there is a shift from their true position. In other words, the existence of longer wave orbital velocity field and SAR velocity causes Doppler shifts. The motion effects cause extremely complex backscatter signals in the SAR image. In addition, Doppler shifts associated with the longer wave orbital velocities can hypothetically be a factor to an extra practical modulation mechanism. This mechanism is well-known as velocity bunching modulation (Fig. 6.14). Velocity bunching is key to determining azimuth shift η_x. The nonlinear wave theory using the Hamiltonian formula of Eq. (6.9) can be used to obtain η_x as follows:

$$\sum_k c_k e^{i(kx - \omega_k t)} \tag{6.38}$$

In this context, the velocity bunching transfer function T_{η_x} can be expressed as:

$$T_{\eta_x}(k, \omega) = \frac{R}{V_s} \omega \left[\sin\theta_0 \frac{k_x}{|k|} + i \cos\theta_0 \right] \tag{6.38.1}$$

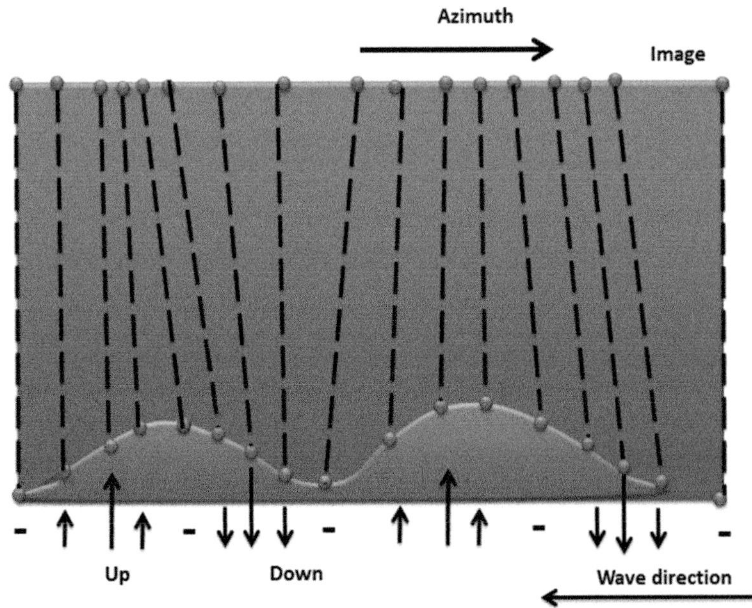

FIG. 6.14 Velocity bunching concept.

Eq. (6.38) reveals that the nature of the surface motion impacts relies on the length of the considered waves. For short waves, the radial velocity is great, as the azimuth displacement equates to the equivalent wavelength, which results in a destructive summation for the image power In contrast, the larger ocean wavelength than the azimuth displacement causing a constructive summation that is known as velocity bunching. According to this understanding, surface motion impacts operate as a low-pass filter in the azimuth direction.

6.7 Relativistic theory of SAR velocity bunching

As noted in Chapter 5, the fundamentals of relativity generally contain two consistent theories obtained by Albert Einstein: special relativity and general relativity. Special relativity applies to essential particles and their interactions, describing all their physical phenomena in the absence of gravity. General relativity, conversely, designates the instruction of gravitation and its relation to dissimilar forces of nature. This leads to the critical question: can relativity theories be used to describe SAR velocity bunching theory? The main cause of velocity bunching is the Doppler frequency shift, which can be easily described by relativity theory. The Doppler effect occurs upon backscattering from the dynamical sea-state, or a change in the frequency of the microwave photon backscattering concerning the incident signal.

Let us assume that $\hbar\omega$ exhibits a single photon of energy and $\hbar\omega c^{-1}$ shows an equivalent momentum of magnitude, upon an oblique incident beam on the dynamical sea surface moving at a radial velocity u_r. Consistent with the quantum depiction of the procedure of backscattering, the microwave photon will be reflected by sea surface particle radial velocity u_r. The dynamical motion of the sea-state will instigate loss or gain of momentum (and, consequently, energy) of the photon upon backscattering, relying on whether the sea surface is traveled along in a positive or negative azimuth direction (x). In the following, let us consider that the motion of the sea-state is nonrelativistic ($u_r \ll$ c), and that the backscattering of the SAR pulse at its surface is perfectly elastic. The pulse is treated as a spherical particle that collides and bounces off a frictionless plane surface.

As a result of the interaction between the SAR signal and the sea surface radial velocity, there will be a shift in the frequency of the photon after the backscattering, and the incident angle θ_0. Let $\hbar\omega'$ and $\hbar\omega' c^{-1}$ signify the particular energy and the momentum magnitude of the backscattered photon. Let us implement the circumstance for the conservation of momentum along the azimuth (x) and range (y) direction, then we obtain:

$$\frac{\hbar\omega}{c}\cos\theta_0 + u_r = -\frac{\hbar\omega'}{c}\cos\alpha + (u_r + \Delta u_r), \tag{6.39}$$

$$\frac{\hbar\omega}{c}\sin\theta_0 = \frac{\hbar\omega'}{c}\sin\alpha, \tag{6.40}$$

Eq. (6.39) demonstrates the radial velocity of the sea surface is reformed after the backscatter by an amount of Δu_r in the azimuth direction. In addition, α is the angle between flight directions of the SAR and wave propagations. The scientific explanation of the law of conservation of energy can mathematically be written as:

$$\Delta u_r(x) = \hbar c^{-1}(\omega \cos\theta_0 + \omega' \cos\alpha), \tag{6.41}$$

In this understanding, the relative radial component of the scatterers velocity can be expressed as:

$$u_{vb} = \frac{R}{V_s}\frac{\hbar c^{-1}(\omega \cos\theta_0 + \omega' \cos\alpha)}{dx_0}, \tag{6.42}$$

Then the Doppler frequency shift ω' is determined from:

$$\omega' = \omega\left(1 - 2u_r \cos\theta_0 c^{-1} + u_r^2 c^{-2}\right)\left(1 - u_r^2 c^{-2}\right)^{-1} \tag{6.42.1}$$

The interested debate is that the resulting formulas (6.41) and (6.42) are essentially nonrelativistic theory, which is also effective in the circumstance when the sea surface is moving at relativistic speeds. The relativity of the Doppler shift depends on whether the ocean wave waves are moving toward the SAR sensor or away from it.

In quantum mechanics, the Doppler shifts transform the energy and momentum of the quanta and it swaps the wavenumber and frequency, but it doesn't swap the ratio of the energy to the frequency, or the momentum to the wavenumber. If the ocean wave is moving toward the SAR sensor, the backscattered pulse is "compressed," meaning that the wavelength of the pulse becomes smaller and then the imaged ocean wavelength would be compressed too. On the contrary, if the ocean wave is moving away from the SAR sensor, the backscattered pulse is "expanded," and the imaged ocean wavelength in SAR data is increased.

Generally, the relativistic velocity bunching impact is the exchange in frequency (and wavelength) of the imaged ocean, triggered through the relative motion of the SAR antenna and the observer (as in the classical Doppler effect), when taking into account consequences described through the distinctive principle of relativity. The relativistic velocity bunching impact is unique from the nonrelativistic velocity bunching impact, as the equations encompass the time dilation impact of special relativity and no longer contain the medium of propagation as a reference point. They describe the complete dissimilarityin determining frequencies and possess the required Lorentz symmetry.

6.8 Relativistic theory of the ocean wavelength in SAR images

As noted in Chapter 5, the principal association of relativistic quantum field theory to SAR is across scattering theory, i.e., the speculation of the backscatter of the signal from the rough sea surface. In the circumstance of relativity, the ocean wavelength in a moving frame would seem foreshortened or contracted in the direction of SAR motion. In other words, the physical characteristics of wave spectra would be transferred using Lorentz symmetry, as this relies on the $\frac{R}{V_S}$ and traveling SAR pulse in the frame of the speed of light. The amount of deep wavelength contraction can be calculated from:

$$L' = 1.56T^2\sqrt{1-\left(\frac{R}{V_S}\right)^2 c^{-2}} \tag{6.43}$$

Eq. (6.43) demonstrates the degree of deep water wavelength as a function of the ratio between slant range and satellite velocity that relies on the speed of the light c. The relativity of the wavelength contraction would be modified as new approach of the imaged shallow water in SAR data as:

$$L' = T(gd)^{0.5}\sqrt{1-\left(\frac{R}{V_S}\right)^2 c^{-2}} \tag{6.44}$$

here d is the shallow water depth and T is the wave period. Since the wavenumber $k=\frac{2\pi}{L_0}$, then the three modulation transfer factors—tilt, hydrodynamic, and velocity bunching—would contain the ocean wavelength contractions as follows:

$$T_I(\theta_0, L') = 4i\frac{\cot\theta_0}{1+\sin^2\theta_0}2\pi\left[1.56T^2\sqrt{1-\left(\frac{R}{V_S}\right)^2 c^{-2}}\right]^{-1} \tag{6.45}$$

$$T_h(\omega, L') = 4.5\omega\frac{L_x'^2}{|L'|}\frac{\omega - i\mu}{\omega^2+\mu^2} \tag{6.46}$$

$$T_{\eta_x}(L', \omega) = \frac{R}{V_s}\omega\left[\sin\theta_0\frac{2\pi\left[1.56T^2\sqrt{1-\left(\frac{R}{V_S}\right)^2 c^{-2}}\right]^{-1}}{\left|2\pi\left[1.56T^2\sqrt{1-\left(\frac{R}{V_S}\right)^2 c^{-2}}\right]^{-1}\right|} + i\cos\theta_0\right] \tag{6.47}$$

The wavelength contractions as a component of relativity theory can impact the behavior of tilt, hydrodynamic, and velocity bunching modulation transfer functions. This can bridge the gap between relativistic quantum theory and the imaged ocean wave mechanism in SAR data. The relativity of the imaged ocean wave spectra in SAR data is presented and modified from deep water to shallow water.

In other words, the wavelength in a moving SAR frame will appear foreshortened or contracted in the direction of motion. This contraction (also known as Lorentz contraction or Lorentz-Fitzgerald contraction after Hendrik Lorentz and George Francis Fitzgerald) is usually only noticeable at a substantial fraction of the speed of light. Length contraction is only in the direction in which the body is traveling. For standard ocean wavelength, this consequence is negligible at everyday speeds and can be ignored for all regular purposes, only becoming significant as the pulse approaches the speed of light relative to the SAR. In general, this contraction only occurs along the line of SAR motion, i.e., the azimuth direction. Therefore, orbital motions of ocean waves cause countless azimuth shifts, which creates nonlinear modulations in the SAR image plane. In this circumstance, the nonlinearity of velocity bunching relies on the nonlinear parameter, which has been established theoretically by Alpers et al. [22] as:

$$N_{nl} = \frac{R}{V_S}\varsigma\left[2\pi\left[1.56T^2\sqrt{1-\left(\frac{R}{V_S}\right)^2 c^{-2}}\right]^{-1}\right]\omega\cos\alpha\sqrt{\sin^2\theta_0\sin^2\alpha+\cos^2\theta} \tag{6.48}$$

The nonlinearity involves the length contraction term due to azimuth shift. The angular frequency ω also contributes to the nonlinearity of SAR wave spectra image as:

$$N_{nl} = \frac{R}{V_S}\xi\left[2\pi\left[1.56T^2\sqrt{1-\left(\frac{R}{V_S}\right)^2 c^{-2}}\right]^{-1}\right]\left[\frac{\omega'\left(1-u_r^2c^{-2}\right)}{\left(1-2u_r\cos\theta_0c^{-1}+u_r^2c^{-2}\right)}\right]\cos\alpha\sqrt{\sin^2\theta_0\sin^2\alpha+\cos^2\theta} \tag{6.49}$$

Eq. (6.49) demonstrates the implementation of relativity theory to SAR Doppler that causes such wavelength contraction along the azimuth. The fluctuation of wave parameters from offshore to shallow water changes the relativistic of SAR Doppler frequency shift, i.e., SAR Doppler frequency shift proves the relativity theory as expressed by:

$$N_{nl} = \frac{R}{V_S}\xi\left[2\pi T(gd)^{0.5}\sqrt{1-\left(\frac{R}{V_S}\right)^2 c^{-2}}\right]^{-1}\left[\frac{\omega'\left(1-u_r^2c^{-2}\right)}{\left(1-2u_r\cos\theta_0c^{-1}+u_r^2c^{-2}\right)}\right]\cos\alpha\sqrt{\sin^2\theta_0\sin^2\alpha+\cos^2\theta} \tag{6.50}$$

The smaller value of the nonlinearity index N_{nl} denotes that the SAR image intensity generates weakly nonlinear patterns. In other words, the relativity of the weak nonlinearity generates a wave-like pattern. The strong Doppler frequency shift across the SAR image owing to an increase of $\frac{R}{V_S}$ causes greater displacement of the wavelength along the azimuth direction.

6.9 Relativistic theory of incidence angle in SAR wave images

Let us assume a regular wave is traveling to the range −90 degrees and the azimuth directions 180 degrees. The simulated SAR image intensity in Fig. 6.15 shows the wave pattern associated with the range propagation wave. In contrast, the traveling azimuthal wave does not show a wave-like pattern (Fig. 6.16) owing to the relativity of nonlinearity as a function of velocity bunching.

The local incident angles are shown in the range traveling wave (Fig. 6.17), while Fig. 6.18 shows the azimuth traveling wave. Both figures show the dissimilar variations of the local incident angles in range and azimuth directions, respectively. The smaller variance of incidence angles in azimuth direction than the range direction causes the wave-like pattern, which does not appear in the azimuth direction due to the nonlinearity of the velocity bunching alliance with the smallest incidence angles. The SAR ocean wave imaging can be thought of as a wave packet, having wave-like properties and also the single position and size we associate with a particle.

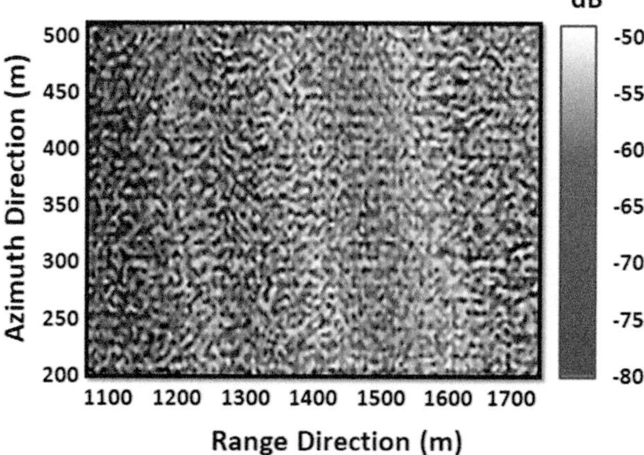

FIG. 6.15 Wave-like pattern range traveling waves.

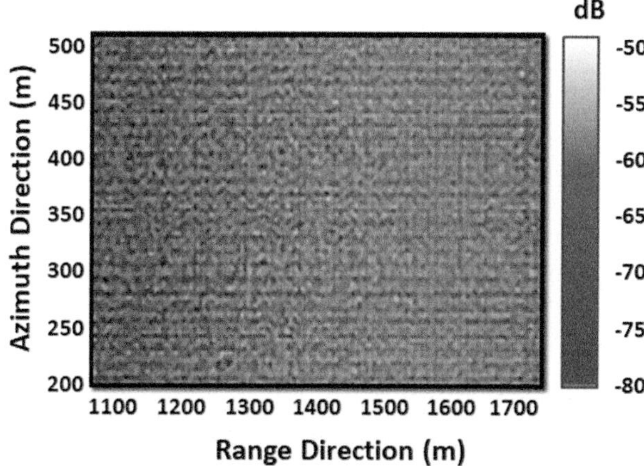

FIG. 6.16 Nonlinearity causing relativity of wave propagation in the azimuth direction.

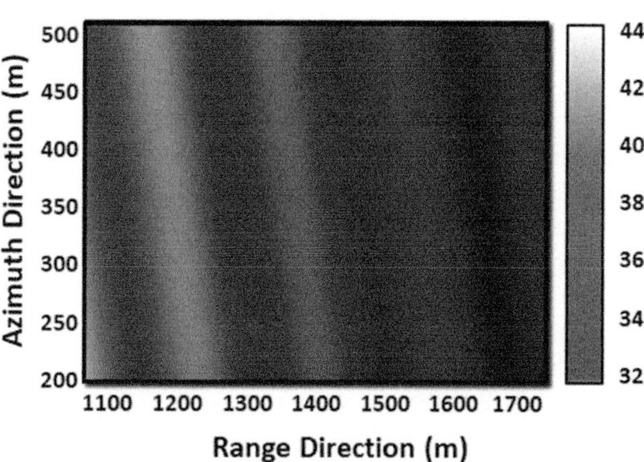

FIG. 6.17 The local incidence angle of range traveling wave.

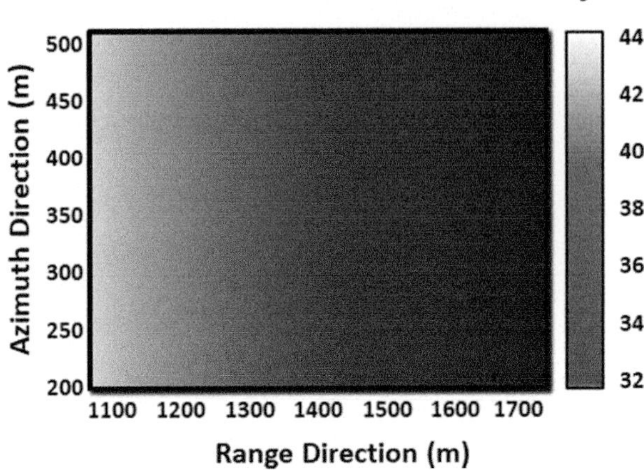

FIG. 6.18 The local incidence angle of azimuth traveling wave.

There are some slight problems, such as that the wave packet doesn't stop at a finite distance from its peak; it also goes on indefinitely. In this understanding, the wavelength of an ocean particle is correlated to the momentum of the orbital particle. Accordingly, energy is also connected to the ocean wave properties such as elevation and frequency. The momentum of the wave is determined by its wavelength. An ocean traveled wave packet is a composite of countless waves.

This indicates that azimuth waves have a weaker backscattering intensity than the range backscattering. The relativity caused by incidence local angle causes hard imaging of ocean waves in SAR images, as the variations of the backscattering intensity are smaller than speckle or white noise [23,24].

Figs. 6.19 and 6.20 show the range wave traveling and azimuth wave traveling, respectively. The range of wave traveling has sharper brightness backscatter than the azimuth one. Moreover, the azimuth of wave traveling is missing more clear wave packets than the range one. In the range wave traveling an obvious ocean wave, a diffraction pattern occurs (Fig. 6.19). In contrast, the azimuth wave traveling captures only a slight wave diffraction pattern (Fig. 6.20).

Both Figs. 6.19 and 6.20 are the proof of wave-particle duality speculation, but the distribution of wave-particle interferences are governed by relativistic of SAR Doppler frequency shift that occurs between range and azimuth directions, respectively, for ocean wave imagine in the SAR satellite data. In azimuth traveling waves dominated by the darkest shadow zone, which would appear as a low wind speed zone or shelter wind zone behind the island as shown in Fig. 6.20. In this understanding, the wave-like properties become less intense due to the shifting of single position and size that are associated with particles.

In general, the imaged ocean surface in SAR data depends on the very high sensitivity of incident angle and radar backscatter signal which are associated with fluctuations on both the local geometry and the spectral density distribution of short gravity and gravity-capillary waves. The resonant surface waves are much shorter at more oblique incidence angles. In other words, the ocean backscatter returns decrease with the increase of incidence angles. Consequently, the large oblique SAR angles view smaller amplitude of Bragg waves, leading

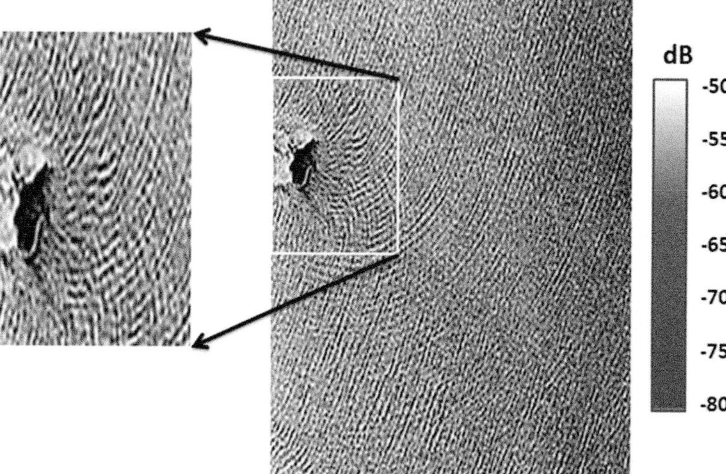

FIG. 6.19 Range wave traveling.

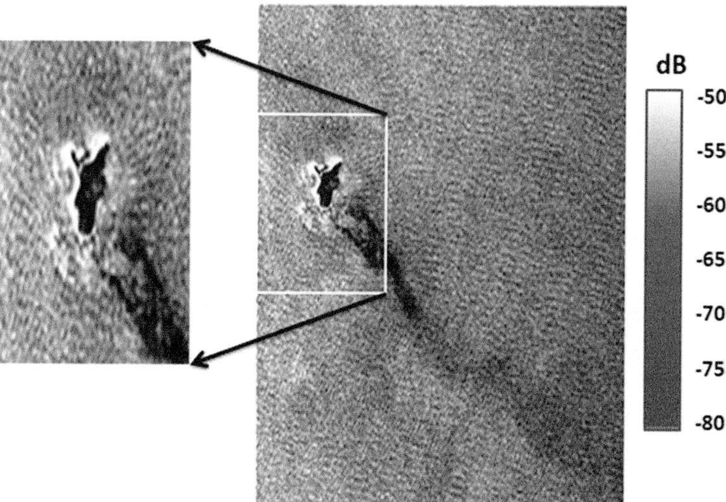

FIG. 6.20 Azimuth wave traveling.

to lower backscatter. The relativistic concept of incidence SAR angles can also explain the wavelength contraction in SAR data.

6.10 Relativistic theory of range bunching

It seems that a SAR image can demonstrate relativity from a different point of view. The scanning distortion, for instance, can be considered with airborne SAR images rather than for satellite-borne SARs. According to Lyzenga [25] and Vachon et al. [26], the long wave speed is

10%–15% higher than airborne SAR, which causes an apparent rotation of waves on a SAR image as a result of the scanning distortion. This phenomenon is neglected in satellite-borne SARs.

The relativity theory also can be proved by the local fluctuation of the sea surface elevation, in which can modify its geometry shapes in SAR images as appears extremely in range direction. In line with Chapter 5, the computation of the scatterers' range position is performed by measuring the time change of a round trip of transmitted and received pulse. In this circumstance, the range relativity appears as the sea surface elevation causes a misplacing of the scatterers in the range direction. Fig. 6.21 shows the range misplacement owing to local variation of the surface height. This can be formulated mathematically as:

$$\eta_R = -\xi(\tan\theta_0)^{-1} \tag{6.51}$$

The relativity impact θ_0 is revealed in Eq. (6.51) as the range distortion increases as θ_0 decreases. Conversely, the range-bunching modulation is commonly linearized and assimilated into the RAR MTF by adding the transfer function term, and can then be given by:

$$\eta_R(L) = -i\left[2\pi\left[1.56T^2\sqrt{1-\left(\frac{R}{V_S}\right)^2 c^{-2}}\right]^{-1}\right]\cot\theta_0 \tag{6.52}$$

However, range bunching becomes obvious with a large incidence angle as it can cause wide distortion for a long wave traveling in range direction [27–29]. A shallow water wave would have a different formula for range bunching than deep water. In this regard, the range bunching for a shallow wave traveling is formulated as:

$$\eta_R(L) = -i\left[2\pi\left[T(gd)^{0.5}\sqrt{1-\left(\frac{R}{V_S}\right)^2 c^{-2}}\right]\right]\cot\theta_0 \tag{6.53}$$

Eqs. (6.52) and (6.53) reveal the range bunching causing wavelength distortion in SAR image that is also a function of the incidence angle. In general, the range-bunching consequence develops significantly when imaging very tall objects, such as mountain ranges (Fig. 6.22).

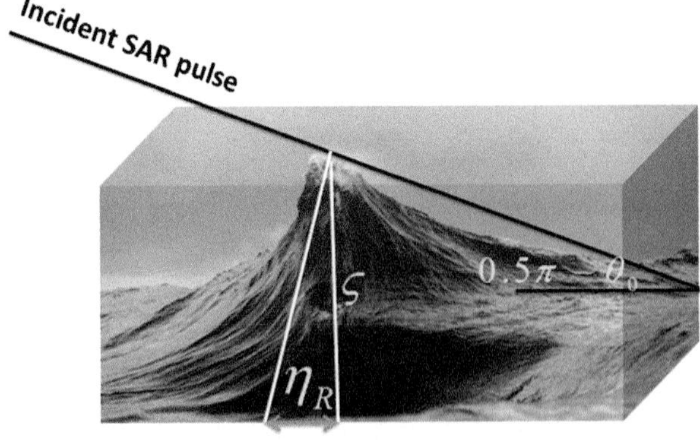

FIG. 6.21 Range bunching due to sea surface variation.

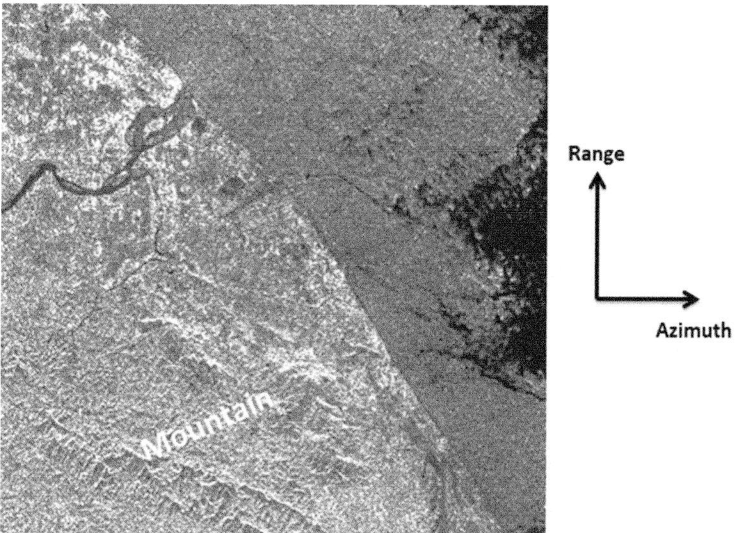

FIG. 6.22 Range bunching along a mountain in the ERS-1 SAR image.

At a low incidence angle of less than 30 degrees, radar cross-section relies on a smaller wave than SAR resolution of 10 m and larger than the resonant Bragg wavelength of 8–10 cm [27,29]. Therefore, the tilt modulation relies on the shape of the elevation of the surface wave spectrum about the Bragg resonant wavenumber as the small waves are fully saturated [28]. Spinning wind speed particles increase the tilt modulation values.

In general, relativity theories are the keystone to understanding the mechanism of the imaged ocean wave in SAR sensors. In relativity, the reference frames can be used to itemize the relationship between a moving SAR and the ocean wave phenomenon or phenomena under observation.In this context, the novel articulation of the imaged ocean wave in the SAR data can come to be "the imaged frame of ocean wave in SAR reference" (or "the imaged reference of ocean wave in SAR frame"), which suggests that the SAR is at rest in the frame, although not necessarily located at its origin. A relativistic SAR reference frame comprises (or indicates) the coordinate time as explained previously, which does not resemble diagonally dissimilar frames transferring relative to each other. The ratio of the range to satellite velocity $\frac{R}{V_S}$ is constant for each SAR sensor, but varies from sensor to sensor. For instance, ERS-1 has $\frac{R}{V_S}$ a value of 120 s while ENVISAT has a value of 114 s, which can exhibit the relativity theory in the imaged ocean wave in SAR frame reference owing to the Doppler frequency shift.

The next chapter will demonstrate the algorithms used to retrieve ocean wave spectra from the SAR image. Advanced quantum computing will be addressed in the following chapters.

References

[1] Fournier A, Reeves WT. A simple model of ocean waves, In: Proceedings of the 13th annual conference on computer graphics and interactive techniques 1986 Aug 31; 1986. p. 75–84.

[2] Donelan MA, Hamilton J, Hui W. Directional spectra of wind-generated ocean waves. Philos Trans R Soc A Math Phys Eng Sci 1985;315(1534):509–62.

[3] Pelinovsky E, Kharif C, editors. Extreme ocean waves. Berlin: Springer; 2008.
[4] Sterl A, Caires S. Climatology, variability and extrema of ocean waves: the web-based KNMI/ERA-40 wave atlas. Int J Climatol J R Meteorol Soc 2005;25(7):963–77.
[5] Hinsinger D, Neyret F, Cani MP. Interactive animation of ocean waves, In: Proceedings of the 2002 ACM SIGGRAPH/Eurographics symposium on computer animation 2002 Jul 21; 2002. p. 161–6.
[6] Pierson WJ. Practical methods for observing and forecasting ocean waves by means of wave spectra and statistics. US Government Printing Office; 1955.
[7] Salart D, Baas A, Branciard C, Gisin N, Zbinden H. Testing the speed of 'spooky action at a distance'. Nature 2008;454(7206):861–4.
[8] Nelson DL, McEvoy CL, Pointer L. Spreading activation or spooky action at a distance? J Exp Psychol Learn Mem Cogn 2003;29(1):42.
[9] Hardy L. Spooky action at a distance in quantum mechanics. Contemp Phys 1998;39(6):419–29.
[10] Tolman HL, Chalikov D. Source terms in a third-generation wind wave model. J Phys Oceanogr 1996;26 (11):2497–518.
[11] Hasselmann S, Hasselmann K, Allender JH, Barnett TP. Computations and parameterizations of the nonlinear energy transfer in a gravity-wave spectrum. Part II: parameterizations of the nonlinear energy transfer for application in wave models. J Phys Oceanogr 1985;15(11):1378–91.
[12] Abdullah SS, Al-Mahdi AA, Mahmood AB. Breaker wind waves energy at Iraqi coastline. Mesopot J Mar Sci 2009;24(2):112–21.
[13] Fouques S. Lagrangian modelling of ocean surface waves and synthetic aperture radar wave measurements. [Dr. Ing. thesis]Norwegian University of Science and Technology; 2005226.
[14] Yigit E, Demirci S, Ozdemir C, Kavak A. A synthetic aperture radar-based focusing algorithm for B-scan ground penetrating radar imagery. Microw Opt Technol Lett 2007;49(10):2534–40.
[15] Zhu D, Li Y, Zhu Z. A keystone transform without interpolation for SAR ground moving-target imaging. IEEE Geosci Remote Sens Lett 2007;4(1):18–22.
[16] Marghany M. Automatic detection algorithms of oil spill in radar images. CRC Press; 2019.
[17] Hasselmann K, Hasselmann S. On the nonlinear mapping of an ocean wave spectrum into a synthetic aperture radar image spectrum and its inversion. J Geophys Res Oceans 1991;96(C6):10713–29.
[18] Mastenbroek CD, De Valk CF. A semiparametric algorithm to retrieve ocean wave spectra from synthetic aperture radar. J Geophys Res Oceans 2000;105(C2):3497–516.
[19] Robinson IS. Discovering the ocean from space: The unique applications of satellite oceanography. Springer Science & Business Media; 2010.
[20] Robinson IS. Measuring the oceans from space: The principles and methods of satellite oceanography. Springer Science & Business Media; 2004.
[21] Schulz-Stellenfleth J, Hoja D, Lehner S. Global ocean wave measurements from ENVISAT ASAR data using a parametric inversion scheme, In: IGARSS 2003. 2003 IEEE international geoscience and remote sensing symposium. Proceedings (IEEE Cat. No. 03CH37477) 2003 Jul 21vol. 5. IEEE; 2003. p. 3111–3.
[22] Alpers WR, Ross DB, Rufenach CL. On the detectability of ocean surface waves by real and synthetic aperture radar. J Geophys Res Oceans 1981;86(C7):6481–98.
[23] Yoshida T, Rheem CK. Research on SAR imaging mechanism for ocean wave propagating in azimuth direction by using numerical simulation, In: 2013 MTS/IEEE OCEANS-Bergen 2013 Jun 10IEEE; 2013. p. 1–6.
[24] Yoshida T. SAR signal simulation in time domain for velocity bunching by ocean wave, In: EUSAR 2012; 9th European conference on synthetic aperture radar 2012 Apr 23VDE; 2012. p. 466–9.
[25] Lyzenga DR. An analytic representation of the synthetic aperture radar image spectrum for ocean waves. J Geophys Res Oceans 1988;93(C11):13859–65.
[26] Vachon PW, Krogstad HE, Paterson JS. Airborne and spaceborne synthetic aperture radar observations of ocean waves. Atmosphere-Ocean 1994;32(1):83–112.
[27] Budge MC, German SR. Basic RADAR analysis. Artech House; 2015.
[28] Anderson S, Anderson C, Morris J. Nonlinearity and multiscale behaviour in ocean surface dynamics: an investigation using HF and microwave radars. Complex computing-networks. Berlin, Heidelberg: Springer; 2006185–96.
[29] Jakowatz CV, Wahl DE, Eichel PH, Ghiglia DC, Thompson PA. Spotlight-mode synthetic aperture radar: A signal processing approach: A signal processing approach. Springer Science & Business Media; 2012.

CHAPTER

7

Quantum nonlinear techniques for retrieving ocean wave spectral parameters from synthetic aperture radar

7.1 Simplification of the magic concept of SAR Doppler shift frequency

The utilization of remote microwave for measurement of the ocean surface was initiated with radar sensors arrayed far closer to Earth than orbit. Consequently, radar signal interference from the ocean surface influenced military maneuvers, particularly throughout World War II when radar technology was in its early days. Subsequently, Kerr [1] and Rice [2] had a strong interest in understanding such effects. In this regard, they tracked the concept of backscattered intensity based on the Bragg resonance theory [3]. In this understanding, the ocean surface is magnified when the wavelength of the radar coordinated the ocean surface wavelength, and become equivalent to Bragg resonance deliberate from scattering off of crystal structures. Bragg scattering would prove to be the keystone to comprehending how to simulate winds over the ocean using SAR technology. This concept is governed by radar frequencies being amplified and wavelengths reduced; a similar phenomenon was observed in microwave radar (>1 GHz frequency) regimes [4].

The principal idea behind radar is simple, as discussed in previous chapters. A radar antenna emits a burst of short pulses of microwave radiation, which is dignified in nanoseconds, or billionths of a second; within microwave wavelengths of a few centimeters. The same antenna that emits the pulse receives the echo of that pulse from the target after a time delay that depends on the distance to the target. The distance is equal to the delay multiplied by the constant speed of light: 300,000 km/s.

Synthetic aperture radar (or SAR) scans a two-dimensional strip of the ocean with a spatial resolution considerably superior in its forward direction than is achievable with a scanning

Nonlinear Ocean Dynamics
https://doi.org/10.1016/B978-0-12-820785-7.00015-0

altimeter. The European ocean satellite ERS-1, for instance, conveyed a radar map of the waves in a 5×10 km area at intervals of 200 km along the satellite's path.

In this context, ERS-1 accomplished a resolution of 30 m in the forward direction and 100 km in the transverse direction. This is due to the fact that ERS-1 irradiates separately every spot in the target area continually at various dissimilar angles of incidence, with pulses of a stable microwave frequency, which allows for better resolution. Therefore, the changing angles of incidence make every backscattered pulse dominated by the Doppler frequency shift, and the Doppler shifts measurably alter as the satellite passes over the spot. Consequently, these shifts efficiently label the spot's position in the target area. The satellite's electronics, therefore, exploit these labels, together with the powers of all the backscatters from that spot, creating a crisp, high-resolution image.

Let us assume that a SAR satellite is traveling in the interval from location 1 to location 2 along a precise orbit that serves as a reference frame from which distances to the sea can be measured (Fig. 7.1). SAR radiates a microwave photon wave-like particle with a frequency, for instance, of 13 GHz and 2.3 cm wavelength. In this circumstance, it can be stated that the radar beam has the shape of a pyramid, whose base is a rectangular footprint on the sea. Each pulse travels to the sea, where the beam reaches the surface, illuminates the waves there, and is reflected.

Each ocean wave in the rectangular footprint of the beam reflects a part of the return pulse, and its part is Doppler shifted in frequency, relative to the incident frequency of 13 GHz. The Doppler shift relies on the position of the wave in the footprint. The shift is proportional to the component of the satellite's velocity along the line of sight from the satellite to the wave. The shift equals zero for a wave directly under the satellite, it is negative for waves further toward the front edge of the rectangular footprint, and positive for waves toward the rear edge. In this way, the location of every wave in the footprint is categorized by a unique Doppler shift.

For instance, the SAR moves from location 1 to location 2, every wave in the footprint is illuminated continually, and for each time it is distinguishing the exchange in the Doppler

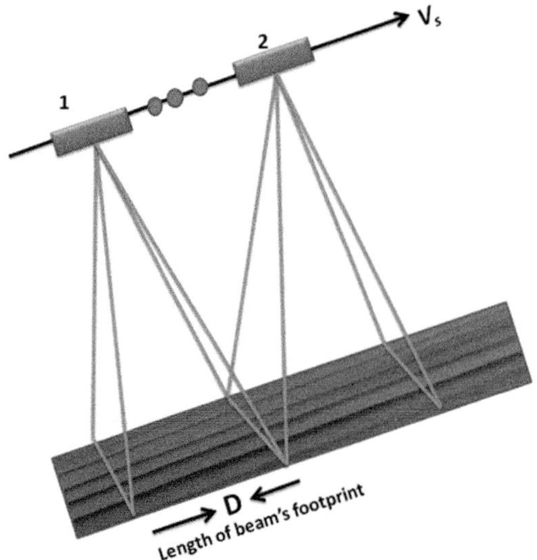

FIG. 7.1 Concept of SAR length beam's footprint.

shift probably. The SAR electronics exploit the Doppler mark to gather and complement the backscattered powers from a specified wave in the footprint. Furthermore, the electronics process all the distance measurements of this wave to define the height of its crest. Evidently, many thousands of backscattered radar signals are gathered for every wave approaches in the footprint. The consequence is a high-resolution two-dimensional microwave SAR image of the sea along the path of the satellite.

SAR has operated since the radar pulse travels at the speed of light, which indicates a faster traveling signal.

Doppler shifts are specifically interrelated with a wave's position within the footprint. Nonetheless, the waves move randomly, and these motions create random Doppler shifts (noise) in the coding of the waves as the thousands of radar spreads are emitted and received. Fig. 7.2 demonstrates an example: an SAR image of coastal water off Tioman Island, Malaysia. The consequence is a marginally fuzzier SAR image than would be created if the waves were immobile (Fig. 7.2).

7.2 SAR sensors for ocean wave simulation

Satellites with synthetic aperture radar (SAR) orbit the Earth in a sun-synchronous low Earth orbit (LOE). LOE is defined as any orbit around the Earth with an orbital period of less than about 2 h. LEO polar orbit and data acquisitions can be made at any time of day or night and independent of cloud coverage, collecting both amplitude and phase data.

As stated by Moore [5], the initial spaceborne active radar for Earth remote sensing was RADSCAT, which operated aboard Skylab. The combined 13.9GHz radiometer and scatterometer quantified the ocean at four polarizations, HH, VV, VH, and HV, respectively, and considered nadir incident angles of 0, 15, 30, 40, and 50degrees, separately. It demonstrated the important ability to simulate both wind and wave patterns. Subsequently, Seasat

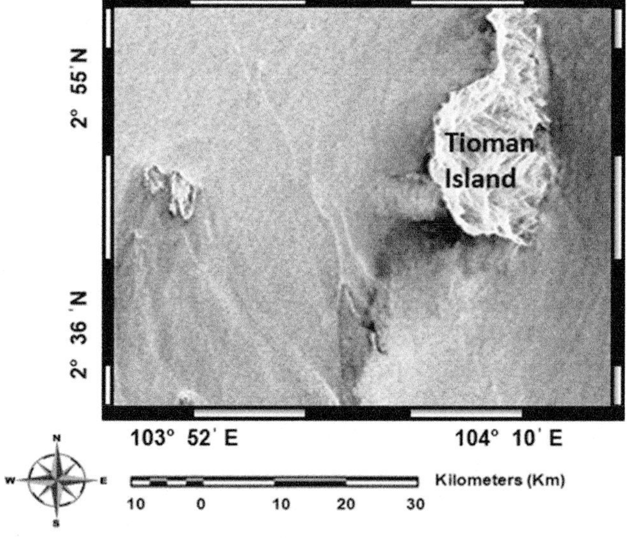

FIG. 7.2 Fuzzy SAR image of coastal water off Tioman Island, Malaysia.

followed quickly in 1978. Seasat carried a suite of new sensors designed to measure various aspects of the ocean surface. There was a radar altimeter that could measure sea surface height, wind speed, and significant wave height; a microwave radiometer for (among other things) wind speed measurements; an active microwave scatterometer for wind speed; and an L-band synthetic aperture radar for ocean surface wave spectra. In addition, the Seasat SAR imagery showed subkilometer-resolution features associated with wind field variability [4].

The SAR satellites have repeating paths which, using two-phase datasets for the same location at different times, allow for interferometric SAR (InSAR) showing relative ground displacements between the two datasets along the direction of the radar beam. The SAR satellites operate at designated frequencies with L-band, C-band, and X-band being the predominant wavelengths. Below is a chart of past, present, and projected SAR satellite missions that is being referred to the bullet list.

Various agencies support dissimilar SAR operations (Table 7.1):

- European Space Agency (ESA): ERS-1, ERS-2, Envisat, Sentinel-1.
- Japan Aerospace Exploration Agency (JAXA): JERS-1, ALOS-1, ALOS-2.

TABLE 7.1 List of SAR satellite sensor characteristics.

Satellites	Acquisition period	Band frequency	Polarization mode	Spatial resolution (m)	Revisit time (days)	Scene cover (km)
Seasat	July–September 1978	L	HH	25	3	100 × 100
ERS-1/2	1991–2011	C	VV	20	35	185 × 185
JERS	1992–98	L	HH	20	44	75 × 75
RADARSAT	1995–2013	C	HH	10–100	24	35 × 500
ASAR	2002–13	C	1 or 2 pol. HH/HV/VV	30–1000	Few–35	100 × 500
PALSAR	2007–11	L	Polarimetric HH/HV/VV	10–100	Few–24	100–500
RADARSAT-2	2007	C	Polarimetric HH/HV/VV	1–15	5–10	NA
TerraSAR-X	2007	X	1 or 2 pol. HH/HV/VV	1–20	Few–11	5–100
Cosmo-Skymed	2007	X	1 or 2 pol HH/HV/VV	1–100	12	10–200
ALOS-2	2013	L	Polarimetric HH/HV/VV	3–100	Few–14	25–350
Sentinel1	2014	C	1 or 2 pol HH/HV/VV	5–100 m	Few–12	80–400
SAOCOM	2015	L	Polarimetric HH/HV/VV	7–100	Few–16	60–320

- Canadian Space Agency (CSA): Radarsat-1, Radarsat-2, Radarsat constellation.
- Deutsches Zentrum für Luft- und Raumfahrt e.V. (DLR): TerraSAR-X, TanDEM-X.
- Indian Space Research Organization (ISRO): RISAT-1, NISAR (w/NASA).
- Comision Nacional de Actividades Espaciales: SAOCOM.
- Italian Space Agency (ASI): COSMO-Skymed.
- Instituto National de Técnica Aeroespacial (INTA): PAZ.
- Korea Aerospace Research Institute (KARI): KOMPSat-5.
- National Aeronautics and Space Administration (NASA): NISAR (w/ISRO) [6].

Since the launch of Seasat in 1978, there has been a fixed development of civilian SAR satellites in space. These include SIR-A, SIR-B, SIR-C/X-SAR, ERS-1/2, JERS-1, ENVISAT ASAR, RADARSAT-1&2, SRTM, COSMO-SkyMed, TerraSAR-X, and Advanced Land Observing Satellite (ALOS)/PALSAR. Nowadays, there are a number of SAR configurations and satellite platforms offering a variety of SAR data with more scheduled for the near future, ensuring a permanent resource of spaceborne SAR data for oceanography applications (Table 7.1). Conversely, most of these SAR sensors deliver a variety of beam modes from 1 m spotlight modes (with limited swath coverage) to low spatial resolution ScanSAR modes (50–100 m resolution) with large swath coverages. Each of these SAR satellite sensors has a web page or other information sources, where detailed information on the sensor specifics, data processing, and products can be acquired [7].

7.3 Sea surface backscatter based on the Kirchhoff approximation

The Kirchhoff approximation model is a cornerstone for comprehending the reflection of the sea surface. It is used when the incident pulse wavelength is shorter than the horizontal roughness scale, which is a function of the large average radius of curvature. In this view, the Green's theorem is essential for understanding the Kirchhoff approximation (KA), which designates the backscatter from the sea surface as tangential fields on the surface.

Let $(\omega, \boldsymbol{r}_r|\boldsymbol{r}_s)$ be the sea surface reflectivity, and ω an angular frequency, where the SAR sensor at height of $\boldsymbol{r}_s = (x_s, z_s)$ and the SAR satellite sensor at $\boldsymbol{r}_r = (x_r, z_r)$, so the sea surface reflectivity, which is coded by SAR single, is formulated as [8]:

$$R_{sea}(\omega, \boldsymbol{r}_r|\,\boldsymbol{r}_s) = -\int_{S_{fs}} \left[G(\omega, \boldsymbol{r}_r|\,\boldsymbol{r}_{fs})\left[\nabla P(\omega, \boldsymbol{r}_{fs}|\,\boldsymbol{r}_s)\cdot \boldsymbol{n}_{fs}\right]\right] dS_{fs}, \tag{7.1}$$

where, $G(\omega, \boldsymbol{r}_{fs}|\boldsymbol{r}_r)$ is the free-space Green functions with sources at $\boldsymbol{r}_{fs} = (x_{fs}, z_{fz})$ on the sea surface and SAR at $\boldsymbol{r}_r$, and $(\omega, \boldsymbol{r}_{fs}|\boldsymbol{r}_s)$ and $\boldsymbol{n}_{fs}$ are the pressure gradient and the normal vector at the sea surface S_{fs}, respectively.

The statistical parameters of the sea surface are required from remote sensing to compute the plane wave reflection coefficient matrix. This technique is measured on account of the difficulty of determining surface height from SAR data. Consistent with Thorsos [9], a Gaussian distribution can be implemented to designate sea surface height. With zero mean and standard deviation of σ, the coherent plane wave reflection coefficient matrix at the mean sea level based on the KA is formulated as:

$$\widehat{R}(\omega, \mathrm{K}_r|\theta_0) = \left\langle \widehat{R}(\omega, \mathrm{K}_r|\theta_0) \right\rangle = e^{\left\{-2\left[k_z^s\sigma\right]^2\right\}\widehat{R}^{Flat}_{coef}(\omega, \mathrm{K}_r|\theta_0)} \tag{7.2}$$

where $\langle\rangle$ is an anticipation operator, K_r is the wavenumber of the SAR, θ_0 is the incident angle, $\overset{\frown}{R}{}^{Flat}_{coef}$ is a finite length flat sea surface plane wave reflection coefficient matrix, and k_z^s is sea surface wavenumber.

In this regard, the Kirchhoff approximation considers the quasispecular reflection; however, it ignores polarization. The limitation of geometrical optics leads the incident wavelength to be zero. Under this circumstance, the Kirchhoff approximation is precise. If the SAR pulse points are far from the reflection of the sea surface, the absolute amplitudes of the reflection are very nearly the same. Consequently, the small-perturbation method (SPM) is used if the wavelength is larger than both the standard deviation and the correlation length of surface heights. Nonetheless, SPM neither explicate for long-scale ocean features in the surface spectrum nor for specular scattering that received by the radar antenna. Both approximations are considered to be rough surface, which has to be either large or small compared with the incidence wavelength. For a sea surface, the slopes are usually small, excluding steep breaking waves, which constitute a distinctly small percentage and arise most effectively at consistently high wind speeds. The small-slope approximation is the consequence of a Taylor expansion with reference to the influences of surface slopes. Finally, the strength of the KA is retraced in expressions of the specular and nonspecular reflections from the rough sea surfaces.

At the near-vertical angles, the impact of wind and waves in a modulation of the smooth sea surface is trivial. The reflectivity of the smooth sea surface, conversely, can be amplified as the wind speed increases. The increase of reflectivity, i.e., backscatter is due to the wide-ranging variation of backscatter intensity in conjunction with azimuth angle that depicts any dielectric impacts. The smooth sea surface would appear to be very dark near the nadir. Nevertheless, any wave propagation rapidly diminishes the reflectivity at large incidence angles. Under this circumstance, a roughened sea seems much darker in the direction of the horizon than a smooth one.

7.4 Imaging Ocean wave parameters in single polarization SAR data

In Chapter 5, it was recognized that tilt modulation, hydrodynamic modulation, and velocity bunching are the main modulation mechanisms for imaging ocean surface waves in SAR data. Briefly, tilt modulation occurs due to variations in the local incidence angle instigated by the surface wave slopes [10]. Tilt modulation, therefore, is robust for waves traveling in the range direction. Hydrodynamic modulation, on the contrary, occurs due to the hydrodynamic exchanges between the long-scale surface waves and the short-scale surface (Bragg) waves that provide the majority of the backscatter at sensible incidence angles [11]. On the other hand, velocity bunching is a modulation process that is unique to SAR imaging systems [12]. It is a consequence of the azimuth shifting of scatterers in the SAR image plane, owing to the mobility of the scattering surface. Velocity bunching is paramount for azimuthally traveling waves.

Numerous nonlinear inversion procedures have been established for simulating wave spectra parameters from SAR image spectra. Most of these practices are grounded on a technique created by Hasselmann and Hasselmann [13]. This innovative technique deployed an iterative technique to guesstimate the wave spectrum from the SAR image spectrum. Original computations are attained by exploiting a linear transfer function [14]. These algorithms are exploited as contributions in the presumptuous SAR imaging model, and the modified SAR image spectrum is exploited to rectify iteratively the preceding approximation of the wave spectra. The specifics of this approach, conversely, rely on the particular SAR imaging model. Consequently, perfections to this algorithm [15] have assimilated closed-form elucidations of the nonlinear transfer function, which transmits the wave spectrum into the SAR image spectrum. Nonetheless, this transfer function also has to be appraised iteratively. Additional enhancements to this technique have been instituted by Engen and Johnsen [16] and Lehner et al. [17]. In this technique, a cross-spectrum is created between dissimilar images of the same SAR ocean wave scene. The principal improvement of this technique is that it settles the uncertainty of 180-degree ambiguity of the wave direction. This technique also diminishes the impact of speckle noises in the SAR spectrum image. Moreover, algorithms that integrate further posterior clues about the wave field, which recover the precision of these nonlinear techniques, have also been established in recent decades [18]. In the suggested slope retrieval techniques, the one nonlinear mechanism that perhaps fully abolishes the wave spectrum pattern is velocity bunching [13]. Velocity bunching, subsequently, is a consequence of scatterers' movements on the ocean surface, either bunching or distending in the SAR image frame. The impacts of velocity bunching would be existed, but are not extremely sufficient to capture the slope of the ocean spectrum retrieval process. It can be reasoned that for velocity bunching to influence the slope approximations, the $\frac{R}{V_s}$ ratio has to be clearly larger than 100s. In other words, if the $\frac{R}{V_s}$ ratio is preciously enlarged to large values, which can trash the wave spectra pattern in the slope approximations. In this circumstance, the shifting of the scatterers in the azimuth direction is perhaps a consequence of the wave spectra pattern destruction in the SAR image. In this regard, the wave slope pattern is well-preserved in the existence of reasonable velocity bunching modulation.

7.5 How to relate wave fields to SAR images

A significant question that may arise is: how do we map the real ocean wave spectra into SAR images? To this end, the main procedure is based on the transfer function. In this regard, it is considered as a mathematical function relating to the real ocean wave spectra or the response of a system such as the SAR frequency domain. The mathematical description of mapping nonlinear RAR to SAR transform domain can be formulated as:

$$\hat{I}_{SAR}(k) = \lim_{x\to\infty} |x|^{-1} \int_x e^{[-ik(u_r + \eta_x(u_r))]} \psi_{RAR}(I_{RAR}) u_r du_r \tag{7.3}$$

Eq. (7.3) reveals that the quantum mechanics of the azimuth shifts η_x owing to radial velocity u_r.

$$\psi|(k)\rangle = \lim_{x\to\infty} |x|^{-2} \int_x e^{[-ik(u_r - u_{r_0})]} \psi_{RAR}|(u_r, u_{r_0}, k_a)\rangle u_r du_0 \tag{7.4}$$

$$\psi_{RAR}|(u_r, u_{r_0}, k_a)\rangle = E\left[I_{RAR}(u_r) e^{\{-ik_a[\eta_x(u_r) - \eta_x(u_{r_0})]\}}\right] \tag{7.4.1}$$

Eq. (7.4) shows a quasilinear expression that is obtained for the first-order part of this formula. The dominant modulation mechanism is predicted to be due to the velocity bunching—constructive and destructive—phenomena. This formula also does not comprise the hydrodynamic modulation definition. Hydrodynamic modulation linearly can describe the small scale scatterers' nonuniform distribution over the long wave profiles, in theory, such as a modulation transfer function that can be governed from the formulation of the energy balance equation of short waves.

7.6 SAR wave retrieval algorithms

A retrieval algorithm commonly endeavors to reform the ocean wave spectrum by diminishing the dissimilarity between its consistent speculative SAR spectrum, which is acquired with a presumptuous transformation, and the satellite observation. The precise derivative of the nonlinear transform ensuing also awkward to perform, the greatest of the prevailing inversion patterns partially disregard the comprehensive nonlinear wave spectral mapping. Besides, nonlinear transform approximation frequently operates as the abridging gradient of well-known optimized SAR quasilinear transform that matches the full nonlinear transform precisely. The simplest algorithm is known as direct quasilinear inversion. The keystone of direct quasilinear inversion is a cutoff parameter, which is based on the modulation transfer functions, i.e., tilt, hydrodynamic, and velocity bunching. Such a technique delivers practical consequences for weak cutoff parameter cases. In the circumstances of small $\frac{R}{V_s}$, low wind velocity conditions (relatively low sea state conditions), a weak cutoff parameter can be used to retrieve very long wave spectral information. The disadvantage of the weak cutoff algorithm is that lower wavenumber spectral features are dominant features through the model. Nonetheless, a quasilinear transform is not able to retrieve lower wavenumber spectra. On the contrary, the full nonlinear algorithm can be used to simulate the capillary wave in the SAR image. The inversion technique is a prerequisite as a priori knowledge of the swell propagation direction.

Another simple technique is a mixed inversion, which is based on an optimized quasilinear SAR spectral mapping. This quasilinear transform is used both to compute the gradient of the cost function and to perform the forward mapping between the iterative steps. This algorithm was first described by Krogstad and Vachon [19], and involves an initial guess spectrum. In this context, the minimization is carried out by stating an initial guess wave spectrum contained by the constituencies expressed by the combination of the SAR and the first guess spectral domain. The cost function can comprise a term allied with the azimuth cutoff. This method will also undergo lower wavenumber features. Correspondingly, in circumstances of opposite wave transmitting schemes as anticipated by a first guess spectrum, one of them would be preserved unaltered throughout the inversion. A partitioning technique is formerly needed to improve the calculation load.

Improved mixed inversion is used to compile the complete nonlinear forward SAR transformation with the use of the gradient of an optimized quasilinear one. This algorithm necessitates an a priori first guess spectrum. This method was first introduced by Hasselmann and Hasselmann [13]. Foremost, the inclusive degree of the unmeasured high wavenumber fragment of the spectrum is someway enhanced, regulated through the presence of an obvious cutoff consequence term in the cost function. Subsequently, the inversion technique merely transforms the comprehensive procedure of the spectrum in the spectral domain for which SAR spectral information is accessible. This algorithm would lead to improbable incoherence in the transition interval extrication of the dissimilar spectral regions of impact. To avoid being overwhelmed by such complexity, a spectral partitioning is presented in an extra iteration loop that describes the input spectrum [20,21].

This inversion technique is by far the most complex scheme. Nonetheless, the scheme operates ocean WAve Model (WAM) as a first guess and a spectral partitioning, the ambiguity is in most circumstances appropriately disentangled for all modes. WAM is also well-known as the approximation technique for retrieving ocean wave spectra parameters; for instance, significant wave height, direction and spectra period. Enhancements in this inversion algorithm have been accomplished by expanding the existing algorithm with a cross-spectral analysis to develop ambiguity removal and to improve handling of low signal-to-noise ratio (SNR) spectra [22,23]. This involves the implementation of a precise technique to approximate the azimuth cutoff parameter, which varies from the utmost of azimuth cutoff approximation procedures.

The first guess wind sea spectrum can be simulated by a two-step inversion scheme. This technique is grounded on the postulation that the wind sea fragment of the spectrum provides the most to the nonlinear imaging mechanism. The inversion is detached into dual phases: first, a nonlinear minimization concerning the wind mode, and second, a linear minimization about the swell mode. The wind mode minimization is achieved concerning the wind, sea peak direction, and wave duration using the full nonlinear SAR forward transformation. The residual SAR spectrum, i.e., the practical SAR spectrum minus the predictable wind sea influence, is then minimized regarding slight swell mode, expending a linear SAR guess. In this phase, the impact of the ocean swell to the cutoff consequences is verified to the extra enhancement of the furthermost expected wave age restriction: if compulsory, a new iteration is required to monitor the accuracy of cutoff retrieving parameters. This method greatly depends on reliable a priori information relating to the local wind sea circumstances. This algorithm has primarily been developed (ARGOOS, IFREMER) using cosited wind vector scatterometer approximations. Enhancing the exact SAR wind vector guess would then further improve this technique, but the examination of a wave prediction model (e.g., WAM) can also be exploited to modify the wind sea modes for the first step. The cross-spectral analysis helps to eliminate the swell ambiguity propagation problem. The algorithm approved by the ESA to deliver the so-called level 2 products belongs to this class of approaches [24].

7.7 Quantum spectra estimation using quantum Fourier transform

Since the wave modifies its direction and wavelength as it propagates, the two-dimensional (2D) discrete Fourier transform (DFT) is used to derive the wave quantity spectra from SAR satellite data. First, select a window kernel length of 512×512 pixels and lines with the pixel size identical to ΔX (Fig. 7.3). It is impossible to acquire wave spectra parameters using a kernel window size of less than 512×512 pixels and lines (Fig. 7.4). The example SAR image is ENVISAT ASAR C_{VV}-band satellite data along the coastal waters off Johor, Malaysia. These data were acquired on December 28, 2010 (Fig. 7.5). The white boxes in

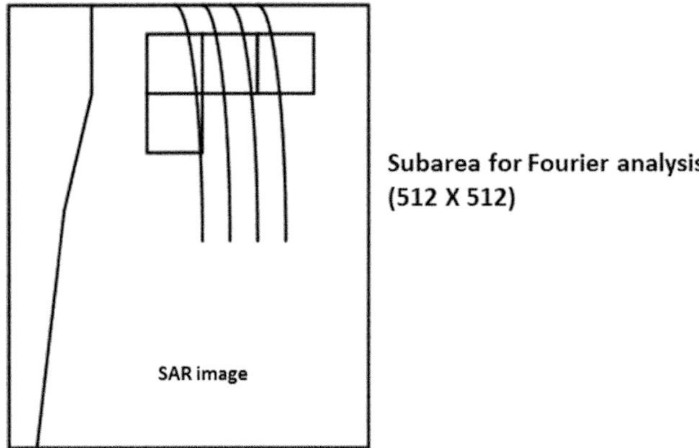

FIG. 7.3 Kernel windows of 2-DFFT.

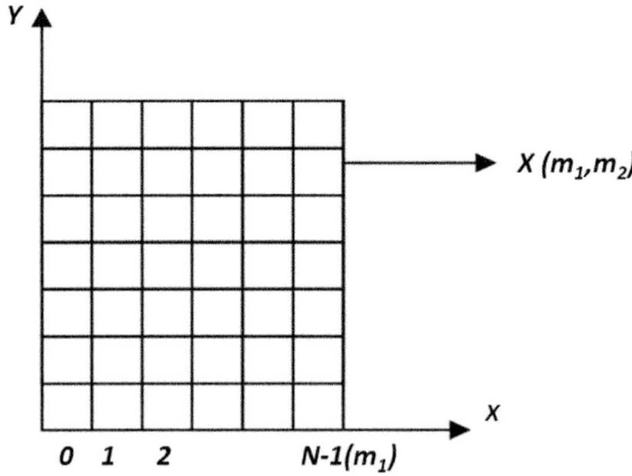

FIG. 7.4 Fourier analysis to extract wave spectral from an SAR image.

Fig. 7.5 show the selected kernel window size of 512×512 pixels and lines to wave spectral parameters through 2-DFFT.

Following Populus et al. [25], let $X(m_1, m_2)$ represent the digital count of the pixel used to perform DFT, which is given by:

$$F(k_a, k_r) = N^{-2} \sum_{m_2=0}^{N-1} \left[\sum_{m_1=0}^{N-1} X(m_1, m_2) \cdot e^{-ik_a \cdot m_1 \cdot \Delta X} \right] \cdot e^{-ik_r \cdot m_2 \cdot \Delta X} \tag{7.5}$$

where, n_1 and $n_2 = 1, 2, 3, \ldots, N$ and k_a and k_r are the wavenumbers in the azimuth and range directions, respectively.

As stated by Populus et al. [25], 2-DFFT transfers the SAR data into the frequency spectral-domain (k_a and k_r) along the azimuth and range directions, respectively. The frequency spectral-domain (Fig. 7.6) is implemented to transfer the image from the spatial domain to the frequency domain.

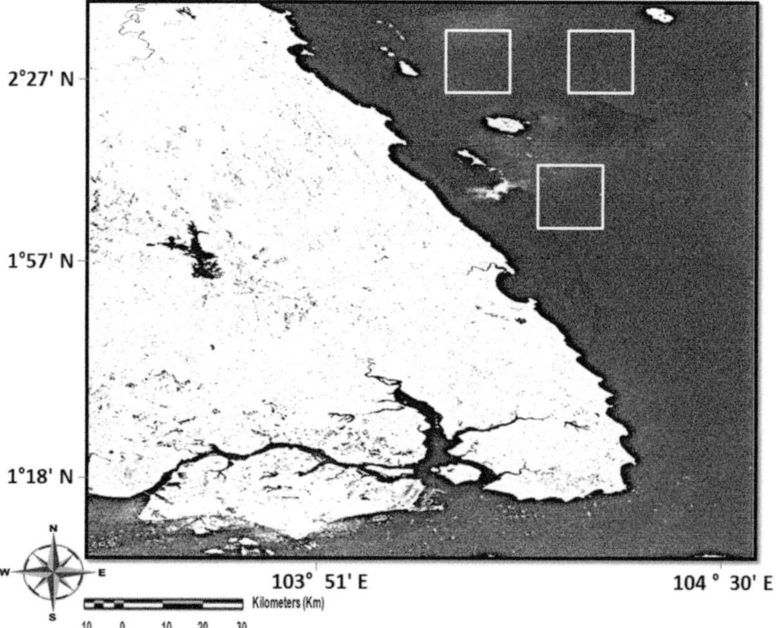

FIG. 7.5 ENVISAT ASAR C_{VV}-band.

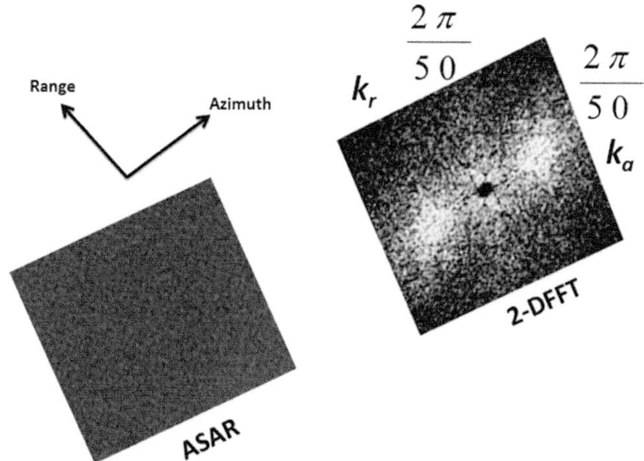

FIG. 7.6 Spectra frequency of ASAR backscatter using 2-DFFT.

Shor [26] and Galindo et al. [27] introduced quantum Fourier transform (QFFT). Let us assume that an orthonormal basis is identified as $|0\rangle, |1\rangle, |2\rangle, \ldots, |N-1\rangle$. Therefore, the linear operator of QFFT can be formulated as:

$$QFFT|j\rangle = \frac{1}{\sqrt{N}} \sum_{k=0}^{N-1} e^{\frac{2\pi ijk}{N}} |k\rangle \tag{7.6}$$

QFFT has a time complexity of $O(\log^2 N)$. The 2DQFFT can be constructed as follows:

Create the subsequent data structures (DS) of pixels and lines of SAR image and unitary operations:

DS1: all of the line vectors are stored at $\vec{k}_a$ and all column vectors are stored in the matrix $\vec{k}_r$.

DS2: Create seven registers that have a data format $\left|\left\|\frac{\vec{k}}{N}\right\|^2\right\rangle\left|\left\|\vec{k}\right\|^2\right\rangle|a\rangle|r\rangle\left|\vec{k}_a\right\rangle\left|\vec{k}_r\right\rangle\left|\left(\vec{k}_a\cdot\vec{k}_r\right)^2\right\rangle$.

A unitary operation U_L is required to load all records that are stored in a classical database into a quantum state. The function of unitary operation U_L can be designated as:

$$\frac{1}{\sqrt{N}}\left(\sum_{a=0}^{N-1}|a\rangle|0\rangle\right)|ancilla\rangle\xrightarrow{U_L}\frac{1}{\sqrt{N}}\left(\sum_{a=0}^{N-1}|a\rangle|record\rangle\right)|ancilla\rangle \tag{7.7}$$

Eq. (7.7) reveals that Grover's algorithm has the function that locates the index a_0 of a special database record, $record_{a_0}$, from the index superposition of state $\frac{1}{\sqrt{N}}\left(\sum_{a=0}^{N-1}|a\rangle\right)$. However, the corresponding record, $record_{a_0}$, cannot be measured out unless the one-to-one mapping relationship between the index a and the corresponding record, $record_{a_0}$, is bound in the entangled state $\frac{1}{\sqrt{N}}\left(\sum_{a=0}^{N-1}|a\rangle|record\rangle\right)$. Then dual U_L is applied to load vectors $\vec{k}_a$ and $\vec{k}_r$ into registers from the classical database, respectively, which are formulated as:

$$\left|\left\|\frac{\vec{k}}{N}\right\|^2\right\rangle\left|\left\|\vec{k}\right\|^2\right\rangle|a\rangle|r\rangle|0\rangle|0\rangle|0\rangle\xrightarrow{U_{L_1}}\left|\left\|\frac{\vec{k}}{N}\right\|^2\right\rangle\left|\left\|\vec{k}\right\|^2\right\rangle|a\rangle|r\rangle\left|\vec{k}_a\right\rangle|0\rangle|0\rangle \tag{7.8}$$

$$\left|\left\|\frac{\vec{k}}{N}\right\|^2\right\rangle\left|\left\|\vec{k}\right\|^2\right\rangle|a\rangle|r\rangle|0\rangle|0\rangle|0\rangle\xrightarrow{U_{L_2}}\left|\left\|\frac{\vec{k}}{N}\right\|^2\right\rangle\left|\left\|\vec{k}\right\|^2\right\rangle|a\rangle|r\rangle|0\rangle\left|\vec{k}_r\right\rangle|0\rangle \tag{7.9}$$

DS4: Design oracle B_{inner} to calculate squared inner product, which is given by:

$$\left|\left\|\frac{\vec{k}}{N}\right\|^2\right\rangle\left|\left\|\vec{k}\right\|^2\right\rangle|a\rangle|r\rangle\left|\vec{k}_a\right\rangle\left|\vec{k}_r\right\rangle|0\rangle\xrightarrow{B_{inner}}\left|\left\|\frac{\vec{k}}{N}\right\|^2\right\rangle\left|\left\|\vec{k}\right\|^2\right\rangle|a\rangle|r\rangle\left|\vec{k}_a\right\rangle\left|\vec{k}_r\right\rangle\left|\left(\vec{k}_a\cdot\vec{k}_r\right)^2\right\rangle \tag{7.10}$$

DS5: Design oracle O_F', which is given by:

$$\left|\left\|\frac{\vec{k}}{N}\right\|^2\right\rangle\left|\left\|\vec{k}\right\|^2\right\rangle|a\rangle|r\rangle\left|\vec{k}_a\right\rangle\left|\vec{k}_r\right\rangle\left|\left(\vec{k}_a\cdot\vec{k}_r\right)^2\right\rangle\xrightarrow{O_F'}(-1)^{F'(k_a,k_r)}\left|\left\|\frac{\vec{k}}{N}\right\|^2\right\rangle\left|\left\|\vec{k}\right\|^2\right\rangle|a\rangle|r\rangle\left|\vec{k}_a\right\rangle\left|\vec{k}_r\right\rangle\left|\left(\vec{k}_a\cdot\vec{k}_r\right)^2\right\rangle \tag{7.11}$$

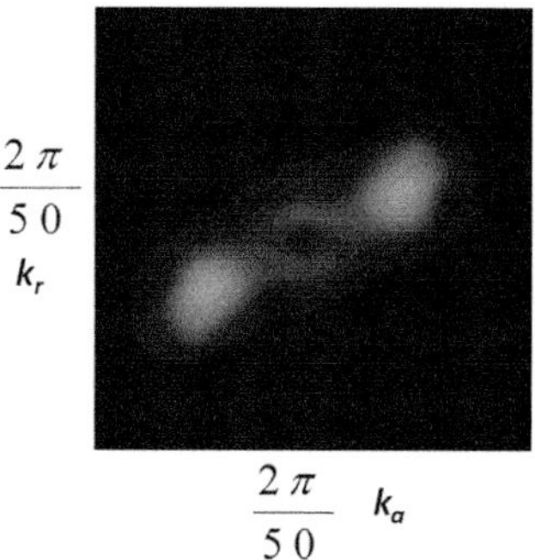

FIG. 7.7 ASAR wave spectra derived by 2DQFFT.

where:

$$F'(k_a, k_r) = \begin{cases} 1 \;\; if \left\langle \left\| \frac{\vec{k}}{N} \right\|^2 \right\rangle \leq \left(\vec{k}_a \cdot \vec{k}_r\right)^2 \leq \left\langle \left\| \vec{k} \right\|^2 \right\rangle \\ 0 \qquad\qquad otherwise \end{cases} \tag{7.12}$$

The final stage is to generate 2D QDFT iteration, which is formulated as [28]:

$$F_{2DQFFT} = \left(2 \left| \frac{1}{N} \sum_{a=0}^{N-1} \sum_{r-0}^{N-1} |k_a\rangle |k_r\rangle \right\rangle \left\langle \frac{1}{N} \sum_{a=0}^{N-1} \sum_{r-0}^{N-1} |k_a\rangle |k_r\rangle \right\rangle - I \right) (U_{L_1})^{\dagger} (U_{L_2})^{\dagger} (B_{inner})^{\dagger} \\ O'_f B_{inner} U_{L_2} U_{L_1} \tag{7.13}$$

The perfect consequence of the 2DQFFT is demonstrated in Fig. 7.7. 2DQDFT delivers sharper spectrum's peaks than 2DFFT. 2DQFFT relies on the redundancy (or smooth property) of the signal sequences. The higher the redundancy is, the faster the 2DQFFT runs.

Pang and Hu [28] state that, compared with 2DFFT, QDFT has two advantages. One of the advantages is that 1D and 2D QDFT has time complexity O $((\sqrt{N}))$ and O(N), respectively. Besides, the other advantage is that QDFT can handle a precise long SAR signal sequence at a time.

7.8 Multilooking and cross-spectral analysis

Complex SAR images contain information about the evolution of the radar backscatter in a short duration since they are created by assimilating the backscattered signal within an approximately interval time of 0.6 s. The "multilooking" technique is applied in which the integration window is divided into dual or extra subwindows. In this circumstance, the azimuth integration is accomplished to acquire numerous "sublooks."

This technique is mainly of significance for the SAR imaging of ocean waves. Presuming that dual sublooks have been extracted from a complex SAR image, they are parted by an approximate interval time of ~0.3 s, and they deliver information on the wave transmission direction without the 180-degree ambiguity [20,22,29]. In other words, an integral transform recitation of the ocean wave spectrum F_k to the SAR crosses the spectrum of dual looks l^i with separated time ∂t, which is given by:

$$\psi^{\partial t}_{l^1,l^2}\left|\vec{K}\right\rangle = \frac{1}{4\pi^2} e^{\left(\left|-k_a^2\right\rangle\left(\frac{R}{V_s}\right)^2 f^v(0)\right)} \cdot \int_{R^2} e^{\left|-ik_a\right\rangle} e^{\left(\left|-k_a^2\right\rangle\left(\frac{R}{V_s}\right)^2 f^v(\vec{x})\right)} \cdot \left\{1 + f^R\left|\vec{x}\right\rangle + \left|-ik_a\right\rangle \frac{R}{V_S}\left(f^{Rv}\left|\vec{x}\right\rangle\right)\right.$$
$$\left. - \left(f^{Rv}\left|-\vec{x}\right\rangle\right) + \left|k_a^2\right\rangle\left(\frac{R}{V_S}\right)^2 \left(f^{Rv}\left|\vec{x}\right\rangle\right) - \left(f^{Rv}\left|0\right\rangle\right)\left(\left(f^{Rv}\left|-\vec{x}\right\rangle\right) - \left(f^{Rv}\left|0\right\rangle\right)\right)\right\} d^2\vec{x} \tag{7.14}$$

Here $\psi^{\partial t}_{l^1,l^2}\left|\vec{K}\right\rangle$ signifies the cross-spectrum guessed from the normalized dual looks of SAR complex data, which is defined as QFFT of the cross-covariance functions f^R, f^{Rv}, and f^v, respectively, which are formulated as follows:

$$f^R\left|\vec{x}\right\rangle = 0.5\int_{R^2} \psi\left|\vec{K}\right\rangle \cdot \left|T_k^R\right\rangle e^{\left|i\omega\partial t\right\rangle} + \psi\left|-\vec{K}\right\rangle \cdot \left|\left|T_{-k}^R\right|^2\right\rangle e^{\left|-i\omega\partial t\right\rangle} e^{\left|i\vec{k}\vec{x}\right\rangle} d^2\vec{K} \tag{7.15}$$

$$f^{Rv}\left|\vec{x}\right\rangle = 0.5\int_{R^2} \psi\left|\vec{K}\right\rangle \cdot \left|T_k^R\right\rangle\left|T_k^{\eta_x}\right\rangle e^{\left|i\omega\partial t\right\rangle} + \psi\left|-\vec{K}\right\rangle \cdot \left|\left|T_{-k}^R\right|^2\right\rangle\left|T_{-k}^{\eta_x}\right\rangle e^{\left|-i\omega\partial t\right\rangle} e^{\left|i\vec{k}\vec{x}\right\rangle} d^2\vec{K} \tag{7.16}$$

$$f^v\left|\vec{x}\right\rangle = 0.5\int_{R^2} \psi\left|\vec{K}\right\rangle \cdot \left|T_k^{\eta_x}\right\rangle e^{\left|i\omega\partial t\right\rangle} + \psi\left|-\vec{K}\right\rangle \cdot \left|\left|T_{-k}^{\eta_x}\right|^2\right\rangle e^{\left|-i\omega\partial t\right\rangle} e^{\left|i\vec{k}\vec{x}\right\rangle} d^2\vec{K} \tag{7.17}$$

Following Hasselmann and Hasselmann [13], Eq. (7.14) can be expanded to first-order concerning the wave spectrum. F_k yields the linear approximation as:

$$\psi^{\partial t}_{l^1,l^2}\left|\vec{K}\right\rangle = 0.5\left(\left|\left|T_k^s\right\rangle\right|^2 e^{\left|i\omega\partial t\right\rangle}\psi\left|k\right\rangle + \left|\left|T_k^s\right\rangle\right|^2 e^{\left|-i\omega\partial t\right\rangle}\psi\left|-k\right\rangle\right) \tag{7.18}$$

Eq. (7.18) involves the transfer function modulation T_k^s as explained and discussed in Chapter 6, which is defined by:

$$\psi\left|T_k^s\right\rangle = \left|T_k^R\right\rangle + i\left(\frac{R}{V_s}\right)\left|k_x\right\rangle\left|T_k^{\eta_x}\right\rangle \tag{7.19}$$

The quasilinear transform can be obtained by expanding an integral part of Eq. (7.14) to linear order and keeps the leading exponential factor as formulated by:

$$\psi^{\partial t}_{l^1,l^2}\left|\vec{K}\right\rangle = 0.5 e^{\left(\left|-k_a^2\right\rangle\left(\frac{R}{V_s}\right)^2 f^v(0)\right)}\left(\left|\left|T_k^s\right\rangle\right|^2 e^{\left|i\omega\partial t\right\rangle}\psi\left|k\right\rangle + \left|\left|T_k^s\right\rangle\right|^2 e^{\left|-i\omega\partial t\right\rangle}\psi\left|-k\right\rangle\right) \tag{7.20}$$

Concerning Eq. (7.20), the quasilinear forward model is useful as it produces a simple transform to retrieve the two-dimensional ocean wave spectrum from the SAR cross-spectrum [16,30].

Fig. 7.8 reveals the two sublooks mined from the same complex ASAR image to perform the cross-spectrum technique. The second plot represents the real part (Fig. 7.9) and the imaginary part (Fig. 7.10) of the cross-spectrum computed from two sublooks extracted from the

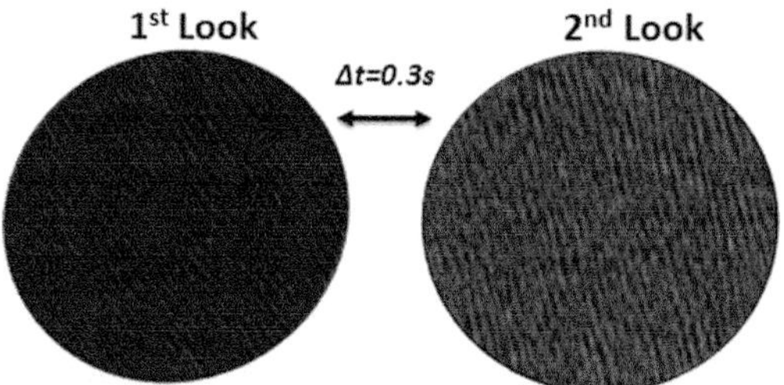

FIG. 7.8 Two selected sublooks from complex ASAR data.

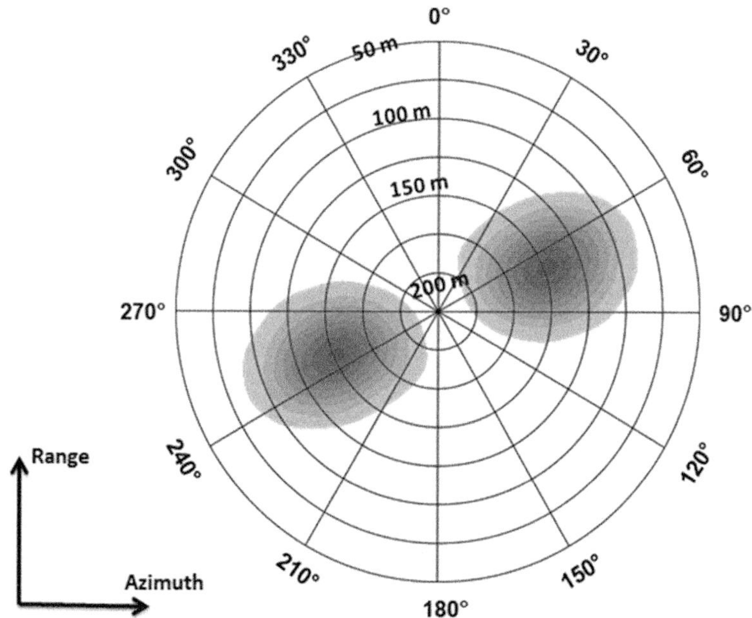

FIG. 7.9 Cross-spectrum results of the real part.

same complex SAR image (Fig. 7.8). The positive imaginary part of the cross-spectrum indicates that the imaged swell system propagates from the ocean toward the Johor coastal waters. Another advantage of SAR cross-spectra is that they have lower noise levels than the SAR image spectrum computed from a single look.

In other words, the real part of the spectrum is symmetric, whereas its imaginary part is antisymmetric. On the contrary, the positive imaginary part of the cross-spectrum specifies the actual propagation direction of the waves in which the wave propagates from the northeast direction by about 60 degrees. This indicates that the cross-spectra technique can be used to remove 180-degree ambiguities. Resolving the 180-degree ambiguities is extremely important when we deal with the simulation of the wave refraction pattern, as will be discussed in the next chapter.

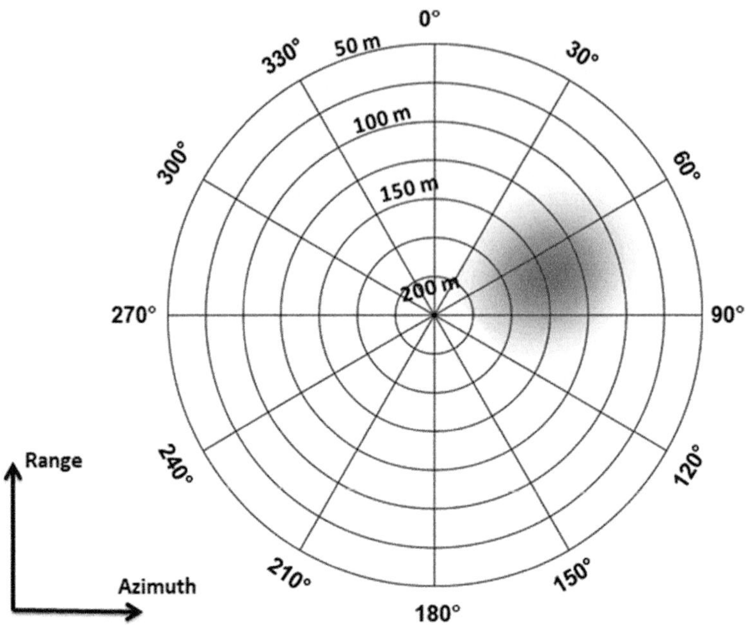

FIG. 7.10 The positive imaginary part of the cross-spectrum.

7.9 Quantum Monte Carlo wave spectral simulation

The term "Quantum Monte Carlo" (QMC) comprises all the Monte Carlo approaches intended for the investigation of quantum systems, by resolving the quantum anticipation values of SAR spectral parameters. In QMC, the multidimensional integrals that develop from transferring ocean wave field characteristics into SAR images are handled via the Monte Carlo method. QMC techniques are the most accurate tool to deal with ground-state properties. In the case of nonlinear wave spectra extraction from SAR images, these methods can produce essentially exact results for the equation of state and structural properties, which are, in both cases, in close agreement with ocean wave field data. On the other hand, they provide an approximate but very accurate description of fermionic systems. This approximation of the fermionic systems is due to the sign problem: the wave function is not positively defined as it must be antisymmetric under particle permutations. As it is not positive, it fails to be used as a probability distribution that could be sampled via Monte Carlo methods.

Let us assume that the $3N$ dimensional vector $K=(k_1,k_2,\ldots,k_N)$, which has a probability distribution $P(K)$. Consequently, the integral of $P(K)$ in the space K is defined as:

$$\int d\vec{K}\,P\left(\vec{K}\right)=1 \tag{7.21}$$

Diffusion quantum Monte Carlo (DQMC) is implemented to retrieve precise ocean wave spectra from the SAR image. It is considered as a stochastic projection method that projects out the ground state by acting the imaginary-time projector $e^{[-\tau \hat{H}]}$ repeatedly on a trial wave function ψ_T. The imaginary-time Schrödinger equation is given by:

$$-\partial_\tau \psi_T\left(\vec{K},\tau\right) = \left(\hat{H} - \frac{\langle \psi_T|\hat{H}|\psi_T\rangle}{\langle \psi_T|\psi_T\rangle}\right)\psi\left(\vec{K},\tau\right). \tag{7.22}$$

Here $\vec{K}$ is the wavenumber vector of SAR wave spectra that is estimated using the cross-spectrum technique and $\hat{H}$ is Hamiltonian. The main concept of the diffusion quantum Monte Carlo technique depends on the Green function:

$$G\left(\vec{K} \leftarrow \vec{K}',t\right) = \left\langle \vec{K}\middle| e^{\left[-\omega t\left(\hat{H} - \frac{\langle \psi_T|\hat{H}|\psi_T\rangle}{\langle \psi_T|\psi_T\rangle}\right)\right]}\middle|\vec{K}'\right\rangle = \sum_i \psi^*\left(\vec{K}\right) e^{\left[-\omega t\left(\hat{H}\left(\vec{K}\right) - \frac{\langle \psi_T|\hat{H}|\psi_T\rangle}{\langle \psi_T|\psi_T\rangle}\right)\right]} \psi\left(\vec{K}\right) \tag{7.23}$$

Let us consider the kinetic energy for 3N-dimensional space. This differential equation describes the diffusion stochastic process, and the Green function has a Gaussian expression with a variance τ, which is given by:

$$G\left(\vec{K} \to \vec{K}',\tau\right) = (2\pi\tau)^{\frac{-3N}{2}} e^{\left[\frac{-\left|\vec{K}-\vec{K}'\right|^2}{2\tau}\right]} \tag{7.24}$$

This green function is interpreted as a transition probability density DMC of walkers that moving from $\vec{K}$ to $\vec{K}'$, which can be simulated by the stochastic process. The probability density regulates the number of walkers that continue to the succeeding step. The parameter $\frac{\langle \psi_T|\hat{H}|\psi_T\rangle}{\langle \psi_T|\psi_T\rangle}$ is accustomed throughout the algorithm to validate that the total number of walkers is steady.

7.10 SAR wave spectra simulated using diffusion quantum Monte Carlo

As stated above, diffusion quantum Monte Carlo is a quantum Monte Carlo technique that makes use of a Green's function characteristic to resolve the Schrödinger equation. DMC is probably a numerically precise algorithm, which means that it can locate the specific principle phase power within a given error for any quantum machine. Fig. 7.11 demonstrates the implementation of DMC to simulate the wave spectra, which is grounded on the Pierson-Moskowitz (PM) spectra. The wave spectra travel northeast toward the coastal water of Johor, Malaysia. This wave spectrum represents the ocean wave characteristics during the northeast monsoon period. This monsoon wave spectra is dominated by a narrow peak with a significant wave height of 3 m and the largest wavelength of 200 m.

Fig. 7.12 reveals the simulated sea surface elevation using DMC. The DMC simulation reveals a clearer sea surface wave elevation (Fig. 7.12B) than the ASAR (Fig. 7.12A). DMC simulation determines the clear pattern of ocean wave propagation into a realistic stage. Fig. 7.12B agrees with PM wave spectra that the swell wave propagation is in the northeast

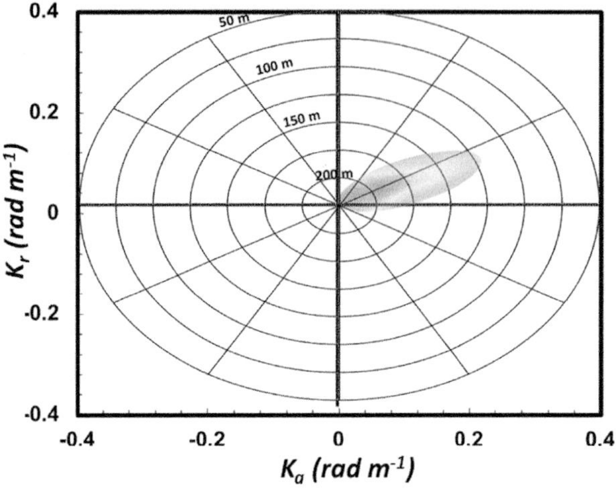

FIG. 7.11 Pierson-Moskowitz (PM) spectra along the coastal waters off Johor, Malaysia.

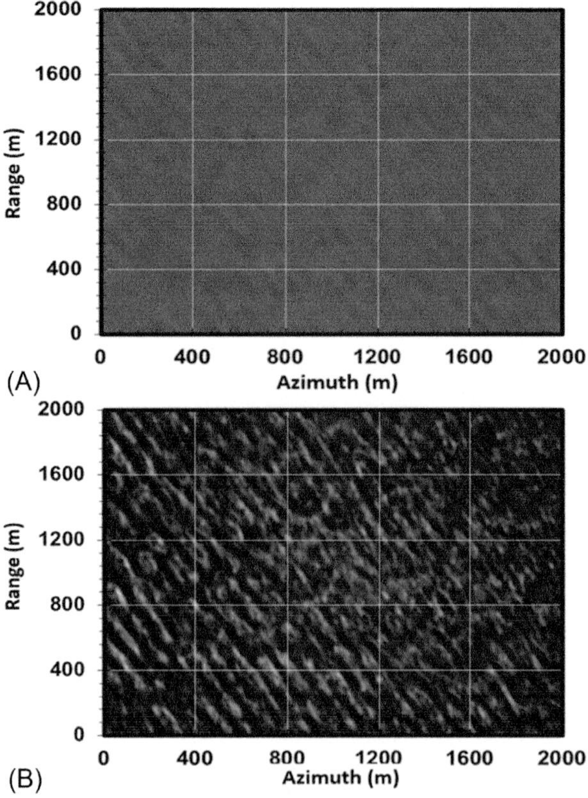

FIG. 7.12 Sea surface elevation simulated using (A) ASAR data and (B) DMC algorithm.

direction with r^2 of 0.85 (Table 7.2). Fig. 7.13 shows the cross-spectrum obtained by DMC as the real part. In this regard, the correlation DMC and cross-spectrum of the real part is 0.86 (Table 7.2); likewise, the real part of the cross-spectrum is the integral transform part [31]. Similarly, the imaginary part of the cross-spectrum obtained by DMC (Fig. 7.14A) is matched

TABLE 7.2 Correlation of different DMC techniques.

Technique	r^2
DMC-cross spectral (real part)	0.86
DMC-integral transforms	0.88
DMC-ASAR (swell propagation)	0.85

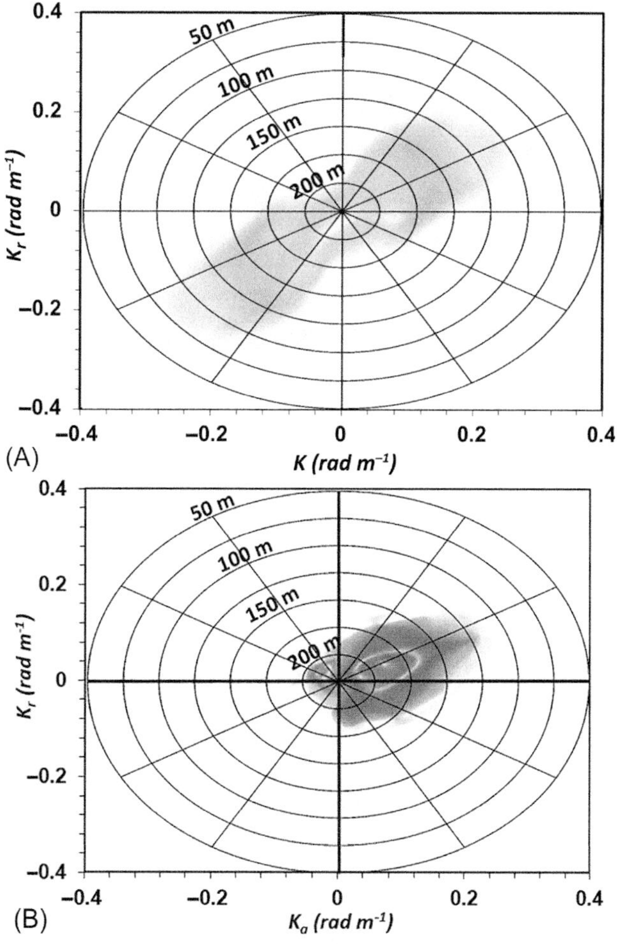

FIG. 7.13 DMC cross-spectrum (A) real and (B) integral transform of real part.

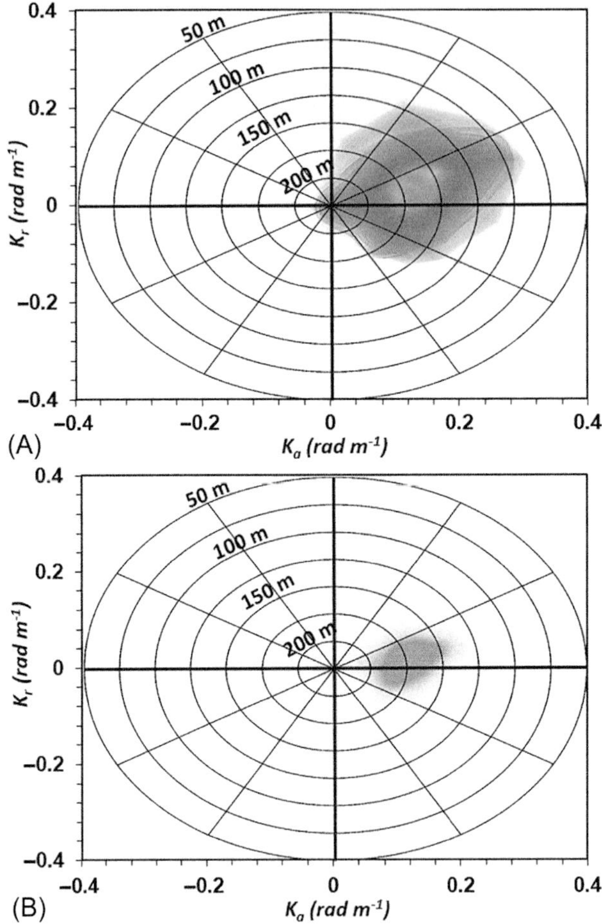

FIG. 7.14 DMC cross-spectrum (A) real and (B) integral transform of the imaginary part.

with the integral transform of the imaginary part of the spectrum (Fig. 7.14B). It can be said that the cross-spectra obtained from DMC simulations match the ones from the integral transform extremely accurately, indicated by r^2 of 0.88 (Table 7.2).

In addition, the dominant traveling wave is the azimuth wave, due to the strongest impact of the nonlinearity associated with the relativistic quantum state, as discussed in Chapter 6. Moreover, DMC delivers a well-constructed RAR image (Fig. 7.15). DMC declines the wave energy along the azimuth direction. Fig. 7.16 shows a simulated ASAR image using DMC in which the azimuth modulation is accurately reconstructed specifically with a long wave of 150m. Likewise, the distortion of the wave field in the ASAR image is extremely effective along the azimuth direction owing to the impact of the nonlinearity of velocity bunching. Lastly, cross-spectra simulated by DMC reveals a harmonic spectral pattern (Fig. 7.17A), in which the spectral peaks are perfectly located away from the azimuth direction and match with PM spectrum peak (Fig. 7.17B).

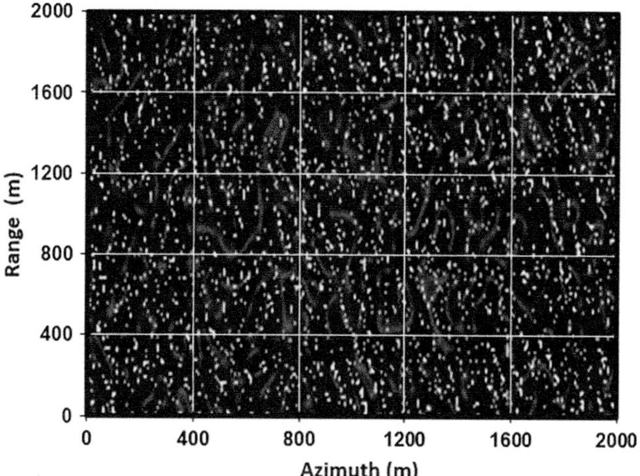

FIG. 7.15 DMC RAR simulated image.

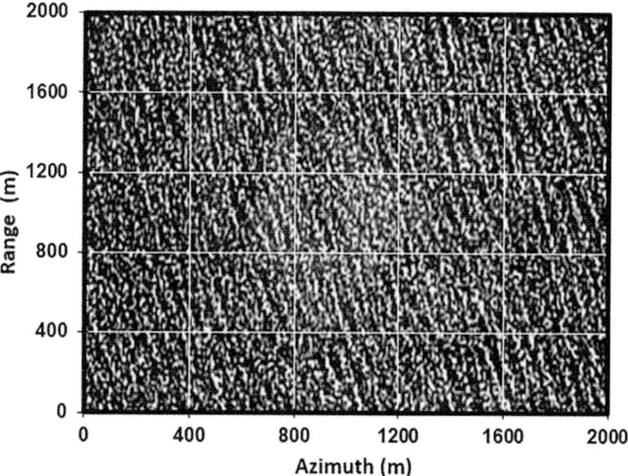

FIG. 7.16 DMC ASAR simulated sea surface preparation.

The advantage of such DMC simulations is their flexibility, since it is conceivable to instigate more sophisticated wave and backscatter models, to the extent that the integral transform approach is improved using quantum Monte-Carlo simulation, and articulate the ocean-to-SAR mapping in a perfectly matched form. The computing of SAR wave spectra using DMC reveals a good estimator, due to the trial wave function being a good approximation. On this understanding, DMC can construct fluctuations of the ocean wave surface as a function of the variations image spectral density owing to the interaction between radar signal energy and sea surface dynamic energy [20–23,29,31]. The greatest common operation of DMC diffusion is matchless controls a renewed experimental function which is also

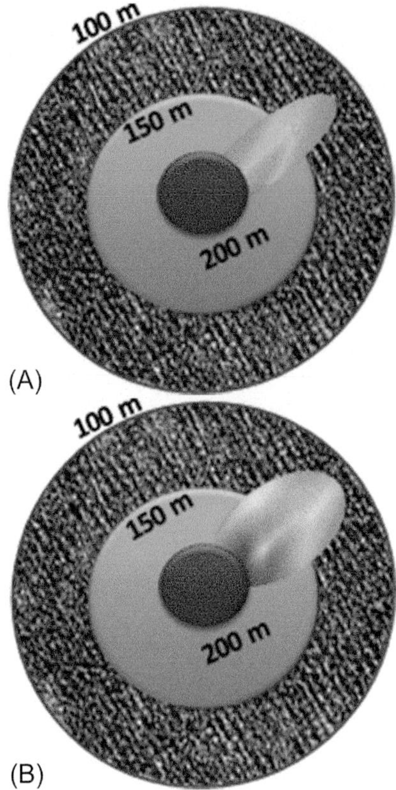

FIG. 7.17 DMC simulated (A) PM spectra and (B) cross-spectrum.

considered as the original state in simulating ocean wave spectral patterns. Finally, for a large projection time, the wave function converges to the lowest state that has an overlap with trail function. As an enhancement phase of the cross-spectrum approximations, the quick random walk of DMC can be used to determine the exact spectrum peak direction. The exact wave spectrum retrieved in SAR images by removing the 180-degree ambiguity can assist in simulating accurate wave refraction patterns, as will be discussed in the next chapter.

References

[1] Kerr DE. The propagation of short radio waves. MIT radiation laboratory series, vol. 13. New York: McGraw-Hill; 1951.

[2] Rice SO. Reflection of electromagnetic waves from slightly rough surfaces. Commun Pure Appl Math 1951;4 (2–3):351–78.

[3] Bragg WL. The diffraction of short electromagnetic waves by a crystal. Proc Camb Philos Soc 1913;17:43–57.

[4] Monaldo FM, Jackson CR, Pichel WG. Seasat to Radarsat-2: research to operations. Oceanography 2013;26 (2):34–45.

[5] Moore RK, Claassen JP, Cook AC, Fayman DL, Holtzman JC, Sobti A, et al. Simultaneous active and passive microwave response of the earth—the Skylab radscat experiment. In: Remote sensing of environment. Elsevier; 1974. p. 189–217.

[6] UNAVACO, 2019. Synthetic aperture radar (SAR) satellites. https://www.unavco.org/instrumentation/geophysical/imaging/sar-satellites/sar-satellites.html. [Accessed 12 May 2020].
[7] Scheer J, Holm WA. Richards MA, Melvin WL, editors. Principles of modern radar. SciTech Pub.; 2010
[8] Asgedom EG, Cecconello E, Orji OC, Söllner W. Rough sea surface reflection coefficient estimation and its implication on hydrophone-only pre-stack deghosting, In: 79th EAGE conference and exhibition 2017; 2017.
[9] Thorsos EI. The validity of the Kirchhoff approximation for rough surface scattering using a Gaussian roughness spectrum. J Acoust Soc Am 1987;83:78–92.
[10] Valenzuela GR. Theories for the interaction of electromagnetic and oceanic waves—a review. Bound-Layer Meteorol 1978;13(1–4):61–85.
[11] Keller WC, Wright JW. Microwave scattering and the straining of wind-generated waves. Radio Sci 1975;10 (2):139–47.
[12] Alpers W, Rufenach CL. The effect of orbital motions on synthetic aperture radar imagery of ocean waves. IEEE Trans Antennas Propag 1979;27(5):685–90.
[13] Hasselmann K, Hasselmann S. On the nonlinear mapping of an ocean wave spectrum into a synthetic aperture radar image spectrum and its inversion. J Geophys Res [Oceans] 1991;96(C6):10713–29.
[14] Lyzenga DR. An analytic representation of the synthetic aperture radar image spectrum for ocean waves. J Geophys Res [Oceans] 1988;93(C11):13859–65.
[15] Hasselmann S, Brüning C, Hasselmann K, Heimbach P. An improved algorithm for the retrieval of ocean wave spectra from synthetic aperture radar image spectra. J Geophys Res [Oceans] 1996;101(C7):16615–29.
[16] Engen G, Johnsen H. SAR-ocean wave inversion using image cross spectra. IEEE Trans Geosci Remote Sens 1995;33(4):1047–56.
[17] Lehner S, Schulz-Stellenfleth J, Schattler B, Breit H, Horstmann J. Wind and wave measurements using complex ERS-2 SAR wave mode data. IEEE Trans Geosci Remote Sens 2000;38(5):2246–57.
[18] Dowd M, Vachon PW, Dobson FW, Olsen RB. Ocean wave extraction from RADARSAT synthetic aperture radar inter-look image cross-spectra. IEEE Trans Geosci Remote Sens 2001;39(1):21–37.
[19] Krogstad HE, Samset O, Vachon PW. Generalizations of the non-linear ocean-SAR transform and a simplified SAR inversion algorithm. Atmosphere-Ocean 1994;32(1):61–82.
[20] Thomas ML, Ralph WK, Jonathan WC. Remote sensing and image interpretation. New York: John Willey & Sons; 2000.
[21] Freeman A, Durden SL. A three-component scattering model for polarimetric SAR data. IEEE Trans Geosci Remote Sens 1998;36(3):963–73.
[22] Brisco B. Mapping and monitoring surface water and wetlands with synthetic aperture radar. Remote sensing of wetlands: Applications and advances. CRC Press; 2015119–36.
[23] Zhang WG, Zhang Q, Yang CS. Improved bilateral filtering for SAR image despeckling. Electron Lett 2011;47 (4):286–8.
[24] Chapron B, Johnsen H, Garello R. Wave, and wind retrieval from SAR images of the ocean. In: Annales des telecommunications. vol. 56, No. 11–12. Springer-Verlag; 2001. p. 682–99.
[25] Populus J, Aristaghes C, Jonsson L, Augustin JM, Pouliquen E. The use of SPOT data for wave analysis. Remote Sens Environ 1991;36(1):55–65.
[26] Shor PW. Algorithms for quantum computation: discrete logarithms and factoring, In: Proceedings 35th annual symposium on foundations of computer science 1994 Nov 20IEEE; 1994. p. 124–34.
[27] Galindo A, Martin-Delgado MA. Information and computation: classical and quantum aspects. Rev Mod Phys 2002;74(2):347.
[28] Pang CY, Hu BQ. Quantum discrete Fourier transform with classical output for signal processing. arXiv; 2007 preprint arXiv:0706.2451.
[29] Shi J, Dozier J. Measurements of snow- and glacier-covered areas with single-polarization SAR. Ann Glaciol 1993;17:72–6.
[30] Li XM. Ocean surface wave measurement using SAR wave mode data. Germany: Department Geowissenschaften der Universität Hamburg; 2010 [Ph.D. thesis].
[31] Chen L. Quantum Monte Carlo study of correlated electronic systems [Doctoral dissertation]. University of Illinois at Urbana-Champaign; 2017.

CHAPTER

8

Polarimetric synthetic aperture radar for wave spectra refraction using inversion SAR wave spectra model

8.1 What is meant by polarimetric synthetic aperture radar?

Polarimetric synthetic aperture radar (POLSAR), also identified as quad-polarization SAR, determines a target's reflectivity with quad-polarizations. In other words, quad-polarization involves horizontal transmitting and receiving (HH), horizontal transmitting and vertical receiving (HV), vertical transmitting and horizontal receiving (VH), and vertical transmitting and vertical receiving (VV). This is accomplished by consecutively transmitting horizontal (H) and vertical (V) polarization radar pulses and receiving both H and V polarizations of reflected pulses with adequately high pulse recurrence frequencies. Dissimilar single or dual-polarization SAR and POLSAR data can be exploited to synthesize reactions from an amalgamation of transmitting and receiving polarizations. This delivers information to portray scattering mechanisms of various terrain covers, for instance, forest, ocean surface, city blocks, grassland, crops, etc. Accordingly, polarimetric SAR has an exclusive supreme perception proficiency by using the single or dual-polarization SAR. A countless variety of POLSAR data is established from space-borne (SIR-C) and airborne (NASA/JPL AIRSAR, DLR E-SAR, EMISAR, PISAR, RAMSES, etc.) instruments, which creates possible precise considerations of a specified constituency of interest in delivering the wide range of polarizations, wavelengths, and interferometric baselines for radar platform data, i.e., airborne and satellite.

8.2 Polarimetric matrix formulations and SAR data representation

Keys to understanding the multilook or spatially averaged POLSAR data are the Stokes reflection, Kennaugh radar backscattering matrix $\boldsymbol{K}$, or regularly by the covariance (lexicographic coherency) matrix $\boldsymbol{C}$, or the Pauli coherency matrix $\boldsymbol{T}$. For single-look complex data,

https://doi.org/10.1016/B978-0-12-820785-7.00008-3

the coherent Sinclair scattering matrix S is utilized. The Sinclair scattering matrix S (for single-look complex data) can be described by:

$$S = \begin{bmatrix} S_{HH} & S_{HV} \\ S_{VH} & S_{VV} \end{bmatrix} \tag{8.1}$$

Eq. (8.1) demonstrates that the orthonormal antenna basis {HV} for the 2×2 coherent Sinclair scattering matrix S is a function of frequency. For backscattering from reciprocal media, S is symmetric: $S_{HV} = S_{VH}$. Concerning other orthonormal polarization bases $\{AB\}$, where $B = A^{\perp}$ with the superscript $^{\perp}$ signifying orthogonality, the demonstration of the Sinclair matrix is reformed consistent with the transformation $S \rightarrow \boldsymbol{U}^T S \boldsymbol{U}$, where $\boldsymbol{U}$ is the unitary modification-of-basis matrix. In this regard, Kennaugh [1] developed the transformation $S \rightarrow \boldsymbol{U}^T S \boldsymbol{U}$, which is commonly termed as a unitary consimilarity transformation or unitary congruence transformation by referring to the bilinear voltage equation [2]. It is very simply obtained, however, by comprehending that the Sinclair matrix S is the matrix part or antimatrix of the antilinear backscatter operator S_a, where the antilinearity results from the time-reversal operation. Accordingly, $S_a = SC$ (or CS) where C signifies complex conjugation with the physical fundamental characteristic of any polarization basis $B = \{b_1, b_2\}$, in which there be existent the associations of $Cb_{1,2} = b_{1,2}$. At that moment, for somewhat unitary transformation which is formulated as:

$$S_a \rightarrow \boldsymbol{U} S_a \boldsymbol{U}^{-1} = \boldsymbol{U} S C \boldsymbol{U}^{*T} = \left(\boldsymbol{U} S \boldsymbol{U}^T\right) C \Rightarrow S \rightarrow \boldsymbol{U} S \boldsymbol{U}^T \tag{8.2}$$

Consistent with Kennaugh [1], unitary consimilarity sustains the regularity of the Sinclair matrix. Consequently, the length and absolute value of the determinant of Sin this transformation (Eq. 8.2) persist symmetrically, which is castoff as:

$$span(S) = \left\{ |S_{HH}|^2 + 2|S_{HV}|^2 + |S_{VV}|^2 \right\} = span\left(\boldsymbol{U}^T S \boldsymbol{U}\right), \tag{8.3}$$

$$|\det S| = \left| S_{HH} S_{VV} - (S_{HV})^2 \right| = \left|\det\left(\boldsymbol{U}^T S \boldsymbol{U}\right)\right| \tag{8.4}$$

Eqs. (8.3) and (8.4) present the coherent radar in the form of the matrix S, which is produced by Sinclair. In this view, coherent radar is a kind of radar that obtains further information about a target through measurement of the phase of echoes from a sequence of pulses. In fact, the phase information is used to improve the signal-to-noise ratio. On this understanding, the Kennaugh backscattering matrix K can be allied to S by:

$$K = 2\left(A^T\right)^{-1} (S \otimes S^*) A^{-1} \quad \text{with} \quad A = \begin{bmatrix} 1 & 0 & 0 & 1 \\ 1 & 0 & 0 & -1 \\ 0 & 1 & 1 & 0 \\ 0 & j & -j & 0 \end{bmatrix}. \tag{8.5}$$

Here A denotes the Kronecker extension matrix. The symbol $\otimes$, therefore, is the standard Kronecker matrix product, the superscript ^{T} signifies the transpose, $^{-1}$ the inverse, and * complex conjugation. Consequently, the matrix A is unitary. In this regard, incoherent scattering of the ensemble or time average, designated by $\langle S \otimes S^* \rangle$, has to be considered.

8.3 The coherency matrix T_{HV} (for single-look or multilook)

The coherence matrix permits investigating single scattering mechanisms within a pixel as well as their contribution to the total signal, while the covariance matrix tolerates the analysis of single-polarization channels as well as their correlations. The coherency matrix T_{HV} is grounded in the development of the matrix S into the Pauli matrices. In this regard, the Pauli-based covariance matrix involves three components, which are specified by:

$$k = \frac{1}{\sqrt{2}} \begin{bmatrix} S_{HH} + S_{VV} \\ S_{HH} - S_{VV} \\ 2S_{HV} \end{bmatrix} \tag{8.6}$$

Eq. (8.6) reveals that the three components of the coherence matrix are based on $\{HV\}$, which is grounded for symmetric random Sinclair matrix on the so-called Pauli target feature vector. Consequently, the implementation of the Pauli vector algorithm delivers excellent and more informative expositions by conveying for $|S_{HH} - S_{VV}|$, $|S_{HV}|$, and $|S_{HH} + S_{VV}|$, respectively. This can be attributed to the fact that the information of phase differences between HH and VV is also incorporated in this display. Expanding Eq. (8.6) with the hypothesis that $S_{HV} = S_{VH}$, the coherency matrix for the reciprocal symmetric case can be obtained by:

$$T_{HV} = \langle kk^{*T} \rangle$$
$$= \frac{1}{2} \begin{bmatrix} \left\langle |S_{HH} + S_{VV}|^2 \right\rangle & \langle (S_{HH} + S_{VV})(S_{HH} - S_{VV})^* \rangle & 2\langle (S_{HH} + S_{VV})S_{HV}^* \rangle \\ \langle (S_{HH} - S_{VV})(S_{HH} + S_{VV})^* \rangle & \left\langle |S_{HH} - S_{VV}|^2 \right\rangle & 2\langle (S_{HH} - S_{VV})S_{HV}^* \rangle \\ 2\langle S_{HV}(S_{HH} + S_{VV})^* \rangle & 2\langle S_{HV}(S_{HH} - S_{VV})^* \rangle & 4\left\langle |S_{HV}|^2 \right\rangle \end{bmatrix} \tag{8.7}$$

Eq. (8.7) denotes that the coherency matrix T_{HV} is more suitable than the covariance matrix with lexicographical ordering for polarimetric analyses. In fact, T_{HV} conserves the relative co/cross-polarization phase performance, as is recognizable from Eq. (8.7).

8.4 Circular polarization-based covariance matrix C_{RL}

The coherent Sinclair scattering matrix S in Eq. (8.1) can be used to derive the circular polarization basis. In this sense, the orthogonal unit vectors for right-handed and left-handed circular polarization, respectively, are formulated as [3–5]:

$$\vec{P}_R = \frac{1}{\sqrt{2}} \begin{bmatrix} 1 \\ -j \end{bmatrix}, \text{ and } \vec{P}_L = \frac{1}{\sqrt{2}} \begin{bmatrix} -j \\ 1 \end{bmatrix}. \tag{8.8}$$

As always, the Jones (Sinclair) vectors recounting the circumstance of polarization are regulated merely to fit for a phase term. Utilizing a unitary consimilarity transformation, the Sinclair matrix in the {R, L} polarization basis becomes:

$$S(RL) = \begin{bmatrix} S_{RR} & S_{RL} \\ S_{LR} & S_{RR} \end{bmatrix} = \frac{1}{\sqrt{2}} \begin{bmatrix} 1 & j \\ j & 1 \end{bmatrix} \begin{bmatrix} S_{HH} & S_{HV} \\ S_{VH} & S_{VV} \end{bmatrix} \frac{1}{\sqrt{2}} \begin{bmatrix} 1 & j \\ j & 1 \end{bmatrix} = U^T S(HV) U \tag{8.9}$$

Here, $\boldsymbol{U}$ is the unitary transition matrix from the $\{HV\}$-origin for the right-left circular $\{RL\}$-origin. The three circular components are formulated as:

$$\begin{aligned} S_{RR} &= (S_{HH} - S_{VV} + j2S_{HV})/2 \\ S_{LL} &= (S_{VV} - S_{HH} + j2S_{HV})/2 \\ S_{RL} &= S_{LR} = j(S_{HH} + S_{VV})/2 \end{aligned} \tag{8.10}$$

The circular covariance matrix is correspondingly expressed as:

$$C_{RL} = \langle cc^{*T} \rangle \text{ with } c = \begin{bmatrix} S_{RR} \\ \sqrt{2}S_{RL} \\ S_{LL} \end{bmatrix}. \tag{8.11}$$

Eq. (8.11) establishes a correlation between the polarization orientation angle and the circular covariance matrix. Indeed, all the matrices C_{HV}, T_{HV}, and C_{RL} are correlated to each other by unitary correspondence transformations. C_{HV}, T_{HV}, and C_{RL} are demonstrations of the similar usual operator with dissimilar polarization grounds, and accordingly have similar eigenvalues.

8.5 Estimation of azimuth slopes using orientation angle

Due to reflection symmetry, the orientation angle measurement is ignored. Nonetheless, the orientation angle is an extremely valuable parameter, which can be retrieved from polarimetric SAR data to measure wave spectra slope. For instance, when polarimetric radar images a sea surface with significant elevation, the change in the polarization orientation angle θ is geometrically related to sea surface slopes and the radar look angle by:

$$\tan\theta = \frac{\tan\omega}{-\tan\gamma\cos\varphi + \sin\varphi} \quad (8.12) \tag{8.12}$$

where φ represents the radar look angle, $\tan\omega$ is the azimuth slope, and $\tan\gamma$ is the slope in the ground range direction.

This equation shows that the orientation angle shift is predominantly suggested by the azimuth slope; nonetheless, it is correspondingly a function of the range slope and the radar look angle. The orientation angle can be mathematically expressed as:

$$-4\phi_{SAR} = Arg\left(\langle S_{RR}S_{LL}^{*}\rangle\right) = \tan^{-1}\left(\frac{-4\,\mathrm{Re}(\langle (S_{HH} - S_{VV})S_{HV}^{*}\rangle)}{-\langle |S_{HH} - S_{VV}|^{2}\rangle + 4\langle |S_{HV}|^{2}\rangle}\right) - \pi \tag{8.13}$$

The orientation angle is calculated by:

$$\phi_{SAR} = \begin{cases} \eta, & if\ \eta \le \pi/4 \\ \eta - \pi/2, & if\ \eta > \pi/4 \end{cases} \tag{8.14}$$

The arctangent in Eq. (8.14) is estimated in the range of $(-\pi, \pi)$. Generally, the orientation angle can be extracted successfully using the phase difference between the most sensitive circular polarization estimator, which contains the right-hand transmit, right-hand receive

(RR), and the left-hand transmit, left-hand receive (LL). On this understanding, azimuth slopes can be obtained from orientation angles by means of L-band and P-band POLSAR data, but less successfully from C-band or higher frequency data. Higher frequency POLSAR responses are less sensitive to azimuth slope variations. The dynamic range and polarization channel isolation of the radar receiver are critical to the success of the orientation angle estimation. The success of the circular polarization methods relies on the accuracy of measuring the $\langle (S_{HH}-S_{VV})S_{HV}^{*}\rangle$ term, which is smaller than $\langle |S_{HH}|^2\rangle$ or $\langle |S_{VV}|^2\rangle$. Coherence between different polarizations of $HH-VV$ is sensitive to surface parameters and can be obtained by:

$$\phi_{HHVV} := \frac{|\langle S_{HH}S_{VV}^{*}\rangle|}{\sqrt{\langle S_{HH}S_{HH}^{*}\rangle\langle S_{VV}S_{VV}^{*}\rangle}} \tag{8.15}$$

The difference between polarizations of $HH-VV$ has been evaluated concerning its sensitivity to dielectric constant and/or root mean square error of the heightened state of rough surfaces [6]. In contrast, the correlation coefficient between **(*HH*+ *VV*)** and **(*HH*− *VV*)** has been determined to correlate with surface roughness [7]. On this understanding, a wide class of natural surface scatterers is described by secondary and/or multiple scattering impacts. With a growing surface roughness, relative to the wavelength, the consequence is that multiple scattering develops more strongly, creating a weighty $|HV|$ scattering constituent. For instance, dihedral scattering owing to minor relationship spans considered by $|HH|>|VV|$, and/or diffuse scattering ($|HV|$ contribution) imitates the backscattered signal.

The SAR polarimetric backscattering, consequently, as a function of the orientation angle at point x_0 is determined from:

$$\phi_{SAR}(x) = \frac{1}{4}\left[\pi + \arctan\left[\frac{-4\,\mathrm{Re}(\langle (S_{HH}(x_0)-S_{VV}(x_0))S_{HV}^{*}(x_0)\rangle)}{-\left\langle |(S_{HH}(x_0)-S_{VV}(x_0))|^2\right\rangle + 4\left\langle |S_{HV}^{*}(x_0)|^2\right\rangle}\right]\right] \tag{8.16}$$

Here, S is the backscattering coefficient, "*" is the conjugate operator, and H and V represent the polarization sorts. Scattering coefficients fulfill the recognized reciprocal relation.

On this understanding, the backscattering coefficient of the ocean surface can be presumed in most circumstances to be Bragg and homogeneous, which delivers superlative circumstances for the orientation angle guess. In an investigation of convergent current front within the Gulf Stream, Lee et al. [8] realized the potential to use the orientation angle in estimating small ocean surface slopes to the precision of a fraction of a degree. In this regard, they found that a sudden change occurred on slopes from positive to negative across the front with an extreme slope swap smaller than 2 degrees. Subsequently, Schuler et al. [9] and Kasilingam et al. [10] have widened this analysis to compute the ocean wave slope spectra, and to investigate ocean surface features such as internal waves.

8.6 Alpha parameter sensitivity to the range of traveling waves

For estimation of sea surface slope along with the azimuth and range directions, alpha angle, $\overline{\alpha}$, is a useful parameter, which is derived from the eigenvector in the $\{H,V\}$-basis.

Subsequently, the eigenvectors are regulated merely up to phase factors, which can be resolved if the first component of each eigenvector e_i ($i=1, 2, 3$) is real, which is described as:

$$e_i = \left[\cos\alpha_i \ \sin\alpha_i \cos\beta_i e^{i\delta_i} \ \sin\alpha_i \sin\beta_i e^{i\gamma_i}\right]^T \tag{8.17}$$

Eq. (8.17) indicates that for surface scattering, $\boldsymbol{\alpha} = \mathbf{0\,degrees}$, for dipole scattering, $\boldsymbol{\alpha} = \mathbf{45\,degrees}$, and for even bounce scattering, $\boldsymbol{\alpha} = \mathbf{90\,degrees}$, generally, the averaged alpha angle $\overline{\alpha}$ is depleted to signify the averaged scattering mechanism, as given by:

$$\overline{\alpha} = P_1\alpha_1 + P_2\alpha_2 + P_3\alpha_3 \tag{8.18}$$

The sensitivity of alpha values to incidence angle is changed, as a function of the modulation transfer function (MTF) for range slope approximation and is reliant on the derivative of α concerning θ. The mathematical relationship of the fluctuations of α concerning θ is identified as:

$$\frac{\partial\alpha}{\partial\theta} = \sin 2\phi\left[1 + \sin^4\phi\right]^{-1} \tag{8.19}$$

Eq. (8.19) is considered for the dielectric infinity of the sea surface, as it is dominated by the high dielectric. Cloude showed that a general covariance matrix [T] can be decomposed as follows:

$$[\mathbf{T}] = \lambda_1 \mathbf{k}_1 \cdot \mathbf{k}_1^\dagger + \lambda_2 \mathbf{k}_2 \cdot \mathbf{k}_2^\dagger + \lambda_3 \mathbf{k}_3 \cdot \mathbf{k}_3^\dagger + \lambda_4 \mathbf{k}_4 \cdot \mathbf{k}_4^\dagger \tag{8.20}$$

Here, λ_i, $i=1, 2, 3, 4$ are the eigenvalues of the covariance matrix, $\mathbf{k}_i$, $i=1, 2, 3, 4$ are its eigenvectors, and $\mathbf{k}_i^\dagger$ signifies the adjoint (complex conjugate transposed) of $\mathbf{k}_i$. In the monostatic (backscatter) case, the covariance matrix has one zero eigenvalue, and the decomposition results in, at most, three nonzero covariance matrices. As such, the average covariance matrix for azimuthally symmetrical sea surface has the general form of:

$$[\mathbf{T}] = \mathrm{C}\begin{pmatrix} 1 & 0 & \rho \\ 0 & \eta & 0 \\ \rho^* & 0 & \zeta \end{pmatrix} \tag{8.21}$$

where:

$$\mathrm{C} = \langle S_{hh} S_{hh}^* \rangle, \tag{8.21.1}$$

$$\rho = \langle S_{hh} S_{vv}^* \rangle \Big/ \langle S_{hh} S_{hh}^* \rangle, \tag{8.21.2}$$

$$\eta = 2\langle S_{hv} S_{hv}^* \rangle \Big/ \langle S_{hh} S_{hh}^* \rangle \tag{8.21.3}$$

$$\zeta = \langle S_{vv} S_{vv}^* \rangle \Big/ \langle S_{hh} S_{hh}^* \rangle \tag{8.21.4}$$

The superscript * signifies complex conjugate, and total measures are amalgamation averages. The parameters C, η, ζ, and ρ rely on the dimension, form, and electrical properties of the scatterers, i.e., dielectric, in addition to their statistical angular distribution.

Let us count Bragg scattering for a Bragg scattering surface. Cloude and Pottier hosted a reviewed parameterization of the eigenvector $\hat{u}_{3P_i}$ of $[C_{3p}]$ which is identified as:

$$\hat{u}_{3P} = \left[\cos\alpha \quad \sin\alpha\cos\beta\exp(j\delta) \quad \sin\alpha\sin\beta\exp(j\gamma)\right]^T \tag{8.22}$$

So that with a revised parameterization of the Pauli Coherency Matrix [C], one obtains:

$$\langle [C] \rangle = [U_{3P}] \begin{bmatrix} \lambda_1 & 0 & 0 \\ 0 & \lambda_2 & 0 \\ 0 & 0 & \lambda_3 \end{bmatrix} [U_{3P}]^{*T} \tag{8.23}$$

It is a peculiarity of a complex covariance matrix [C] that the orthonormal eigenvectors [C] are regulated merely up to a phase term, which obeys, from substituting into Eq. (8.23), the expression $\hat{u}_{3P_i} \cdot \hat{u}_{3P_i}^{*T} = \left(e^{j\phi}\hat{u}_{3P_i}\right) \cdot \left(e^{j\phi}\hat{u}_{3P_i}\right)^{*T}$. In particular, the first component of all eigenvectors can be selected as real, where:

$$[U_{3P}] = \exp(j\varphi) \begin{bmatrix} \cos\alpha_1 & \cos\alpha_2\exp(j\varphi_2) & \cos\alpha_3\exp(j\varphi_3) \\ \sin\alpha_1\cos\beta_1\exp(j\delta_1) & \sin\alpha_2\cos\beta_2\exp(j\delta_2) & \sin\alpha_3\cos\beta_3\exp(j\delta_3) \\ \sin\alpha_1\sin\beta_1\exp(j\gamma_1) & \sin\alpha_2\sin\beta_2\exp(j\gamma_2) & \sin\alpha_3\sin\beta_3\exp(j\gamma_3) \end{bmatrix} \tag{8.24}$$

The parameterization of the 3×3 unitary matrix $[U_{3p}]$ in expressions of column vectors of dissimilar α_i, β_i, ϕ_i, δ_i, and γ_i $i=1, 2, 3$ φ signifying a complete phase permits a probabilistic clarification of the scattering process. However, the quantities of α_1, α_2, α_3; β_1, β_2, β_3; etc.; are not reciprocal, which requires further investigation. In the first instance, the introduction of a three-symbol Bernoulli process, according to Cloude and Pottier [11], initiated a scattering model based on terms of three distinct Sinclair matrices $[S_i]$, one each to correspond to a column of $[U_{3p}]$, which ensue within probabilities P_i, so that $\sum_{i=1}^{3} P_1 = 1$. Cloude and Pottier [11] subsequently developed a mean parameter of random arrays allied with the Bernoulli process. In this regard, the dominant scattering matrices from the 3×3 Pauli coherency matrix [C] can be demarcated as a mean target vector $\hat{u}_{3P_m}$:

$$\hat{u}_{3m} = \left[\cos\overline{\alpha} \quad \sin\overline{\alpha}\cos\overline{\beta}\exp(j\overline{\delta}) \quad \sin\overline{\alpha}\sin\overline{\beta}\exp(j\overline{\gamma})\right]^T \tag{8.25}$$

This approach is firmly model-reliant on, and comparable to, the [S] matrix decomposition into the Pauli matrix set $\psi_P\{[\sigma_i], i=0,1,2,3\}$, not exclusive. This requires additional investigation for the development of other model-dependent decompositions of the eigenvectors $\hat{u}_{3P_i}$. Consequently, the direct relationship of the respective set of [S] and [C] model-dependent Pauli spin matrix decompositions are pursued and perhaps shed light on the "polarimetric scatter dichotomy," which forms an indispensable piece of "polarimetric radar theory." The alpha parameter developed from the Cloude-Pottier polarimetric scattering decomposition theorem, which has desirable directional measurement properties. It is roll-invariant in the azimuth direction, and in the range direction, it is highly sensitive to wind-induced wave.

8.7 Examined POLSAR and AIRSAR data

The Jet Propulsion Laboratory (JPL) Airborne Synthetic Aperture Radar (AIRSAR) data were attained on December 6, 1996 and September 19, 2000 from the coastline of Kuala Terengganu, Malaysia between 103°4′E to 103°15′E and 5°20′N to 5°24′N (Fig. 8.1). In addition, the sea wave actual data were collected by wave rider buoy from the Malaysian Meteorological Service between latitudes of 5°18′N and 5°26′N and longitudes of 103°32′E 103°40′E on December 6, 1996 and September 19, 2000 (during that time, the airborne AIRSAR and POLSAR were flown over the study area, respectively). The in situ observation data included wave height and wave direction, which were used for wave spectra modulation with AIRSAR and POLSAR data. The wind data were collected at the meteorological station at Sultan Mahmud Airport, Kuala Terengganu, at latitudes of 5°23′N and longitude of 103°06′E and obtained by the Malaysia Meteorological Service in Kuala Terengganu. AIRSAR is a NASA/JPL multifrequency instrument package aboard a DC-8 aircraft, operated by NASA's Ames Research Center at Moffett Field. AIRSAR flies at 8 km over the average terrain height at a velocity of 215 m s^{-1}. The AIRSAR system is premeditated to have hovered on a minor and big airplane. The system necessitates a scanner port (18 cm × 36 cm) on the underside of the airplane. JPL's airborne synthetic aperture radar (AIRSAR) is a unique system, encompassing three radars at HH-, VV-, HV-, and VH-polarized signals from 5 m × 5 m pixels recorded for three wavelengths: C band (5 cm), L band (24 cm), and P band (68 cm) [12]. AIRSAR data assemblies are involved fully polarimetric SAR data (POLSAR); which can be composed at entirely three frequencies, while cross-track interferometric data (TOPSAR) and along-track interferometric (ATI) data can be collected at C- and L-bands [12]. In this chapter, VV-, HH-, and VH-polarized are selected for simulating ocean wave refraction from C and L bands, respectively (Fig. 8.2).

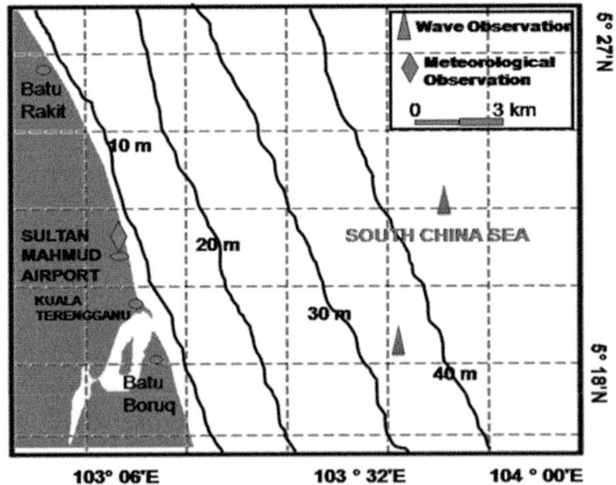

FIG. 8.1 The geographical location of the AIRSAR/POLSAR acquisition data.

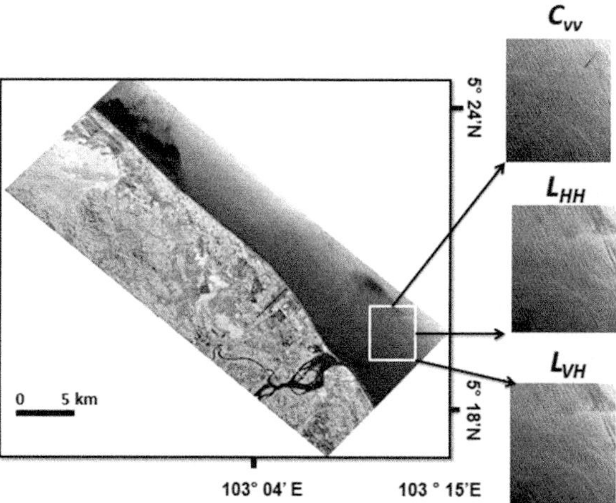

FIG. 8.2 Selected and polarized bands for wave refraction simulation.

8.8 Wave spectra model

Consistent with Pierson and Moskowitz [13], the significant wave height spectrum H_s can be related to the one-sided directional wave number spectra density $S\left(\vec{K}, \varphi\right)$ by the following formula:

$$H_s = 4\left[\iint S\left(\vec{k}, \varphi\right) dk_i dk_j\right]^{0.5} \tag{8.26}$$

where k_i and k_j are the wavenumbers in the azimuth and range directions, respectively, and $\vec{K}$ is the wave-number magnitude of k_i and k_j in the azimuth and the range directions, respectively. The ocean wave spectrum $S\left(\vec{K}, \varphi\right)$ was obtained by the input of in situ wave parameters, which involved dominant wave number, wave propagation direction φ, and significant wave height H_s. The actual ocean wave spectrum density $S\left(\vec{K}, \varphi\right)$ can be calculated from the following equation [13]:

$$S\left(\vec{K}, \varphi\right) = \frac{\sqrt{S(k_i, k_j) d\vec{K}}\, e^{\left[-2\left(5.88U^{-2}/\vec{K}\cos\varphi\right)^2\right]}}{\left[\vec{K}^2 - \left(0.5gU^{-2}\right)\vec{K}\cos\varphi + 2.4U^{-2}\right]^2} \tag{8.27}$$

where U is the wind speed and φ is the wave direction.

The real wave spectra contours have been determined using Eq. (8.27). Furthermore, Eq. (8.27) has been used with a quasilinear model to map the AIRSAR/POLSAR wave spectra onto ocean wave spectra.

8.9 Two-dimensional quantum Fourier transform for retrieving SAR wave spectra

To this end, the ocean surface wave spectra and wave parameters are retrieved from AIRSAR/POLSAR data as follows: (1) select 512×512 pixel size subimages. Subsequently, the geophysical information and SAR system information from AIRSAR/POLSAR, for instance, latitude, longitude, incident angle, and other information for each pixel, radar look angle, and track angle are retrieved. The complex backscattering coefficients S_{VV}, S_{HH}, and S_{VH} for each pixel of AIRSAR/POLSAR data are then computed. The azimuth slope and the ground slope can be estimated using the orientation angle in Eq. (8.16). The 2-QDFFT (recalled from Chapter 7) is used to retrieve slope spectra (Fig. 8.3) from a smoothed image using a Gaussian filter. In the quantum context, the linear transformation can be represented as follows: given an input image state, $|f\rangle$ the output image spectra state $|G\rangle = U_U|f\rangle$. The corresponding unitary operator U_U is identified as:

$$U_U = QFT_N = \frac{1}{\sqrt{N}}\begin{bmatrix} 1 & 1 & \cdots & 1 \\ 1 & e^{\frac{-2\pi}{2^n}} & \cdots & \left[e^{\frac{-2\pi}{2^n}}\right]_n^{N-1} \\ \vdots & \vdots & \vdots & \vdots \\ 1 & \left[e^{\frac{-2\pi}{2^n}}\right]_n^{N-1} & \cdots & \left[e^{\frac{-2\pi}{2^n}}\right]_n^{(N-1)(N-1)} \end{bmatrix} \tag{8.28}$$

The 2-DQFFT spectra derived from POLSAR data are represented as a quantum state, and encoded in n qubits. The spectra quantum image transformation is performed by unitary operator U_U. The 2-DQFFT delivers well-constructed spectra peaks, which are useful to estimate the sea surface slope along the azimuth direction. The cluster center of the spectrum, as represented in Fig. 8.2, is assumed to be $K_s = (k_{na}, k_{nr})$ of $N \times N$the image spectrum. Let us

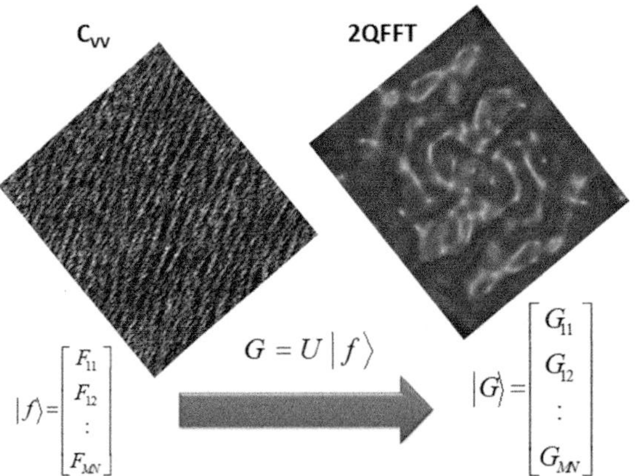

FIG. 8.3 The demonstrated wave spectra retrieved by using 2-QDFFT.

assume the image spectrum $X=\{X|x_i\rangle|x_i\in G, 1\le i\le N^2\}$, which is formulated using a weighting average approach in the circumstance of symmetric property, as:

$$|K_s\rangle = \frac{1}{\sum_{i}^{0.5N^2} X|x_i\rangle} \sum_{i}^{0.5N^2} X|x_i\rangle \times |x_i\rangle \tag{8.29}$$

The wavelength can be estimated using:

$$\lambda = \frac{\Delta x \cdot lag(pixels)}{\sqrt{\left(k_x^{\,2} + k_y^{\,2}\right)}} \tag{8.30}$$

8.10 Quasilinear transform

To represent the observed POLSAR spectra into the real ocean wave spectra, a quasilinear model is implemented. The simplified quasilinear theory is explained below. Along with the Gaussian linear theory, the relationship between ocean wave spectra $S\left(\vec{K},\varphi\right)$ and AIRSAR/POLSAR image spectra $S_Q\left(\vec{K}\right)$ could be designated by tilt and hydrodynamic modulation (real aperture radar (RAR) modulation), as discussed in Chapters 6 and 7 [14]. The tilt modulation is linear to the local surface slope in the range direction, i.e., in the plane of radar illumination. The tilt modulation in general is a function of wind stress and wind direction for ocean waves and AIRSAR/POLSAR polarization. Consistent with Vachon et al. [15,16], the tilt modulation is the largest for HH polarization. In addition, hydrodynamic interaction between the scattering waves (ripples) and longer gravity waves creates a focus of the scatterer on the upwind face of the swell [17,18]. Following Marghany [12], AIRSAR image spectra can be presented into ocean wave spectra under the assumption of the quasilinear modulation transfer function $S_Q\left(\vec{k}\right)$, which is identified as:

$$S_Q\left(\vec{K}\right) = R(K)H(k_i;K_c)\cdot\left[\frac{S\left(\vec{K},\varphi\right)}{2}\left|T_{lin}\left(k_{i,j}\right)\right|^2 + \frac{S\left(-\vec{K},\varphi\right)}{2}\left|T_{lin}\left(-k_{i,j}\right)\right|^2\right]. \tag{8.31}$$

where $H(k_i;K_c)$ is an azimuth cutoff function that is contingent upon azimuth wave number k_i and range wave number k_j, the cutoff azimuth wave number is K_c, and $R(\vec{K})$ is the AIRSAR point spread function.

The AIRSAR/POLSAR point spread function is a function of the azimuth and the range resolutions [12]. Consistent with Vachon et al. [16], T_{lin} is a linear modulation transfer function, which is composed of the RAR (the tilt modulation and hydrodynamic modulation) and the velocity bunching modulation. The RAR modulation transfer function (RAR MTF) is the coherent sum of the transfer function associated with each of these terms, i.e.:

$$T_{lin}(\theta_0, L') = M_t(\theta_0, L') + M_d(\theta_0, L') + M_v(\theta_0, L'). \tag{8.32}$$

The relativity tilt modulation $T_l(\theta_0, L')$ in shallow water, which is modified after Hasselmann and Hasselmann [14], can be formulated as:

$$M_t\left(\theta_0, \vec{L'}\right) = \left[T(gd)^{0.5}\sqrt{1-\left(\frac{R}{V_S}\right)^2 c^{-2}}\right]^{-1} \frac{4\cot\theta}{1 \pm \sin^2\theta} e^{\left(\frac{i\pi}{2}\right)} \tag{8.33}$$

where the first term is wavelength relativity in shallow water as the inverse of the range wave number and θ is the local incident angle of the radar beam.

Modifying Vachon et al.'s [15] work as a function of the relativity of wavelength contraction, the hydrodynamic modulation transfer function can be given by:

$$M_d\left(\theta_0, \vec{L'}\right) = 4.5\omega\left[T(gd)^{0.5}\sqrt{1-\left(\frac{R}{V_S}\right)^2 c^{-2}}\right]^{-1} \frac{\omega - i\mu}{\omega + \mu^2}\sin^2\varphi. \tag{8.34}$$

where i is $\sqrt{-1}$, ω is the angular frequency of the long waves, φ is the azimuth angle, and μ is the relaxation rate of the Bragg waves, which is 0.5/s.

According to Hasselmann and Hasselmann [14] and Vachon et al. [15,16], velocity bunching can contribute to Eq. (8.32) (linear MTF) based on the following equation, which was given by Vachon et al. [16]:

$$M_v\left(\vec{K}\right) = \frac{R}{V_S}\left[\omega\left(1 - 2u_r\cos\theta_0 c^{-1} + u_r^2 c^{-2}\right)\left(1 - u_r^2 c^{-2}\right)^{-1}\right]\left[k_i\left[T(gd)^{0.5}\sqrt{1-\left(\frac{R}{V_S}\right)^2 c^{-2}}\right]^{-1} \sin\theta + i\cos\theta\right]. \tag{8.35}$$

where k_i is the azimuth wavenumber, $\frac{R}{V_S}$ is the scene range to platform velocity ratio, which is approximately 32 s in the case of AIRSAR data [19], and u_r is the wave radial velocity as discussed in Chapter 6.

The SAR signal travels at the speed of light c, which is considered as the relativity source of radiated and received SAR pulses.

The wave spectral information has been acquired from the compositing of four subimages, and each subimage was 512 by 512 pixels (Fig. 8.4). The regular subimages spectral information was exploited with the quasilinear model. In ease of depiction, the AIRSAR and POLSAR imagery wavelength spectra are demonstrated in polar plots by circles that designate the wavelength spectra disparity. The wavelength values diminish from the outer to the inner circle. The space of the peaks from the center is inversely proportional to its wavelength. In addition, the angular position of the peaks specifies the wave propagation direction, (Figs. 8.5 and 8.6). The two-dimensional quantum Fourier transform changes the SAR data to the domain frequency. The largest wavelength is found in outer circles and decreased to inner circles (Figs. 8.5 and 8.6). This confirms the study by Marghany [20].

The wave spectra peaks fluctuate from C_{VV}, L_{HH}, and L_{VH} bands. The spectral peaks of the C_{VV} band match with the spectral peak of the in situ wave measurements. However, the retrieved spectrum from HV does not fit accurately with the one measured in situ (Figs. 8.5 and 8.6).

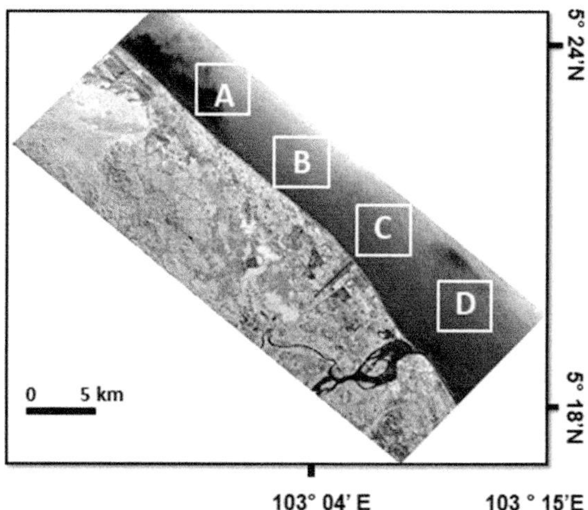

FIG. 8.4 Location of wave spectra window selections.

Therefore, similar results are obtained from the POLSAR data (Figs. 8.7 and 8.8). The L_{VH} band illustrates a poor match with in situ wave measurements than C_{VV} and L_{HH} bands. However, POLSAR data also prove that the C_{VV} band matches better with in situ measured spectrum physical properties, for instance, peak locations, wavelength, and direction of propagation.

Close inspection of spectra plots indicated the dominant wavelengths of 20, 50, and 100 m, which have been resolved by spectral analysis. The largest average wavelength occurred in areas A and B, with a value of 75 m, compared to areas C and D. Note that the wave refraction can be observed as the wave directions changed dramatically in areas A and B as they approached the shallow water in areas C and D (Figs. 8.5 and 8.7). It could further be seen that the offshore wave spectra travel at approximately 50° and approach the nearshore zone with an incident angle of about 45 degrees.

The retrieval of wave spectra using the quasilinear algorithm is in good agreement with in situ measurements (Fig. 8.5C and D). AIRSAR and POLSAR data present different retrieval wave refraction patterns (Fig. 8.7). The wavelength retrieved from POLSAR data is about 80 m and propagated at about 120°. It is interesting to find that the quasilinear model can retrieve a decreasing wavelength of 25 m, with direction change of roughly 140°, in the shallow water areas of A and B (Fig. 8.7B). This means that the retrieval wave refraction pattern from both airborne radar data agrees with the simulated one using in situ wave measurements. Fig. 8.7D, nevertheless, does not agree with Fig. 8.7B, where a sharp wave refraction pattern is presented in comparison to the quasilinear model for areas A and B.

Generally, the quasilinear model predicts only the wave spectra peak, which rotates toward the azimuth direction. This means that the quasilinear model corrects the distortion that occurs due to the Doppler shift frequency on the SAR image. It was noticed that there is a similar pattern of peak locations between observed wave spectra (in situ measurements and AIRSAR and POLSAR spectra) and the retrieval spectra using the quasilinear model as a result of low $\frac{R}{V_S}$, i.e., 32 s. The retrieval spectra, being characterized by relatively small values for $\frac{R}{V_S}$, were less susceptible to the azimuth cutoff, and provide good spectral fidelity. Thus, airborne imagery of ocean waves may be much less susceptible to imaging nonlinearity. This confirms the studies of Forget et al. [17] and Hasselmann et al. [21].

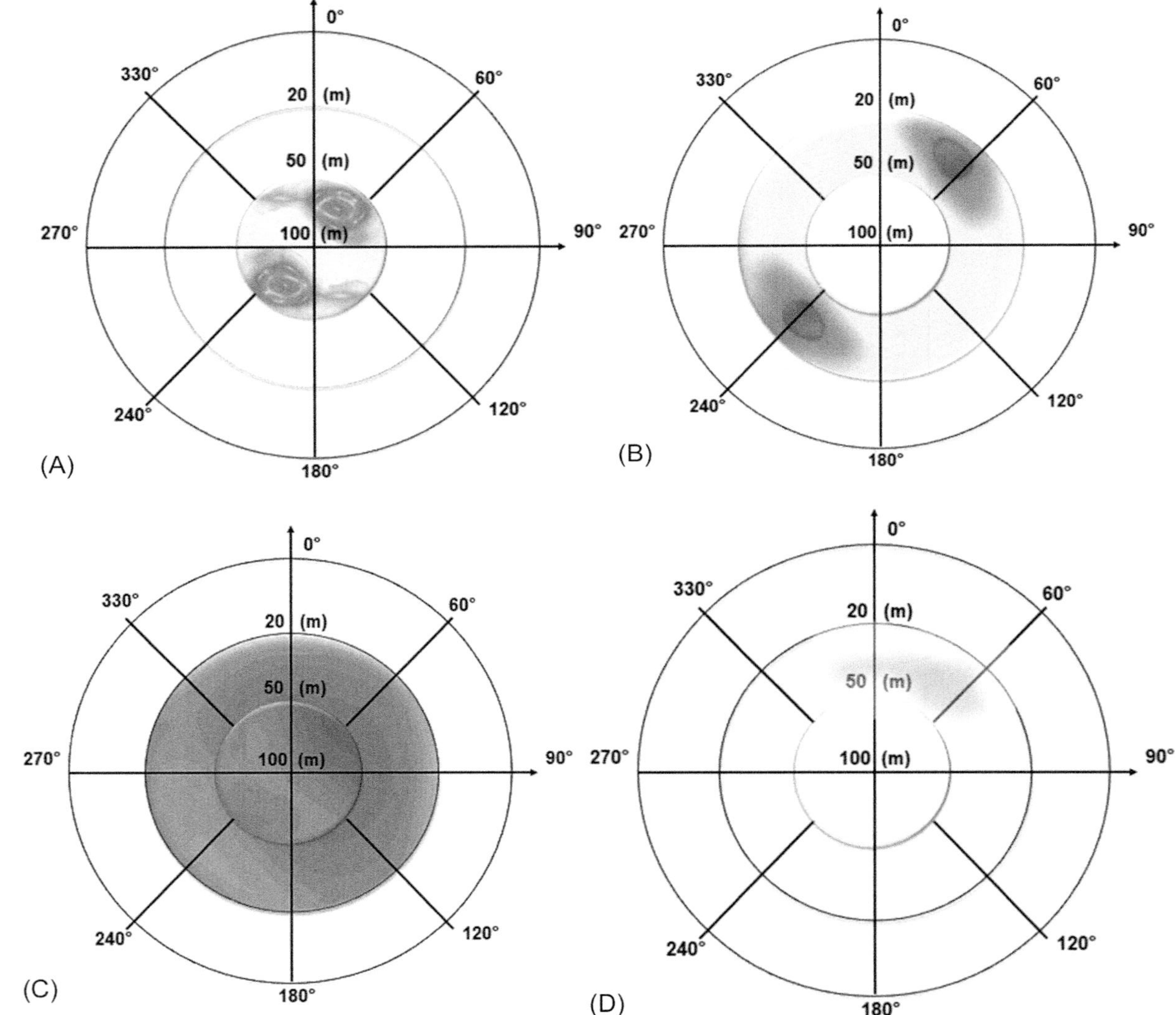

FIG. 8.5 Average AIRSAR wave spectra from C_{VV} band, (A) areas A and B, (B) areas C and D, (C) simulated in situ offshore waves, and (D) in situ onshore waves.

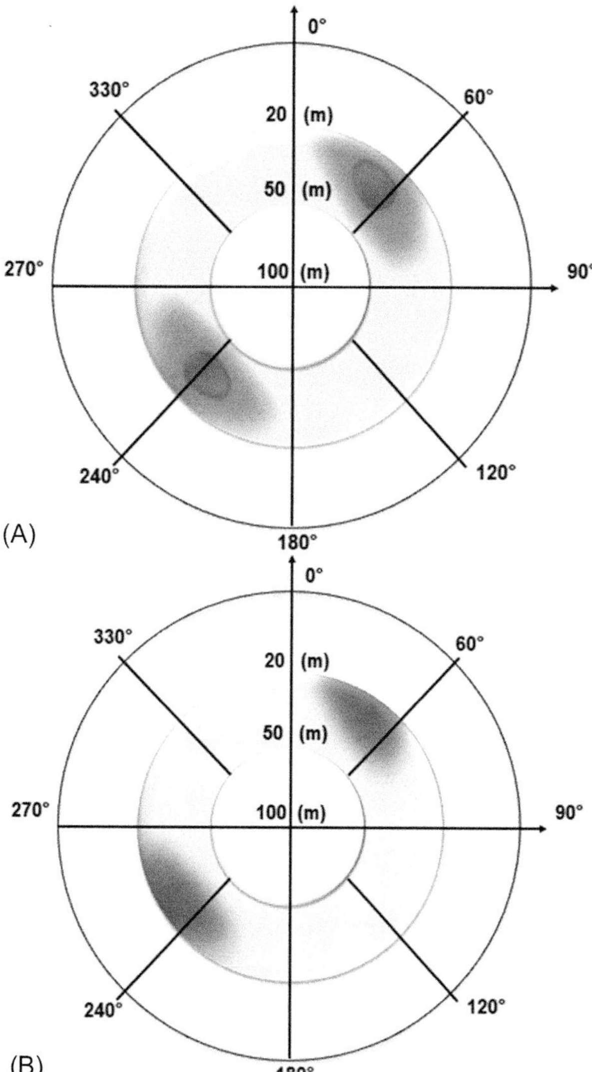

FIG. 8.6 Average AIRSAR wave spectra from areas A and B for different polarizations of (A) L_{HH} and (B) L_{VH} bands.

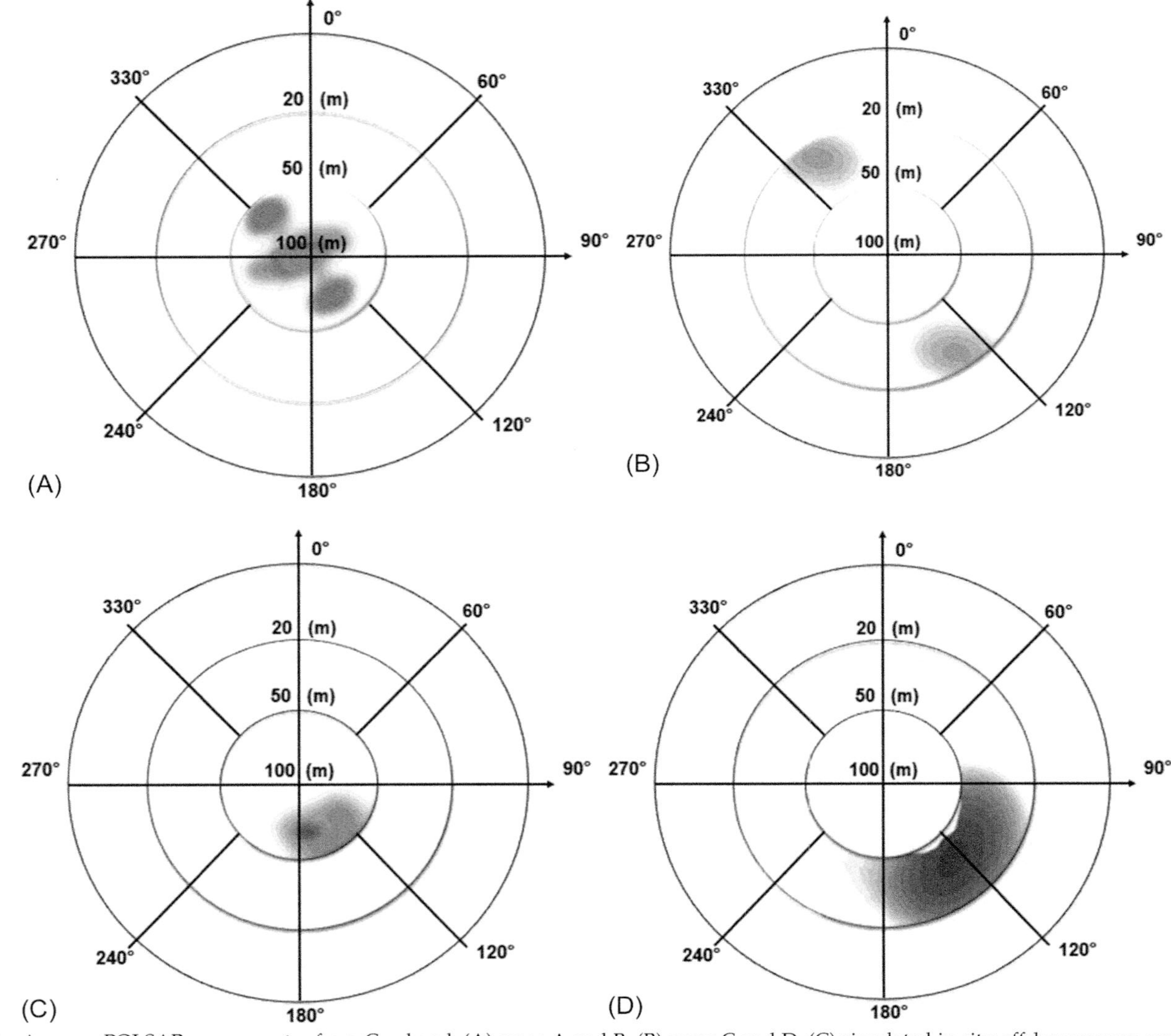

FIG. 8.7 Average POLSAR wave spectra from C_{VV} band, (A) areas A and B, (B) areas C and D, (C) simulated in situ offshore waves, and (D) in situ onshore waves.

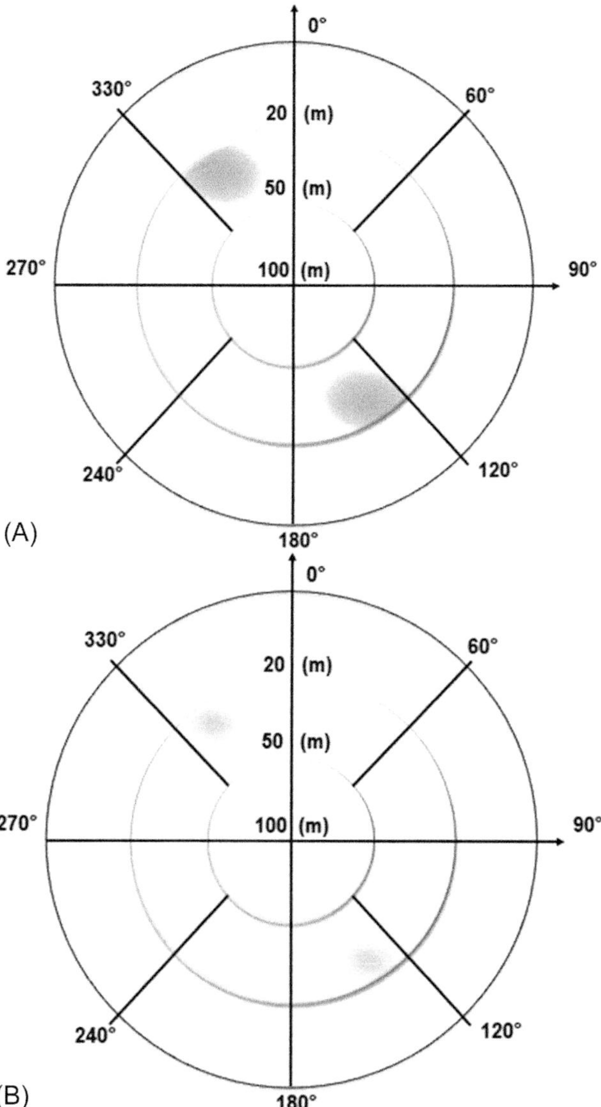

FIG. 8.8 Average POLSAR wave spectra from areas A and B for different polarizations of (A) L_{HH} and (B) L_{VH} bands.

8.11 Modeling significant wave height using azimuth cutoff model

The azimuth cutoff is the degree to which the SAR image spectrum is constrained in the azimuth direction. The azimuth cutoff is shaped by the wind and wave conditions in a quasilinear forward-mapping model. To approximate the significant wave height from the quasilinear transform, we adopted the algorithm that was given by Marghany [20] as appropriate for the geophysical conditions of tropical coastal waters:

$$\lambda_c = \beta\left(\int_{H_{s0}}^{H_{sn}} \sqrt{H_s} dH_s + \int_{U_0}^{U_n} \sqrt{U} dU\right). \tag{8.36}$$

where $\lambda_c$is the cutoff azimuth wavelength, and H_s and U are the in situ data of significant wave height and wind speed, respectively, along the coastal waters of Kuala Terengganu, Malaysia.

The measured wind speed was estimated at 10 m height above the sea surface. The changes of significant wave height and wind speed along the azimuth direction are replaced by dH_s and dU, respectively. The subscript zero refers to the average in situ wave data collected before flight pass over by 2 h, while the subscript n refers to the average of in situ wave data during flight pass over the study area. β is an empirical value, which results from R/V multiplied by the intercept of azimuth cutoff (c) when the significant wave height and the wind speed equal zero. A least-squares fit was used to find the correlation coefficient between cutoff wavelength and the one calculated directly from the AIRSAR/POLSAR spectra image by Eq. (8.36). The following equation was then adopted by Marghany [12,20] to estimate the significant wave height (H_{sT}) from the AIRSAR/POLSAR images:

$$H_{sT} = \beta^{-2} \int_{\lambda_{c0}}^{\lambda_{cn}} (\lambda_c)^2 d\lambda_c. \tag{8.37}$$

where β is the value of $\left(c\frac{R}{V}\right)^{-1}$ and H_{sT} is the significant wave height simulated from AIRSAR images.

A comparison of the retrieval of significant wave heights with in situ measurements indicates a positive linear relationship, as shown in Table 8.1. This indicates an r^2 value of 0.63, and the probability (P) is less than .05. The retrieval of significant wave height ranges between 0.56 and 1.34 m. This agrees with the studies of Marghany [19,20].

The retrieved significant wave height ranges between 0.56 and 1.34 m. Both AIRSAR/POLSAR data reveal that the C_{VV} band (Fig. 8.9A) has the highest r^2 value of 0.63 and lowest RMSE of ±0.3 compared to L_{HH} (Fig. 8.9B) and L_{HV} bands. On the contrary, the L_{VH} band demonstrates (Fig. 8.9C) the weakest correlation with in situ wave measurements and the highest RMSE of ±0.57 compared to C_{VV} band and L_{HH}. A similar scenario is shown with POLSAR data as C_{VV} band (Fig. 8.10A) provides better significant wave height fit with in situ measurements than L_{HH} band (Fig. 8.10B) and L_{VH} band (Fig. 8.10C).ạ

TABLE 8.1 Regression parameters for retrieval of significant wave height from AIRSAR and POLSAR data.

Statistical parameters	Airborne data	
	AIRSAR	POLSAR
r^2	0.63	0.62
P	<.05	<.05

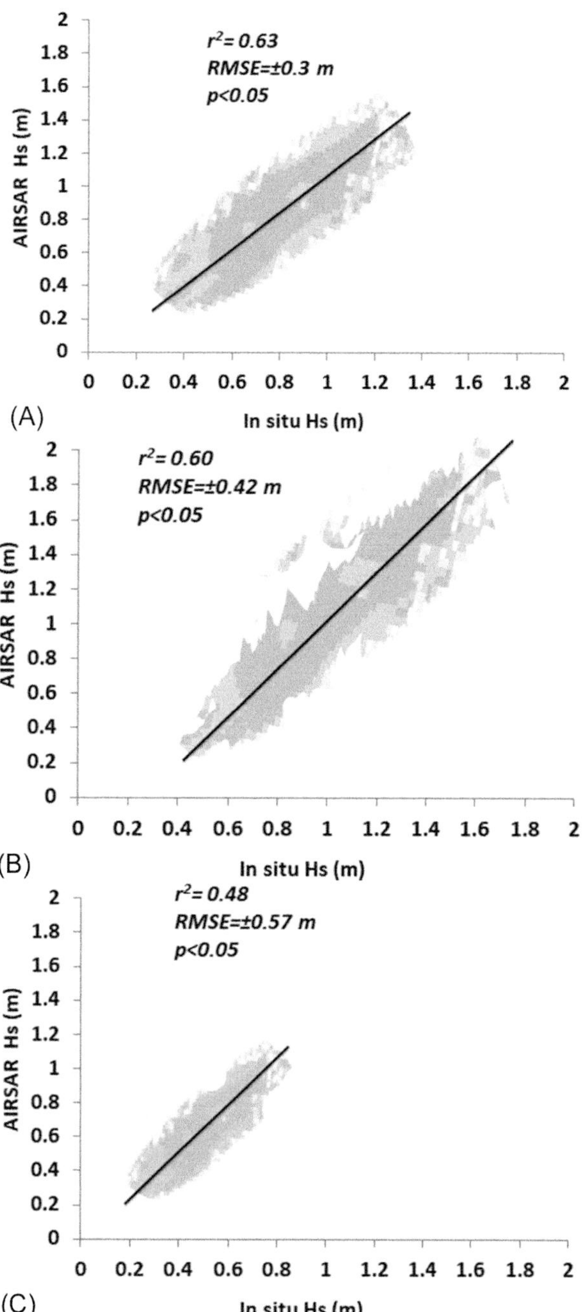

FIG. 8.9 AIRSAR significant wave height reveals from different polarization data sing azimuth cutoff model: (A) C_{VV} band, (B) L_{HH} band, and (C) L_{VH} band.

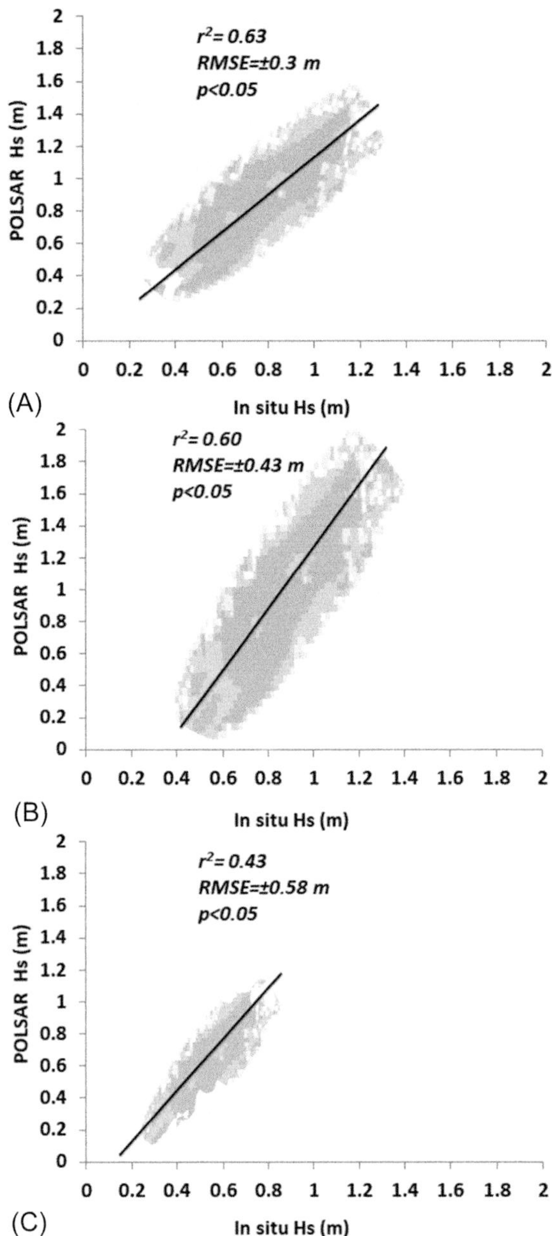

FIG. 8.10 POLSAR significant wave height reveals from different polarization data sing azimuth cutoff model: (A) C_{VV} band, (B) L_{HH} band, and (C) L_{VH} band.

8.12 AIRSAR/POLSAR cross-spectrum inversion

The recovery of the wave spectrum $S\left(\vec{k}\right)$ from the SAR cross-spectrum throughout the mathematical model presented in Eq. (7.14) is an inversion task to be instigated as an optimization problem. This comprises the explanation of an objective function $F[S]$ to be diminished beginning from a practicable guess $S^o\left(\vec{k}\right)$. Accordingly, the most naive objective function would contain a foremost correction term. It uses the reduction of the deviation between the observed SAR spectrum and the simulated one. Hasselmann et al. [21] suggested an inversion scheme grounded on the utilization of the SAR image spectrum ($\tau=0$) to determine the inherent 180° ambiguity and to deliver the lost wave information that is not exposed by SAR imaging at high azimuth wavenumbers. In this regard, the objective function comprises dual further regularization terms other than the principal term. The first term corrects the deviations from a "first guess" spectrum. Therefore, the second term matches the differences between the observed cutoff term and the simulated one. Consequently, the VV inversion scheme necessitates a consistent guess spectrum to operate accurately and to be referred to. On the other hand, the SAR image cross-spectrum characteristics that preserve information about the transmission direction of the wave systems could be suitably exploited and a simplified inversion scheme could be envisaged [22]. To achieve the mission of the inversion scheme, the gradient-descending technique has been preferred though inhibited nonlinear programming algorithms would be demoralized to acquire the optimum solution. Owing to the physical denotation of $S\left(\vec{k}\right)$, the examination of the minimum for the objective function would be reserved to nonnegative solutions. Subsequently, the gradient-descending scheme is an unconstrained technique of optimization of ocean wave spectra parameters. In this view, a modification is introduced by De Carolis et al. [22], with the intention of merely nonnegative quantities of the restated ocean spectrum are reserved at individually iteration step. The modified scheme at the n-th iteration step can be expressed as:

$$\begin{cases} S^n\left(\vec{k}\right) = S^{n-1}\left(\vec{k}\right) + \wp^n W^n\left(\vec{k}\right) \dfrac{\partial F\left[S^{n-1}\left(\vec{k}\right)\right]}{\partial S^{n-1}\left(\vec{k}\right)}, \quad \wp^n > 0 \\ \text{if } S^n\left(\vec{k}\right) < 0 \text{ then } S^n\left(\vec{k}\right) = 0 \end{cases} \tag{8.38}$$

where the objective function is demarcated as [22]:

$$F\left[S^{n-1}\left|\vec{K}\right\rangle\right] = \int \left|\psi_{l^1,l^2}^{n-1}\left|\vec{K}\right\rangle|\tau\rangle - \psi_{l^1,l^2}^{\partial t}\left|\vec{K}\right\rangle|\tau\rangle\right|^2 d\vec{k}' \tag{8.39}$$

Eq. (8.39) is implemented through diffusion quantum Monte Carlo (DQMC), which is fully described by Eqs. (7.22) and (7.23).

Fig. 8.11A reveals C_{VV} band inverse spectra with a clearer spectra peak than the L_{HH} (Fig. 8.11B) band and L_{VH} (Fig. 8.11C) band for AIRSAR data. Like AIRSAR data, POLSAR data proves that the C_{VV} band (Fig. 8.12A) has better sensitivity to image the ocean spectra peak than the L_{HH} (Fig. 8.12B) band and L_{VH} band (Fig. 8.12C).

Here, $\psi_{l^1,l^2}^{\partial t}\left|\vec{K}\right\rangle|\tau\rangle$ is the measured SAR image cross-spectrum. The weight function $W^n\left(\vec{k}\right)$ permits spectral growths both to be improved where SAR information is applicable and to be

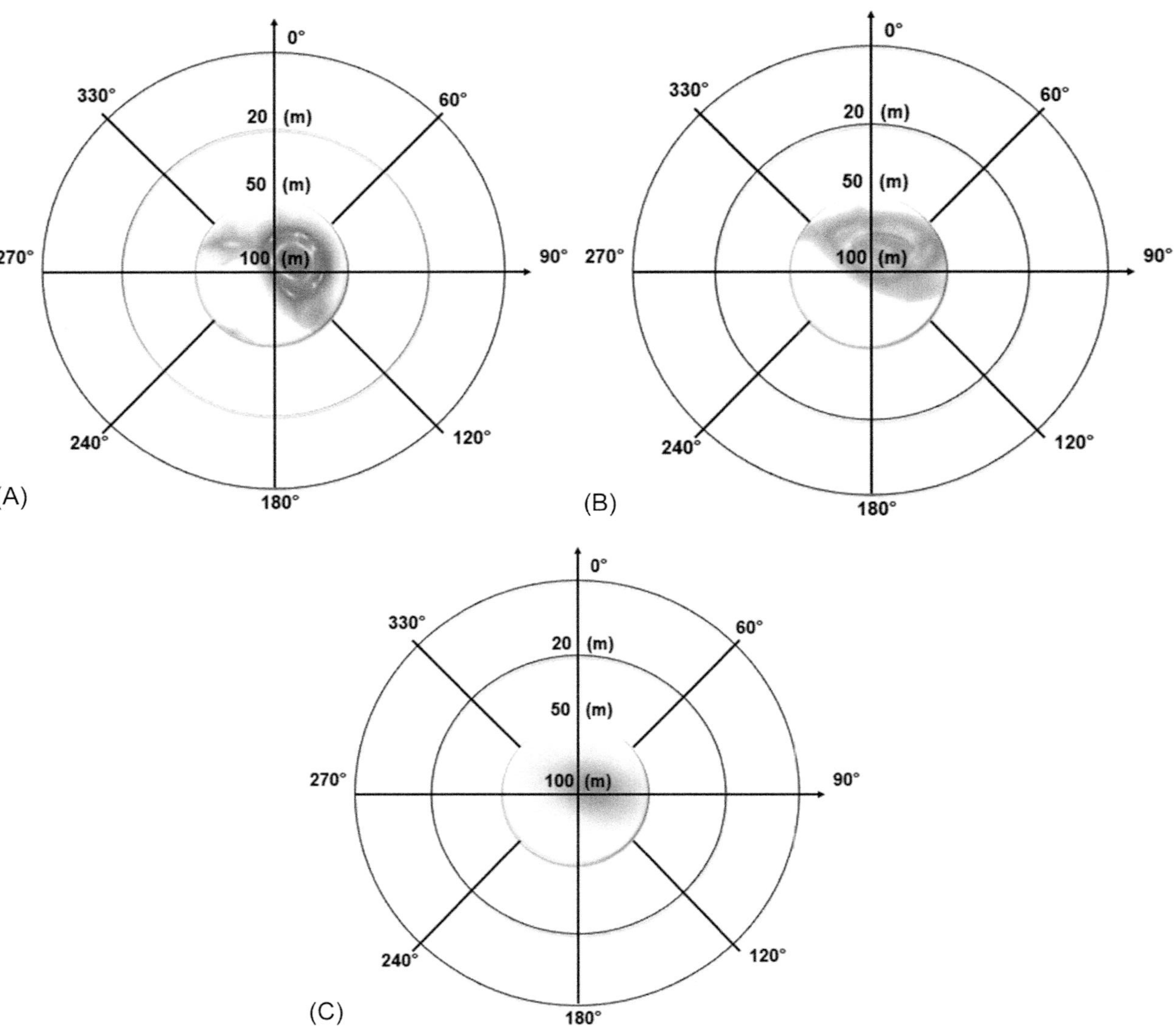

FIG. 8.11 Average AIRSAR image spectra inversion algorithm for (A) areas A and B of Fig. 8.4 for (A) C_{VV}, (B) L_{HH}, and (C) L_{VH} bands.

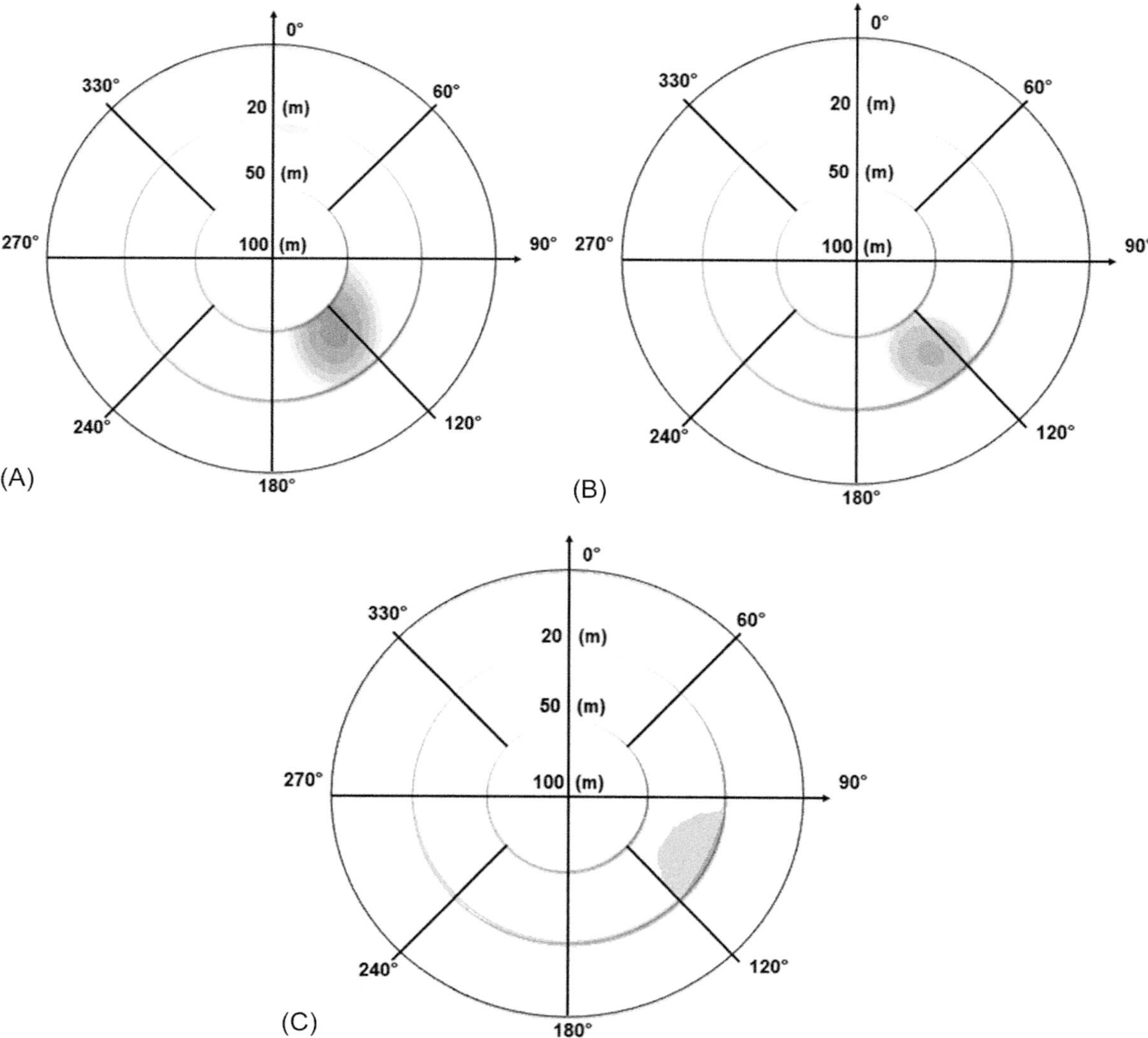

FIG. 8.12 Average POLSAR image spectra inversion algorithm for (A) areas A and B of Fig. 8.4 for (A) C_{VV}, (B) L_{HH}, and (C) L_{VH} bands.

disciplined at high wavenumbers [23]. In this understanding, the simplest optimal consists of a bell-shaped or Gaussian form for $W^n\left(\vec{k}\right)$ whose width is recursively examined for each iteration step. The Gaussian width value for which the objective function proceeds, the minimum quantity is neglected. Lastly, the parameter $\wp^n$ is estimated by first considering the linear contribution of the simulated SAR cross-spectrum concerning $\wp$ and then diminishing the objective function F [22].

8.13 Differences between deep and shallow water waves

Prior to simulating the wave refraction spectra from AIRSAR/POLSAR data, it is critical to answer the above question to understand perfectly the mechanisms of wave refraction. Following on from Chapter 7, the deep water waves, as the name suggests, have their basis where the depth of the water in the ocean is large, and there is no shoreline to provide any obstacle to their transfer [24]. Technically stated, they are twisted in spaces where the depth of the water is more than half of the wavelength of the wave. On the contrary, shallow waves have their origin where the depth of the water is much less. They usually propagate in waters that have depths less than $\frac{1}{20}$th of the wavelength of the wave. The wave's velocity is a function of its wavelength: waves with a longer wavelength transfer at greater velocities than waves with a shorter wavelength. Nonetheless, unlike deep water waves, the velocity of the wave has nothing to do with the wavelength of the wave and is only a function of the depth of water. Consequently, waves in shallow waters propagate faster than waves in deeper waters. More specifically, the wave's velocity C_v is equal to the square root of the product of the depth of water (d) and the acceleration due to gravity (g), i.e., $C_v = \sqrt{gd}$.

Deep water is also dominated by many waves of dissimilar wavelengths, which are superimposed upon one another to create an amalgamated superior wave. They are elongated and propagate in straight lines, and have adequate power to travel considerably greater distances than other waves, such as breaking waves. The foremost force of causality is wind stress, which can be from confined or widespread wind flows. They are also recognized as Stokesian waves or short waves. The length of these waves is less than the depth of the water they enter, which diminishes the velocity of the waves. This results in the reduction of the wavelength and growth in the height, ultimately breaking the waves, which drain from the beach as a backwash. Therefore, refracted waves move into shallow water when they come close to the shore, and the shallowness reduces the wave's power and causes a curved pattern. These are usually observed near headlands and bays.

8.14 Quantum of wave refraction

The significant question is: can we quantize the wave refraction spectra? Let us assume that a wave traveling with wave frequency ω, wavenumber k, and velocity potential ϕ, as the surface displacement η_s is given by:

$$\eta_s = A\cos(kx - \omega t) \tag{8.40}$$

$$\phi = \omega A \frac{\cosh(k(z+d))}{k\sinh(kh)} \sin(kx - \omega t) \tag{8.41}$$

where A is amplitude.

The dispersion relation for the small-amplitude surface water waves as well as their phase velocity v and group velocity v_g are formulated as follows:

$$\omega^2 = gk\tanh(kd) \tag{8.42}$$

$$v = \sqrt{\frac{g}{k}\tanh(kd)} \tag{8.43}$$

$$v_g = 0.5\sqrt{\frac{g}{k}\tanh(kd)}\left[1 + \frac{2kd}{2\sinh(2kd)}\right] \tag{8.44}$$

Consistent with Eq. (8.44), the ratio of the water depth to wavelength $\frac{d}{\lambda}$ involves two conditional phases: a shallow wave and deep wave. The shallow wave occurs when $\frac{d}{\lambda} \ll 1$ while deep wave occurs when $\frac{d}{\lambda} \gg 1$. For the case of shallow-water waves, the dispersion relationship can be approximated by:

$$\omega = k\sqrt{gd}\left[1 - \frac{k^2h^2}{6} + \dots\right]; v_0 = \sqrt{gd}. \tag{8.45}$$

Eq. (8.45) reveals that for very long shallow-water waves, $\omega = kv_0; v = \omega k^{-1} \approx v_0; v_g = \frac{\partial \omega}{\partial k} \approx k_0$. In contrast, for deep-water waves, the approximation for the dispersion relation is $\omega = \sqrt{gd}$ and the phase and group velocity are $v = \sqrt{gk^{-1}}; v_g = \sqrt{g(2k)^{-1}}$, respectively. Therefore, the group velocity is smaller and half the phase velocity.

Generally, ocean wave systems contain ordered ensembles of tiny particle moments—spins—coupled by the exchange interaction. On a quasiclassical basis, the particle moments are discretely capable of precessing approximately their equilibrium alignment. Owing to the pairing between spins, it is conceivable to stimulate phase-coherent precessional wave spectra spinning—the average particle moment per unit volume. These waves are termed "spin waves," which are described by their amplitude, phase, frequency, wave vector, and group and phase velocities, each representing a resource for wave propagation.

The key feature that makes spin waves unique is their dispersion, which can be peculiarly anisotropic depending on the dominant interaction between wind stress and water or wave-particle moments. There are dualistic foremost interfaces to deliberate. The quantum-mechanical exchange interaction, which is responsible for the wind spin ordering, dominates over centimeter wavelengths and gives rise to an isotropic, parabolic dispersion of so-called exchange spin waves.

Spin waves are keystones to transfer the energy along the direction and speed of wave propagation, which is determined by their group velocity, demarcated as the incline of the angular frequency in reciprocal space. As well as the bias wind field, the spin-wave group velocity relies on the angle between the spin wave's wave vector and the wind stress over the ocean. In this regard, the dispersion is different for each superimposed wave and relies on the water depth, which suggests continuous variation of the depth and/or width as a means by which to control the spin-wave propagation from deep water to shallow water.

The dispersion of spin waves in the ocean is also sensitive to the variation of the water depth. Moreover, the finite depth leads to quantization and thus the appearance of several dispersion branches for spin waves (of any sort) propagating within the plane of the ocean.

The relation of the angles between the incident and refracted waves is called Snell's law. Refraction arises owing to the alteration in the phase velocity between different water depths, characterized by dispersion relation. Spin waves in the small wavenumber regime have an anisotropic dispersion relation owing to the anisotropic nature of wave-particle and bottom interactions.

Consequently, the phase velocity of spin waves relies on the angle of deep and shallow wave propagation. In ordinal refraction between two different media with positive refraction indices, the refracted wave has a wave vector directing forward along the boundary. The divergence refraction of spin waves is created from the anisotropy of the dispersion relation and consequent variance in the direction of the wave vector and group velocity.

Note that the spin-wave ray represents the energy flow owing to spin waves, divergence, and convergence of wave refraction in both cases because refracted spin waves have the opposite sign of the group velocity components compared to the incident spin waves. The relation between refraction and incident angle is modeled by considering the anisotropic dispersion relation of spin wave transfers and the wavenumber conservation of spin waves among the domain waves.

8.15 Wave refraction graphical method

The wave refraction model of the AIRSAR and POLSAR images is articulated based on the wave number and wave energy conversation principle, gentle bathymetry slope, steady wave conditions, and refractive depth (Fig. 8.13). Consistent with Herbers et al. [18], wave refraction can be formulated mathematically as:

$$\frac{\partial}{\partial x}\left(H_s{}^2 c_g \cos\varphi\right) + \frac{\partial}{\partial y}\left(H_s{}^2 c_g \sin\varphi\right) = 0. \tag{8.46}$$

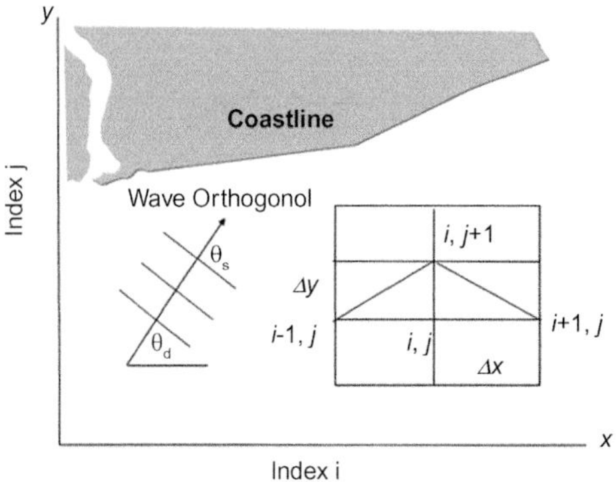

FIG. 8.13 Sketch of wave refraction over gentle bathymetry slope.

where the coordinates and the wave angle φ are orientated according to the notation of Fig. 8.13.

Eq. (8.46) is a first-order partial differential equation (PDE) in the unknown variables $\varphi(x,y)$ and $H_s^2(x,y)$; the group velocity c_g is a known function of the wave period T and the known local depth $h(x,y)$. Following Herbers et al. [18], in the notation of Fig. 8.13, the explicit finite difference scheme, centered in x, proposed for the solution of Eq. (8.46) takes the following form:

(1) Wave angle equation: solved for φ_{ij+1}:

$$\varphi_{ij+1} = \arccos\left[\left(\frac{\Delta y_j}{2}\left(\left(\frac{\sin\varphi_{i+1j}}{c_{i+1j}} - \frac{\sin\varphi_{ij}}{c_{ij}}\right)\Delta x^{-1} + \left(\frac{\sin\varphi_{ij}}{c_{ij}} - \frac{\sin\varphi_{i-1j}}{c_{i-1j}}\right)\Delta x^{-1}{}_{i-1}\right) + \frac{\cos\varphi_{ij}}{c_{ij}}\right)\cdot c_{ij+1}\right]. \tag{8.47}$$

(2) Significant wave height equation: solved for Hs_{ij+1}:

$$H_{s_{ij+1}}{}^2 = \frac{1}{c_{gij+1}\sin\varphi_{ij+1}}\left[\left(H^2{}_s c_g \sin\varphi\right)_{ij} - 0.5\Delta y_j\left(\frac{\left(H^2{}_s c_g\cos\varphi\right)_{i+1j}}{\Delta x_i} - \frac{\left(H^2{}_s c_g\cos\varphi\right)_{ij}}{\Delta x_i} + \frac{\left(H^2{}_s c_g\cos\varphi\right)_{ij} - H^2{}_s c_g\cos\varphi)_{i-1j}}{\Delta x_i - 1}\right)_j\right]. \tag{8.48}$$

The boundary conditions completing the model are:

(i) It is assumed that the parallel depth contours are as shown in Fig. 8.13.
(ii) The φ and H_s values are given as initial conditions on the open sea boundary ($j=1$).
(iii) The computation is terminated on the coastal boundaries ($h=0$). The wave breaking criterion is applied in shallow waters. The computed significant wave height H_s is compared to:

$$0.78h_{ij};\text{if}\,H_{s_{ij+1}} > 0.78h_{ij},\text{then}\,H_{s_{ij+1}} = 0.78h_{ij} \tag{8.49}$$

The spectra energy of significant wave height distribution due to wave refraction is then estimated by using the following formula adapted from Hasselmann and Hasselmann [21]:

$$E\left(\vec{K}, H_S\right) = S(k_x, k_y)p(H_S). \tag{8.50}$$

where $S(k_x, k_y)$ is the distribution for the wave number and $p(H_S)$ is the probability distribution of the significant wave height in the convergence and divergence zone.

Consistent with Herbers et al. [18], the refraction index (K_r) for a straight coastline with parallel contours can be estimated using the following equation:

$$K_r = \sqrt{\frac{\cos\theta_d}{\cos\theta_r}}. \tag{8.51}$$

where θ_d and θ_r are the deep and shallow waves incidence angles, respectively.

To this end, diffusion quantum Monte Carlo (DQMC) is implemented to retrieve precise inversion wave cross-spectra, which is fed into the wave refraction model to simulate AIRSAR/POLSAR wave refraction. The details of the implementation of DQMC are given in Chapter 7, Section 7.9.

Fig. 8.14 shows the wave refraction pattern modeled from the image spectra inversion of quasilinear model and in situ wave data. The input image spectra inversion wavelength is 80 m. Both AIRSAR and POLSAR wave refraction pattern results indicate the refractive index is 2.60 and 2.54 at the Sultan Mahmud Airport station and the location of Batu Rakit station, respectively, showing the convergence of wave energy (Figs. 8.14A and 8.15A) in C_{VV} band for both AIRSAR and POLSAR data, respectively. In contrast, the convergence and divergence zones are not well-recognized in the L_{VH} band (Figs. 8.14C and 8.15C) for both AIRSAR and POLSAR, respectively. However, the L_{HH} (Figs. 8.14B and 8.15B) band shows a better wave refraction pattern than the L_{VH} band; however, L_{HH} cannot reveal a wave refraction pattern as well as the C_{VV} band. These can also be realized by refraction index values. For instance, at the Batu Rakit station, which is close to the river mouth of Kuala Terengganu, the refractive index values are less than 1.00, suggesting divergence of wave energy. In other locations, the refractive index values are close to 0.99, indicating no change in the concentration of wave energy at the coastline. Although the refractive index values for the quasilinear model differed with those of the in situ wave spectra refraction, the same trend of wave energy dispersion and concentration occurs at the coastline. This means that the wave refraction pattern simulated by using the inverse SAR spectra of the quasilinear model is similar to the wave refraction simulated from the in situ wave data. The largest refractive index value is observed at the Sultan Mahmud Airport station. This could be attributed to the slightly concave shoreline profile, which made the incoming north wave energy converge.

Fig. 8.16 shows the average wave refraction spectra energy because of convergence and divergence simulated from C_{VV} of AIRSAR/POLSAR data. The convergence spectrum has a sharper peak compared to divergence spectra. The sharp peak of the convergence spectrum is 0.84 m^2 s (Fig. 8.16A) while the divergence spectrum peak is less than 0.4 m^2 s (Fig. 8.16B). The convergence spectrum peak is located along the azimuth direction. It can be explained that the highest spectra energy propagated close to the azimuth direction is caused by the influence of the Doppler frequency shift, which is produced by convergence. This result agrees with the studies of Vachon et al. [15,16] and Marghany [19]. Consequently, the Doppler frequency shift leads to the conclusion of the impact of the relativity in SAR imaging wave spectra traveling. The SAR imaging mechanism of ocean wave not only established as the real spectra wave imagine either along range or azimuth directions, but just as relativity backscatter at the background of SAR image owing to the impact of the relativity of the Doppler frequency shift. Conversely, the changing of spectra physical parameters such as wavelength and peak directions from band to another (band C and band L) and among different polarization also creates SAR wave spectra relativity as it noticed through the VV, HH, and VH polarizations. The spectra pattern is more distorted than for the C_{VV} and L_{HH} bands. Moreover, velocity bunching and range bunching across VH cause wavelength distortion in the SAR image that is also a function of the incidence angle.

Generally, the exact wave spectrum retrieved in the SAR image by removing the 180° ambiguity can assist to simulate accurate wave refraction patterns. This is demonstrated using the inverse cross-spectra algorithm. Generally, comparing the three modeled inverse cross-spectra, it is clear that the HH spectrum reveals the largest spectral energy at a given

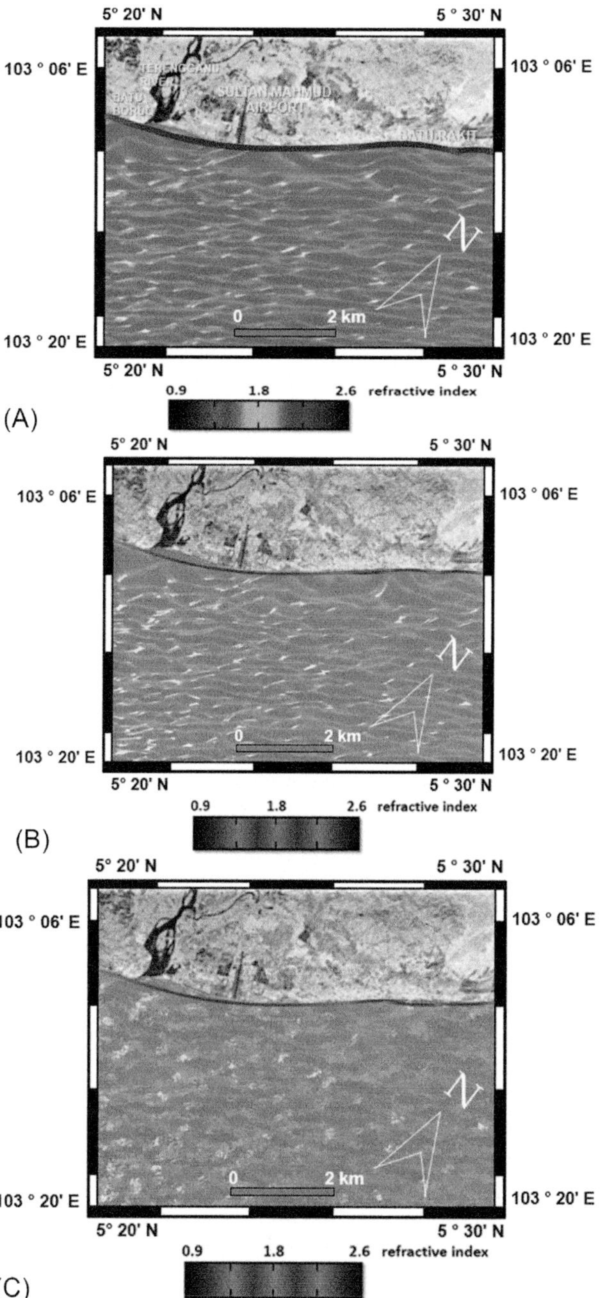

FIG. 8.14 Wave refraction pattern from AIRSAR for (A) C_{VV}, (B) L_{HH}, and (C) L_{VH} bands.

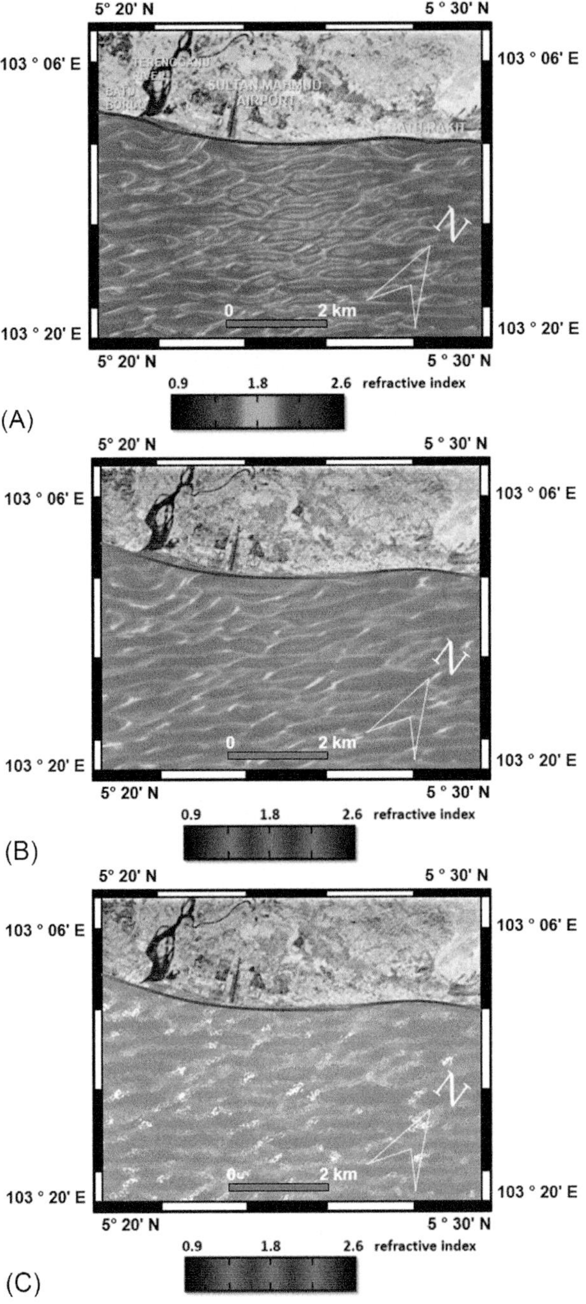

FIG. 8.15 Wave refraction pattern from POLSAR for (A) C_{VV}, (B) L_{HH}, and (C) L_{VH} bands.

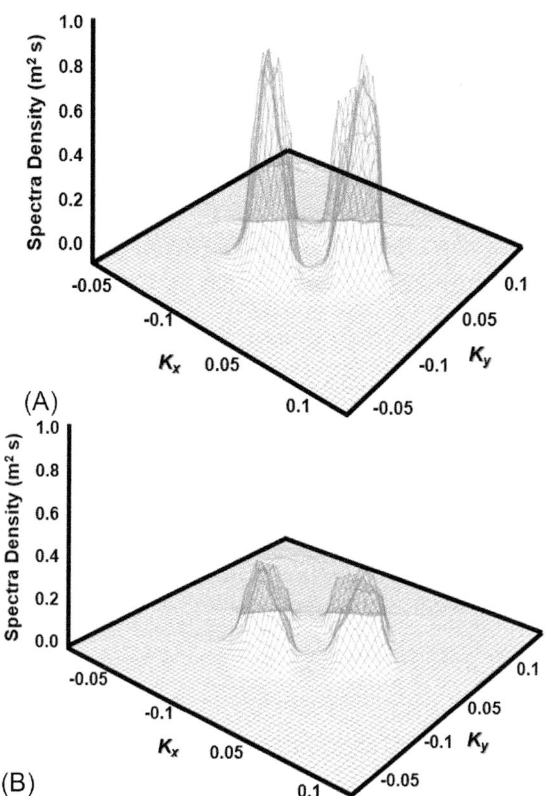

FIG. 8.16 Wave refraction spectra (A) convergence and (B) divergence.

wavenumber, in line with the expectation, owing to the largest RAR MTF in HH polarization. In other words, the wave spectra pattern on the VH band for both AIRSAR and POLSAR data is more blurred than that of VV and HH polarization. The realistic wave refraction pattern is also delivered in the C_{VV} band owning to VV polarization being more sensitive to the ocean surface than HH and VH. Moreover, diffusion quantum Monte Carlo (DQMC) assists in reconstructing real wave refraction spectra in both AIRSAR and POLSAR data. DQMC can be interpreted, however, as describing the behavior of a system of particles, each of which performs a particle random walk, i.e., diffuses isotropically, and at the same time is subject to multiplication, which is determined by the value of the point function. Indeed, the DQMC technique is a simple game of chance involving the random walks of wave particles through space and their occasional multiplication or disappearance, as illustrated in the simulated wave refraction pattern in AIRSAR/POLSAR data.

References

[1] Kennaugh EM. Polarization properties of radar reflections. [Doctoral dissertation]Columbus, OH: The Ohio State University; 1952.

[2] Lüneburg E. Principles of radar polarimetry. IEICE Trans Electron 1995;78(10):1339–45.

[3] Boerner WM, Mott H, Luneburg E. Polarimetry in remote sensing: basic and applied concepts, In: IGARSS'97. 1997 IEEE international geoscience and remote sensing symposium proceedings. Remote sensing—A scientific vision for sustainable development 1997 Aug 3vol. 3. IEEE; 1997. p. 1401–3.
[4] Mott H. Antennas for radar and communications: A polarimetric approach. Wiley; 1992.
[5] Krogager E. Aspect of polarimetric radar imaging. [Ph. D. dissertation]Danish Defence Research Establishment; 1993.
[6] Borgeaud M, Noll J. Analysis of theoretical surface scattering models for polarimetric microwave remote sensing of bare soils. Int J Remote Sens 1994;15(14):2931–42.
[7] Hajnsek I, Pottier E, Cloude SR. Inversion of surface parameters from polarimetric SAR. IEEE Trans Geosci Remote Sens 2003;41(4):727–44.
[8] Lee JS, Jansen RW, Schuler DL, Ainsworth TL, Marmorino GO, Chubb SR. Polarimetric analysis and modeling of multifrequency SAR signatures from Gulf Stream fronts. IEEE J Ocean Eng 1998;23(4):322–33.
[9] Schuler DL, Kasilingam D, Lee JS, Jansen RW, De Grandi G. Polarimetric SAR measurements of slope distribution and coherence changes due to internal waves and current fronts, In: IEEE international geoscience and remote sensing symposium 2002 Jun 24vol. 1. IEEE; 2002. p. 638–40.
[10] Kasilingam D, Schuler D, Lee JS, Malhotra S. Modulation of polarimetric coherence by ocean features, In: IEEE international geoscience and remote sensing symposium 2002 Jun 24vol. 1. IEEE; 2002. p. 432–4.
[11] Cloude SR, Pottier E. An entropy based classification scheme for land applications of polarimetric SAR. IEEE Trans Geosci Remote Sens 1997;35(1):68–78.
[12] Marghany MM. TOPSAR wave spectra model and coastal erosion detection. Int J Appl Earth Obs Geoinf 2001;3 (4):357–65.
[13] Pierson Jr. WJ, Moskowitz L. A proposed spectral form for fully developed wind seas based on the similarity theory of SA Kitaigorodskii. J Geophys Res 1964;69(24):5181–90.
[14] Hasselmann K, Hasselmann S. On the nonlinear mapping of an ocean wave spectrum into a synthetic aperture radar image spectrum and its inversion. J Geophys Res Oceans 1991;96(C6):10713–29.
[15] Vachon PW, Krogstad HE, Paterson JS. Airborne and spaceborne synthetic aperture radar observations of ocean waves. Atmosphere-Ocean 1994;32(1):83–112.
[16] Vachon PW, Liu AK, Jackson FC. Near-shore wave evolution observed by airborne SAR during SWADE. Atmosphere-Ocean 1995;2:363–81.
[17] Forget P, Broche P, Cuq F. Principles of swell measurement by SAR with application to ERS-1 observations off the Mauritanian coast. Int J Remote Sens 1995;16(13):2403–22.
[18] Herbers TH, Elgar S, Guza RT. Directional spreading of waves in the nearshore. J Geophys Res Oceans 1999;104 (C4):7683–93.
[19] Marghany M. Velocity bunching model for modelling wave spectra along east coast of Malaysia. J Indian Soc Remote Sens 2004;32(2):185–98.
[20] Marghany M. ERS-1 modulation transfer function impact on shoreline change model. Int J Appl Earth Obs Geoinf 2003;4(4):279–94.
[21] Hasselmann S, Bruning C, Hasselmann K, Heimbach P. An improved algorithm for the retrieval of ocean wave spectra from synthetic aperture radar image spectra. Oceanogr Lit Rev 1997;3(44):280.
[22] De Carolis G, Parmiggiani F, Arabini E, Trivero P. Wave and wind field extraction from ERS SAR imagery. In: SAR image analysis, modeling, and techniques III. vol. 4173. International Society for Optics and Photonics; 2000. p. 64–74.
[23] Cloude S.R. Potential new applications of polarimetric radar/SAR interferometry. Proc. on advances in radar methods—With a view towards the twenty first century, EC-JRC/SAI (ISPRA) Hotel Dino, Baveno, Italy, 1998 July 20–22; 1998. (Proc. to be published in Fall 98).
[24] Boerner WM. Basics of SAR polarimetry I. Illinois Univ at Chicago; 2005.

CHAPTER

9

Wavelet transform and particle swarm optimization algorithms for automatic detection of internal wave from synthetic aperture radar

9.1 Introduction

The previous chapters have tackled the SAR ocean wave imaging mechanisms from the point of view of quantum mechanics and relativity theories, especially due to the relativity of the Doppler frequency shift. In previous chapters, the mechanism of ocean surface waves was discussed as well as the implementation of the SAR wave spectra retrieving algorithm. This chapter demonstrates a certain sort of ocean wave, which is known as the internal wave. Following on from previous chapters, SAR data have proved to be very promising for sea surface phenomena detections. Can SAR data deliver precise clues about the internal wave?

9.2 What is meant by internal wave?

Beneath the ocean surface, hidden from human eyes, lie "internal" waves. They occur due to the deep waters of the ocean being denser than the surface waters. When a parcel of deep (dense) water rises closer to the surface, gravity forces it to sink. Correspondingly, if you upraised a parcel of seawater mass into the air; it would free fall down. By the same principle, the effective "buoyancy" of surface waters, means they would return to the surface if they were temporarily sunk.

In this regard, restoring force governs the expansion of compressed water and the sinking of raised water. In other words, restoring force acts on vertical variants of density, which generates internal waves. This means that, in every phase, there are dual layers of fluid with a curved interface between them, as demonstrated in Fig. 9.1.

https://doi.org/10.1016/B978-0-12-820785-7.00005-8

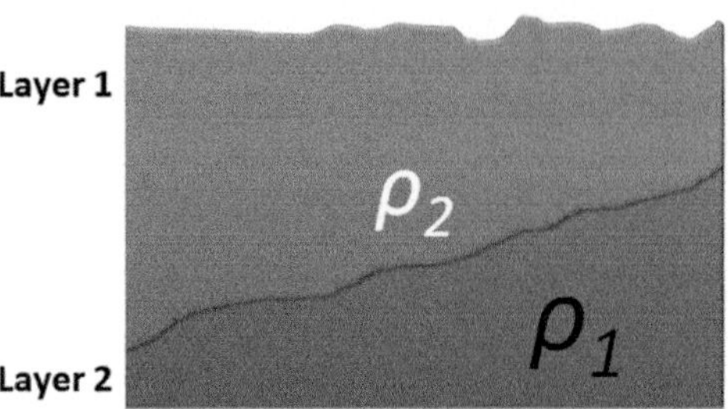

FIG. 9.1 A simple explanation of internal wave generation.

Let us assume that the difference in density between these two, $\rho_1 - \rho_2$, in the notation illustrated in Fig. 9.1, is about **1000 kg/m³**. In this regard, the density variations within the ocean are relatively small, and as a consequence, the restoring force for internal waves is weak compared with that of surface waves. This contributes to the slow movement of the internal wave within the ocean body. The critical question is: how slowly, precisely? Some mathematical work is required to answer this question. Let us consider the specific density of surface seawater to be $\rho_1 = \mathbf{1024\,kg/m^3}$ while the deep density is usually of value $\rho_2 = \mathbf{1025\,kg/m^3}$. The simple mathematic formula to describe the difference between the two layers of water density is:

$$\Delta\rho = \rho_1 - \rho_2 \tag{9.1}$$

It is also appropriate to designate the mean density, and for that it is conventional to use a subscript zero, as given by:

$$\rho_0 = 0.5(\rho_1 + \rho_2) \tag{9.2}$$

Consequently, the relative change in density is $\frac{\Delta\rho}{\rho_0}$. It would be an excellent application to declare whether the comparative density change is an large number or a small one. In this regard, $\frac{\Delta\rho}{\rho_0}$ is a function of gravity, or the "buoyancy." Let us assume that g signifies the acceleration owing to gravity, which is approximately $10\,\mathrm{m\,s^{-2}}$. Thus the mathematical description of an *effective* gravity felt by internal waves is formulated as:

$$g' = g\frac{\Delta\rho}{\rho_0} \tag{9.3}$$

Eq. (9.3) denotes the reduced gravity g'. The simple mathematical formula can be exploited to deliver the speed of the internal wave as a function of the water depth d as:

$$V_I = \sqrt{gd} \tag{9.4}$$

If the water approaches the shallow water depth of about 1–10 m deep, with the intention of the speed of waves V_I, substituting into Eq. (9.4), its velocity would be approximately $3\text{–}10\,\mathrm{m\,s^{-1}}$. In this view, this wave speed is similar to one can run at $3\,\mathrm{m\,s^{-1}}$, but only an Olympic athlete can run at $10\,\mathrm{m\,s^{-1}}$, and then only for a few tens of seconds. Consequently, internal waves would propagate more slowly than surface waves. In this circumstance, an exact value of g' is $0.01\,\mathrm{m\,s^{-2}}$, and consequently, the speed would be abridged by a factor of approximately $\frac{1}{30}$, in which 30 is about equal to $\sqrt{1000}$. In this regard, we assume that the surface/internal waves of waters with irregular depth, for an area from a kilometer to a few hundred kilometers from shore, is $d = 100\,\text{m}$ [1]. Consequently, the surface waves travel at a speed of $30\,\text{m}\,\text{s}^{-1}$, while the internal waves travel at a relatively slow $1\,\text{m}\,\text{s}^{-1}$.

The predication time for internal wave traveling can be simulated by:

$$t = \frac{L}{4\sqrt{gd\frac{\Delta\rho}{\rho_0}}} \tag{9.5}$$

The time variation of the internal wave propagation is inversely proportional to the root square of $\frac{\Delta\rho}{\rho_0}$ as a function of buoyancy and directly with its wavelength L. Subsequently, the slope S_l of internal wave traveling can be identified by:

$$S_l = \frac{L}{4\sqrt{8\times10^{-4}gdS‰}} \tag{9.6}$$

here $S‰$ is seawater salinity, the square root of which is inversely proportional with the internal wave slope. Indeed, the sharp slope of internal wave occurs due to the low fluctuation of water salinity time the gravity through the water depth d.

9.3 Simplification of internal wave generation mechanisms

Internal waves are also known as gravity waves, which oscillate within, as opposed to on the surface of, a water body. They are generated by relationships within a dynamical stratified ocean system. In other words, internal waves are gravity waves that oscillate inside a fluid medium, instead of on its surface, and the ocean must be well-stratified to generate internal waves. In this regard, the density ought to decrease continuously or discontinuously with depth because of physical characteristic fluctuations, for instance, in temperature and/or salinity and density (Fig. 9.2).

Vertical mixing occupies the water column as result of vertical velocity shear impact that is caused by internal waves. External forces, for instance, the tide, can create internal waves in highly stratified waters—tide relocates a water parcel, which is reinstated by buoyancy forces. Then the restoration force perhaps surpasses the equilibrium stage and generates an oscillation, which is generally known as an internal wave. An internal tide, consequently, arises from the interface between irregular topography and the barotropic tide. In other words, numerous varieties of internal waves develop in the plume zone and buoyant waters.

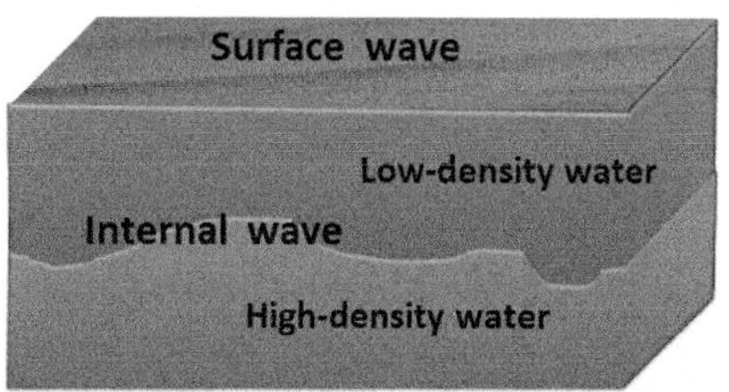

FIG. 9.2 Physical circumstances for the creation of internal waves.

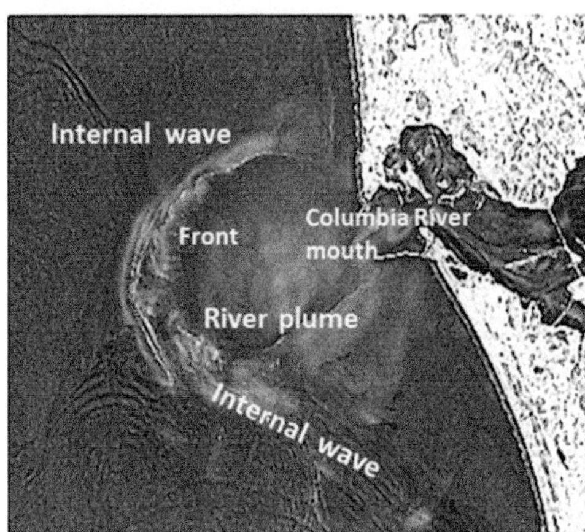

FIG. 9.3 Internal waves induced along the Columbia River plume.

One is created by interface between the tides and the sharp topography of the continental shelf, roving shoreward, which is a usual mechanism for generating the internal waves that end at the continental shelf. In this view, the plume fronts also concern barotropic tide, increasing internal tides on the plume flow. Consequently, the internal waves are allied with drastic escalations in phytoplankton and zooplankton aggregates.

Consequently, solitons are regularly seen on the leading edge of a semidiurnal internal tide bore. These solitary waves, therefore, reveal a second-order Korteveg-de Vries (KDV) configuration: the Comb-KDV model with minor variation in soliton width and amplitude. A different sort of internal wave is caused at the front of the river plume, traveling off shoreward. For instance, Nash and Moum [2] explored the creation mechanism of the internal waves induced at the Columbia River plume front, presuming that the wave separation from the braking plume front is accountable for the internal wave creation (Fig. 9.3). Nonetheless, the dynamic sorts and influences of internal waves roving away from the plume expanse have been little explored owing to a lack of in situ measurements.

This is characterized by horizontal length scales of the order 1–100 km and its horizontal velocity of 0.05–0.5 $\mathrm{m\,s^{-1}}$. Finally, its time generation scale ranges from a few minutes to days. On the shelf, therefore, the dynamic of the internal wave is strongly nonhydrostatic and thus cannot be well resolved in ocean general circulation models that frequently use a hydrostatic approximation [1–3].

Similarly to surface waves, internal waves change as they approach the shore. As the ratio of wave amplitude to water depth becomes such that the wave "feels the bottom," water at the base of the wave slows down due to friction with the seafloor. This causes the wave to become asymmetrical and the face of the wave to steepen, and finally, the wave will break, propagating forward as an internal bore [1, 4].

9.4 Mathematical description of internal waves

Let us assume two water layers have different densities, which are separated into two mean flows, U_1 and U_2, respectively, by an interface $z=\zeta$ (Fig. 9.4). A perturbation in the

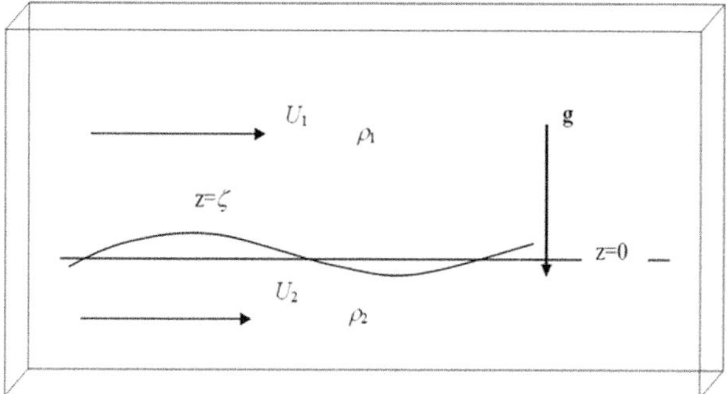

FIG. 9.4 Two different layers of densities and separated interfacing mean flows.

different layers of water densities is presented into the water body by perturbing the interface a small quantity to the interface proceeds from the $z=0$ surface by an amount $\zeta(x,t)$. At every layer the density is persistent, friction is discounted as being insignificant and, in the absence of planetary rotation, the movement taking place from relaxation will be, and continue to be, irrotational. Consequently, every layer above and below the interface is identified as [5, 6]:

$$\nabla^2\varphi_j=0,\ \ j=1,2 \tag{9.7}$$

where the index j equals 1 for the layer above the interface and equals 2 for the layer below the interface.

In every fluid, at the interface the Bernoulli equation yields:

$$\frac{\partial\varphi_j}{\partial t}+0.5\left|\nabla\varphi_j\right|^2+g\zeta=-\frac{p_j(x,\zeta,t)}{\rho_j} \tag{9.8}$$

and the kinematic condition, spread on every side of the interface is identified by:

$$\frac{\partial\zeta}{\partial t}+\nabla\varphi_j\cdot\nabla\zeta-\frac{\partial\varphi_j}{\partial z}=0 \tag{9.9}$$

Combining Eqs. (9.8) and (9.9) leads to small perturbations to the streaming movement in every layer and can be linearly formulated as:

$$\begin{aligned}\left(\frac{\partial}{\partial t}+U_j\frac{\partial}{\partial x}\right)\varphi_j+g\zeta&=-\frac{p_j}{\rho_j}\\ \left(\frac{\partial}{\partial t}+U_j\frac{\partial}{\partial x}\right)\zeta&=\frac{\partial\varphi_j}{\partial z}\end{aligned} \tag{9.10}$$

Boundary conditions can be implemented into the intact free surface location at $z=0$. For simple explanation, let us assume each flow layer to be semiinfinite in the z direction, in which there are cross boundaries that are considerably spread away from the interface than a wavelength of the disruption. In this circumstance, a wave-like solution of Laplace's equation in every layer, which fulfills the circumstance that the disturbance be finite as $|z|$ grows precisely into a great depth and can be given by [7, 8]:

$$\varphi_1=\mathrm{Re}A_1e^{i(kx-\omega t)}e^{-kz}, \tag{9.11}$$

$$\varphi_2 = \mathrm{Re} A_2 e^{i(kx-\omega t)} e^{kz}, \tag{9.12}$$

$$\eta = \mathrm{Re} N_o e^{i(kx-\omega t)} \tag{9.13}$$

Implementing Eqs. (9.11)–(9.13) into the kinematic boundary conditions yields:

$$i(Uk-\omega)\left[\frac{ik}{U_1k-\omega}A_1\right] = -kA_1, \tag{9.14}$$

$$i(Uk-\omega)\left[\frac{ik}{U_1k-\omega}A_1\right] = kA_2 \tag{9.15}$$

The pressure variation through the water layers is required to determine the dynamic boundary condition, which is applied to the interface. In this regard, the Bernoulli equation is modified as a function of the pressure variation:

$$-p_1 = \rho_1 g\eta + \rho_1\left(\frac{\partial}{\partial t} + U_1\frac{\partial}{\partial x}\right)\varphi_1 = \rho_2 g\zeta + \rho_2\left(\frac{\partial}{\partial t} + U_2\frac{\partial}{\partial x}\right)\varphi_2 + p_2 = 0 \tag{9.16}$$

If $z=0$, then Eq. (9.16) becomes:

$$(\rho_2-\rho_1)g\zeta = \rho_1\left(\frac{\partial}{\partial t} + U_1\frac{\partial}{\partial x}\right)\varphi_1 - \rho_2\left(\frac{\partial}{\partial t} + U_2\frac{\partial}{\partial x}\right)\varphi_2 \tag{9.17}$$

Based on solutions for the potential velocities and the interface, a second algebraic equation for A_1 and A_2 is formulated as:

$$-\frac{(\rho_2-\rho_1)gk}{i(U_1k-\omega)}A_1 = i\rho_1(U_1k-\omega)A_1 - i\rho_2(U_2k-\omega)A_2 \tag{9.18}$$

The kinematic condition can be used to eradicate A_2 from Eq. (9.18), which develops into:

$$A_1\left[(\rho_2-\rho_1)gk - \rho_1(U_1k-\omega)^2 - \rho_2(U_2k-\omega)^2\right] = 0 \tag{9.19}$$

The mathematical expression of the internal wave packet speed can be written as:

$$c \equiv \frac{\omega}{k} = \frac{\rho_1U_1+\rho_2U_2}{\rho_1+\rho_2} \pm \sqrt{\left[\frac{(\rho_2-\rho_1)g}{(\rho_2+\rho_1)k} - \frac{(\rho_1\rho_2)}{(\rho_2+\rho_1)^2}\{U_1-U_2\}^2\right]} \tag{9.20}$$

Eq. (9.20) reveals that the wavenumber k is the keystone to determine the internal wave structure in space and time. In this sense, some special phases can deliver:

$$\text{(i)}\ \ U_1 = U_2 = 0,\ \frac{\rho_1}{\rho_2} \to 0, \tag{9.21}$$

In this regard, the shear is fixed to zero and the upper layer has a negligible density concerning the lower layer, so the water body should mimic the simple gravity wave, which diminishes to:

$$\omega = \pm\sqrt{(gk)} \tag{9.22}$$

Eq. (9.22) presents precisely the dispersion relationship for an individual layer in the short wave limit concerning the depth:

$$\textbf{(ii)}\ U_1 \neq 0,\ U_2 \neq 0,\ \frac{\rho_1}{\rho_2} \to 0, \tag{9.23}$$

Consequently, the upper layer's density is trivial concerning the lower layer's density. In this context, even though there is a shear flow diagonally across the interface there is no dynamical interaction of the two layers and the frequency is given by [8]:

$$\omega = U_2 k \pm \sqrt{(gk)} \tag{9.24}$$

Eq. (9.24) shows a short wave in a large depth. In this circumstance, the gravity wave Doppler shifted by the mean current is formulated by:

$$\textbf{(iii)}\ U_1 = 0,\ U_2 = 0,\ 0 < \rho_1 < \rho_2 \tag{9.25}$$

In the circumstance of the absence of shear across the interface, there is rapid fluctuation in density, less than ρ_2, and the frequency is:

$$\omega = \pm\sqrt{\left[\left(\frac{\rho_2 - \rho_1}{\rho_2 + \rho_1}\right) gk\right]} \tag{9.26}$$

The configuration of the dispersion relationship is similar to that for the individual fluid under the circumstances, if the two density layers are approximately similar. In this regard, the frequency of the internal wave is considerably less than the solitary layer model, since in that circumstance [6–8]:

$$\left(\frac{\rho_2 - \rho_1}{\rho_2 + \rho_1}\right) g \equiv g' \ll g \tag{9.27}$$

Formula (9.27) introduces what is known as the *reduced gravity*, g', to replace the gravity, g, which leads to frequency reduction. In this regard, the free waves have comparatively low frequencies concerning the solitary layer model and the resulting wave seems to the observer stunningly twisting. The velocity potentials, therefore, in the two layers have opposite signs, and thus the x-velocity in the two layers is decaying away from the interface. Through a wavelength, the motion exponentially a minor growth. In the water body with a free upper surface along with this interface, the movement is formed owing to these wave occurrences on the interface, which will tend to be imperfect to layers close to the interface layer and be approximately invisible at the surface. For that reason, these waves are known as internal waves [6].

9.5 Kelvin-Helmholtz instability

Let us assume that the velocities of the mean flow are different in each layer, i.e., $U_1 \neq U_2$ with the similar densities of $\rho_1 = \rho_2$. The mathematical description of this water flow is given by:

$$\sigma = \frac{U_1 + U_2}{2} k \pm i \frac{|U_1 - U_2|}{2} k \tag{9.28}$$

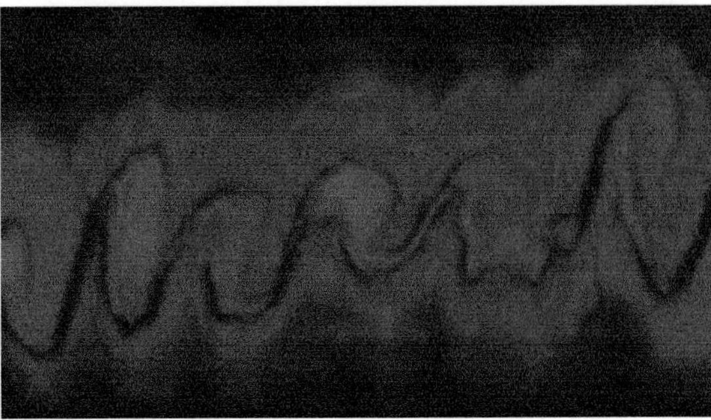

FIG. 9.5 Example of the shape of Kelvin-Helmholtz instability.

The first term on the right-hand side of Eq. (9.28) is the Doppler shift of the frequency by an amount that is contingent on the mean flows of the two streams. The second term yields an imaginary influence to the frequency and the corresponding growth rate is larger the smaller the wavelength of the disturbance. This *shear instability is* generally known as a *Helmholtz instability*. Strong shears give rise to these small-scale instabilities and they are an essential mechanism for minor-scale turbulence in the atmosphere and the oceans. Fig. 9.5 demonstrates the *Kelvin-Helmholtz instability across the Atlantic Ocean*, which occurs 500 m below the sea surface.

In the circumstance of $U_1 \neq U_2$ and $\rho_2 > \rho_1$, the density structure is stable and by itself can sustain stable internal waves owing to the gravitational restoring force that arises when heavy fluid is lifted into lighter fluid. On the other hand, the shear can influence to destabilize the flow. Along with Eq. (9.20), this instability is known as *Kelvin-Helmholtz instability*, which can occur whenever the radicand in that equation turns out to be negative or whenever the shear is sufficiently strong, and is given by:

$$\frac{(\rho_2 - \rho_1)g}{(\rho_2 + \rho_1)k} < \frac{(\rho_1 \rho_2)}{(\rho_2 + \rho_1)^2}\{U_1 - U_2\}^2 \tag{9.29}$$

Consistent with Eq. (9.29), short unsteady waves can always be occurred. For extremely small wavelengths, the full structure of the shear zone cannot be disregarded. On the contrary, at the same order wavelength as the width of the shear zone, the buoyancy forces can stabilize the shear layer [6, 8].

9.6 Internal wave imaging in SAR

Internal waves can be monitored from space. Synthetic aperture radar (SAR) can detect the radar backscatter signals from the ocean surface under all weather and night conditions, with a spatial resolution as fine as a few meters. Currents can be generated by an internal wave, which forms a convergence zone on one side of the internal wave and divergence on the other side. Consequently, the convergence roughs the sea surface by escalating surface waves, although the divergence can flatten the sea surface by destroying the surface waves. Previous chapters have revealed that a rough surface has a greater radar backscattering section, while a smoother one yields less backscatter. Consequently, in the SAR image, one can see bright and dark textures allied with internal waves below the ocean surface (Fig. 9.6).

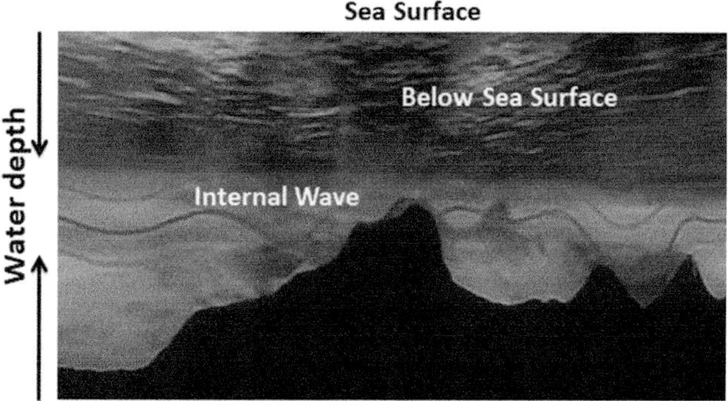

FIG. 9.6 Internal wave generation below the sea surface.

The internal wavelength persistently diminishes from the front to the rear of the wave, highlighting the nonlinearity of these internal waves. These internal waves are traveling off shoreward from the plume front based on their spatial shapes as discussed in Section 9.3 [9].

Consistent with Da Silva et al. [9], SAR data provides excellent confirmation for the existence of shorter period internal solitary waves (ISWs) in the ocean. This is due to a mechanism whereby horizontally-spreading internal waves are highlighted on the thermocline, usually some tens of meters below the surface, which can generate a signature in the surface roughness field because of the modulating consequence of convergence and divergence in the near-surface currents allied with the internal waves. This modulation is extremely operative for a short period (30 min or shorter) of ISWs since the strain of short (Bragg) surface waves (or ripples) is robust in these periods. The wind, tidal flow and internal wave interactions during a large tidal period of 12.4 h cause extremely strongest radar backscatter than wind-tide interactions. Moreover, the occurrence of internal waves in the SAR image is exhibiting by the existing of a surface film [10].

Da Silva et al. [11] are pioneers who provided the scientific explanation of SAR imaging of internal waves. They revealed that sea surface signatures of mode 1 and mode 2 ISWs can be explicitly recognized in SAR images, providing the waves are traveling in relatively deep water when the lower layer is significantly deeper than the mixed layer depth. If the mixed group of waves in two layers traveling along the pycnocline are waves of depression (which is often the case in deep waters), then mode 1 ISWs would be revealed in SAR images by bright bands leading to darker ones in their direction of travel. In this view, depression soliton waves mean that solitary waves maintain a constant shape as they travel with little dissipation, but a solitary wave—or soliton—travels through a water body as only an "up" or a "down" wave. Nonetheless, mode 2 ISWs would have precisely the opposite contrast in the SAR, since their radar signature consists of dark bands preceding bright bands in their direction of propagation (Fig. 9.7).

Fig. 9.7 reveals shows ISWs traveling from left to right. From top to bottom, the horizontal profiles signify the succeeding structures: SAR intensity image forms the tilted shape of the ISWs, with bright boosted backscatter, which is forgoing dark abridged backscatter in the bearing of the internal wave traveling. Surface roughness, consequently, demonstrates how rough or smooth the surface is along the ISW wave packet. Likewise, isopycnical

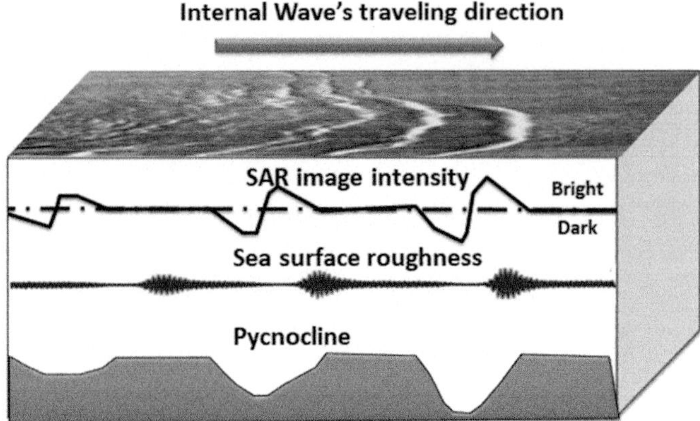

FIG. 9.7 Internal wave imagined in SAR data.

dislocations are twisted by ISW transmission. This is because surface velocity fields formed by traveling ISWs of dissimilar modes can generate diverse convergence and divergence patterns, which then modify the surface roughness and thus the power of the radar backscatter signal. Particularly, ISWs move along the thermocline and create a divergence pattern on the surface, tailed by a convergence pattern in the transmission path. This dynamical variant generates the characteristic of the black bands preceding the bright bands in their direction of travel (Fig. 9.7).

9.7 Internal wave radar backscatter cross-section

Internal wave revealing in SAR images is grounded in the integration of hydrodynamic interaction theory and Bragg scattering theory. The hydrodynamic modulation transfer function (MTF) delineates the variations in the backscattering cross-section since the hydrodynamic modulation of the ripple energy spectral density at the mean (nonmodulated) Bragg wave number by the long gravity waves. Hydrodynamic interaction leads to a nonuniform distribution of the Bragg waves. The Bragg waves, consequently, ride the longest wave and are subject to the current stresses of the long wave orbital motion.

According to Plant [12], the radar backscatter cross-section per unit area, σ_0 for the ocean surface is formulated as

$$\sigma_0(\theta)_{ij} = 16\pi k_0^2 |p_{ij}(\theta)|^2 \Psi(0, 2k_0 \sin\theta). \tag{9.30}$$

where θ is the incident angle, k is wave number, and the high-frequency capillary-gravity wave spectrum is Ψ.

Accordingly, the horizontal ($p_{HH}(\theta)$) and vertical polarization ($p_{VV}(\theta)$) states, respectively, are given by:

$$p_{HH}(\theta) = \frac{(\varepsilon_r - 1)\cos^2\theta}{\left[\cos\theta + (\varepsilon_r - \sin^2\theta)^{1/2}\right]^2}, \tag{9.31}$$

and:

$$p_{VV}(\theta) = \frac{(\varepsilon_r - 1)\left[\varepsilon_r(1-\sin^2\theta) - \sin^2\theta\right]\cos^2\theta}{\left[\varepsilon_r\cos\theta + (\varepsilon_r - \sin^2\theta)^{1/2}\right]^2}, \tag{9.32}$$

where ε_r is the relative dielectric constant of seawater.

In agreement with Zheng et al. [13], the high-frequency capillary-gravity wave spectrum Ψ is estimated by:

$$\Psi = m_3{}^{-1}\left[m\left(\frac{W_*}{c}\right)^2 - 4\gamma^2\varpi^{-1} - S_{\alpha\beta}\frac{\partial U_\beta}{\partial x_\alpha}\omega^{-1}\right]k^{-4}, \tag{9.33}$$

where m is a dimensionless constant (=0.04), W_* is the wind speed friction, ϖ is the angular frequency, c is the phase speed, γ represents the viscosity, U_β represents the velocity components of the large-scale current field, $S_{\alpha\beta}$ represents the excess moments flux tensor, and subscripts α and β represent different horizontal coordinates (x or y).

$$S_{\alpha\beta}\frac{\partial U_\alpha}{\partial x_\beta} = \frac{1}{2}\left[\frac{\partial u}{\partial x}\cos^2\varphi + \left(\frac{\partial u}{\partial y} + \frac{\partial v}{\partial x}\right)\cos\varphi\sin\varphi + \frac{\partial v}{\partial y}\sin^2\varphi\right]. \tag{9.34}$$

Eq. (9.34) establishes that the gradient current of $\frac{\partial u}{\partial y}$ and $\frac{\partial v}{\partial x}$ is interrelated to internal wave and upper ocean constraints, for instance, amplitude and wavelength of the internal wave, along with water column stratification. On this understanding, the greater the amplitude of the internal wave, the more the gradient current escalates. Eq. (9.34) shows that the normalized radar cross-section grows into greater convergent flow layers when $\frac{\partial U_\alpha}{\partial x_\beta} < 0$. On the contrary, if $\frac{\partial U_\alpha}{\partial x_\beta} > 0$, the divergent flow is generated and then the normalized radar cross-section becomes less.

Thus, the radar backscatter cross-section per unit area has a form of:

$$\sigma_0(\theta)_{ij} = 16\pi k_0{}^2 m_3{}^{-1}|g_{ij}(\theta)|^2\left[m\left(\frac{W_*}{c}\right)^2 - 4\gamma^2\varpi^{-1} + \frac{1}{2}\frac{\eta_0 c_0}{h}a\left(1+\frac{1}{a^2b^2}\right)\cos^2\phi\frac{\sinh 2a\xi}{\left(\cosh^2 a\xi + \frac{1}{a^2b^2}\sinh^2 a\xi\right)^2}\right]k^{-4}. \tag{9.35}$$

The soliton-caused radar backscatter cross-section $\sigma_{0IS}(\theta)_{ij}$ is the last term in the right-hand side of Eq. (9.35), which is expressed as:

$$\sigma_{0IS}(\theta)_{ij} = 8\pi k_0{}^2 m_3{}^{-1}|p(\theta)|^2 k^{-4}\frac{\eta_0 c_0}{d}a\left(1+\frac{1}{a^2b^2}\right)\cos^2\phi\left[\frac{\sinh 2a\xi}{\left(\cosh^2 a\xi + \frac{1}{a^2b^2}\sinh^2 a\xi\right)^2}\right] \tag{9.36}$$

Eq. (9.36) reveals the vertical displacement η_0 of the interface between two layers of water, where c_0 is soliton phase speed, d is water depth, and ϕ is propagation direction. The nonlinear least square algorithm must be applied to obtain the parameters a, b, and sea surface elevationξ. Consequently, parameter b is computed based on the water depth layer d and vertical displacement η_0 as:

$$b = \frac{4d_1^{\ 2}}{3\eta_0} \tag{9.36.1}$$

whereas a is at wavenumber like parameter sustaining the relationship between water depth; vertical displacement and surface height (H):

$$ab\tan(aH) = 1, \tag{9.36.2}$$

As stated by Pan and Jay [14], the normalized formula of the backscatter cross-section instigated by an internal wave is calculated as:

$$\hat{\sigma}_{0IS}(\theta)_{ij} = \frac{1}{M}\frac{\sinh 2a\xi}{\left(\cosh^2 a\xi + \frac{1}{a^2b^2}\sinh^2 a\xi\right)^2}, \tag{9.37}$$

where M is the maximum rate of the function $\frac{\sinh 2a\xi}{\left(\cosh^2 a\xi + \frac{1}{a^2b^2}\sinh^2 a\xi\right)^2}$.

Eq. (9.37) shows that $\xi = x - ct$, where x is the horizontal space of the internal wave transmission, c is the phase velocity, and t is time. ξ shapes the radar backscatter, which relies on radar wavelength, incidence angle, and relaxation rate. The relaxation rate, therefore, is reasonably flexible, depending on wind speed and direction. In addition, the presumptions necessitate the wind momentums to exceed the threshold for Bragg wave creation, which is approximately 2–3 $\mathrm{m\,s^{-1}}$, and be below approximately of 10 $\mathrm{m\,s^{-1}}$, at which point the wind-created roughness and the internal wave-created roughness patterns can no longer be discriminated from each other.

9.8 Internal wave detection using two-dimensional wavelet transform

The wavelet transform tool is predominantly exploited for investigating time-varying signals. This practice engenders spectral decomposition throughout the scale mathematical model. In SAR image analysis, the two-dimensional wavelet transform (2-DWT) works up as an highly effective band-pass filter. In this respect, 2-DWT are ruled to discrete frequency channels with transformed scales. Further than, the strong ability of wavelet transforms in achieving the time localization for a one-dimensional performance, it can provide a precise indication of the feature description in SAR data. These precise functions of 2-DWT suggest that the wavelet transform is a robust algorithm for extracting features "physical properties" precisely in SAR data. "physical properties precisely in SAR data. The continuous wavelet transform of a signal $f(t)$ is then identified as:

$$CWT_f^{\psi}(s,\tau) = \int_{-\infty}^{\infty} f(t)\psi_{s,\tau}(t)dt \tag{9.38}$$

The 2-DWT function, having an oscillation in x-direction, is expressed by:

$$\Psi_{s,\tau}(t) = \frac{1}{\sqrt{s}}\Psi\left(\frac{t-\tau}{s}\right) \tag{9.39}$$

$\Psi(t)$ is the transforming function, and is known as the mother wavelet. The dual novel variables s and τ are the scale and translation of the daughter wavelet, respectively. The expression $\sqrt{s}$ normalizes the energy for dissimilar scales, although the other terms describe the width and the translation of the wavelet. The continuous wavelet transform (CWT) is demarcated as:

$$\gamma(s,\tau) = \int f(t)\Psi^*_{s,\tau}(t)\mathrm{d}t \tag{9.40}$$

The asterisk signifies a complex conjugate function. There are, nevertheless, two conditions that the wavelet has to achieve: the admissibility condition and the regularity condition. The admissibility condition is achieved by:

$$\int_{-\infty}^{\infty} \frac{\left|\hat{\Psi}(\omega)\right|^2}{\omega} d\omega < \infty \tag{9.41}$$

where ω is the angular frequency.

Eq. (9.41) must achieve the following rules: (i) the transformation is invertible—all square-integrable functions that fulfill the admissibility condition can be used to examine and recreate the signal somewhat; consequently, (ii) the function must have a value of zero at zero requency. Thus, the wavelet itself has an average value of zero. This implies that the wavelet produces an oscillatory signal, where the positive and negative values terminate each other. This offers the 'wave' form in wavelet.

The symmetry characteristics necessitate that the 2-DWT have to be smooth in the vicinity and robust in both the time and frequency domains.In this circumstance, vanishing moments can be acquainted with a new approach in dealing with symmetry occurrences in 2-DWT computing algorithm. If the Fourier transform of the wavelet is M times differentiable, the wavelet has M vanishing moments.

$$\int x^n \Psi(x) dx = 0 \quad n \in [0, M] \tag{9.42}$$

where n is the role of the shift parameter.

For SAR data analysis, the elementary wavelets commonly castoff are the Gaussian modulated sine and cosine wave packet (the Morlet wavelet) and the second derivative of a Gaussian (the Mexican-hat wavelet). In this chapter, the examining of wavelet is demarcated as the Laplacian of a Gaussian algorithm as:

$$\omega(t) = \frac{1}{\sqrt{2\pi\sigma}} e^{\frac{-t^2}{2\sigma^2}} \tag{9.43}$$

The mother wavelet can be specified by:

$$\psi(t) = \frac{1}{\sqrt{2\pi\sigma^3}} \left(e^{\frac{-t^2}{2\sigma^2}} \cdot \left(\frac{t^2}{\sigma^2} - 1 \right) \right) \tag{9.44}$$

The Morlet wavelet is expressed as:

$$\omega(t) = e^{iat} \cdot e^{-\frac{t^2}{2\sigma}} \tag{9.45}$$

where a is the modulation parameter and σ is the scaling parameter that determines the width of the kernel window.

Each high-pass filter generates a comprehensive version of the imaginative signal, and the low-pass a smoothed form. Three bands are tangled in Morlet wavelets: (i) low-pass band S_jf; (ii) horizontal high-pass band S_j^Hf; and (iii) vertical high-pass band S_j^Vf. In this framework, the edge detection of internal wave based on both S_j^Hf and vertical high-pass band S_j^Vf is formulated as:

$$S_jf = \sqrt{\left|S_j^Hf\right|^2 + \left|S_j^Vf\right|^2} \tag{9.46}$$

Then the gradient angle ϑ_j of the internal wave edge domain is shaped from:

$$\vartheta_jf = \tan^{-1}\left(\frac{S_j^H}{S_j^V}\right) \tag{9.47}$$

Eq. (9.47) also defines 1-D local maxima in the direction of the gradient, which is counted as edge pixels of internal wave in SAR data. In this framework, the local maxima is a function of internal wave boundary morphology in the SAR image.

The algorithm is considered as a conventional edge detector where edges are operated on a discrete pixel level, which is frequently disagreeable when treating images contaminated comprehensively with speckles, for example, SAR data. Fluctuating scaling factor σ would emphasize structures with changed spatial scales [15]. The constraint scale in the wavelet analysis is comparable to the scale used in maps. As in the example of maps, large scales give a nondetailed global view (of the signal), and small scales give a detailed view. Correspondingly, in terms of frequency, low frequencies (high scales) give global information of a signal (that usually spans the entire signal), whereas high frequencies (low scales) provide comprehensive information of a hidden pattern in the signal (that usually lasts a relatively short time). In practical applications, low scales (high frequencies) do not last for the entire duration of the signal, but they regularly perform from time to time as short bursts or spikes. High scales (low frequencies) generally last for the whole period of the SAR signal [15–17].

9.9 Particle swarm optimization (PSO) algorithm

PSO, which is an optimization method based on the social manner of bird clustering, was articulated by Edward and Kennedy in 1995 [18]. This method is comparable to the infinite genetic algorithm (GA) in that it instigates with a casual population matrix. Unlike the GA, PSO has no evolution operators, for instance, crossover and mutation. The PSO optimum solution is originated by sharing statistic probability estimation in feature detection. This is a population-based search algorithm that is initialized with the population of random solutions, known as particles, and the population is recognized as a swarm [18]. The cornerstone strength of PSO is that it is precise to instigate and debauched convergent of a specific feature searching. PSO has come to be widely applied in continuous and discrete optimization for engineering applications [19].

Let assume N particles, in which each particle exchanges almost its cost function with a velocity $v_i(t)$ as a function of the time t. In this view, the particles renovate their velocities

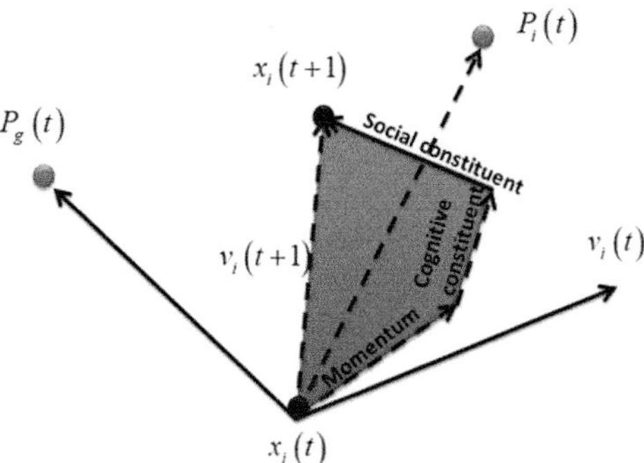

FIG. 9.8 A single particle's movement.

and locations grounded in the local and global finest clarifications. Fig. 9.8 spectacles the interchange of a single particle (i) at the time step t in space search. At time step (t), the location, speed, individual finest and global finest are specified as $x_i(t)$, $v_i(t)$, $p_i(t)$, and $p_g(t)$, respectively. The velocity $v_i(t)$ operates as a memory of the earlier movement direction, can be grasped At time step ($t+1$), the new location $x_i(t+1)$ can be computed founded on three constituents, which are momentum, cognitive and social component.

Two techniques are involved in the application of PSO: initialization and velocity update. The first step comprises the creation of a random population, which is a prerequisite for possible solutions. In this regard, particles are denoted as a random population. Lastly, the stimulated particles flood from side to side of SAR data as a function of apprising their rapidity. In this instance, the particles can comprehend their earlier flood through SAR image and those of their assembly exchange for the requested feature detection [20].

After acquiring the discrete best and global optimum, the particle is then speeded up toward those binary optimum amounts by apprising the particle location and speed for the next iteration exercising the succeeding set of equations which are specified as:

$$p_{m,n,l} = p_{m,n,l} + v_{m,n,l}, \tag{9.48}$$

where $p_{m,\ n,\ l}$ represent a particle and $v_{m,\ n,\ l}$ is the velocity of the 3-D particle in the i, j, and k agents, respectively.

$$v_{m,n,l} = v_{m,n,l} + c_1 r_1 (lbest_{m,n,l} - p_{m,n,l}) + c_2 r_2 \left(gbest_{m,n,l} - p_{m,n,l}\right) \tag{9.49}$$

where $lbest_{m,\ n,\ l}$ is the local best particle, $gbest_{m,\ n,\ l}$ is the global best particle, r_1 and r_2 are random variables, and c_1 and c_2 are the swarm system variables.

After each iteration, the global best g_{best} particle and the agent local best l_{best} particle are evaluated based on the maximum fitness functions of all particles in the solution space. Then Eq. (9.48) can be expressed as follows [21]:

$$v_{m,n,l} = w \cdot v_{m,n,l}(t-1) + c_1 \cdot r_1 (p_{m,n,l}(t-1) - \psi_{m,n,l}(t-1)) + c_2 \cdot r_2 (p_{m,n,l}(t-1) - \psi_{m,n,l}(t-1)) \tag{9.50}$$

$$\psi_{m,n,l} = \psi_{m,n,l}(t-1) + v_{m,n,l}(t) \tag{9.51}$$

In Eq. (9.51), $\psi_{m,\ n,\ l}$ is the location of the particle of wavelet transformation, $v_{m,\ n,\ l}$ is the current velocity of the particles in the SAR image. The velocity is measured by a set of rules that influence the dynamics of the swarm. Furthermore, numerous parameters must be determined, for instance, initial population, representation of position and velocity strategies, fitness function identification, and the limitation [21]. These constraints are for PSO routines. Following Ibrahim et al. [22], the initial swarm particles suggests PSO is modified to comprise 3000 plugs of particles for $\psi_{m,\ n,\ l}$ and velocity $v_{m,\ n,\ l}$. The plugs are arbitrarily designated in the azimuth and range directions of the wavelet transform for SAR data. Conversely, according to Kennedy and Eberhart [18] and Marghany [23], particle swarm optimization (PSO) is a population-based random searching process. It is assumed that there are N "particles," i.e., physical properties of an internal wave and its surrounding environments: bright and dark backscatter variations, linearity, depressions, and direction of propagation in SAR data. These internal wave structures intrusive associates arbitrarily with each other and appear in a "clarification space" [18]. Consequently, the optimization difficulties can be resolved for data assembling; there is always a criterion, for instance, the squared error function, for every single particle at their location in the solution space [20]. The N particles will maintain poignant and computing the criteria in every pixel remains in SAR image, which is named as fitness in PSO that approaching the criteria of the satisfied threshold. Consequently, each internal wave facet (particles) maintains its matches in the clarification pixels of SAR image, which is merged with the best fitness that has extremely achieved by requested the matched physical features, i.e., particles.

The simplification of the above procedures and mathematical equations can be articulated from the point of view of probability. In this sense, computers do not "think" the same way humans do. How do we tell a computer to send an *agent* toward the most likely solution? An understanding of probability can be used as well as randomly generated numbers to accomplish the task. In this context, the probability is the likelihood that something will happen or be true, which is mathematically formulated as:

$$P(E) = \frac{\text{Number of Elements in} \text{``}E\text{''}}{\text{Size of the Sample Space}} \tag{9.52}$$

Eq. (9.52) demonstrates that the sum of all probabilities for all events in the sample space must be equal to 1. On this understanding, PSO involves **a random number in which** a number is drawn from a set of numbers where each number is equally likely to be drawn. Consequently, **random number generators are** a deterministic algorithm that generates a string of random numbers (better-called pseudorandom) (Fig. 9.9).

For instance, a random number between 0 and 1 is drawn to be placed in a bin. However, if the number is between 0 and 0.5, it will be placed in Bin#1. If it is greater than 0.5 and less than 1, it will be placed in Bin#2.

9.10 Tested SAR data

Advanced synthetic aperture radar (ASAR) is a ground-range projected detected image in zero-Doppler SAR coordinates, with a 12.5 m pixel spacing and nominal resolution at a range and an azimuth of 30 m × 30 m. It has four overlapping looks in Doppler covering a total bandwidth of 1000 Hz, with each look covering a 300 Hz bandwidth. Sidelobes reduction is applied

```
Equation (9.49)
v[] = c0 *v[]
   + c1 * rand() * (pbest[] - present[])
   + c2 * rand() * (gbest[] - present[])
(in the original method, c0=1, but many
   researchers now play with this parameter)

Equation (9.51)
present[] = present[] + v[]

For each particle
   Initialize particle
END

Do
   For each particle
      Calculate fitness value
      If the fitness value is better than its peronal best
         set current value as the new pBest
   End

   Choose the particle with the best fitness value of all as gBest
   For each particle
      Calculate particle velocity according equation (a)
      Update particle position according equation (b)
   End
While maximum iterations or minimum error criteria is not attained
```

FIG. 9.9 Pseudo-code of PSO.

to achieve a nominal PSLR of less than −21 dB. ASAR data are radiometrically calibrated and are numerically scaled such that a beta0 value of 0 dB corresponds to a product digital number (DN) value of 682.3. ENVISAT's mission was ended on May 9, 2012, but for this study, one image was selected: ASAR data acquired on December 31, 2004 (Fig. 9.10) with HH polarization (Table 9.1).

9.11 Backscatter distribution along with internal wave in SAR data

It is interesting to observe the occurrence of internal waves after the tsunami hit Asia on 31 December 2004. ENVISAT SAR data were acquired in orbit 148 along with the Indian Andaman Islands and the Ritchie's Archipelago (Fig. 9.11). The water depth around Andaman and Nicobar Islands is 200–3000 m.

The normalized radar cross-section ranges from −24 to −4 dB. The lowest normalized radar cross-section of −28 dB is designated as the low window zone shelter around Andaman and Nicobar Islands (Fig. 9.12). Nonetheless, the highest backscatter of −4 dB defines the existence of a whirlpool in the east of the Andaman Sea. This whirlpool is positioned between the latitudes of 14°N to 15°N and longitudes of 94°E and 96°E. The whirlpool has a radius of 1.9 km and is situated above a water depth gradient of 1000 m. It rotates in an anticlockwise direction (Fig. 9.13).

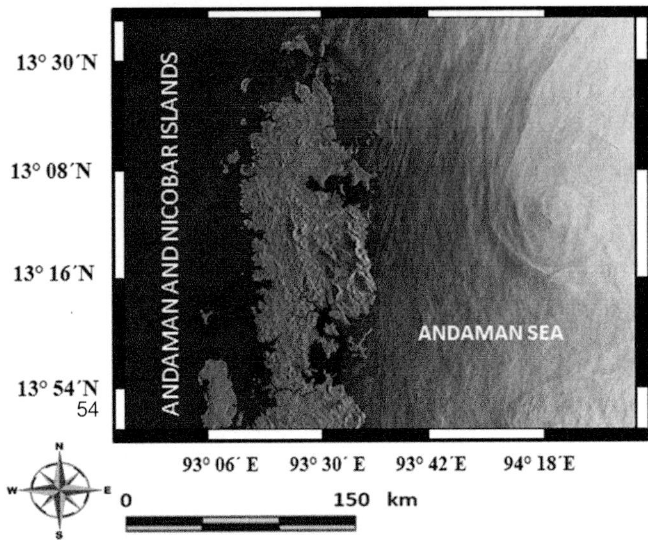

FIG. 9.10 ENVISAT data acquired December 31, 2004.

TABLE 9.1 ENVISAT data characteristics.

ENVISAT	Characteristics
Date	December 30, 2004
Polarization	HH
Product	ASAR-IMP
Incident Angle (degrees)	15–45
Band	C
Swath (km)	58–110
Resolution (m)	30–150

Whirlpools are associated with the occurrence of internal waves (Fig. 9.14); in this case, the internal wave traveling parallel to the coastal waters, for 72 km, of two Indian Andaman Islands and the Ritchie's Archipelago. The internal wave, consequently, has an approximate maximum wavelength of 1000 m and shifts westward to Andaman and Nicobar Islands. It travels in an irregular parallel-packet and phase speed vectors, analogous to propagation inside a waveguide owing to the sharp slope of an irregular bathymetry pattern gradient of about 1000 m. These waves are internal waves, and they run through the lowest layers of ocean water, in no way swelling the surface.

It can be seen that these internal waves have higher backscatter of −4 dB than the surrounding sea environment. The internal wave crests dominate by higher backscatter while the troughs dominate by lower backscatter of −10 dB (Fig. 9.14). Therefore, the internal wave

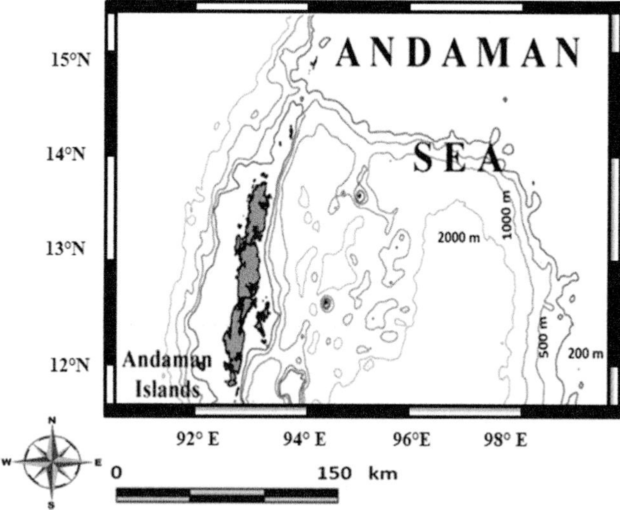

FIG. 9.11 Bathymetry map of the Andaman Sea.

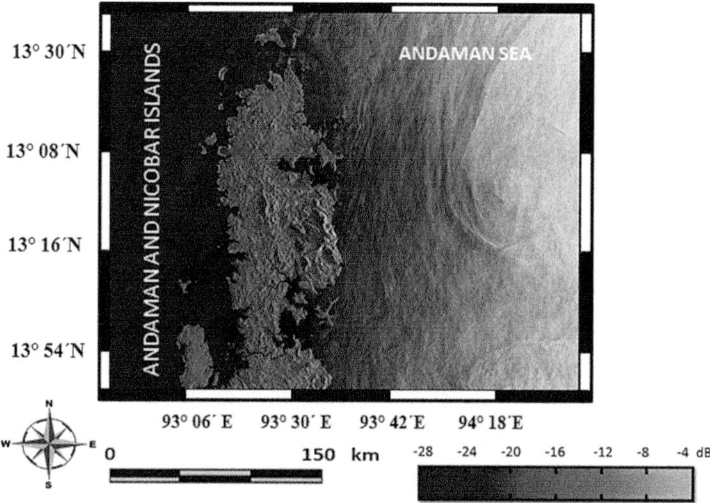

FIG. 9.12 Backscatter variations in ENVISAT of Andaman and Nicobar Islands.

edges appear as discontinuities in the backscatter. In this circumstance, it is barely possible to detect objects that have intensity discontinuities in SAR data because of the multiplicative speckle noise. The signal backscattered from the sea surface might be comparable to one from the land.

The internal waves modulate the overlying surface waves, and the sea surface manifestations of internal waves are readily observed in synthetic aperture radar (SAR) images. They appear as alternating bright and dark bands corresponding to regions of increased surface

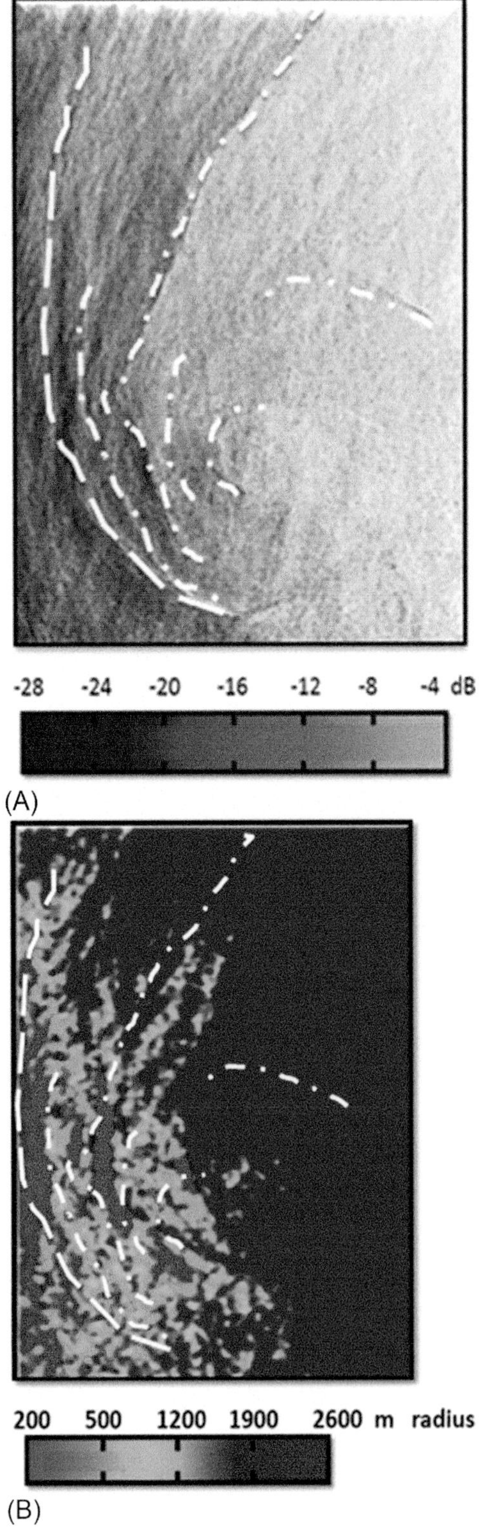

FIG. 9.13 Whirlpool characteristics (A) backscatter and (B) radius.

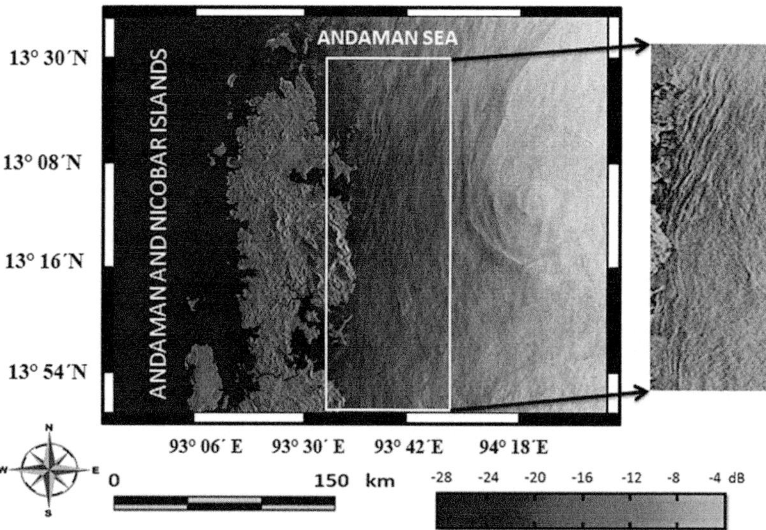

FIG. 9.14 An internal wave traveling along the Andaman Islands in ASAR data.

roughness with maximum normalized radar cross-section of −4 dB alternating with smooth regions with a minimum of −10 dB (Fig. 9.14).

It is worth mentioning that the Bragg scattering is the main concept of the scattering mechanism for ocean surface. In this view, the backscatter variations are mainly a function of incidence angle and wavelength as well as the sea surface conditions such as wind speed. For ASAR, wavelength is 5.3 cm, which indicates that the Bragg scattering is approximately equivalent to its wavelength. Further, current shear, wind speed, long gravity waves associated with whirlpool, and internal waves created modulation in the short waves. Under these circumstances, the ASAR data show high variability of backscatter as a function of surface roughness. The large incidence angle of ASAR, which ranges from 15° to 45°, allows for extreme discrimination between water/land boundary as a function of a large radar backscatter contrast.

9.12 Automatic detection of internal wave using two-dimensional wavelet transform

The automatic detection of the internal wave can be performed using a 2-DWT algorithm. However, the 2-DWT is not able to distinguish between internal wave and whirlpool features. It may be that both features have a similar shape, especially along the edges. In addition, 2-DWT also detected the edges of look-alikes and current shear boundaries. Look-alikes and current shear occasionally have quasilinear periodic lines, which are similar to the signature of internal waves in SAR data (Fig. 9.15). However, 2-DWT produces a false alarm by delineating the pattern of the whirlpool pattern. In this understanding, the 2-DWT is not able to identify automatically the lineament pattern of internal waves from the whirlpool pattern.

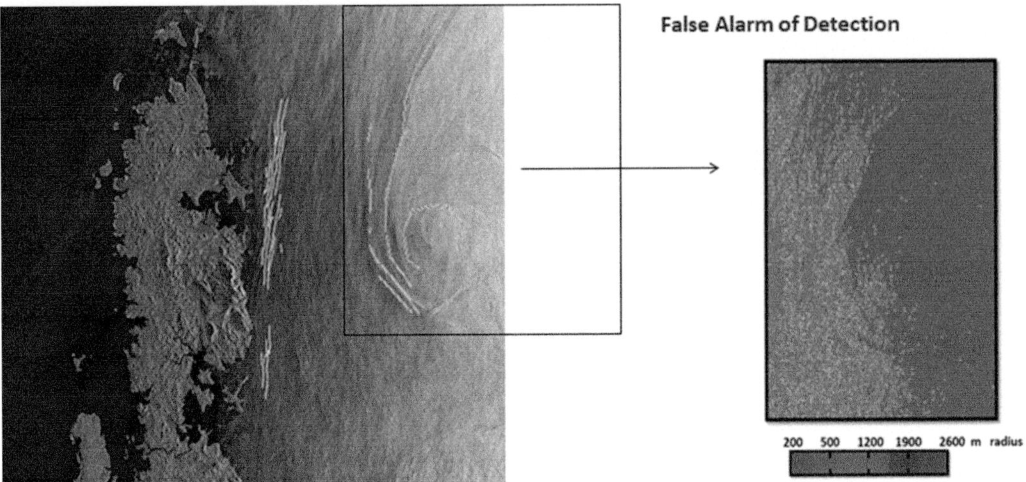

FIG. 9.15 Internal wave detection using 2-DWT.

This may explain the difficulties that arise for automatic detection of internal waves in SAR data. There are numerous features allied with internal wave look-alikes: upwelling, eddies, oil slicks, current shear zones, wind sheltering by land, natural film, and threshold wind speed zones. These are well-thought-out as secondary information in SAR data as well as the impact on the sea surface roughness, which do not allow 2-DWT for accurate detection of internal waves in ASAR data (Fig. 9.16).

Ultimately, Fig. 9.16 suggests that the intense packets of internal waves are related to natural slicks. While internal waves of higher magnitudes will frequently smash after crossing over the shelf break, smaller internal wave trains will continue steadily throughout the

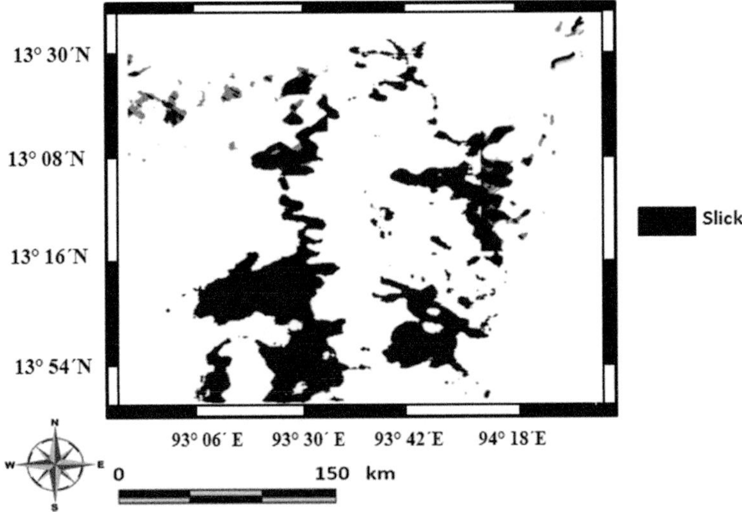

FIG. 9.16 Slicks associate with internal waves.

shelf. At low wind speeds those internal waves are evidenced utilizing the formation of extensive surface slicks, orientated parallel to the bottom topography, which develop shoreward with the internal waves. Waters above an internal wave converge and sink in its trough and upwelling, and diverge over its crest. The convergence zones related to internal wave troughs frequently acquire oils and flotsam that occasionally develop shoreward with the slicks (Fig. 9.16) [20,24].

9.13 Internal wave packet detection by PSO

The performance of PSO algorithm is achieved through different numbers of particles, fitness, and clamping velocity. In this investigation, the *gbest*-to *lbest* PSO is implemented for ASAR data. In this algorithm, *lbest* begins with a zero-radius neighborhood, then *gbest* is achieved as the *a* neighborhood radius is linearly increased to avoid being trapped in local optima. The algorithm converges into the best solution by using *gbest* approach. Fig. 9.17 reveals the initial stage of PSO operation in the ASAR image within the maximum iteration of 100 and root mean square error (RMSE) of ±70. The initial PSO result reveals some information about pixel brightness and dark variation patterns.

The increment of particle sizes of 400 can reconstruct coastline information about the Andaman Islands with a reduction of RMSE to ±66 (Fig. 9.18). It is worth noting that the whirlpool pattern morphology structures are well organized in 600 particle sizes with RMSE ±54 (Fig. 9.19). As long as the number of particle swarms increases to 1000 particle sizes, the worthy organization of internal wave morphology structures is precisely achieved (Fig. 9.20). This is proved by best fitness function of 95 and RMSE of ±0.43.

Fig. 9.21 shows the best performance for automatic detection of the internal wave with RMSE of ±0.39. This occurred within clamping of velocity updates and swarm particle sizes

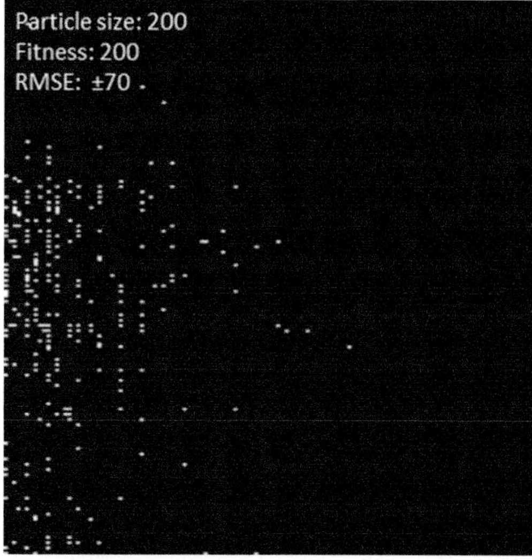

FIG. 9.17 PSO initial image processing within 200 particle sizes.

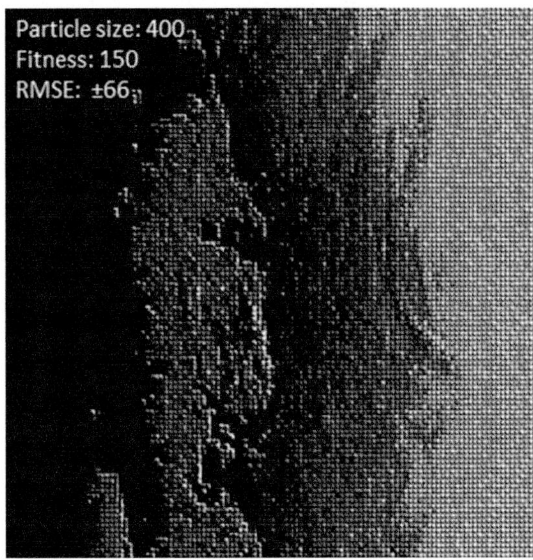

FIG. 9.18 Coastline detection using PSO 400 particle sizes.

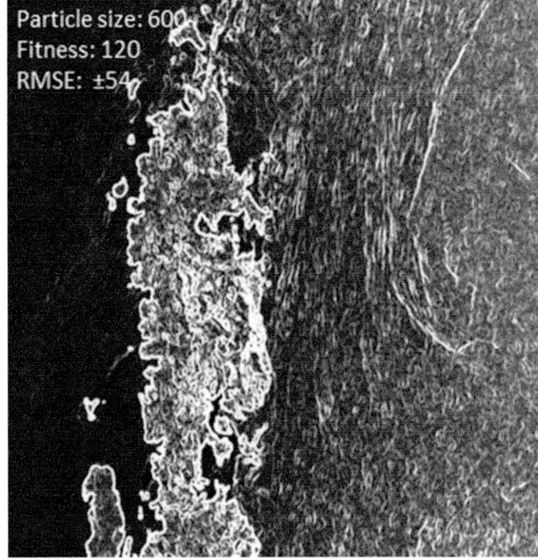

FIG. 9.19 Whirlpool detection within PSO 600 particle sizes.

of 1050 with a fitness function of 73, the internal waves are distinguished precisely from the surrounding sea features and land too.

Generally, growth in the number of swarm particles enlarges diversity, in this manner limiting the weights of initial conditions and decreasing the probability of being blocked in local minima. Furthermore, less reducing the velocity of swarm particles appraises the precise

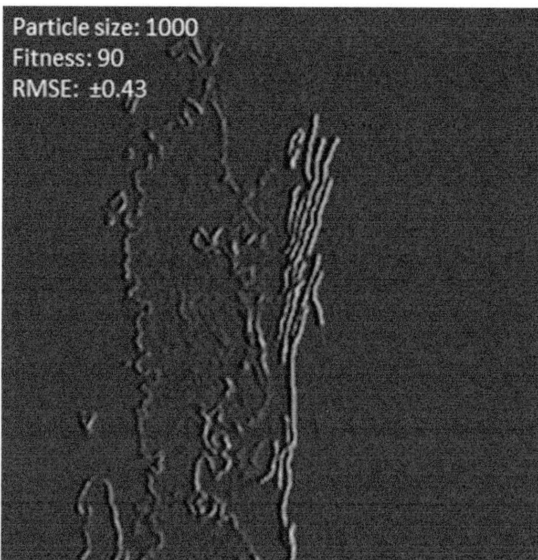

FIG. 9.20 Internal wave well detection using 1000 particle sizes.

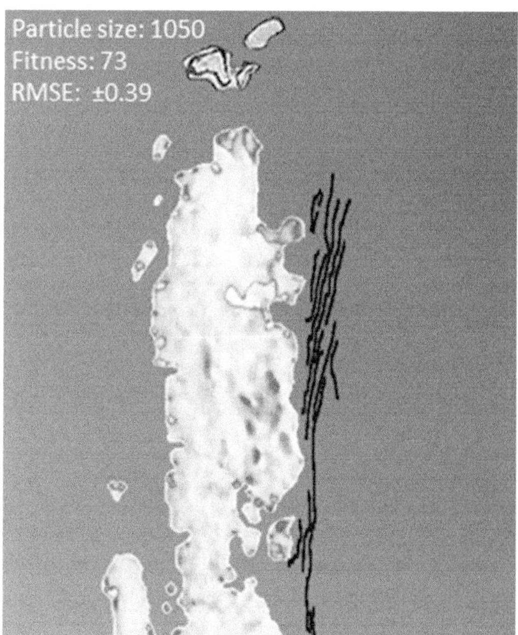

FIG. 9.21 Internal wave fine structure using 1050 particle sizes.

structure of the internal wave in SAR data. Indeed, this tolerates the swarm to be further accelerated in the analytical domain in a SAR image for automatic detection of an internal wave.

In fact, the PSO circumvents a decreasing resolution by making a weighted combination of running average with the neighboring surrounding pixels. This reduces the noise in the features' edge areas without losing edge sharpness. The implementation of PSO helps to determine optimal growth regions across the continuing and discontinuing internal wave edges. In fact, in the PSO system, particles are impartial of every other and their movements are governed by a set of rules. With this in mind, PSO synchronizes sequence sides of the internal wave edge. The PSO algorithm commences by creating random locations for the particles, within an initialization pixel of the internal waves in the ASAR image. Particles in the PSO algorithm can also be modified to zero or two minor random values to avoid particles' withdrawal from the search space of internal wave pixels during the first iterations. Throughout the core loop of the algorithm, the velocities and the locations of the particles are iteratively rationalized until an end condition is encountered. Besides, PSO algorithm constructed the discontinuity of the internal wave pixels in quality order. This is appropriate where a high-intensity line or curve of fixed length and locally low curvature boundary is known to exist between edge elements and high noise levels in ASAR data.

9.14 Why do internal waves occur in the Andaman Sea?

The automatic detection of 10 packets of the internal wave was performed using a PSO algorithm. Interestingly, the PSO algorithm delivers automatic detection of the full length of internal wave packets (Fig. 9.22). Moreover, these internal waves have approximately 10 packets. These internal waves occurred with an ocean bathymetry gradient of 1000 m. The PSO analysis picks out several curves that are the most prominent boundaries in the image of this scale [20].

Fig. 9.22 shows the bathymetry variations of the Andaman and Nicobar Islands in which the water depth along Andaman and Nicobar Islands is between 200 and 3000 m. The Andaman and Nicobar Islands on the western side of the Andaman Sea are volcanic in origin.

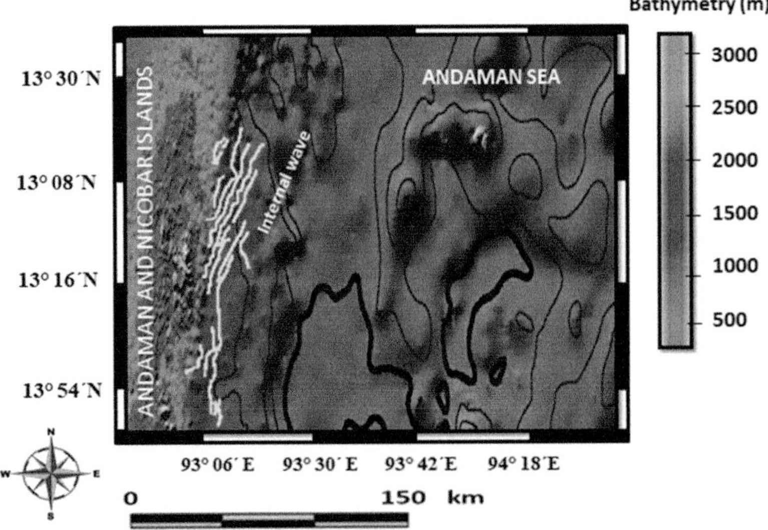

FIG. 9.22 Internal wave automatic detection with bathymetry.

The sills between the islands, as well as several underwater volcanic seamounts, are all potential sources of internal waves. The result is an area rich in internal wave excitations and complex soliton-soliton interaction. It is well known that an exceptionally large range of internal waves occur within the Andaman Sea of the Indian Ocean. The internal waves propagate beneath the Andaman Sea surface in the water layers whose density fluctuates as a function of the water depth. In addition, a large packet of internal wave groups have been found in the western verge upon the Strait of Malacca between Phuket, Thailand, and the northern coast of Sumatra.

Marghany [20] stated that internal waves happen because the ocean is layered. Deep water is cold, dense, and salty, while shallower water is warmer, lighter, and fresher. The differences in density and salinity cause the various layers of the ocean to behave like different fluids. When tides drag the ocean over a shallow barrier such as a ridge on the ocean floor, it creates waves in the lower, denser layer of water. These waves—internal waves—can be tens of kilometers long and can last several hours.

References

[1] Kelley D. Waves beneath the sea. Assoc Sci Teach J 2002;3(1):1–8.

[2] Nash JD, Moum JN. River plumes as a source of large-amplitude internal waves in the coastal ocean. Nature 2005;437(7057):400–3.

[3] Staquet C, Sommeria J. Internal gravity waves: from instabilities to turbulence. Annu Rev Fluid Mech 2002;34 (1):559–93.

[4] Orr MH, Mignerey PC. Nonlinear internal waves in the South China Sea: observation of the conversion of depression internal waves to elevation internal waves. J Geophys Res Oceans 2003;108(C3):1–16.

[5] Helfrich KR, Melville WK. On long nonlinear internal waves over slope-shelf topography. J Fluid Mech 1986;167:285–308.

[6] Pedlosky J. Geophysical fluid dynamics. Springer Science & Business Media; 2013.

[7] Helfrich KR, Melville WK. Long nonlinear internal waves. Annu Rev Fluid Mech 2006;38:395–425.

[8] Pedlosky J. Ocean circulation theory. Springer Science & Business Media; 2013.

[9] Da Silva JC, New AL, Magalhaes JM. On the structure and propagation of internal solitary waves generated at the Mascarene Plateau in the Indian Ocean. Deep-Sea Res I Oceanogr Res Pap 2011;58(3):229–40.

[10] Ermakov SA, da Silva JCB, Robinson IS. Role of surface films in ERS SAR signatures of internal waves on the shelf. 2. Internal tidal waves. J Geophys Res 1998;103:8032–43.

[11] da Silva JCB, Magalhaes JM. Internal solitons in the Andaman Sea: a new look at an old problem. In: SPIE remote sensing. International Society for Optics and Photonics; 2016. p. 9999071–99990713.

[12] Plant WJ. Bragg scattering of electromagnetic waves from the air/sea interface. In: Surface waves and fluxes. Dordrecht: Springer; 1990. p. 41–108.

[13] Zheng Q, Clemente-Colón P, Yan XH, Liu WT, Huang NE. Satellite synthetic aperture radar detection of Delaware Bay plumes: jet-like feature analysis. J Geophys Res Oceans 2004;109(C3):1–11.

[14] Xie J, He Y, Chen Z, Xu J, Cai S. Simulations of internal solitary wave interactions with mesoscale eddies in the northeastern South China Sea. J Phys Oceanogr 2015;45(12):2959–78.

[15] Ghobadi M. Wavelet-based coding and its application in JPEG2000, webhome.cs.uvic.ca/~pan/csc461s06/.../csc561-monia-wavelet-v2.doc; 2017. [Accessed 16 February 2017].

[16] Mallat S. A wavelet tour of signal processing. 2nd ed. Academic Press; 1998.

[17] Robi P. The engineering ultimate guide to wavelet analysis: The wavelet tutorial. Rowan University, College of Engineering; n.d.

[18] Kennedy J, Eberhart R. Particle swarm optimization, In: Proceedings of ICNN'95—International conference on neural networks Nov 27, 1995, vol. 4. IEEE; 1995. p. 1942–8.

[19] Bien D. Dynamic modeling and control of a flexible link manipulators with translational and rotational joints. VNU J Sci Math Phys 2018;34(1):52–66.

[20] Marghany M. Advanced remote sensing technology for tsunami modelling and forecasting. CRC Press; 2018.

[21] Dorigo M, de Oca MA, Engelbrecht A. Particle swarm optimization. Scholarpedia 2008;3(11):1486.

[22] Ibrahim S, Khalid NE, Manaf M. Computer aided system for brain abnormalities segmentation. Malays J Comput 2010;386:22–39.

[23] Marghany M. Copper mine automatic detection from TerraSAR-X using particle swarm optimization, In: CD of 36th Asian conference on remote sensing (ACRS 2015), Manila, Philippines, 24–28 October 2015; 2015a-a-r-s.org/acrs/administrator/components/com.../files/.../TH3-2-1.pdf.

[24] Wu SY, Liu AK. Towards an automated ocean feature detection extraction and classification scheme for SAR imagery. Int J Remote Sens 2003;24(5):935–51.

CHAPTER

10

Modeling wave pattern cycles using advanced interferometry altimeter satellite data

One limitation of SAR is the coverage area, as it cannot cover large areas such as the Indian Ocean, whereas a satellite microwave altimeter can offer a larger coverage scale than SAR satellite data. Furthermore, SAR satellite data cannot determine annual wave pattern cycles. Interferometry is a well-recognized technique that can only operate with SAR data, although there is a possibility of interferometry operation of satellite altimeter data. So how can satellite altimeter involve interferometry techniques?

This chapter covers advanced interferometry technology that can operate in satellite altimeter data. The west coast of Australia is examined as an interesting area due to southern Indian Ocean dynamic fluctuations. The annual significant wave pattern and power are also investigated.

10.1 Microwave altimeter

An altimeter is also referred to as an altitude meter; it is a device to compute an object's height above a stable point. The estimation of altitude is known as altimetry. Bathymetry, which is the computation of depth beneath the sea surface, is associated with altimetry. The height is usually measured between the altimeter platform, i.e., satellite or aircraft, and the Earth's surface (Fig. 10.1). In other words, altimeters are nadir-looking pulse-radars; they transmit short microwave pulses and measure the round trip time delay to targets to determine their distance from the air- or spaceborne sensor.

10.2 Principles of altimeters

Like a synthetic aperture radar (SAR), an altimeter emits a radar signal and then records the backscattered signal from objects. Unlike SAR, the altimeter emits and receives the radar

Nonlinear Ocean Dynamics
https://doi.org/10.1016/B978-0-12-820785-7.00003-4

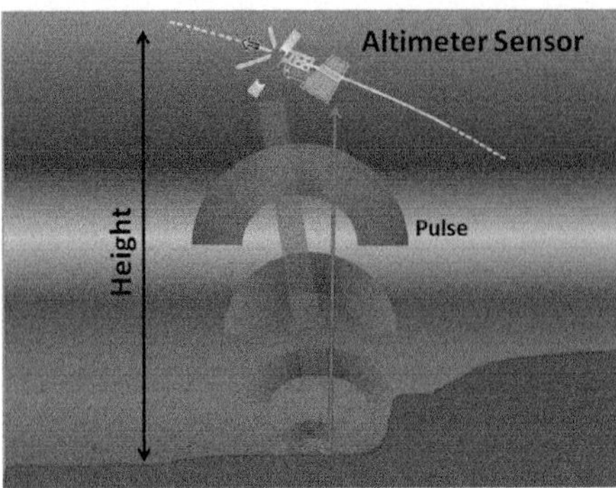

FIG. 10.1 The concept of the altimeter.

waveform perpendicular to the object. This mechanism allows estimation of the object height from the inverted backscatter signal. In this context, ocean wave height can be inverted more easily than with SAR as a function of perpendicular backscatter signal amplitude [1]. The main signal bands used with altimeters are E-band, K_a-band, and S-band. Advanced sea-level retrieving parameters are easily made by S-band. Consequently, more reliable and precise ocean wave height is provided by altimeter than by SAR sensors.

10.3 Types of radar altimeter frequencies

There are two main components of radar altimeters: (i) frequency modulated continuous wave (FMCW) and (ii) pulse altimeters, which are a function of radar signals used. Two sorts of FMCW altimeters are generally used: broad-beamwidth and narrow-beamwidth types. Both FMCW altimeters are a function of antenna beamwidth. In contrast, the pulse altimeters are also known as short-pulse altimeters or pulse-compression altimeters, which are a function of intrapulse modulation. In addition, altimeters can also operate in optical bands, for instance, laser altimeters.

In other words, a simple continuous-wave radar device without frequency modulation has the disadvantage that it cannot determine target range because it lacks the timing mark necessary to allow the system to time accurately the transmit and receive cycle and to convert this into range. Such a time reference for measuring the distance of stationary objects can be generated using frequency modulation of the transmitted signal. In this method, a signal is transmitted, which increases or decreases in frequency periodically. When an echo signal is received, that change of frequency produces a delay Δt (by runtime shift), similar to the pulse radar technique. In pulse radar, however, the runtime must be measured directly. In FMCW radar, the differences in phase or frequency between the actual transmitted and the received signal are measured instead (Fig. 10.2).

For lower height measurements, a frequency modulated continuous wave radar (FMCW) can be used. This technology provides a better resolution but a lower maximum unambiguous height measurement at the comparable wiring effort, material 1 cavity of individual components in these systems. Often both technologies are used simultaneously [1–3].

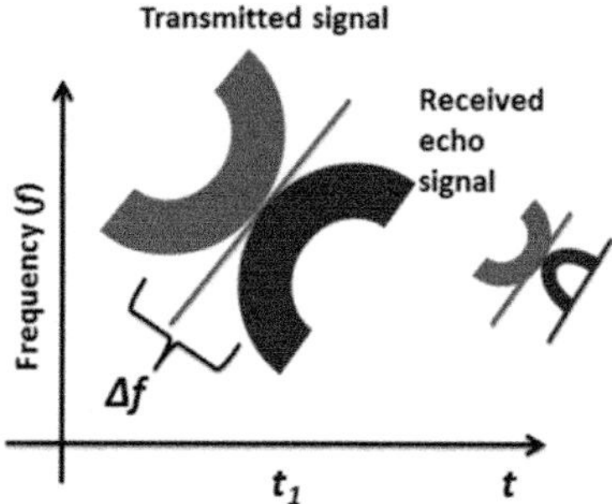

FIG. 10.2 Ranging from an FMCW system.

10.4 How does a radio altimeter work?

Radio altimeters are simpler and work in a similar way to radar (the system that planes, ships, and other vehicles use to navigate). In other words, they just radiate a beam of radio waves down from a satellite or airplane and wait for the reflections to return. Since radio waves travel at the speed of light (300,000 km or 186,000 miles per second), it takes only a few hundredths of a second for a radio beam to make the 20,000-m or so round trip to Earth's surface and back. The altimeter times the beam and calculates its altitude in kilometers by multiplying the time in seconds by 150,000 (300,000 divided by two—the beam has traveled twice as far as its altitude going to the ground and back again). Radio altimeters are much quicker and more precise than pressure instruments and are widely used in high-speed airplanes or ones that need to fly at particularly low altitudes, such as jet fighters [4–6].

In general, the speed of light is about a million times faster than the cruising speed of a typical plane (v), so a radio signal bounced to the ground and back travels a distance of about twice the plane's altitude ($2h$) (Fig. 10.3). Consequently, the altimeter must multiply the time of the signal, which radiates from the transmitter and reflects to the receiver, by half the speed of light. In theory, the faster the plane travels, the less accurate the measurement, because the radio beam has further to travel; in practice, the speed of light is so much faster than the speed of the plane that any error is minimal [3, 7, 8].

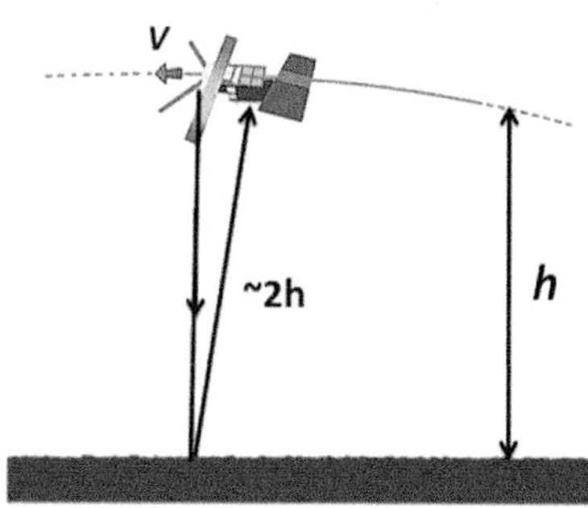

FIG. 10.3 The basic theory of altimeter function.

10.5 How is surface height estimated by radio altimeter?

The transmitted pulse reaches the surface first on a small point. As the pulse advances, the illuminated area grows rapidly from a point to a disk, as does the returned power. The reflecting area depends on the beamwidth of the antenna. The energy from the center of the main beam has to travel a shorter path than the energy from the edges. An annulus is formed and the geometry is such that the annulus area remains constant as the diameter increases [6, 8]. The returned signal strength, which depends on the reflecting area, grows rapidly until the annulus is formed and remains constant until the growing annulus reaches the edge of the radar beam, where it starts to diminish. The simple mathematical formula to identify the surface height (H) is given by:

$$\text{(Corrected) height} = \text{altitude} - \text{(Corrected) variety} \tag{10.1}$$

In the case of sea surface height, several circumstances must be considered. These include ocean surface height, ocean circulation, and other physical parameters, such as wind speed, eddies, and seasonal variations. Ocean surface height is determined without referring to other physical parameters that are associated with the sea surface, for instance, tide, wind speed, and currents. The geoid is governed by the sea surface due to the impact of gravity distribution over the world. Under this circumstance, the geoid fluctuates due to changes in water masses and densities. In other words, a sea bathymetry at the geoid is noticeable as a seafloor has a denser rock zone at the seafloor that would distort sea level by tens of meters. Furthermore, dynamic topography, which is known as ocean circulation, is a function of the Earth's rotation. It is a derived impact of about 1 m. By removing the geoid from sea surface height, the dynamic topography is then computed. In practice, the mean sea level is calculated to yield the variable component (sea degree anomalies) of the ocean signal [9].

10.6 Pulse-limited altimetry

Consider a radar pulse emanating from a radar beacon propagating downwards and interacting with a flat ocean surface. Fig. 10.4 shows an illustration of the vertical cross-section and top-down view of the radar pulse.

Implementing the Pythagorean theory, the leading edge r_p of the pulse is calculated as:

$$r_p = \sqrt{Hct_p} \tag{10.2}$$

where H is satellite height, c is the speed of light, and t_p is pulse time.

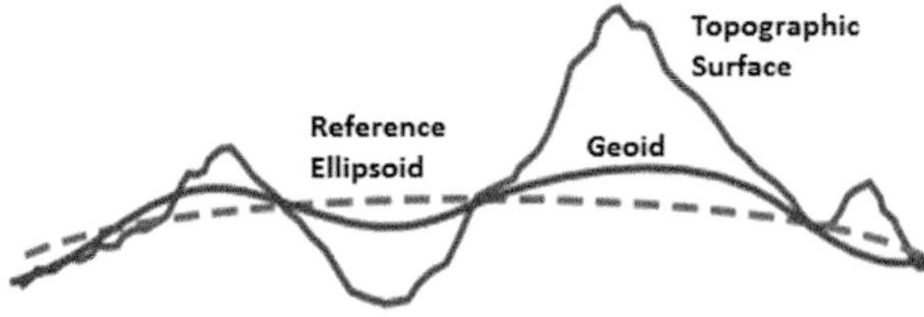

FIG. 10.4 Radar pulse.

Eq. (10.2) demonstrates that the variation of pulse time t_p owing to the fluctuation of the backscatter signal of the ocean or land surfaces are identified as (i) the period before the pulse arrives, (ii) the period after the pulse arrives and before the tail of pulse has been received by the antenna, and after the tail, pulse has been received by the antenna.

In order to determine the power signal of the delay-Doppler radar as a function of time, we will need to assume that the footprint of the pulsed radar is small enough to be considered as two rectangles of width W. This can be expressed by:

$$P(W) = \begin{cases} 0 & t < t_0 \\ 2W_r(t) & t_0 < t < t_0 + t_p \\ 2W_r\left[r(t) - r\left(t - t_p\right)\right] & t > t_0 + t_p \end{cases} \tag{10.3}$$

The altimeter radiates a pulse toward the Earth's surface. The time, which intervenes from the transmission of a pulse to the reception of its backscatter of the Earth's surface, is proportional to the satellite's altitude. Some theoretical details of the principle of radar are applied to the altimeter, which can assist in understanding the different behaviors and characteristics of the pulse in the function of irregularities on the surface encountered. In this regard, the magnitude and shape of the echoes (or waveforms) also contain information about the characteristics of the surface that caused the backscatter. The best results are acquired over the ocean, which is spatially homogeneous and has a surface that conforms with known statistics. In contrast, land surfaces, which are not homogeneous and contain discontinuities or significant slopes, make precise analysis more challenging. Even in the best case (the ocean), the pulse should last no longer than 70 ps to achieve an accuracy of a few centimeters. Technically, this means that the emission power should be greater than 200 kW and that the altimeter would have to switch every few nanoseconds.

These problems are solved by the full deramp technique, making it possible to use only 5 W for emission. The range resolution of the altimeter is about 0.5 m (3.125 ns) but the range measurement performance over the ocean is about one order of magnitude greater than this. This is achieved by fitting the shape of the sampled echo waveform to a model function, which represents the form of the echo [3–7].

10.7 Altimeter sensors

The detailed concurrent measurements of the sea surface height (SSH) and radar backscattering strength at a nadir in deep water are available. These measurements were made with microwave radars on board the Jason-1, Topex/Poseidon, ENVISAT, and Geosat Follow-on (GFO) altimetric satellites (Fig. 10.5) (Table 10.1) [1].

The details of the coastal wave interaction, for the most part, are not accounted for in the propagation models that simulate open ocean wave dynamics. Therefore, coastal measurements are difficult to use for validation and comparison of propagation model data. All four satellites provided unambiguous deep ocean measurements of the wave, to be used in propagation models [8].

FIG. 10.5 Different altimeter satellites.

10.8 Principles of synthetic aperture radar altimeter interferometry

The synthetic aperture (SAR) interferometer radar altimeter SIRAL-2 design is based on existing equipment, but with several major enhancements designed to overcome difficulties associated with measuring ice surfaces. It works by bouncing a radar pulse off the ground and studying the echoes from the Earth's surface. By determining the position of the spacecraft—achieved with an on-board ranging instrument called DORIS (Doppler Orbitography and Radiopositioning Integrated by Satellite)—the signal return time will reveal the surface altitude. Correct antenna orientation is vital for this and is maintained using a trio of star trackers. Moreover, SIRAL is the primary instrument of the mission, designed and developed for ESA by Thales Alenia Space (formerly Alcatel Alenia Space), France. SIRAL is of Poseidon-2 heritage flown on the Jason-1 mission. The objective is to observe ice sheet interiors and ice sheet margins for sea ice and other topography.

Since the launch of Cryosat-2 in 2010, a new technology of altimeters expending Doppler and interferometric competencies have emerged and will most probably become the pattern for upcoming altimeters—at least the Doppler one, as is already the case for the Sentinel-3 altimeter launched in 2016 [3]. In this context, the Delay-Doppler Altimeter (DDA) concept (also acknowledged as an SAR altimeter) was first proposed by Raney [9].

DDAs have an excessive pulse repetition frequency (PRF) to make sure of pulse-to-pulse coherence, leading to a practicable along-track resolution of approximately 300 m, improved signal-to-noise ratio, and improved altimeter ranging performance [10]. For instance, the Cryosat-2 Synthetic Aperture Interferometric Radar Altimeter (SIRAL) exploits the SAR mode over ocean zones.

TABLE 10.1 Altimeter satellite sensors.

Sensors	Band	Radar frequency (GHz)
Geosat	ku	13.5
ERS-1	ku	13.8
ERS-2	ku	13.8
TOPEX	Ku c	13.6 5.3
Poseidon	ku	13.65
GFO	ku	13.5
Jason-1	ku c	13.6 5.3
ENVISAT	ku s	13.6 3.2
Jason-2	ku c	13.6 5.3
Jason-3	ku c	13.6 5.3
SENTINEL-3	ku c	13.575 5.41
Cryosat-2	ku	13.575

The SAR interferometric mode (SARin) is CryoSat's most advanced mode, particularly when used over the ocean surface margins. At this juncture, the altimeter performs synthetic aperture processing and exploits the second antenna as an interferometer to detect the across-track angle to the earliest radar returns. The SARin mode offers the precise surface measurements of the location when the surface is sloping and can be used to find out about extra contrasted sea surface slopes. Over most of the rough ocean, SIRAL operates in the preferred Low Rate Mode (LRM), which is the conventional pulse-limited radar altimeter mode. In this mode, the data rate is considerably lower than the different dimension modes. The SIRAL data provide an excellent opportunity to investigate the competencies and deserves three distinctive altimeter function modes for the detection and estimation of even tiny floating debris characteristics (free-board and surface).

10.9 Altimeter interferometry technique

The concept of the interferometric altimeter was first proposed by Jensen [11] and led to the improvement of the Cryosat mission. A detailed description of the concepts and processing of the Cryosat SARin information is given by Wingham et al. [3]. The primary (left) antenna transmits the radar signal and the two antennas measure the backscattered echo waveform

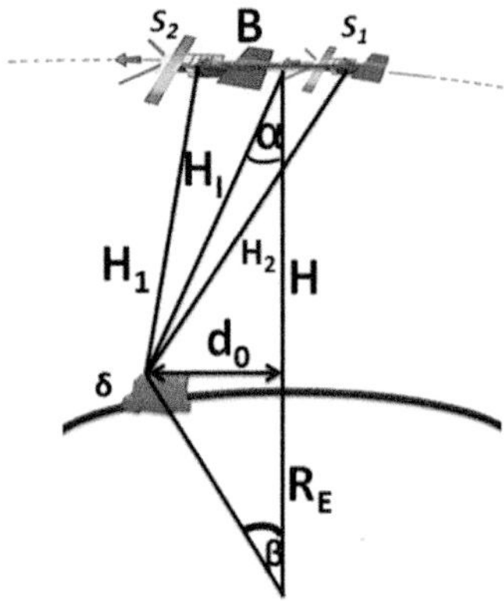

FIG. 10.6 Geometry of altimeter interferometry.

(Fig. 10.6). The primary complex waveform is multiplied with the complex conjugate of the second antenna waveform. The phase of the ensuing cross-channel waveform is then defined as the interferometric phase difference, which results from the slight range distinction of an off-nadir scatterer for the two antennas. The normalized modulus of the conjugate product offers an estimate of the signal coherence. The stacked SAR echoes for each antenna are computed using the SAR mode processing. The SAR echoes, phase, and coherence are furnished with ESA Level-1B products.

In SARin mode, the waveform evaluation window is elevated to 512 bins (240 m) to sample sloping terrains more effectively. In the CryoSAT Baseline-C data; for instance, CryoSAT ocean wave spectra patterns are processed. The use of zero-padding prior to FFT processing, besides, increase the variety of range bins in 1024 except altering the range window. Note that, If you apply a windowing function to your waveform, the windowing function needs to be applied before zero padding the data. This ensures that your real waveform data starts and ends at zero, which is the point of most windowing functions. Each bin corresponds, consequently, to 1.565 ns or 0.23 m along with the range.

Let us assume that the two SAR antennas S_1 and S_2 mounted on a satellite platform are separated by a baseline B with a baseline angle α, and observe the complex response at the point δ with a slant range H_1 (Fig. 10.6). The mathematical relationships between the interferometric phase difference, $\Delta\psi$, and the off-nadir angle, α, can be given by:

$$\Delta\psi = \frac{2\pi B}{\lambda}\sin(\alpha) \tag{10.4}$$

where λ is the radar wavelength and B is an interferometer baseline (distance between the two antennas). Under the small-angle approximation, the off-nadir angle α is expressed by:

$$\alpha = \frac{\lambda \Delta\psi}{2\pi B} \tag{10.5}$$

Galin et al. [12] estimated the across-track distance to nadir, d_0, by:

$$d_0 = \frac{ct_i}{2}\overline{\alpha} = H_i\overline{\alpha} \tag{10.6}$$

Eq. (10.6) demonstrates that range H_i equals half of the multiplication of pulse time t_i by the speed of light c then multiplied by angle scaling $\overline{\alpha}$. Consequently, the freeboard δ due to the earth curvature is defined as:

$$\delta = (H - H_i \cos\alpha_1 + R_E(1 - \cos\beta))\cos\beta \tag{10.7}$$

where $\beta = H/R_E\alpha_1$ and $H_i = ct_1/2$.

The SARin echoes are similar to the SAR ones, except that the number of range bins in the echoes' thermal noise part (TNP) is significantly larger (125×2 vs 50). Altimeters overpass open waters have a detectable noise; which is termed as thermal noise part (TNP). In this view, TNP is accounted as easily detected pulse above the sea surface in high resolution (HR) waveforms of pulse-limited altimeters [12].

The swath over which sea level can be detected, which is of the order of 6 km, is thus significantly increased to 12 km. The SAR detection algorithm can be applied to the SARin waveforms without modification. However, in the echoes, the TNP signals received by the two antennas are by nature random noise and thus incoherent [13].

10.10 InSAR precision procedures altimeter scheme

InSAR altimeter accurate analysis is a function of the specifics of satellite altitude, slant range, baseline length, baseline angle, and phase difference, which are denoted as σ_H, σ_B, σ_{H_1}, σ_α, and σ_β, respectively. In these circumstances, these parameters are independent, and can be expressed mathematically as:

$$\sigma_{R_E} = \sigma_H\left(\frac{\partial R_E}{\partial H}\right) + \sigma_B\left(\frac{\partial R_E}{\partial B}\right) + \sigma_{H_1}\left(\frac{\partial R_E}{\partial H_1}\right) + \sigma_\alpha\left(\frac{\partial R_E}{\partial \alpha}\right) + \sigma_\beta\left(\frac{\partial R_E}{\partial \beta}\right) \tag{10.8}$$

Eq. (10.8) demonstrates that the height approximation accuracy is a function of a decreasing of the look angle α. In this circumstance, α must equal 0.5 degrees to obtain accurate height estimation. In contrast, decreasing of α causes decreasing of the range resolution. In addition, low edge of sea surface can be presented as a quasispecular scattering in altimeter antenna. In this regard, the echo-tracking model has differences in backscattering echoes from the sea surface.

The echo-tracking algorithm is implemented in the slant range enhancement after the phase and geometry computations to deliver a precise slant range estimation. More precisely, Fig. 10.7 demonstrates that the location of the sea surface at the point $\delta\alpha$ is the radar look angle and r is a function of α. In other words, the echo waveform is a function of the plane surface impulse response $\delta_{FS}(\tau, r)$, system point target response $S_r(\tau, r)$, and height probability density function of the sea surface scattering elements $Q(\tau)$. The mathematical expression of the sea surface echoes waveform is based on the convolution [14], which is given by:

$$\delta_r(\tau r) = \delta_{FS}(\tau r) \times S_r(\tau r) \times Q(\tau) \tag{10.9}$$

$$\delta_r(\tau, r) = \int_{SeaSurfaceArea} \frac{\lambda^2 \delta(\tau)\sigma_0(r)}{(4\pi)^3 L_p \|r_1\|^4} \|G(r)\|^2 dA \tag{10.10}$$

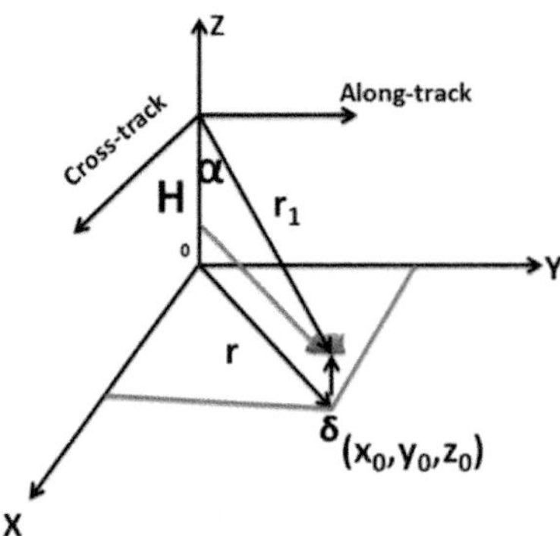

FIG. 10.7 SAR geometry.

Eq. (10.10) is considered an example of the radar equation. In this context, $\delta(\tau)$ is the transferred waveform in the duration period τ with wavelength λ. At the position r over the changing area dA, the sea surface backscatter is σ_0, L_p is the double loss of the atmosphere, and G is the antenna gain. Consequently, system point target response $S_r(\tau, r)$ is formulated as:

$$S_r(\tau,r)=C^2\,\mathrm{sinc}^2\left(\frac{\tau_a}{2\pi}\left(2\pi f_a-2\pi f_d\left(1-\frac{k_r t_0}{f_0}\right)\right)\right)\mathrm{sinc}^2\left(\tau_c\left(\frac{k_r}{cH}\|r\|^2-\tau\right)\right) \tag{10.11}$$

Eq. (10.11) indicates a filter length of along-track and cross-track τ_a and τ_c, respectively. Consequently, f_a, f_d, and f_0 are azimuth, Doppler, and wave frequencies, respectively. Therefore, the linear frequency modulation rate k_r equals the ratio of baseline B and period T, i.e., $k_r=\frac{B}{T}$. Finally, C is a function of SAR along-track and cross-track filter length.

The height probability density function of the sea surface scattering elements $Q(\tau)$ is determined from:

$$Q(\tau)=\frac{1}{\sqrt{2\pi\sigma_s}}\left[1+\frac{\lambda}{6}\left(\frac{\tau^3}{\sigma_s^3}-3\frac{\tau}{\sigma_s}\right)\right]e^{\left(-\frac{\tau^2}{2\sigma_s^2}\right)} \tag{10.12}$$

Eq. (10.12) is the keystone to estimate the significant wave height H_s from the backscatter quantity σ_s, which indicates the root-mean-square wave height as follows:

$$\sigma_s=\frac{1}{2C}H_s \tag{10.13}$$

Fig. 10.8 explains the delay time of the echo waveform model of SAR observation, for instance, at the position of $\delta(\tau)$ (10000,0,1) with H equaling 800 km and H_s at about 2 m, which produces the maximum normalized epoch power of 1 with a minimum delay time less than 0.008 s. The maximum delay time 0.04 s coincides with low normalized echo power closes in 0.

The time delay of the emitting signals derives from the mid-power epoch of the echo under the circumstance of fitting the backscatter signal to the echo waveform model. The precise slant range is obtained via multiplying by the speed of light [14].

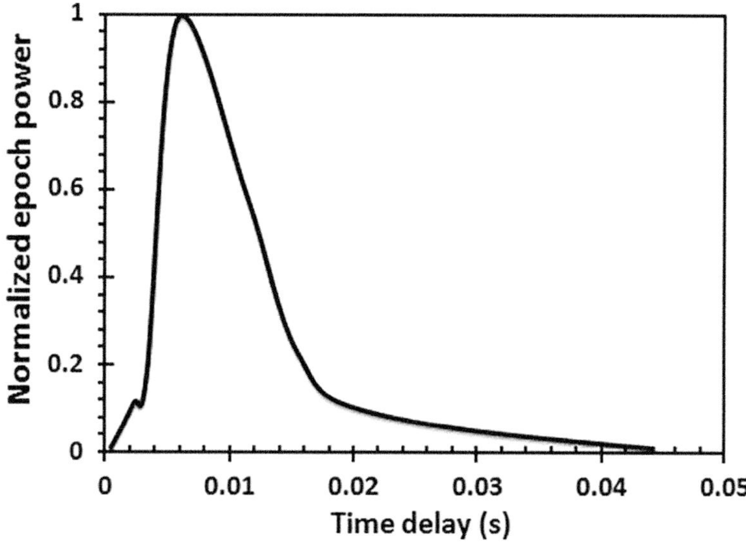

FIG. 10.8 The epoch waveform of near-nadir SAR observation with a certain position.

10.11 Delay-Doppler altimeter

Conventional radar altimeters emit pulses with a long interval period of approximately 500 μs. SIRAL, however, emits a burst of pulses with an interval of only 50 μs between them. The receiving echoes are thus interrelated, and by considering the complete burst of pulses in one campaign, the data processor can separate the echo into the set of strips across the track by manipulating the slight frequency shifts. In this context, the Doppler effect causes a frequency shift in the forward- and aft-looking parts of the beam. In this way, noise reduction is achieved by superimposing and average of strips on each other.

In interferometric mode (SARin), however, a second receiving antenna is triggered to determine the receiving angle. SARin, therefore, is able to sense approximately radar backscatter signals received from a target not precisely positioned beneath the altimeter sensor. Consequently, the variance in the path-length time of the radar echoes is tiny between radar echoes on the track and radar echoes out of the track. In this circumstance, α must be accurate.

The Delay-Doppler Altimeter differs from a common radar altimeter theory in that it utilizes coherent dispensation of sets of transmitting pulses. In this regard, the full Doppler bandwidth is contracted to achieve the most efficient use of the backscatter intensity, which is reflected from the Earth's surface (Fig. 10.9). In contrast, the conventional altimeter system determines the distance between the satellite and the mean ocean surface. In this context, the Delay-Doppler Altimeter method differs from conventional altimeter instruments by two techniques. The primary technique of single antenna involves pulse-to-pulse coherence and full Doppler processing, which permit for the quantity of the along-track spot of the range measurement. The latter technique of altimeter interferometry includes the implementation of two antennas and two receiver channels, which allow for the dimension of the across-track angle of the range quantity.

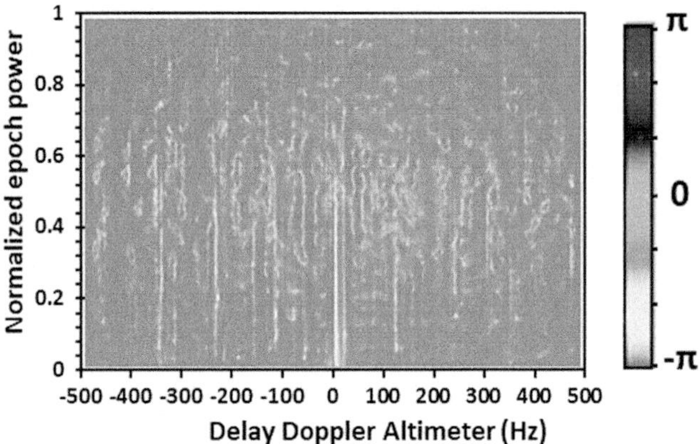

FIG. 10.9 Delay-Doppler Altimeter interferometry.

The Delay-Doppler Altimeter technology has numerous advantages over a conventional altimeter. The sea surface height precision offered by Delay-Doppler Altimeter technology is approximately twice that of the conventional altimeter. Simulations of the associated signal processing concepts have produced 0.5 cm precision in a calm sea, with precision remaining better than 1.0 cm even in significant wave heights as great as 4 m. The Delay-Doppler Altimeter technique reduces the uncertainties caused by ocean waves.

For instance, both Delay-Doppler Altimeter and conventional altimeter experience comparable levels of random noise from a calm sea. Nevertheless, as the waves grow, a conventional altimeter suffers intense noise level growth, whereas the coherent processing of the Delay-Doppler Altimeter experiences only a trivial intensification in random noise with wave height. Consequently, the Delay-Doppler Altimeter is particularly appropriate for geodetic uses. Random error owing to ocean waves is the dominant error source. Consequently, wind speed and wave height retrievals from the Delay-Doppler Altimeter have twice the precision of current sensors.

10.12 CRYOSAT-2 SIRAL data acquisitions

In this investigation, CryoSat-2 SIRAL data are implemented on wave spectra patterns in the Indian Ocean. CryoSat-2 revolves on a nonsun-synchronous polar orbit with an inclination angle of 92 degrees at an altitude of 713 km. Consequently, the repeat cycle for CryoSat is 369 days and 5344 orbits. This cycle provides full coverage of the earth and the cycle number is provided in the CryoSat product headers. The repeat cycle is made up of approximately 30-day subcycles, defined and used primarily for statistic and quality reporting [15]. There are two instruments attached to a board of CryoSat-2's mission. The primary two instruments aboard CryoSat-2 are SIRAL-2; in which SIRAL-2 instruments serve as a backup in case the other fails.

A second instrument is Doppler Orbit and Radio Positioning Integration through Satellite (DORIS), which is deployed to compute precisely the spacecraft's orbit. An array of retroreflectors is additionally conducted on board of the spacecraft and permit quantities of backscattered signals to be made from the ground to insist on the orbital data supplied through DORIS [16]. In LRM the altimeter operates as a conventional pulse narrow altimeter. This mode functions at a pulse repetition frequency (PRF) of 1970 Hz. This PRF is low enough to ensure that the echoes are decorrelated [16]. Consequently, the generated pulses may be incoherently extra to decrease speckle noise by $\frac{1}{\sqrt{M}}$ a factor, where M is the number of being an average of pulses in the selected time interval [16].

In SAR mode, the pulses are emitted in bursts, and correlation between echoes is anticipated. Thus, in SAR mode, the PRF within a burst is higher than the LRM, and equal to 17.8 kHz [3]. At this PRF, every burst emits 64 pulses. Every time the pulses are transmitted, the altimeter waits for the returns and then transmits again for the succeeding burst. Hence, there is not just an "intra-burst" PRF, but a burst repetition frequency (BRF), which, according to SIRAL measurements, is 85.7 Hz.

Both SIRAL LRM and SAR modes transmit pulses of equivalent pulse length. The principal distinction between these modes is the PRF and its related effects. In SAR mode, pulse-to-pulse correlation is a consequence of its excessive PRF, while pulse-to-pulse correlation is not existent in LRM, nor preferred. In this regard, the decorrelation of PRF is obtained by dividing the spacecraft velocity V by the decorrelation distance and correcting for the curvature of the earth R_e [17] as given by:

$$PRF = \frac{V}{0.31\lambda H(r^{-1})\left(\frac{R_e + H}{R_e}\right)} \tag{10.14}$$

where λ is the radar altimeter wavelength, H is the altimeter height, and r is the radius of the circular uniformly illuminated area.

The mathematical relationship between r and significant wave height can be expressed as:

$$r = \sqrt{\frac{(ct + 2H_s)H}{1 + \frac{H}{R_e}}} \tag{10.15}$$

where t is a lag time between the leading and trailing edges of the pulse.

Consequently, Eq. (10.15) demonstrates that the increase of PRF is due to the extreme growth of significant wave height H_s (Fig. 10.10).

Fig. 10.11 shows an example of the simulated H_s with an area of 300 km × 300 km, with a 500 m wavelength and 4 m H_s.

10.13 Cycle of significant wave heights and powers: Case study of west coast of Australia

The west coast of Australia was selected as it is a significant zone for crashing the Malaysian Airlines MH370. This part of the southern Indian Ocean is a dynamical zone, which is required for further investigations. Let us assume that area of the ocean surface A is under the action of incident regular waves of varying wave frequency and/or under the action of irregular waves. The significant wave heights delivered by CryoSat-2 are used to

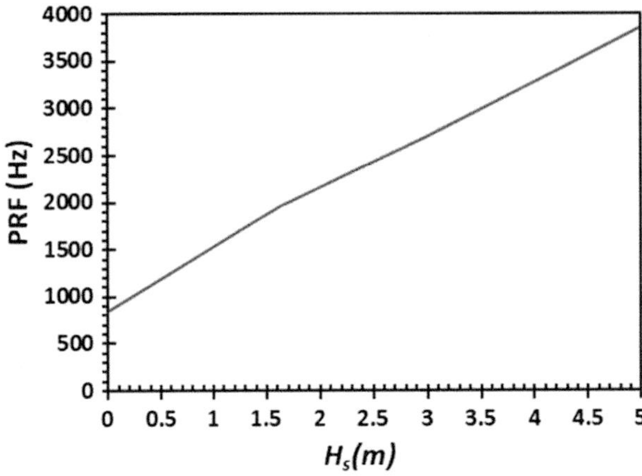

FIG. 10.10 PRF as a function of significant wave height H_s.

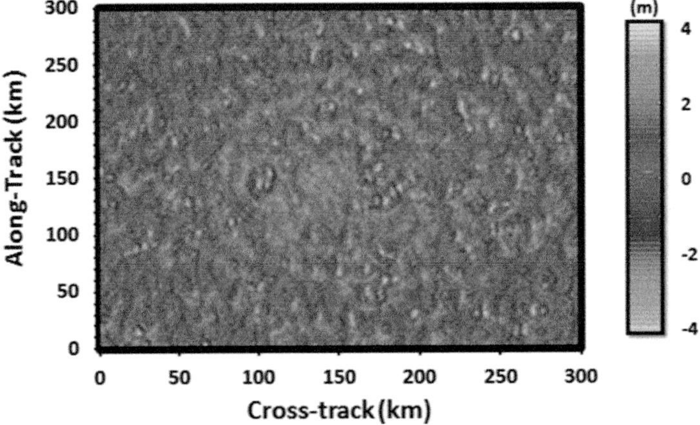

FIG. 10.11 Simulation of H_s using SIRAL LRM data.

compute the irregular wave power spectra by using the Pierson-Moskowitz (P-M) spectrum [17–19], which is given by:

$$S_I = \frac{H_s^2 T_z}{8\pi^2} \left(T_z \frac{\omega}{2\pi} \omega\right)^{-5} e^{\left[-\frac{1}{\pi}\left(T_z \frac{\omega}{2\pi}\right)^{-4}\right]} \tag{10.16}$$

where T_z is the zero upcrossing periods and ω is wave frequency rad/s.

In this regard, ω varies between 0.6 and 7.6 rad/s. Then the averaged produced power in irregular waves is computed as follows [19–21]:

$$S_P(\omega) = \overline{P}(\omega) S_I(\omega) \tag{10.17}$$

In this context, $\overline{P}$ is the total averaged power as a function of significant wave height modeled from altimeter interferometry.

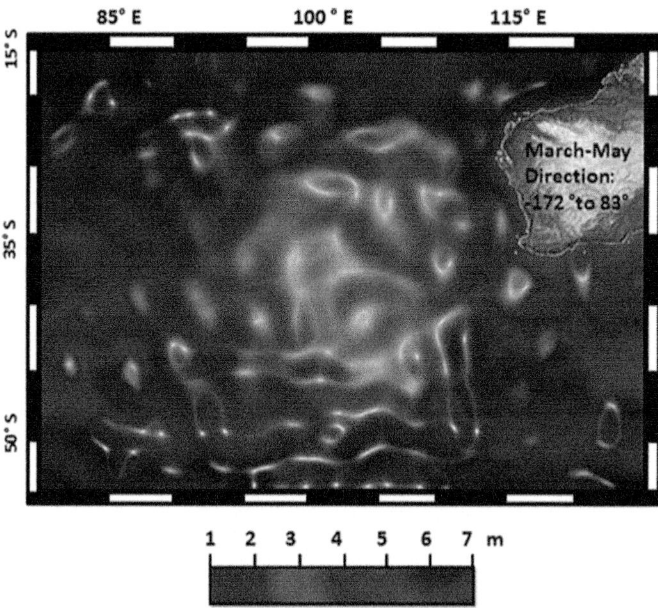

FIG. 10.12 Significant wave height spatial variations from March to May 2014.

The CryoSat-2 coverage cycles cover the southern Indian Ocean between 23°57′26.98″S and 101°18′18.76″E. In other words, CryoSat-2 satellite data can deliver significant wave height cycles in the southern Indian Ocean. In this regard, offshore regions of the southwest Australian shelf are dominated by oceanic swell and seas with an average significant wave height of about 4 m and a period from 5 to 20 s. The western coast of Australia is dominated by a maximum significant wave height of 6 m from March to May. Further north, the significant wave height is 3 m. In other words, there is a swell pool occupying the offshore regions of the southwest of Australia. The southwest significant wave height of 172° propagated toward the northeast's of 83° (Fig. 10.12).

From June to August 2014, the maximum significant wave height of 7 m approached the southwest Australian shelf. In the northwest Australian shelf, the significant wave height fluctuated from 1 to 3 m (Fig. 10.13). Consequently, the swell propagated from 174° to 62°. September to October 2014 was dominated by maximum significant wave height 6 m in the northwest Australian shelf with the dominant direction of 74°. In contrast, the southwest Australian shelf had a swell pool of 7 m significant wave height, which propagated from the southwest with the direction of 171° (Fig. 10.14).

The significant wave height pattern increased dramatically to 7 m, in both the southwest and northeast of the Australian shelf (Fig. 10.15) from November to December 2014. However, the swell continued traveling from southwest to northeast and turned to the northwest of the offshore of the Australian shelf.

For a 1-year significant wave height cycle, the swell remains propagating from the southwest to the northeast from January to March 2015 with a maximum significant wave height of 5 m in the northeast offshore of the Australian shelf and 7 m significant wave height

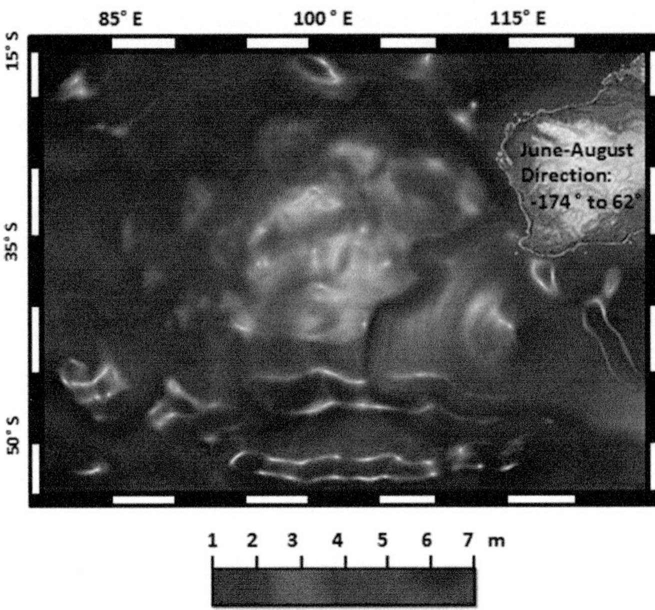

FIG. 10.13 Significant wave height spatial variations from June to August 2014.

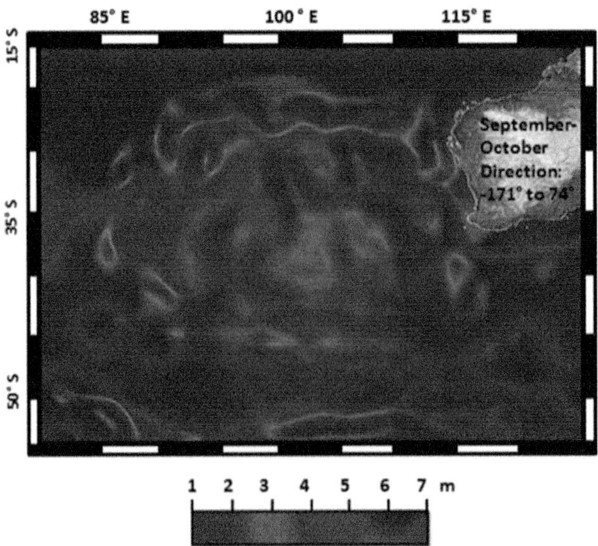

FIG. 10.14 Significant wave height spatial variations from September to October 2014.

southwest of the Australian shelf (Fig. 10.16). In contrast, inshore, the wave climate responds rapidly to the afternoon onset of the sea breeze. Wind waves associated with the sea breeze have a significantly shorter period and length than the swell and are less affected by inshore bathymetry.

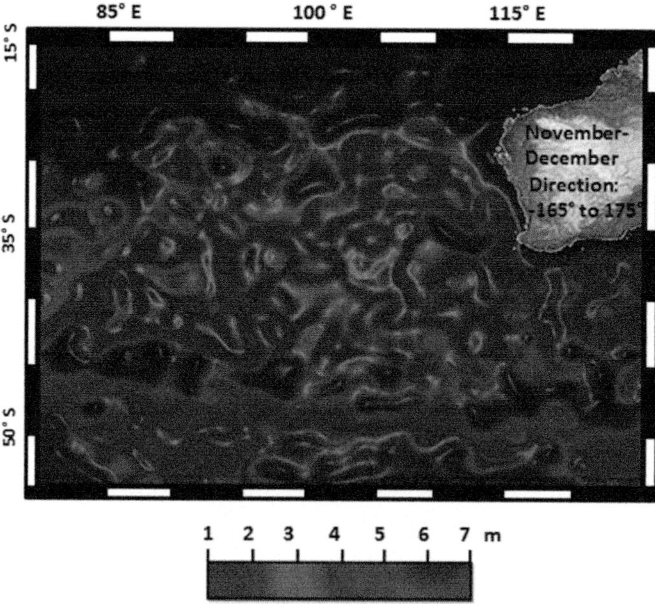

FIG. 10.15 Significant wave height spatial variations from November to December 2014 to May 2015.

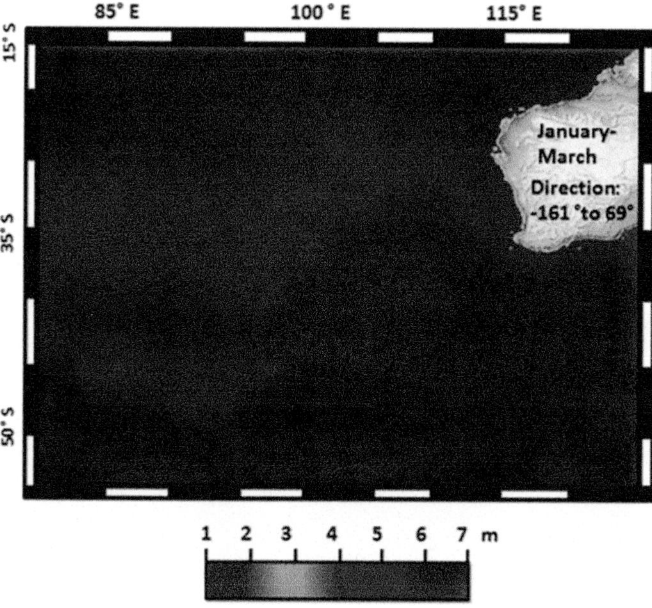

FIG. 10.16 Significant wave height spatial variations from January to March 2015.

In this view, most of the year, the waves arrive from the southwest, though winter storms often result in waves from the west and northwest bringing high energy conditions to the coast for short periods. Inshore, much of the coastline is sheltered from the direct impact of the swell-wave activity by an extensive chain of reefs, which cause significant attenuation of the ocean waves [22, 23]. The search area is dominated by a maximum wave power of 30,000 kJ/m/wave, located southwest of the Australian shelf. However, the northeast and west of the Australian shelf are dominated by a maximum wave power of 20 kJ/m/wave. This quantity fluctuated from month to month (Figs. 10.17–10.21). In the first 3 months, March to May 2014, the highest wave power pool was southwest of the Australian shelf with 30,000 kJ/m/wave (Fig. 10.17). From June to August 2014, there was a slight reduction of the wave power of 25,000 kJ/m/wave (Fig. 10.18). The northwest sector offshore of the Australian shelf was dominated by wave power fluctuating between 5000 and 25,000 kJ/m/wave. From September to October southwest of the Australian shelf was dominated by a maximum wave power of 30,000 kJ/m/wave, while north offshore of the Australian shelf was dominated by wave power of 15,000 kJ/m/wave (Fig. 10.19). This wave power of 30,000 kJ/m/wave remained in the southwest of the Australian shelf and decreased to 20,000 kJ/m/wave in the northeast Australian shelf from November to December 2014 (Fig. 10.20). From January to March 2015, the wave power increased rapidly in the northeast to 25,000 kJ/m/wave (Fig. 10.21).

In general, around Australia, annual mean wave power is greater in the southern waters than in the northern waters [24, 25]. The largest and longest-period powerful waves occur off the west coast of Western Australia. Low mean heights and shorter periods occur on the northwest shelf. This pattern of wave climate is generally consistent with previous studies

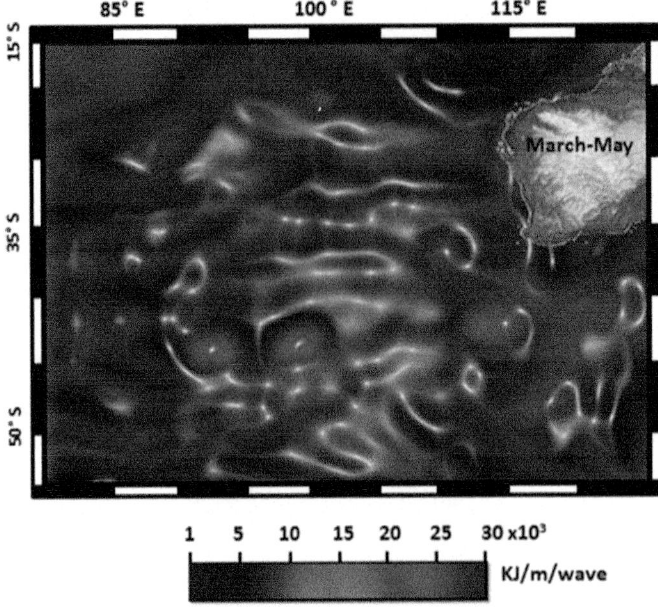

FIG. 10.17 Wave power from March to May 2014.

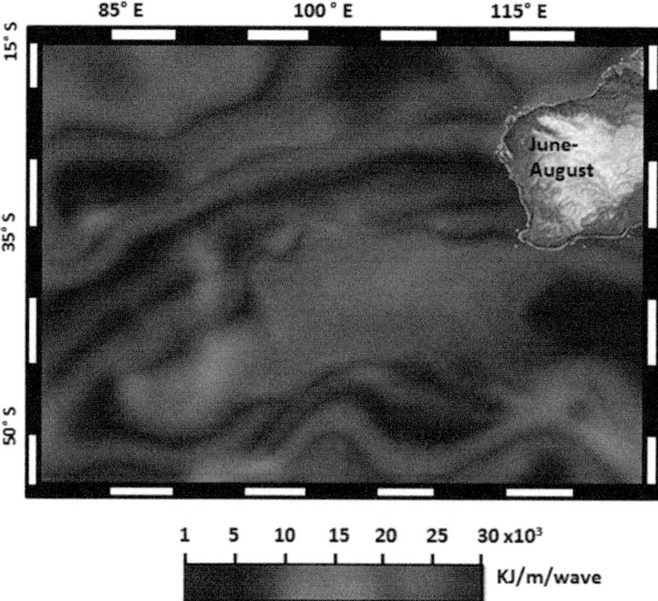

FIG. 10.18 Wave power from June to August 2014.

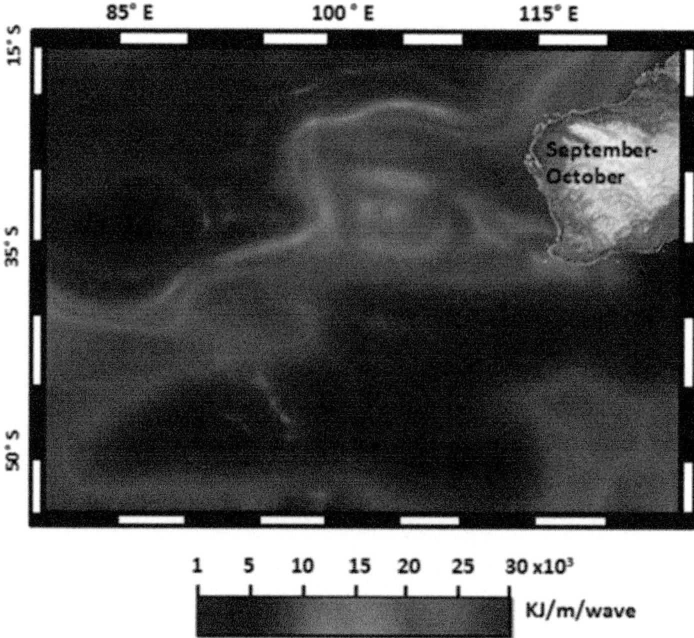

FIG. 10.19 Wave power from September to October 2014.

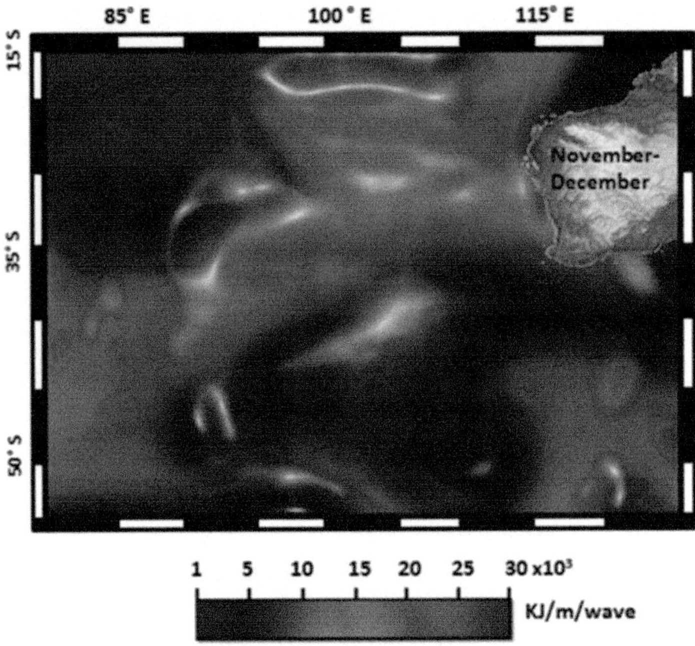

FIG. 10.20 Wave power from November to December 2014.

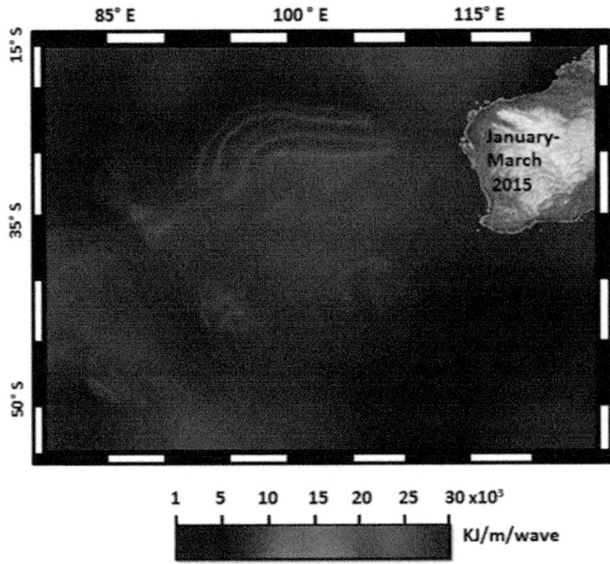

FIG. 10.21 Wave power from January to March 2015.

of waves on the continental shelf around Australia [23–25]. Maximum wave power peaks in the southwest of the Australian shelf are associated with waves generated by tropical cyclone conditions [26].

Furthermore, the offshore wave climate of Australia is dominated by a persistent moderate-energy wave regime with waves of 2–4m in height. It is characterized by south to southwesterly swells along the south and west coasts, and low-energy waves from the west and northwest along the northern coast. The east coast of Australia is characterized by a strong southeasterly swell, interrupted only briefly by a locally generated choppy northeasterly following the passage of an anticyclone. The inshore wave energy along the northern coasts of the continent is generally low since it is dissipated across a broad continental shelf. In late summer and autumn, tropical cyclones may generate high-energy waves during intense but short-lived storms. Consequently, the swells of the west, south, and east coasts are generated in the storm belt of the Southern Ocean, between 50° and 60°S [24–26].

Despite the strong to moderate offshore wave energy of southwestern Australia, the inshore wave energy is considerably less due to dissipation via refraction and diffraction processes around reefs and headlands. This effect is particularly apparent on the west coast, where an extensive reef chain parallels the coast and may attenuate wave energy by up to 50%. Hence, the shoreline of southwestern Australia experiences modally low wave energies. Isolated reefs and offshore islands offer some degree of local protection to the beaches along the southern coast. However, wave dissipation along the southern coast is considerably less than on the west coast and beaches there are subjected to a moderate to heavy southwesterly swell. Refraction around headlands is an important form of wave energy dissipation along the southern and eastern coasts of the continent.

References

[1] Chelton DB, Ries JC, Haines BJ, Lee-Lueng F, Callahan PS. Satellite altimetry. In: International geophysics. vol. 69. Academic Press; 2001. p. 1–ii.

[2] Rosmorduc V, Benveniste J, Bronner E, Dinardo S, Lauret O, Maheu C, Milagro M, Picot N, Ambrozio A, Escolà R, Garcia-Mondejar A, Restano M, Schrama E, Terra-Homem M. Benveniste J, Picot N, editors. Radar altimetry tutorial. 2016. http://www.altimetry.info.

[3] Wingham DJ, Francis CR, Baker S, Bouzinac C, Brockley D, Cullen R, de Chateau-Thierry P, et al. CryoSat: a mission to determine the fluctuations in Earth's land and marine ice fields. Adv Space Res 2006;37(4):841–71.

[4] Hayne GS. Radar altimeter mean return waveforms from near-normal-incidence ocean surface scattering. IEEE Trans Antennas Propag 1980;28(5):687–92.

[5] Robinson IS. Measuring the oceans from space: The principles and methods of satellite oceanography. Springer Praxis Books; 2004. 669 pp.

[6] Chelton DB, Ries JC, Haines BJ, Fu LL, Callahan PS. Satellite altimetry. In: Fu LL, Cazenave A, editors. Satellite altimetry and earth sciences. Academic Press; 2001.

[7] Marghany M. Advanced remote sensing technology for tsunami modelling and forecasting. CRC Press; 2018.

[8] Vaijayanthi S, Vanitha N. Aircraft identification in high-resolution remote sensing images using shape analysis. Int J Innov Res Comput Commun Eng 2015;3(11):11203–9.

[9] Raney RK. The delay/Doppler radar altimeter. IEEE Trans Geosci Remote Sens 1998;36(5):1578–88.

[10] Martin-Puig C, Ruffini G, Marquez J, Cotton D, Srokosz M, Challenor P, Raney K, Benveniste J. Theoretical model of SAR altimeter over water surfaces, In: Geoscience and remote sensing symposium, 2008. IGARSS 2008. IEEE International, vol. 3. IEEE; 2008. p. III–242.

[11] Jensen JR. Angle measurement with a phase monopulse radar altimeter. IEEE Trans Antennas Propag 1999;47 (4):715–24.

[12] Galin N, Wingham DJ, Cullen R, Fornari M, Smith WHF, Abdalla S. Calibration of the CryoSat-2 interferometer and measurement of across-track ocean slope. IEEE Trans Geosci Remote Sens 2013;51(1):57–72.

[13] Tournadre J, Girard-Ardhuin F, Legrésy B. Antarctic icebergs distributions, 2002–2010. J Geophys Res Oceans 2012;117(C5):1–15.

[14] Tournadre J, Bouhier N, Boy F, Dinardo S. Detection of ice berg using Delay Doppler and interferometric Cryosat-2 altimeter data. Remote Sens Environ 2018;212:134–47.

[15] Sui X, Zhang R, Wu F, Li Y, Wan X. Sea surface height measuring using InSAR altimeter. Geod Geodyn 2017;8 (4):278–84.

[16] Martin-Puig C, Marquez J, Ruffini G, Keith Raney R, Benveniste J. SAR altimetry applications over water. arXiv; 2008. preprint arXiv:0802.0804.

[17] CryoSat mission and data description, Doc No. CS-RP-ESA-SY-0059; 2007.

[18] Pandey PC, Gairola RM, Gohil BS. Wind-wave relationship from SEASAT radar altimeter data. Bound-Lay Meteorol 1986;37(3):263–9.

[19] Fujisaki K, Tanaka K, Tateiba M. A simulation of satellite altimeter return pulses from three-dimensional ocean waves. In: IEEE 1999 international geoscience and remote sensing symposium. IGARSS'99 (Cat. No. 99CH36293) Jun 28, 1999, vol. 2. IEEE; 1999. p. 995–7.

[20] Fukuda K, Fujisaki K, Tateiba M. A computer simulation of relative backscattering coefficients of sea surfaces in satellite altimetry. Electron Commun Jpn (Part II: Electronics) 2007;90(12):121–8.

[21] Portilla J, Ocampo-Torres FJ, Monbaliu J. Spectral partitioning and identification of wind sea and swell. J Atmos Ocean Technol 2009;26(1):107–22.

[22] Michailides C, Angelides DC. Optimization of a flexible floating structure for wave energy production and protection effectiveness. Eng Struct 2015;85:249–63.

[23] Steedman R. Collection of wave data and the role of waves on nearshore circulation. Report prepared for the Water authority of Western Australia, Perth; 1993.

[24] Sanderson PG, Eliot I. Compartmentalisation of beachface sediments along the southwestern coast of Australia. Mar Geol 1999;162(1):145–64.

[25] Louis JP, Radok JRM. Propagation of tidal waves in the Joseph Bonaparte Gulf. J Geophys Res 1975;80 (12):1689–90.

[26] Hemer MA, Simmonds I, Keay K. A classification of wave generation characteristics during large wave events on the Southern Australian margin. Cont Shelf Res 2008;28(4–5):634–52.

CHAPTER

11

Multiobjective genetic algorithm for modeling Rossby wave and potential velocity patterns from altimeter satellite data

11.1 What is meant by Rossby wave?

Previous chapters have dealt with wind waves, swell, and internal waves. However, there is a unique sort of wave known as a Rossby wave, which differs from other wave types. A recent chapter delivers a generation mechanism besides simulation of its behavior across the ocean. In this regard, the arising significant question is: what is meant by the Rossby wave?

Waves in the ocean generate numerous unique shapes and dimensions. Slow-moving oceanic Rossby waves are essentially different from ocean surface waves. Unlike waves that break on the shore, Rossby waves are huge, representing the undulating dynamic of the ocean that stretches horizontally throughout the planet for thousands of kilometers in a westward direction (Fig. 11.1). They are so massive that they can alter Earth's local weather conditions. Along with rising sea levels, "king tides," and the consequences of El Niño, oceanic Rossby waves contribute to excessive tides and coastal flooding in some zones of the globe [1, 2].

Rossby waves are not only planetary waves (Fig. 11.2), but inherently occur in gyrating fluids. Within the Earth's ocean and atmosphere, these waves are generated as a consequence of the rotation of the planet [3, 4].

Rossby wave dynamic movement is complex. The horizontal wave velocity of a Rossby (the quantity of time it takes the wave to travel across an ocean basin) is a function of upon the geographical latitude of the wave propogation. In the Pacific, for instance, waves at lower latitudes (closer to the equator) (Fig. 11.3) might take months to a year to travel across the ocean. Waves that build farther away from the equator (at mid-latitudes) of the Pacific can take 10–20 years to make the journey [1–4].

Nonlinear Ocean Dynamics
https://doi.org/10.1016/B978-0-12-820785-7.00014-9

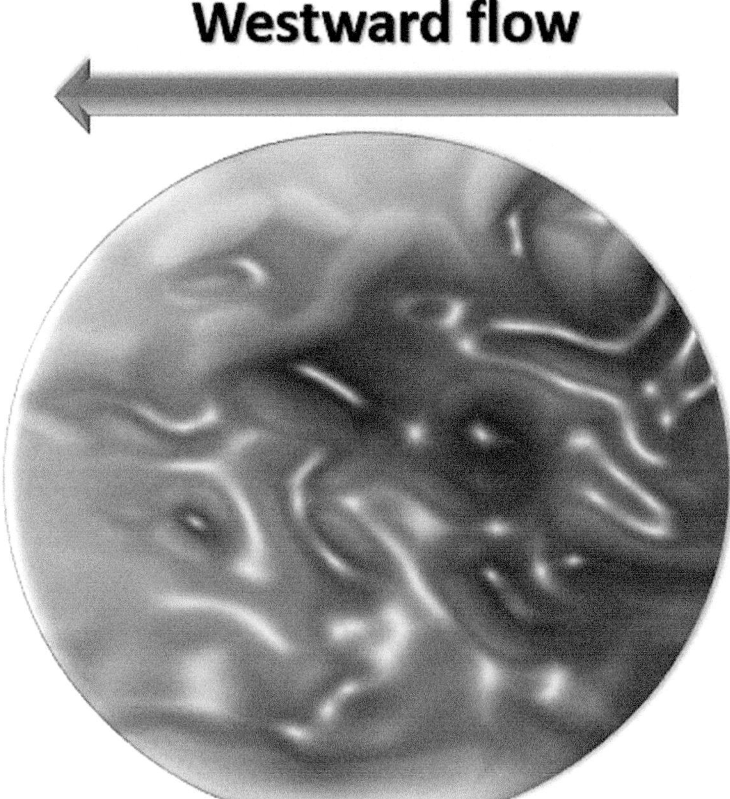

FIG. 11.1 Rossby wave westward flow.

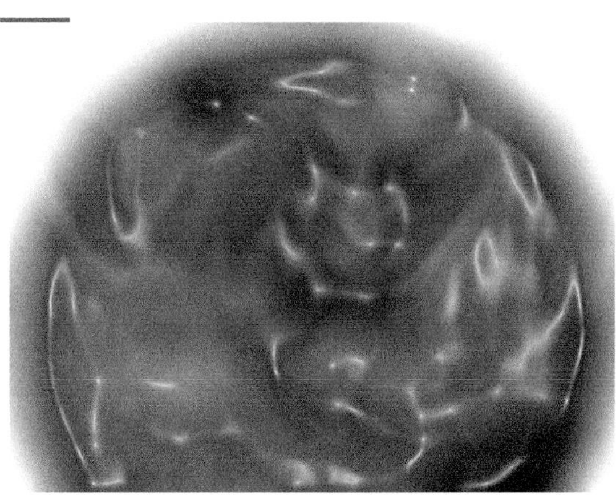

FIG. 11.2 Example of planetary waves.

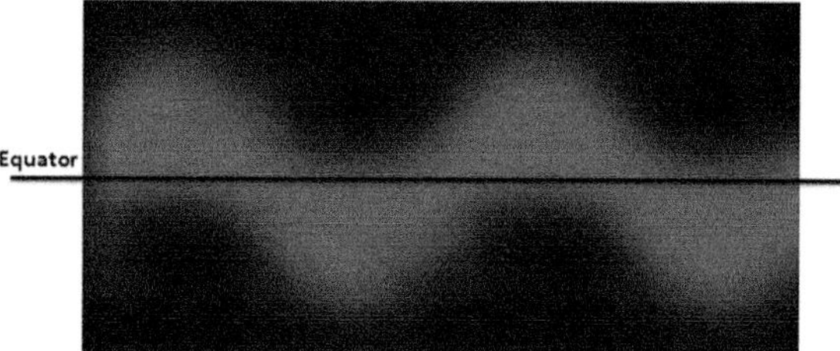

FIG. 11.3 Equatorial Rossby wave.

The vertical dynamic of Rossby waves is small close to the ocean's floor and larger close to deeper thermoclines (Fig. 11.4)—the transitional area between the ocean's warm upper layer and chillier depths. This type of vertical movement of the water's surface can be quite dramatic: the regular vertical dynamic of the water's surface is commonly 4 in. or less, while the vertical motion of the thermocline of the equatorial wave is about 1000 times greater. In other words, for a 4-in. or much less surface displacement close to the ocean surface, there may additionally be greater than 300 ft of corresponding vertical motion in the thermocline a long way beneath the surface [3]. Due to the minimal vertical motion at the ocean's surface, oceanic Rossby waves are undetectable to the human eye. Scientists generally rely on satellite radar altimetry to study these massive waves [5].

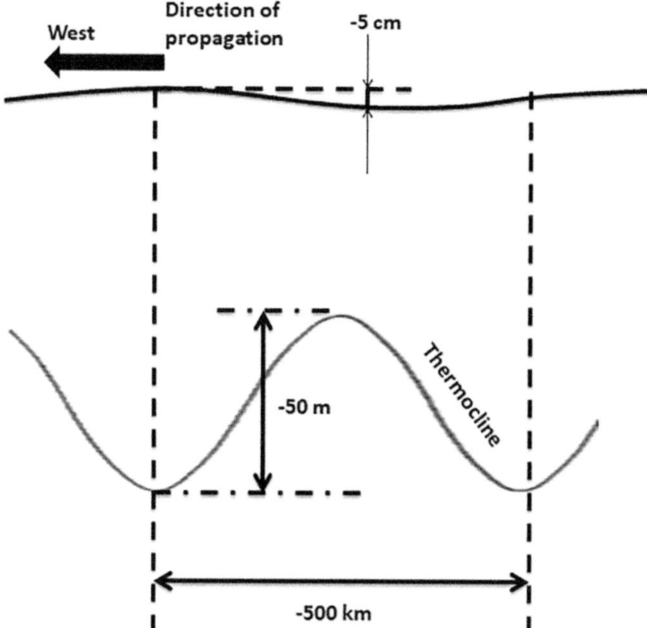

FIG. 11.4 The vertical dynamic of a Rossby wave.

11.2 Rossby waves algebraic portrayal Coriolis

The Coriolis parameter is the key to understanding Rossby waves. The Coriolis force f_{cor} is an imaginary force, which generates due to the Earth's rotation around its axis with constant angular velocity Ω, which equals $7.292 \times 10^{-5}\ \mathrm{s}^{-1}$ (Fig. 11.5). In this sense, the scientific explanation of f can mathematically be written as:

$$f_{cor} = 2\Omega \sin(\phi) \tag{11.1}$$

Eq. (11.1) demonstrates that the Coriolis force equals zero along the equator as its latitude ϕ is zero. In this regard, what is the role of Ω? The fixed frame of the Earth's rotation causes the potential vorticity of the parcel of fluid in the ocean, which is sustained.

It is well known that the latitudinal variation (β) in f_{cor} is simplified using:

$$\beta = \frac{df_{cor}}{d\phi} = 2\Omega \cos(\phi) \tag{11.2}$$

Then the potential vorticity across the shallow water and a homogenous layer of thickness H is defined as:

$$\frac{\partial\left(\xi - \frac{f_{cor}\eta}{H}\right)}{\partial t} + \beta v = 0 \tag{11.3}$$

Eq. (11.3) is considered a continuity equation. In this equation, the vertical component of the potential vorticity can be represented by ξ through the free height of the shallow layer η. Consequently, v is the meridional (north-south) component of the fluid speed.

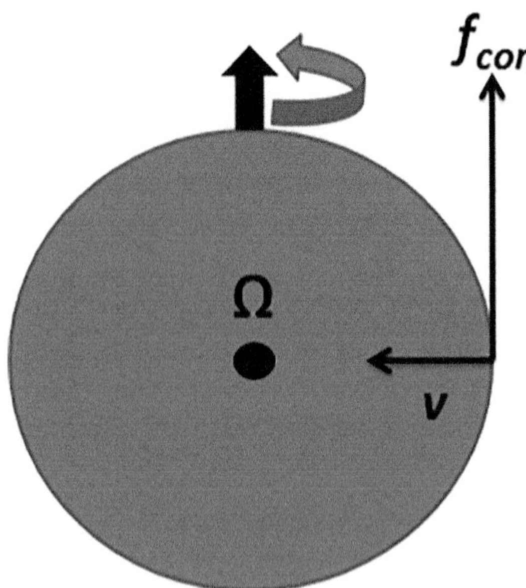

FIG. 11.5 Coriolis force concept.

The potential vorticity is conserved because of a restoring force. In other words, a parcel of fluid latitudinally is a function of a restoring force. A propagation of Rossby wave is the resultant effect of the initial disturbance, the restoring force, and the inertia of the fluid parcel. The resultant effects are caused by atmospheric force or a change in ocean currents. Therefore, the propagation of Rossby waves can be described as:

$$\eta \sim e^{[i(kx+ly-\omega t)]} \tag{11.4}$$

where k and l are the zonal (east-west) and meridional wavenumbers, respectively.

Consequently, the wave frequency (ω) suggests the dispersion relation for zero backgrounds mean flow, which is calculated as:

$$\omega = -\frac{\beta k}{(k^2+l^2+r^{-2})} \tag{11.5}$$

where the local deformation radius of Rossby waves is r, which varies with latitude,ϕ local density stratification, and mode number, and where the minus sign on the right-hand side of Eq. (11.5) [4] indicates that Rossby waves have westward phase velocity.

In this regard, each baroclinic mode has a deformation radius. In other words, mode baroclinic Rossby radius is:

$$r = V_n |f_{cor}|^{-1}, \quad n = 1,2,3,\ldots \tag{11.6}$$

Eq (11.6) suggests that the radius of the baroclinic Rossby wave is a function of the propagation velocity V with the nth mode. The baroclinic Rossby radius is a length scale often associated with boundary currents, fronts, and eddies. In the circumstance of Eqs. (11.5) and (11.6), the longest wavelength of Rossby waves is limited to:

$$k^2 + l^2 \ll r^{-2} \tag{11.7}$$

Eq. (11.7) suggests that Rossby waves are nondispersive. In the circumstances of $l=0$ and $k=\frac{1}{r}$, respectively, Rossby waves have a maximum or cutoff frequency ω_c, which is given by:

$$\omega_c = -0.5(\beta r) \tag{11.8}$$

Rossby waves have a revolving latitude, ahead of which Rossby wave clarifications form farther away from the equator (at mid-latitudes) of the Pacific may take closer to 10–20 years to make the journey. In other words, the Rossby wave propagation rate becomes zero. Therefore, in the dynamic ocean, wind stress curl, buoyancy forcing, coastally trapped waves, and reversals in coastal currents are the forces that generate Rossby waves. Moreover, the dependence of transport on f_{cor} means that on large, "planetary" scales, *variation* in Coriolis causes convergence and divergence to the west and east of eddies such that their pattern propagates westward.

11.3 Rossby waves causing convergence and divergence zones

Rossby waves generate growth of energy in the west of the ocean gyres and create strengthening currents on the western side of the ocean basins. It is necessary to take into account the rate of the thermocline displacement, the convergence, and the divergence processes to calculate the speed of Rossby wave propagations. If we assume that the propagation of the Rossby waves between two latitudes ψ_1 and ψ_2, respectively, in which f_{cor} changes by an amount $\beta\Delta\phi$, i.e., $f_{cor}=f_0+\beta\Delta\psi$ (in other words, the β-plane approximation, which is just a convenient way of representing the variation of f_{cor} with latitude), then β can be identified by:

$$\beta=\frac{\partial f_{cor}}{\partial\psi}=2\Omega\cos\phi\frac{\partial\phi}{\partial\psi} \tag{11.9}$$

where:

$$\frac{\partial\phi}{\partial\psi}=R_{Earth}^{-1} \tag{11.9.1}$$

where R_{Earth} is the radius of the Earth, which is 6371 km.

The β value can be given by:

$$\beta=\frac{2\times 7.292\times 10^{-5}}{6371\times 10^{3\cos\phi}} \tag{11.10}$$

$$\beta=2.28\times 10^{-11\cos\phi}=2.28\times 10^{-11\cos\phi}\,\mathrm{m}^{-1}\mathrm{s}^{-1}$$

For instance, at 20°S, $\beta=2.15\times 10^{-11}\,\mathrm{m}^{-1}\mathrm{s}^{-1}$, while at 40°S, $\beta=1.75\times 10^{-11}\,\mathrm{m}^{-1}\mathrm{s}^{-1}$. Consequently, the net volume convergence N_c between two latitudes ψ_1 and ψ_2 is calculated using:

$$N_c=g\overline{H}\frac{\Delta\rho}{\rho_0}\Delta H\frac{\beta\Delta\psi}{f^2} \tag{11.11}$$

where ρ_0 is seawater density, and g is the gravitational constant.

Eq. (11.11) suggests that the volume convergence must be balanced by a pycnocline that is being driven down at a vertical velocity of $\frac{\partial H}{\partial t}$. Under the circumstance of balancing, Eq. (11.11) becomes:

$$\Delta\psi_1\Delta\psi_2\frac{\partial H}{\partial t}=g\overline{H}\frac{\Delta\rho}{\rho_0}\Delta H\frac{\beta\Delta\psi}{f^2} \tag{11.12}$$

Eq. (11.12) suggests that on the east side, ΔH changes sign and this means that $\frac{\partial H}{\partial t}$ is negative, consistent with a thinning upper layer. In this regard, the vertical displacement of the layer H is calculated as:

$$\frac{\partial H}{\partial t}=\frac{\beta g\overline{H}}{f^2}\frac{\Delta\rho}{\rho_0}\frac{\partial H}{\partial x} \tag{11.13}$$

where x is the longitudinal distance across the ocean.

In this sense, the ratio of $\frac{\partial H}{\partial t}/\frac{\partial H}{\partial x}$ demonstrates the velocity v at which a line of constant H (a wave crest for instance) moves eastward. On the contrary, the planetary eddy pattern moves westward at a speed of $-v$, with:

$$v=\beta\frac{\Delta\rho g H}{\rho_0 f_{cor}^2}\,\mathrm{ms}^{-1} \tag{11.14}$$

Eq. (11.14) demonstrates that the thermocline displacements have small sea surface displacements associated with them, which can be observed using a satellite altimeter. In an interval time of Δt, the Rossby wave travels west with a distance of Δx; consequently, the slope of these lines is calculated via:

$$\frac{\Delta x}{\Delta t} = \frac{\partial H}{\partial t}\left[\frac{\partial H}{\partial x}\right]^{-1} = -v \tag{11.15}$$

For instance, Eq. (11.15) can fit the observations in the Indian Ocean as:

$$v = \frac{\partial x}{\partial t} = \frac{30°\text{lon} \times 111\,\text{km} \times \cos(25)}{60\,\text{cycle} \times 10\,\text{days} \times 86{,}400\,\text{s}}$$
$$= 5.8 \times 10^{-2}\,\text{m}\,\text{s}^{-1}$$

In the circumstance of the Indian Ocean, which has a thermocline depth of 1000 m, an average surface water density ρ_0 of 1026.6, and depth of 1027.8, the Rossby wave velocity is estimated through Eq. (11.14) as $6.3 \times 10^{-2}\,\text{m}\,\text{s}^{-1}$.

The application of Eq. (11.14) shows that the Rossby wave speed grows larger when it is approaching the equator, always with positive (i.e., westward) propagation. Furthermore, this is an approximate equation for very long wavelength, long period (many months) Rossby waves, for the transition area between the ocean's warm upper layer and colder depths.

11.4 Collinear analysis for modeling Rossby wave patterns from satellite altimeter

A significant question is: how can Rossby waves be simulated using a satellite altimeter? The initial investigations of Rossby wave is determined by Geosat altimeter satellite data. In this view, the information the westward propagation of Sea surface Height signals through Geosat data turned into expressions of annual Rossby Waves [6]. However, these consequences had been particularly ambiguous, owing to the aliasing belongings of the Geosat data. However, the westward propagation of signals has been confirmed utilizing numerous studies based on the usage of T/P altimeter and in situ data, as well as with the aid of numerical models [7–9]. These signals originate off the coast due to the annual fluctuations in the alongshore winds. Thus, along the Australia west coast, the annual and interannual variability in the coastal currents and water mass characteristics are transferred westward, into the deep ocean [10].

Let us assume that I_i is the vector of altimeter measurements in the ith and I_{ij} is the measurements at the jth grid point. In this understanding, N is the number of grid points along the pass, where the mean sea surface height h_j is determined with a certain error, e_{ij}. At the jth grid point, $I_{ij} \in h_j \in e_{ij}$. Then the mean sea surface height is simplified using [11]:

$$h_j = I_{ij} - e_{ij} - C_j x_i \tag{11.16}$$

where C_j is the raw of the vector of model coefficients for the time-dependent radial orbital error. Additionally, C_j is a function of travel time t_j to the jth grid point from some reference point along the pass. The traveling time can thus be expressed by:

$$C_j = \left(1, \sin\left(2x\pi\frac{t_j}{T}\right)\right) \tag{11.17}$$

$$C_j = \cos\left(2x\pi\frac{t_j}{T}\right) \tag{11.18}$$

where T is the orbital period of the satellite.

The linear model used to compile all the N repeat passes is formulated as:

$$I = [E_N \otimes I_n, I_n \otimes C]\begin{bmatrix} h \\ x \end{bmatrix} + e = X\begin{bmatrix} h \\ x \end{bmatrix} + e \tag{11.19}$$

where $\otimes$ is the Kronecker product and E_N, which is the N-vector of that element, is a "1."

Eq. (11.19) can be expressed based on the Gauss-Jordan elimination as [11]:

$$X^g = \begin{bmatrix} N^{-1}E_N^T \otimes I_n \\ \left(I_N - N^{-1}E_N E_N^T \oplus (C^T C)^{-1} C^T\right) \end{bmatrix} \tag{11.20}$$

Eq. (11.20) involves the transfer matrix T for the selected parameters, which are stated above in Eqs. (11.16)–(11.19). The sea surface height variability H_j at grid point j can be computed using:

$$\begin{aligned} H_j = N^{-1} l^T \left(I_{nN} - \begin{bmatrix} I_n & N^{-1}E_N^T \oplus C \\ 0 & (I_n - N^{-1}E_N E_N^T \oplus I_r) \end{bmatrix} \right) \\ (I_N \oplus E_{ij}) \left(I_{nN} - \begin{bmatrix} I_n & N^{-1}E_N^T \oplus C \\ 0 & (I_n - N^{-1}E_N E_N^T \oplus I_r) \end{bmatrix} \right) \end{aligned} \tag{11.21}$$

Therefore, Eq. (11.21) is the inverse form of Eq. (11.20) in grid position i, j being E_{ij} is the (n,n) elementary matrix with "1" [12, 13].

11.5 Rossby wave spectra patterns using fast Fourier transform

The Fast Fourier Transform (FFT) is used to estimate the frequency domain and the wavelength of the Rossby wave. To this end, the FFT is a method that is converted into a wave number and frequency space, which highlights the distinct spectral components. The FFT can be expanded to display single foundations of a wave, which may additionally translate to distinct baroclinic modes [14]. Nevertheless, due to the fact it maps a single wave frequency to every component, it requires that the propagation characteristics of the wave continue to be regular for the vicinity and time being studied [15]. The FFT is used to calculate the wavelength, period, and amplitude of the first-mode baroclinic Rossby waves in the zonal gradient of every parameter. The wavelength and period are calculated by taking the inverse of the frequency components of the peak corresponding to the first-mode Rossby wave, and transforming to kilometers and days respectively. The amplitude is calculated with the aid of dividing the absolute cost of the Fourier Transform utilizing half of the product of the length of the x and y dimensions [16].

Consequently, the transformation of H_j from the time domain to the frequency domain is based on the Fourier transform and its inverse, which are defined as:

$$S(\omega) = \int_{-\infty}^{\infty} s(t)e^{j2\pi f_r t}dt \tag{11.22}$$

where:

$$s(t) = \int_{-\infty}^{\infty} s(f_r)e^{j2\pi f_r t}df_r \tag{11.22.1}$$

where $s(t)$, $S(\omega)$, and f_r are the time signal, the frequency signal, and the frequency, respectively, and $j = \sqrt{-1}$.

Physicists and engineers sometimes prefer to write the transform in terms of angular frequency $\omega = 2\pi f_r$. To restore the symmetry of the transforms, the convention is [17–21]:

$$S(\omega) = \frac{1}{\sqrt{2\pi}} \int_{-\infty}^{\infty} s(t)e^{j\omega t}dt \tag{11.23}$$

where:

$$s(t) = \frac{1}{\sqrt{2\pi}} \int_{-\infty}^{\infty} S(\omega)e^{j\omega t}d\omega \tag{11.23.1}$$

The mathematical description of the Rossby wave amplitude "a" and wavelength L in the angular frequency domain is formulated as:

$$\eta = a \cdot \cos(kx - \omega t) \tag{11.24}$$

where η is the FFT parameter, k is the wavenumber, i.e., $k = \frac{2\pi}{L}$, and angular frequency is ω for a Rossby wave traveling over time t across distance x.

For instance, the Rossby wave spectra derived from the FFT suggest westward propagation with a maximum frequency of 0.2 (1/day) and wavenumber of −0.3 (1/degree) (Fig. 11.6).

11.6 Multiobjective algorithm for modeling Rossby waves in altimeter data

Fitness roles $f_1, f_2, \ldots, f_n$ are one of complicated procedures that involve in a multiobjective optimization (MOO). The MOO procedure aims to obtain optimal solutions. On the other hand, the solutions would be at the slightest acceptable, and reliable with all the events concurrently. On the other hand, the solutions would be at the least acceptable, and reliable with

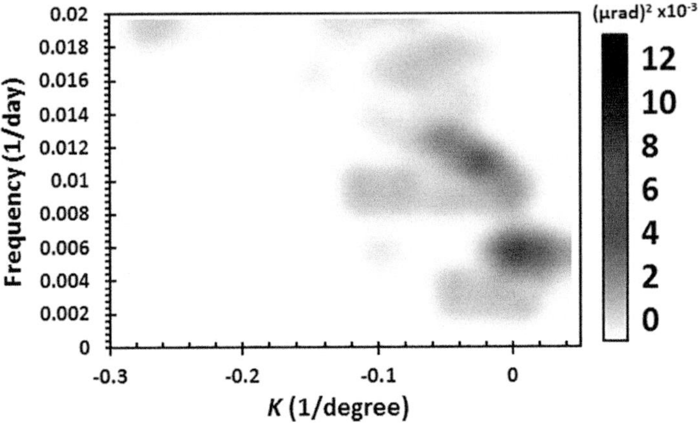

FIG. 11.6 Derived Rossby wave spectra using FFT.

all the events concurrently. In this understanding, the difficulty of altered solutions is one of the furthermost stimulating subjects to be functioned in evolutionary schemes, such as genetic programming.

Generally, the foremost challenge in the multiobjective algorithm is to regulate the optimal solutions into the following mathematical formula:

$$\text{Maximize } f(x) = f_1(x), f_2(x), \ldots, f_m(x) \tag{11.25}$$

$$\text{Subject to } x \in X, \tag{11.26}$$

Eq. (11.26) indicates a search space or *decision space* that presents as x. Therefore, X presents the potential solutions or *decision alternatives*. In this way, there is frequently no additional restraint on the province of the decision alternatives. Let us consider that $Y = f(X)$ presents the *objective space* as the component $y \in Y$ presents an *objective vector* and its constituents, *objective values*. Consequently, the objective function $f: X \mapsto \mathbb{R}^m$, $m \in \mathbb{N}$ is well-known as the practical association between the decision alternatives and the decision criteria. To comprehend thoroughly multiobjective optimization problems (MOOP), algorithms, and principles, particular definitions must be explained.

11.6.1 Decision adjustable and objective space

The variable constraints of an optimization problem bound distinct decision constraints to a lower and upper bound. Alternatively, it presents a space recognized as the decision variable space. In multiobjective optimization, consequently, the number of element functions creates a multidimensional space recognized as an objective space. The individual decision element in element space is thus similar to a target in impartial space.

11.6.2 Feasible and infeasible solutions for Rossby waves

Multiobjective algorithms comprise dual categories of solutions that are feasible and infeasible. A feasible solution fulfills all linear and nonlinear constraints. Infeasible elucidation presents a solution that would fulfill all the constraints and dissimilarity element constraints and equivalence. Conversely, the infeasible solution contains a solution that does not influence all element constraints and bounds. On the other hand, a specific solution may be infeasible that does not involve infeasible problems. Nevertheless, infeasible problems do happen.

11.6.3 Ideal objective vector

Let us assume that v^{*i} is a vector of variables that optimizes, maximizes, or minimizes the ith objective, which accesses a multiobjective optimization problem. This can be expressed mathematically by:

$$\exists v^{*(i)} \in \Omega, v^{*(i)} = \left[v_1^{*(i)}, v_2^{*(i)}, v_3^{*(i)}, \ldots\ldots, v_N^{*(i)}\right] : f_i\left(v^{*(i)}\right) = OPTf_i(v) \tag{11.27}$$

Then, the vector V can be given by:

$$V^* = f^* = \left[f_1^*, f_2^*, f_3^*, \ldots, f_N^*\right]^T \tag{11.28}$$

where f_N^* is the optimum of the Nth objective function, which is the ultimate for a problem of multiobjective optimization.

Moreover, the perfect solution of this vector can be resolved by the object fit in $\Re^N$. Basically, the ultimate objective vector is attained if a vector has slighter constituents than that of a perfect objective vector. On the other hand, the ultimate objective vector is described as:

$$\forall i = 1,2,3,4,\ldots,n \ : V_i^{**} = V_i^* - \varepsilon_i, \ \ \varepsilon_i > 0. \tag{11.29}$$

11.6.4 Domination

Contradictory objects are the foremost concern in most real-world applications. In this way, optimizing a solution approximately, one object cannot be a consequence of a perfect solution regarding the other objects.

Let us assume that $\triangleright$ is the operator between binary solutions s_i and s_j, i.e., $s_i \triangleright s_j$ for N objective MOP. Conversely, it clarifies that a solution s_i is more precise than the solution s_j to determine a confident objective. In this viewpoint, the description of domination for both minimization and maximization MOP can be specified as a feasible solution $s^{(1)}$ is predictable to govern alternate feasible solution $s^{(2)}$ as, $s^{(1)} \leq s^{(2)}$, in the following circumstances:

(i) The solution $s^{(1)}$ is no poorer than $s^{(2)}$ with style to all objectives rate. On the other hand, mathematical manifestation of feasible solution can be specified by:

$$f_j\left(s^{(1)}\right) \triangleright \mathrm{f_j}\left(\mathrm{s}^{(2)}\right) \ \ \text{for all } \mathrm{j} = 1,2,3,\ldots,\mathrm{N} \tag{11.30}$$

(ii) The first feasible solution $s^{(1)}$ is decisively consistent than the second feasible solution $s^{(2)}$ within the tiniest one objective value. This can be formulated precisely as:

$$f_i\left(s^{(1)}\right) \triangleleft \mathrm{f_i}\left(\mathrm{s}^{(2)}\right) \ \ \text{for at least one } \mathrm{i} \in \{1, 2, 3, \ldots, \mathrm{n}\}. \tag{11.31}$$

In these statuses, the solution $s^{(1)}$ governs the solution $s^{(2)}$. Nonetheless, the solution $s^{(1)}$ is not governed by the solution $s^{(2)}$. Alternatively, the solution $s^{(2)}$ is governed by the solution $s^{(1)}$.

From a mathematical point of view, a multiobjective problem involves optimizing. Optimization is minimizing or maximizing numerous objectives synchronously with a quantity of dissimilarity or equality constraints. The Rossby wave problem can be written mathematically as:

$$\text{Find } v=(v_i)\ \forall i=1,2,\ldots,N_{param} \text{ such as} \tag{11.32}$$

$$f_i(v) \text{ is a minimum}, \forall i=1,2,\ldots,N_{obj} \tag{11.33}$$

subject to:

$$v=v_j-\frac{\beta}{R_1^{-2}+R_2^{-2}}+\frac{v_i-v_j}{1+\dfrac{H_2}{H_1}} \tag{11.34}$$

$$R_n=\frac{\sqrt{gH_n}}{f_{cor}} \tag{11.35}$$

where v is the Rossby wave phase propagation in i and j directions as a function of the internal Rossby radius R_n for the nth layer, H_n is the layer thickness, and β is the meridional of the derivative of f_{cor}.

The potential velocity ϕ, which is a result of the Rossby wave propagation from the surface to the layer thickness of H_n, is assumed to accelerate the descent of water mass parcels to Z sinking depth.

$$\phi_j(v)\geq 0\ \ \forall j=1,2,3,\ldots,M \tag{11.36}$$

$$Z_k(v)\geq 0\ \ \forall k=1,2,3,\ldots,K \tag{11.37}$$

Eqs. (11.36) and (11.37) suggest that the Rossby wave propagation increases the potential velocity. In this view, v is a vector containing the N_{param} design, parameters $f_i((v))_{i=1,\ldots,N_{obj}}$ the objective functions, and N_{obj} the number of objectives. In this investigation, inequality constraints are only deliberated and are suggested as bounded domains. In other words, upper and lower limits are executed on all parameters:

$$-Z\in\left[v_{i,\min};v_{i,\max}\right]\ \ i=1,\ldots,N_{param}. \tag{11.38}$$

The evolutionary algorithm is used to solve Eqs. (11.36)–(11.38). To this end, an initial population of individuals (solutions) of retrieving velocity potential for Jason-2 data is generated from the matter domain, which then undergoes evolution by suggesting the reproduction, crossover, and mutation of individuals until an appropriate resolution exists.

Hence, similar to most alternative evolutionary algorithms, the genetic algorithm demands that only the parameters of the matter be quantified. Subsequently, the algorithmic program is applied to attain a solution, which is usually a problem-independent. Genetic algorithms typically represent all solutions within the sort of fastened length character strings analogous to the

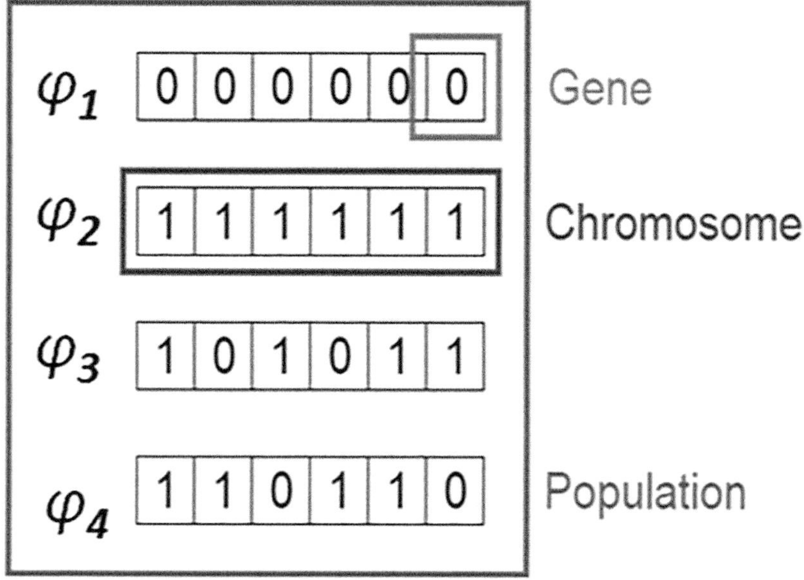

FIG. 11.7 Bit-string GA representation of velocity potential.

DNA found in living organisms. The rationale for the fastened length character strings is to permit easier manipulation, storage, modeling, and implementation of the genetic rule [22].

In this regard, binary numbers jointly afford straightforward conversion to and from the precise solution. Conversely, since there are infinite real numbers between 1 and 2, fixed-length strings cause further weakness for the computer programmer. To unravel this, the real number range should be discretized into a finite variety of constituent real segments, resembling every binary number utilized in the character string. Suppose that the character strings have a length of $n=10$. Then the potential values of the character string of the sinking depth of water mass parcels would be from 0000000000 to 1111111111 (Fig. 11.7) [22].

11.7 Rossby wave population of solutions

A collection of potential solutions is unbroken throughout the life cycle of the genetic algorithmic program. This assortment is mostly called the population since it is analogous to a population of living organisms. The population, typically, is either of fastened or variable size; however, fastened size populations are generally used so that the precise quantity of computer resources will be predetermined. The population of solutions is held in main memory or on external storage, depending on the sort of genetic algorithm program and computer resources accessible.

At the start of the algorithmic program, a population of solutions is generated randomly. In the case of the square root problem, a hard and fast variety of 10 character binary strings are generated randomly. This population is then changed through the mechanisms of evolution

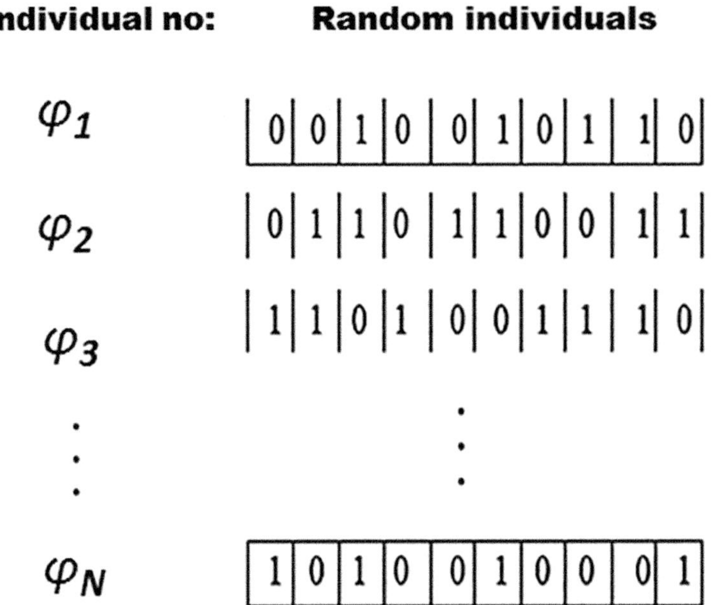

FIG. 11.8 Initial populations.

to result eventually in individuals that are nearer to the solution than these initial random ones (Fig. 11.8).

Fig. 11.8 represents the initial populations of 1000 for velocity potential individual generations from Jason-2 data. A binary number between 0 and 1 is randomly generated and sorted as a string in the computer memory. It can also be seen that they are not sorted in order, but randomly, which represents only the row of the Jason-2 data.

One of the explanations for the exploitation of binary numbers is to forbid incorrectly formatted solutions automatically. In binary, it is easier to visualize some characteristics of water density being present (by a 1) or absent (by a 0). This can also be applicable to nonnumeric problem domains. There are only two potential binary values of velocity (1 and 0). This suggests that every one of the potential binary values is often created by these two values. Consequently, the binary individuals of sinking water depth of any water mass parcels ranging between 0000000000 and 1111111111 contain all the genetic material attainable, i.e., they span the solution space (Fig. 11.8).

11.8 Fitness procedures for simulation of Rossby wave patterns

Consequently, the range of the population is sustained by a widespread fitness sharing function. Satisfying the N determination is used to determine the Pareto solution. To this end, the blended crossover is exercised to generate children on fragments identified via two parents and a specific parameter. In this optimization, new plan variables of Rossby wave potential vorticity have a weight common as given by:

$$Ch_1 = \varpi^* P_1 + (1 - \varpi)^* P_2 \tag{11.39}$$

$$Ch_2 = (1 - \varpi)^* P_1 + \varpi^* P_2 \tag{11.40}$$

where $\varpi = (1 + 2\ell)\_ran_1 - \ell$, Ch_1, and Ch_2 are child 1 and 2, and P_1 and P_2 are parent 1 and 2, which represent programmed scheme variables of the members of the new population and a reproduced pair of the old generation. Therefore, *ran* is a random number that is uniform in [0,1]. When the mutation takes place, Eqs. (11.39) and (11.40) can be given as follows:

$$Ch_1 = \varpi^* P_1 + (1 - \varpi)^* P_2 + \alpha(\text{ran}_2 - 0.5) \tag{11.41}$$

$$Ch_2 = (1 - \varpi)^* P_1 + \varpi^* P_2 + \alpha(\text{ran}_2 - 0.5) \tag{11.42}$$

where ran_2 is a random number that is uniform in [0,1], and α is set to 5% of the given range of each variable (Fig. 11.9).

Subsequently, since the potential vorticity is a feature of Rossby wave propagation, its diagram parameters have to be addressed predictably. Otherwise, the computation deviates and infinite population cannot be weighed. Consequently, if set to 0.0, then mutation takes place at a likelihood of 10% [22].

In line with Sivanandam and Deepa [23], a genetic algorithm is commonly a characteristic of the reproducing step, which entails the crossover and mutation techniques in altimeter data. In the crossover step, the chromosomes interchange genes. A local fitness value results in every gene as given by:

$$f\left(P_i^j\right) = \left|\phi - P_i^j\right| \tag{11.43}$$

where ϕ is the velocity potential, $f(P_i^j)$ is local fitness value for every gene, and P_i^j is a probability variation along i and j, respectively.

In this regard, the fitness values are changed between 0 and 1 (Fig. 11.10) with iteration increments. The highest fitness value of 0.8 does not provide clear information about ϕ

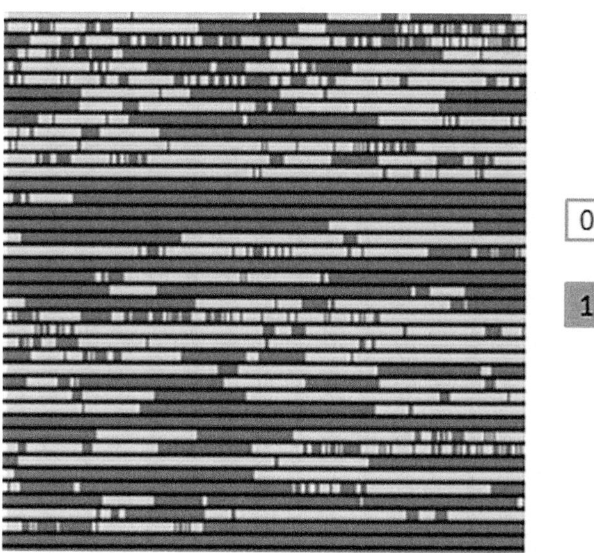

FIG. 11.9 A genetic algorithm for initial velocity potential generation.

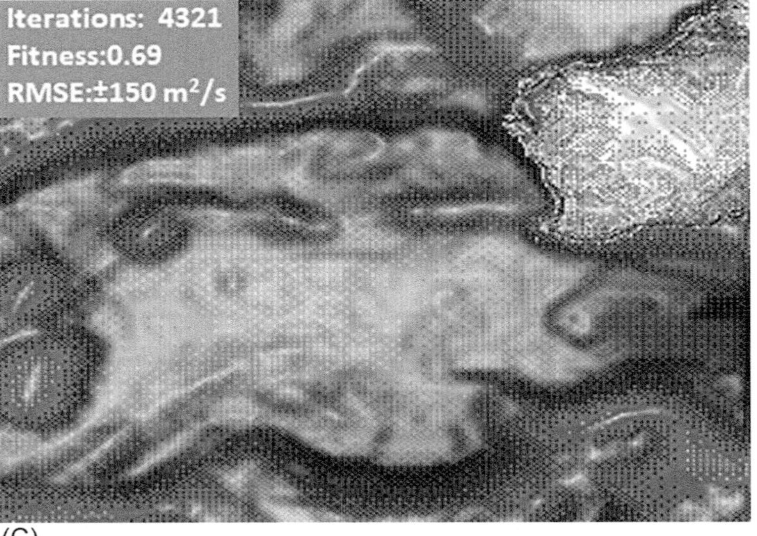

FIG. 11.10 Fitness variations with iterations (A) 0.9, (B) 0.78, and (C) 0.69.

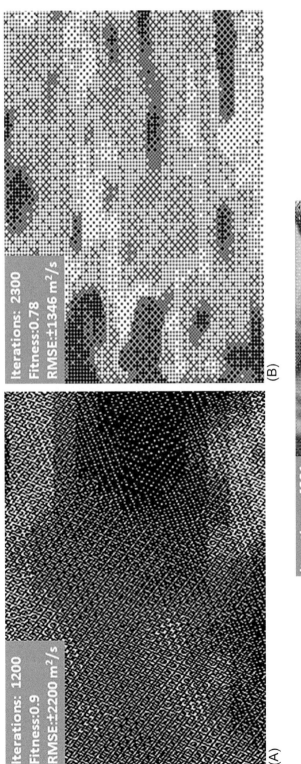

(Fig. 11.10A). As the fitness is gradually decreased with iteration increments, the ϕ is calculated (Fig. 11.10B). The lowest fitness value suggests a clear ϕ pattern (Fig. 11.10C).

Fig. 11.10 shows the fitness of 0.69 and 0.9. In fact, the standardized fitness attempts to restrict the fitnesses to the range of positive real numbers only. The adjusted fitness changes the fitness value so that it lies strictly within the 0–1 range. Furthermore, Fig. 11.10 indicates whether the algorithm is convergent or not. If there is visible convergence and no solution has yet been found, then the algorithm can be extended over more generations. If convergence is not reached, then the parameters of the run can be tweaked to better suit the problem domain.

11.9 Cross-over and mutation for Rossby wave reconstruction from altimeter data

The crossover and mutation procedures that are described in the following sections; which are difficult subsequent steps of a genetic algorithm [22]. The crossover operator is constructed to converge around options with excessive fitness. Thus, the closer the crossover probability is to 1, the faster is the convergence [22–26]. Then the crossfire between two individuals consists of keeping all individual populations of the first parent that have a local fitness greater than the average local fitness $f(P^j_{av})$ and substituting the remaining genes with the corresponding ones from the second parent. Hence, the average local fitness is defined by:

$$f\left(P^j_{av}\right) = \frac{1}{K}\sum_{i=1}^{K} f\left(P^j_i\right) \tag{11.44}$$

Hence, the mutation operator denotes the phenomena of the greatest probability of the evolution process. In this circumstance, the fitness value is reduced to 0.49, which is a clear feature of the west Australian shelf and ϕ is well identified (Fig. 11.11). In addition, this step improves the fitness procedure by showing the lowest RMSE value of $\pm 0.03\,\mathrm{m}^2/\mathrm{s}$ compared to Fig. 11.10.

Some valuable genetic procedures involving the chosen population may be required to replace the duration of the reproducing step. As a result, the mutation operator introduces new genetic facts to the ordinary gene. Generally, the genetic algorithm will take binary match individuals and mate them (a process referred to as crossover). The offspring of the mated pair will acquire some of the traits of the parents. The methods of selection, crossover, and mutation are called genetic operators [22, 26].

11.10 Velocity potential patterns in the southern Indian Ocean from Jason-2

The Jason-2 altimeter satellite data are used to reveal the sea surface height anomaly (SSHA) data. It has a spatial and temporal resolution of 9.9156 days. These data are retrieved with an accuracy of 0.25–0.4 m. The data cover 1 year from March 2014 to March 2015 along the Southern Indian Ocean. The data, consequently, are obtained from "www.aviso.oceanobs.com" These data were used to retrieve the Rossby wave pattern on the western shelf of Australia.

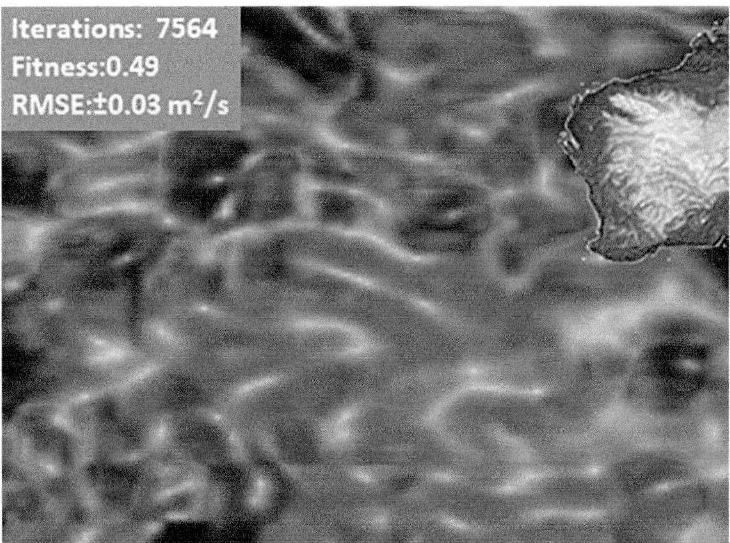

FIG. 11.11 Crossover improves fitness value.

Fig. 11.12 shows the synoptic information retrieved of velocity potential from Jason-2 satellite data, which is appropriately simulated using the multiobjective genetic algorithm. The Rossby wave patterns are dominant in the southern Indian Ocean from March 2014 to March 2015. In this context, the velocity potential suggests rotational vorticity toward the westward with $100 \times 10^5 \mathrm{m}^2 \mathrm{s}^{-1}$ from March to May 2014.

The velocity potential continues with $-80 \times 10^5 \mathrm{m}^2 \mathrm{s}^{-1}$ from June to August 2014. The potential flow rotates anticlockwise toward the west (Fig. 11.13). The distance of 500 km is dominated by a strong potential flow, which increases toward the west.

The velocity potential suggests an irregular irrotational flow from September to October 2014 (Fig. 11.14). The west Australian shelf is dominated by the lowest velocity potential of $-40 \times 10^5 \mathrm{m}^2 \mathrm{s}^{-1}$. On the contrary, the maximum velocity potential records from November to December 2014 where the highest retrieved: ϕ of $120 \times 10^5 \mathrm{m}^2 \mathrm{s}^{-1}$ (Fig. 11.15).

From January to March 2015, the latitude of 25°S to 40°S is dominated by a westward strong velocity potential of $120 \times 10^5 \mathrm{m}^2 \mathrm{s}^{-1}$ (Fig. 11.16). This pattern flow continues the widespread of potential vorticity along the west Australian shelf.

The change in the velocity potential pattern is created due to the propagation of the Rossby wave across the southern Indian Ocean. Indeed, the Hovmöller diagram of the SSH (Fig. 11.17) shows the Rossby waves propagating westward. This requires accurate steric height estimation from satellites because various kinds of studies of the oceans can be performed with these data. Furthermore, SSH and pressure are related, therefore, the height field is also related to the velocity field. Similar to the atmosphere, the balance of pressure gradient force, Coriolis force, and the frictional force is required to determine the kind of circulation that would dominate the ocean depths. Indeed, the time evolution of ocean circulation depends on the external factors, viz., wind stress, heating and cooling, evaporation, and precipitation. Hence, the ocean temperature, salinity, and other properties depend on these three external

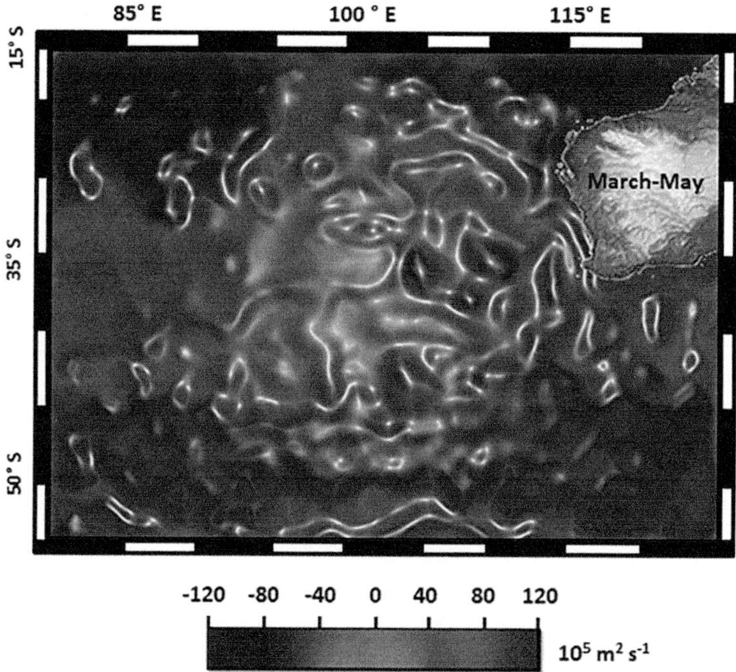

FIG. 11.12 Velocity potential from March to May 2014.

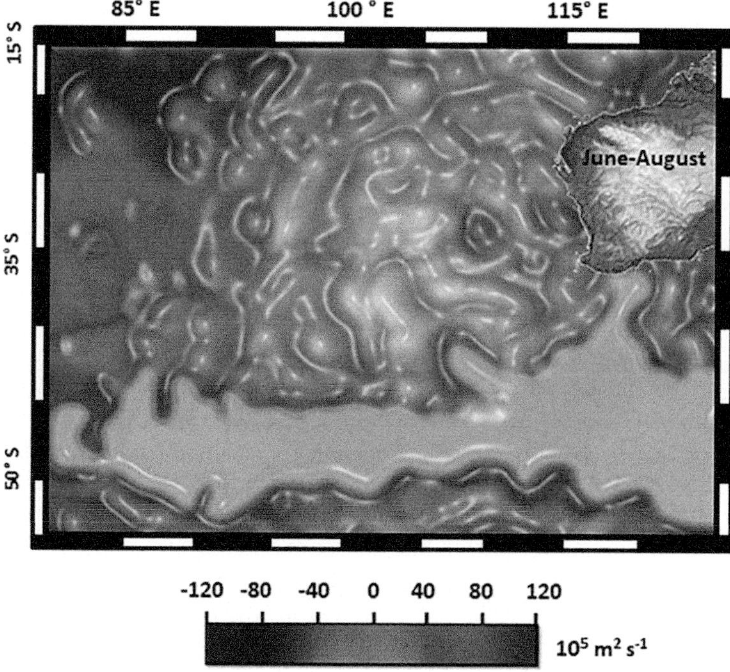

FIG. 11.13 Velocity potential from June to August 2014.

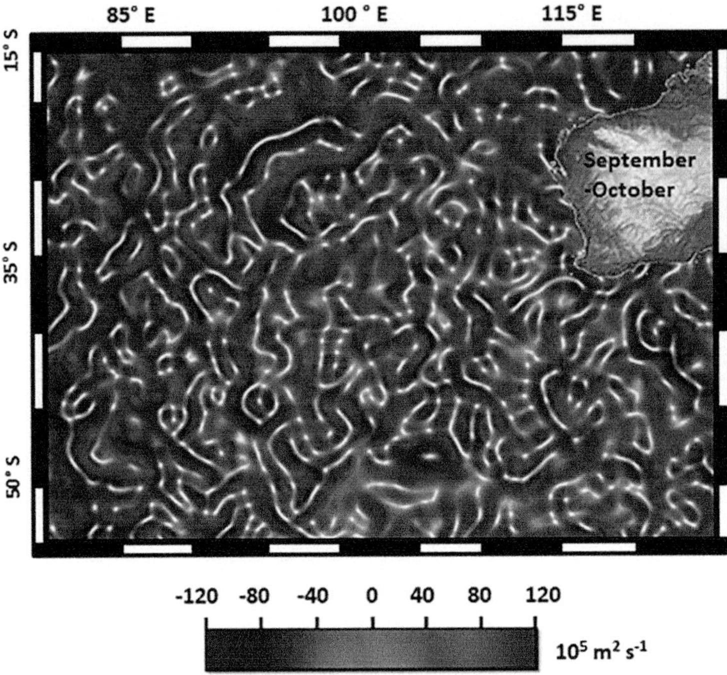

FIG. 11.14 Velocity potential from September to October 2014.

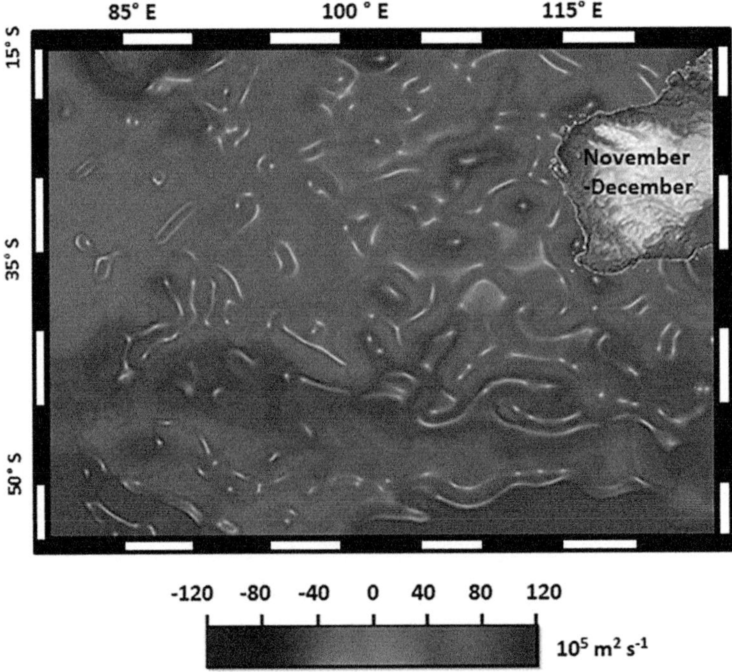

FIG. 11.15 Velocity potential from November to December 2014.

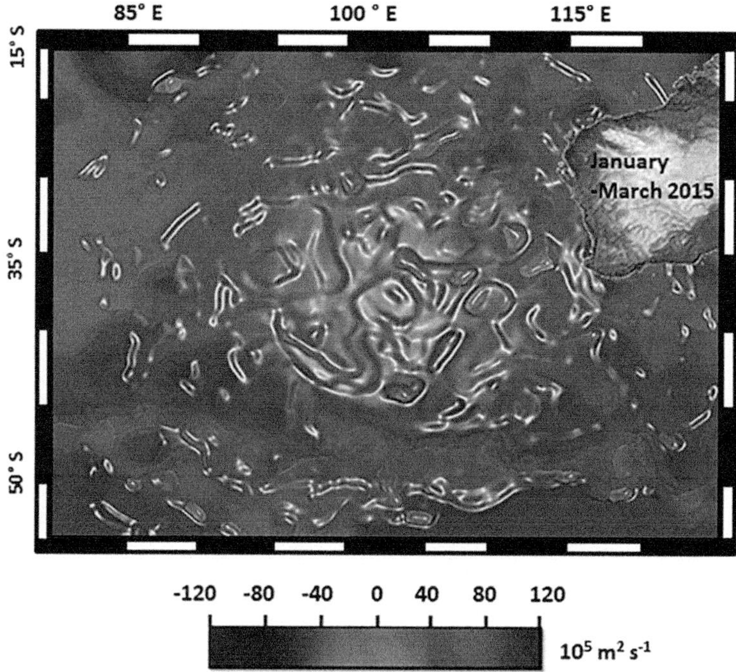

FIG. 11.16 Velocity potential from January to March 2015.

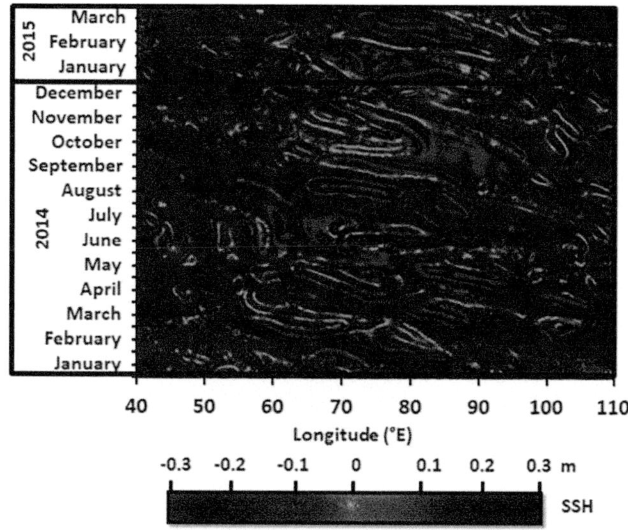

FIG. 11.17 Rossby wave propagation simulated from Jason-2 satellite data.

factors and also on how the moving seawater parcels would balance under the combined action of these forces. The solution of the β-plane linear Rossby wave equation represents long periodic oscillations with periods of several months or longer (Fig. 11.17).

It is worth mentioning that the impact of the rotation vorticity is the dominant feature of Rossby wave propagation under the circumstance of $f_{cor} \rightarrow 0$, $R \rightarrow \infty$. In this context, the vorticity occurs because of the large length scale comparable with R (Figs. 11.12–11.17).

11.11 Pareto algorithm simulation of water parcel sinking due to vorticity potential velocity

In the decision fluctuating space, the solution uncertainty is not governed by any other solution is termed as the Pareto-optimal solution. In this framework, the Pareto-optimal is the well-known (optimal) solution about entire objectives. Additionally, it cannot be finalized in some objective deprived of fading the alternative objective. In this considerate, the group of completely feasible solutions that are nongoverned by slightly extra solution is recognized as the Pareto-optimal or nondominated set. Contrariwise, the overall Pareto-optimal set is predictable when the nondominated set is bordered by the precise feasible solution. The Pareto-optimal (P_o) is accurately formulated by:

$$P_o = \{s \in \Omega \neg \exists s' \in \Omega \;\; f(s') \leq f(s)\}. \tag{11.45}$$

The objective rate functions associated with each clarification of a Pareto-optimal set in objective space are identified as the Pareto-front (Fig. 11.18). In this concern, a representative Pareto-front of a binary objective declining grouping optimization obstacle in detached space.

On the other hand, the Pareto-front (P_F) is mathematically expressed as:

$$P_F := \{f(s) | s \in P_o\}. \tag{11.46}$$

Eq. (11.46) demonstrates that as long as the idea of domination permits evaluation of the solutions regarding multiobjectives, the majority of multiobjective optimization algorithms

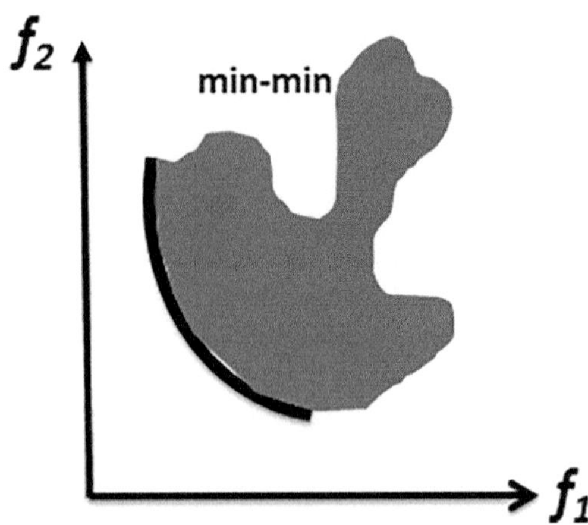

FIG. 11.18 Concept of Pareto-front.

exercise this principle to acquire the nondominated set of solutions, as a result of the Pareto-front.

An important question is: did the Rossby wave strength have an impact on the submerged water mass parcel or any heavy debris? The answer to this question can be provided by the Pareto-optimal solution. In a multiobjective problem, the set of constraints, i.e., the individuals in the evolutionary algorithm terminology, can be compared according to Pareto's rule. In this way, the individual sinking depth Ψ_s dominates the individual velocity potential ϕ, if for at least one of the objectives, $\Psi_s \in H$ and $H \in \phi$. Subsurface water mass Ψ_s such as thermocline layer and deep water mass layertakes a place under the constrain circumstances of:

$$\max \Psi_s = \max \sqrt{2\phi \frac{H}{U}} \tag{11.47}$$

$$\text{Subject to } \Psi_s \in \phi \in H \tag{11.48}$$

$$\begin{aligned} &U(i) < 0 < -1 \quad i = 1,2,3,\ldots k \\ &\Psi_s \leq H \leq 0 \end{aligned} \tag{11.49}$$

Eq. (11.47) can achieve a true multiobjective when the Ψ_s is dominated by other multiobjectives. In other words, the other multiobjectives of velocity potential, the velocity of the Rossby wave, and thickness layer H are considered independent of Ψ_s. Offshore of the west of Australia shelf, the Rossby wave propagates with maximum vorticity velocity of $120 \times 10^5\,m^2\,s^{-1}$ (Fig. 11.19). Under this circumstance, any water mass parcels or heavy debris must sink in a water depth of less than 1000 m (Fig. 11.20).

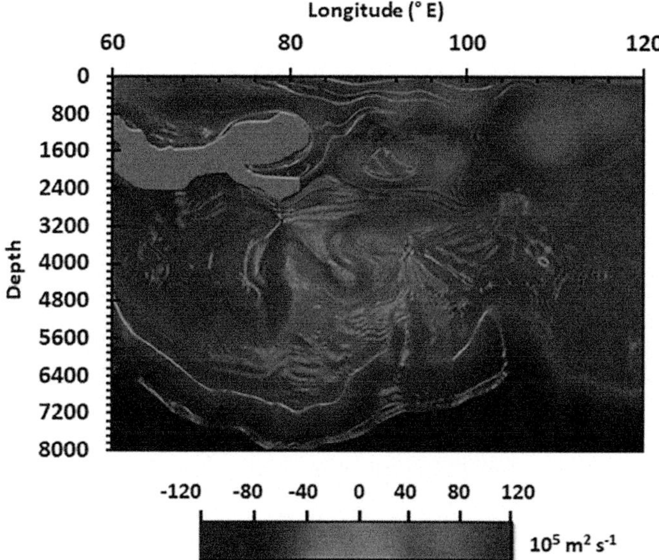

FIG. 11.19 Rossby waves traveling through the deepest water west of the Australian shelf.

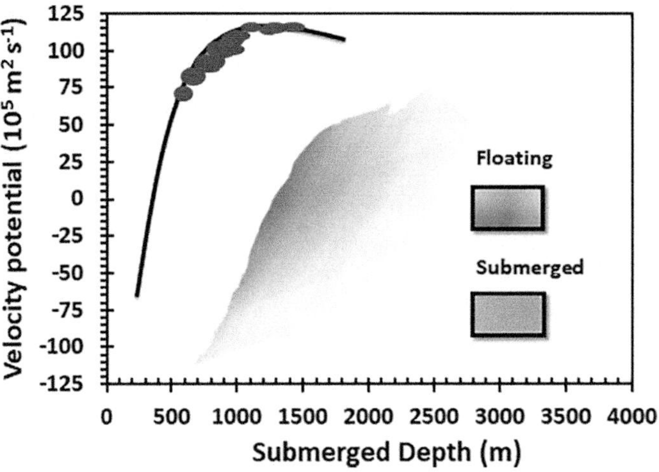

FIG. 11.20 Pareto-optimal for sinking water depth due to Rossby waves.

Pareto front curve confirms that the submerged intensity of water mass will increase as the velocity potential, which generated by Rossby wave in the southern Indian Ocean (Fig. 11.19). In this circumstance, the water mass parcels or any heavy debris should submerge to a water depth of more than 1000m. In this view, the velocity potential of approximately $120 \times 10^5 \text{m}^2 \text{s}^{-1}$ in addition to wave power of 22,000kJ/m/wave must lead to the sinking of any water mass parcels or heavy debris between the longitude of 99°E and 110°E and the latitude of 22°S and 28°S. This proves by using the usage of the submerged water mass parcel is placed absolutely at the Pareto front curve rather than floating object. In this understanding the Rossby wave forms turbulent flow; which cause any object or any water mass parcel to sink through the water column (Fig. 11.20). Consistent with Fig. 11.21, any debris can drift on the surface of the water for no longer than 3months. This is confirmed also by the impact of the velocity potential pattern from the surface to 8000m across the southern Indian Ocean. This reveals that this zone of the southern Indian Ocean is dominated by strong turbulent flow through the water column, which causes downward mixing.

11.12 How can Rossby waves mobilize water mass parcels and heavy debris?

The propagation of the Rossby wave causes the variation of the thermocline. In this regard, the anomalies on the thermocline migrate 1° per 28.01days in the zonal direction. In other words, the changes in thermocline can occur within 3months. At a large scale, Rossby waves are likely to dominate, whereas at a small scale, advection and turbulence dominate. This can break down the thermocline layer [27], which leads to strong large-scale mixing [28]. In this circumstance, the water mass parcels would sink deep due to wave-turbulence crossover. The fluid is swirled at some well-defined scale through the water body, producing an energy input. It is assumed that no energy is lost to smaller scales, i.e., energy cascades to large scales at that same rate. At some point on the scale, the β term in the vorticity equation will start to make its presence felt. By analogy with the procedure for finding the viscous dissipation scale

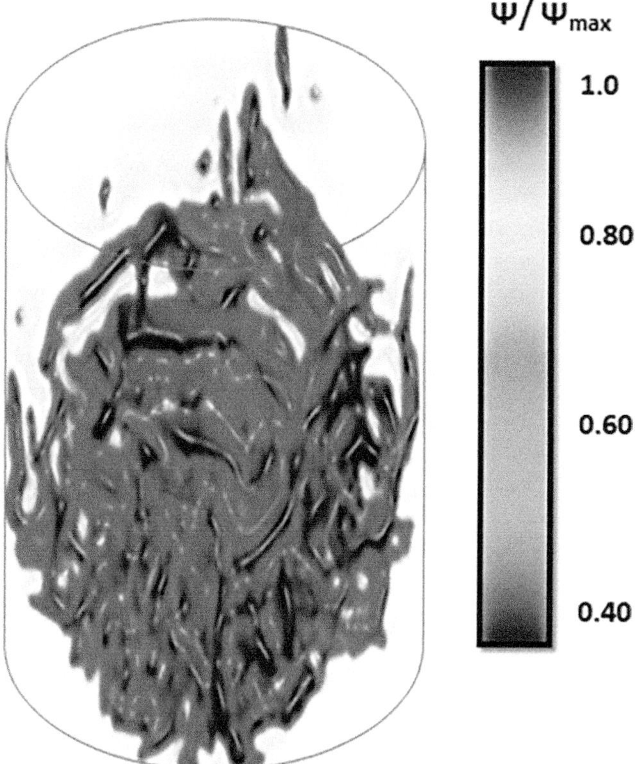

FIG. 11.21 3-D ψ/ψ_{max} associated with the eddy turbulent flow.

in turbulence, we can find the scale at which linear Rossby waves dominate by equating the inverse of the turbulent eddy turnover time to the Rossby wave frequency. Furthermore, the turbulent eddy transfer rate is proportional to $k^{2/3}$ in a $k^{-5/3}$ to the Rossby wave frequency $\beta k^x/k^2$ [29]. In this context, the inverse cascade plus Rossby waves thus leads to a generation of zonal flow (Fig. 11.20). What occurs physically? The region inside the dumbbell shapes in Fig. 11.20 is dominated by Rossby waves, where the natural frequency of the oscillation is higher than the turbulent frequency.

If the flow is stirred at a wavenumber higher than this, the energy will cascade to larger scales, but because of the frequency mismatch, the turbulent flow will be unable to excite modes efficiently within the dumbbell. Nevertheless, there is still a natural tendency of the energy to seek the weightiest mode, and it will do this by cascading toward the $k_x = 0$ axes; that is, toward the zonal flow [30]. Thus, the combination of Rossby waves and turbulence will lead to the formation of zonal flow and, potentially, zonal jets. In other words, at large scales, Rossby waves are likely to dominate whereas at small scales, advection and turbulence dominate, as mentioned above. The combination of small-scale advection and turbulence stirs the water mass parcels or any heavy debris downward along with the mixing layer, which has a thickness of more than 1000 m (Fig. 11.20).

The Antarctic intermediate water inflow creates patches of high potential vorticity at intermediate depths in the southern Indian Ocean, below which the field becomes dominated by planetary vorticity, indicating a weaker meridional circulation and weaker potential vorticity sources [31]. Wind-driven gyre depths have lower potential vorticity gradients, primarily due to same-source waters.

The Indian Ocean adds unique features to the study of potential vorticity with its complicated bathymetry, a prominent link to the Southern Ocean, and reversing (Fig. 11.19) monsoonal currents. In this view, thermocline water masses have unique potential vorticity signals. The changes in the potential vorticity can cause high eddy variability. The active wind-driven currents in the Indian Ocean overlie a strong abyssal circulation, allowing the possibility of comparison between the relative effects of each on potential vorticity. The potential vorticity is mapped on several neutral surfaces to provide a steady-state picture of the potential vorticity as well as to develop a better understanding of the thermohaline circulation's effect on the potential vorticity signature [32–34]. The ratio between vorticity velocity ψ and maximum vorticity velocity ψ_{max} suggests eddy turbulent flow. This eddy turbulent flow has approached 1 at the surface and 0.4 at the bottom. This could be sufficient to submerge any object at a water depth of more than 2500 m across the southern Indian Ocean (Fig. 11.21). The mechanism of potential vorticity impact on the water mass parcels sinking is confirmed by the Pareto front where its ratio velocity of ψ/ψ_{max}, which ranges between 0.6 and 1, pulls the water mass parcel or any heavy debris down to a sinking depth of 3000 m (Fig. 11.22).

Nonetheless, the water mass parcels or heavy debris with different ranges ψ/ψ_{max} are not dominated by the Pareto front. In addition, the sinking domain with ψ/ψ_{max} less than 0.6 is not dominated by the Pareto front, which confirms that $\psi/\psi_{max} \leq 0.6$ can be responsible for the sinking of water mass parcels or heavy debris in a water depth of approximately 3000 m.

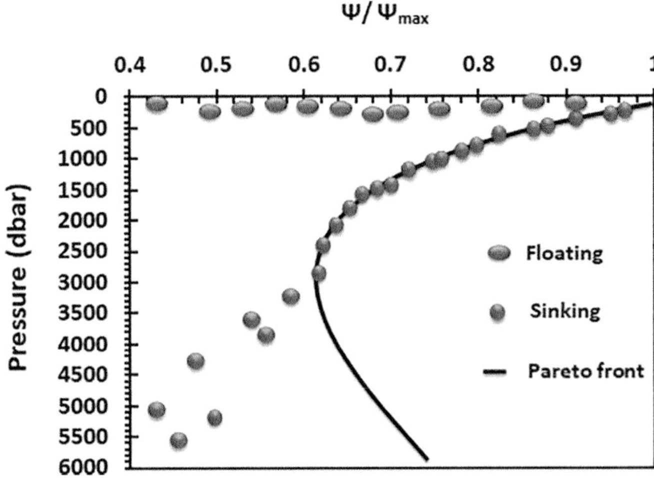

FIG. 11.22 Pareto front for water mass parcels sinking due to potential vorticity.

References

[1] Polvani LM, Alan Plumb R. Rossby wave breaking, microbreaking, filamentation, and secondary vortex formation: the dynamics of a perturbed vortex. J Atmos Sci 1992;49(6):462–76.
[2] Lovelace RVE, Li H, Colgate SA, Nelson AF. Rossby wave instability of Keplerian accretion disks. Astrophys J 1999;513(2):805.
[3] Lorenz EN. Barotropic instability of Rossby wave motion. J Atmos Sci 1972;29(2):258–65.
[4] Woollings T, Hoskins B, Blackburn M, Berrisford P. A new Rossby wave–breaking interpretation of the North Atlantic Oscillation. J Atmos Sci 2008;65(2):609–26.
[5] Ambrizzi T, Hoskins BJ, Hsu H-H. Rossby wave propagation and teleconnection patterns in the austral winter. J Atmos Sci 1995;52(21):3661–72.
[6] White WB, Tai C-K, Dimento J. Annual Rossby wave characteristics in the California Current region from the Geosat exact repeat mission. J Phys Oceanogr 1990;20(9):1297–311.
[7] Vivier F, Kelly KA, Thompson LA. Contributions of wind forcing, waves, and surface heating to sea surface height observations in the Pacific Ocean. J Geophys Res Oceans 1999;104(C9):20767–88.
[8] Strub PT, James C. Altimeter-derived variability of surface velocities in the California Current System: 2. Seasonal circulation and eddy statistics. Deep-Sea Res II Top Stud Oceanogr 2000;47(5–6):831–70.
[9] Kelly KA, Anne Thompson L. Scatterometer winds explain damped Rossby waves. Geophys Res Lett 2002;20:29.
[10] Marchesiello P, McWilliams JC, Shchepetkin A. Equilibrium structure and dynamics of the California Current System. J Phys Oceanogr 2003;33(4):753–83.
[11] van Gysen H, Coleman R, Morrow R, Hirsch B, Rizos C. Analysis of collinear passes of satellite altimeter data. J Geophys Res Oceans 1992;97(C2):2265–77.
[12] Van Gysen H, Coleman R. On the analysis of repeated geodetic experiments. J Geod 1999;73(5):237–45.
[13] Van Gysen H, Coleman R, Hirsch B. Local crossover analysis of exactly repeating satellite altimeter data. J Geod 1997;72(1):31–43.
[14] Subrahmanyam B, Robinson IS, Blundell JR, Challenor PG. Indian Ocean Rossby waves observed in TOPEX/POSEIDON altimeter data and in model simulations. Int J Remote Sen 2001;22(1):141–67.
[15] Cipollini P, Cromwell D, Challenor PG, Raffaglio S. Rossby waves detected in global ocean colour data. Geophys Res Lett 2001;28(2):323–6.
[16] Subrahmanyam B, Heffner DM, Cromwell D, Shriver JF. Detection of Rossby waves in multi-parameters in multi-mission satellite observations and HYCOM simulations in the Indian Ocean. Remote Sens Environ 2009;113(6):1293–303.
[17] Sevgi L, Akleman F, Felsen LB. Groundwave propagation modeling: problem-matched analytical formulations and direct numerical techniques. IEEE Antennas Propag Mag 2002;44(1):55–75.
[18] Sevgi L. Complex electromagnetic problems and numerical simulation approaches. John Wiley & Sons; 2003.
[19] Stoica P, Moses RL. Introduction to spectral analysis. vol. 1. Upper Saddle River, NJ: Prentice Hall; 1997.
[20] Stearns SD, Ahmed N. Digital signal analysis. IEEE Trans Syst Man Cybern 1976;10:724.
[21] Harries D, McHenry M, Jennings P, Thomas C. Hydro, tidal and wave energy in Australia. Int J Environ Stud 2006;63(6):803–14.
[22] Marghany M. Advanced remote sensing technology for tsunami modelling and forecasting. CRC Press; 2018.
[23] Sivanandam SN, Deepa SN. Genetic algorithm optimization problems. In: Introduction to genetic algorithms. Berlin, Heidelberg: Springer; 2008. p. 165–209.
[24] Michalewicz Z. Genetic algorithms+data structures. Evolution programs. New York: Springer-Verlang; 1994.
[25] Mohanta RK, Sethi B. A review of genetic algorithm application for image segmentation. Int J Comput Technol Appl 2011;3(2):720–3.
[26] Deb K, Agrawal S, Pratap A, Meyarivan T. A fast elitist non-dominated sorting genetic algorithm for multi-objective optimization: NSGA-II, In: International conference on parallel problem solving from natureBerlin, Heidelberg: Springer; 2000. p. 849–58.
[27] You Y, Tomczak M. Thermocline circulation and ventilation in the Indian Ocean derived from water mass analysis. Deep-Sea Res I Oceanogr Res Pap 1993;40(1):13–56.
[28] Talley LD. Antarctic intermediate water in the South Atlantic. In: The South Atlantic. Berlin, Heidelberg: Springer; 1996. p. 219–38.
[29] You Y. Seasonal variations of thermocline circulation and ventilation in the Indian Ocean. J Geophys Res Oceans 1997;102(C5):10391–422.

[30] Szoeke d, Roland A, Chelton DB. The modification of long planetary waves by homogeneous potential vorticity layers. J Phys Oceanogr 1999;29(3):500–11.
[31] Dewar WK. On the potential vorticity structure of weakly ventialted isopycnals: a theory of subtropical mode water maintenance. J Phys Oceanogr 1986;16(7):1204–16.
[32] Dewar WK. On "too fast" baroclinic planetary waves in the general circulation. J Phys Oceanogr 1998;28 (9):1739–58.
[33] Holland WR, Keffer T, Rhines PB. Dynamics of the oceanic general circulation: the potential vorticity field. Nature 1984;308(5961):698.
[34] Killworth PD, Chelton DB, de Szoeke RA. The speed of observed and theoretical long extratropical planetary waves. J Phys Oceanogr 1997;27(9):1946–66.

CHAPTER

12

Nonlinear sea surface current mathematical and retrieving models in synthetic aperture radar

12.1 Introduction

To date, there are few theories regarding synthetic aperture radar (SAR) imaging of ocean currents. The existing SAR theories are largely concerned with ocean waves, as discussed earlier, leaving SAR ocean current imaging mechanisms requiring full attention. Ocean currents play a tremendously important role in world climate fluctuations. For instance, currents influence the Earth's climate by creating the warm water flow from the Equator and cold water from the poles around the Earth. The warm Gulf Stream, for example, carries milder winter weather to Bergen, Norway, than that experienced by New York, which lies considerably further south.

The SAR ocean current imaging theory is restricted to the Doppler frequency theory without further development except for along-track interferometry for imaging ocean surface currents. This chapter presents the techniques used for measuring ocean currents, modeling, and retrieving information from SAR satellite data. The chapter demonstrates these techniques using Sentinel-1A satellite data from its overpass of the east coast of Malaysia on November 9, 2019.

12.2 What is meant by ocean current?

It is easy to observe ocean waves moving toward the shore, but we cannot see a current. In the oceans and other water bodies, the motion of the water is defined by currents. From the point of view of wave-particle duality, it only considers a water current as a horizontal particle movement casing by its wave propagation. In this specific view, the water particles spin either horizontally or vertically to represent the current motion. In other words, when measured, the water particles' spin will be "up," or "down," or "mixed," in which case the spin

https://doi.org/10.1016/B978-0-12-820785-7.00016-2

might be 40% probability of "up" and 60% probability of "down." In this regard, the term "current" describes the spin motion of the ocean or water particles. However, what are the forces that spin the water particles to create a current motion?

12.3 Ocean current theory

Newton's laws of motion are the key to understanding the forces that induce an ocean current. It is worth noting that Newton's laws of motion are three physical laws that, taken together, laid the groundwork for conventional mechanics. They designate the association between a body and the forces acting upon it, and its motion in response to those forces. However, they are nonrelativistic equations that do not take relativity into account; and as long as ocean particles have a path from one location to another, relativistic quantum mechanics must be taken into account to describe this motion. On this understanding, ocean current is quantized and signifies a field by imagining each water particle conveyed by a swiftly spinning time. Its influence specifies the particle's phase and the probability of taking a particular path. Space-time, consequently, is considered to designate the forces to induce ocean current, rather than thinking of the ocean current in the space and in time independently [1–4].

Oceanic currents are driven by three main causes, as follows.

(i) The rise and fall of the tides

Tides generate a current in the oceans, which are robust near the shore and in bays and estuaries along the coast. These are known as "tidal currents." Tidal currents vary in a precise, steady pattern and can be forecast for future dates. In some places, robust tidal currents can travel at a velocity of eight knots or more. Tidal current can be tackled with quantum gravity and particle spinning theories owing to the Earth's rotation. In quantum gravity theory, gravity is formed of a composite particle, which acts as force particles. Consequently, tide fluctuation is generated due to acts of force particles between the water body, the moon, and the sun [4–7].

By considering general relativity as an operative field theory, one can essentially create valid expectations for quantum gravity, at least for low-energy phenomena. An example is the well-known calculation of the tiny first-order quantum-mechanical correction to the classical Newtonian gravitational potential between three masses like the moon, the sun, and the water body to cause tides [3, 6, 8]. The tide due to gravitational forces would occur through spin networks, which, signifying the water body and the volume of every surface or space zone equally, has a discrete spectrum. Thus the area of the ocean and the volume of any portion of space is correspondingly quantized, where the quanta are rudimentary quanta of space. Particles across the spin networks have angular momentum, which directs them to travel from one location to another in space-time.

(ii) Wind

Winds induce currents at or near the ocean's surface. Near coastal areas, winds tend to force currents on a restricted scale and can result in phenomena corresponding to coastal upwelling. On a more global scale, in the open ocean, winds create currents that move water for thousands of miles across the ocean basins. Wind particles spin on the sea surface causing stress forces that direct the water particle to spin in a different direction as a function of the Earth's rotation [3–7]. In this regard, quantum water particles' spin is related to angular momentum, the physical property attributed to rotating particles, which induces opposite current motion to wind spinning particles.

(iii) Thermohaline circulation

This is a process caused by density differences in water due to temperature (thermo) and salinity (haline) dissimilarities in different parts of the ocean. Currents caused by thermohaline circulation appear at both deep and shallow ocean stages and travel much slower than tidal or surface currents. For instance, deep **ocean currents** are induced by density and temperature gradients [9]. Thermohaline **circulation** is correspondingly recognized as the **ocean's** conveyor belt (which denotes deep **ocean** density-directed **ocean** basin **currents**). These **currents**, so-called submarine rivers, emanate below the surface of the **ocean** and are hidden from immediate detection or measurement [3, 5, 8, 10].

The quantization of thermohaline circulation is a function of understanding quantum mechanics. In this view, a change in the quantum behavior of water mass particles at different temperatures and salinities produces a current that flows upwards or downwards. In this way, two water particles are mixed in such a way that a change made to the state of one water particle is reflected instantly in the other, however far apart they are separated. It is possible to define this procedure through the water column as **quantum water particle entanglement**. Quantum entanglement can occur on a large scale across the ocean due to the Coriolis effect.

When spinning objects impact with another moving or stationary force, it generates a different movement. The Earth's revolution creates dual currents: one, a clockwise transfer of water in the northern hemisphere; the other, a counter-clockwise movement of water in the southern hemisphere. When these currents rebound from land masses, they generate massive ocean currents, termed "gyres." Quantum entanglement can create two currents spinning in opposite directions to each other owing to the Earth's rotation or spinning.

12.4 Ocean current measurements

12.4.1 Lagrangian method

Direct current measurements fall into two categories: (i) those that monitor a parcel of the moving water and (ii) those that measure the speed and direction of the water as it travels to a stationary location. Moving current can be monitored with buoys, which signal their positions acoustically to a research vessel or shore station. In this way, buoys' routes are determined and their velocities and displacements owing to current movements are gauged.

Consistent with Stewart [6], oceanography and fluid mechanics discriminate between dual current measurement techniques: Lagrangian and Eulerian. Lagrangian techniques track water particles, whereas Eulerian techniques compute the velocity of water particles at a static location.

The Lagrangian technique is a conventional current measurement depleted to gauge the current patterns (speed and direction). In this regard, tracking the changeable of a drifter displacement across space-time either on the surface or deeper water column can assist to calculate the current speed and direction. The mean velocity, therefore, over some period is computed from the distance between sequences of the numerous locations of the drifter at the beginning and the completion of the period allocated by the drifter. This technique involves numerous uncertainties owing to: (i) miscalculations in determining the position of the drifter; (ii) the failure of the drifter to track a parcel of water—for instance, the drifter remains in the ocean, but external forces such as storm acting on the drifter cause it to drift relative to the water; and (iii) sampling uncertainties occur as the drifters undergo chaotic current movements such as they travel into the convergence zone and divergent zone, respectively [6, 7].

12.4.2 How to design a simple drogue for sea surface current tracking

The drogue can be made of dual intersecting sails of 2.5m length and 0.6m width (Fig. 12.1). A weight below it should ensure that it sinks until its tether rope is taut and essentially vertical. The tether rope is polyethylene, 3m in length and 5mm in diameter. The surface float is a Styrofoam board of size 50 × 50 × 4cm. The flag mast can be made from PVC tubing 2m in length and 2cm diameter. A small weight at the lower end of the flag mast stabilizes the mast, which holds a marker flag of 30 × 20cm above the water surface. It also carries a beacon for nighttime observation [11]. A drogue of this design was deployed and its movement observed twice in the study area. The deployment was at latitude 30°30.0′N and longitude 103°54.0′E (about 26 nautical miles from the coastline) on October 30, 2019, at 1400h. The drogue was allowed to drift for 12days and was recovered on November 9, 2019, at 1800h. The drogue traveled southward very fast immediately after release with a speed of approximately 0.86ms^{-1}, which was associated with the flood tide of 1.5m.

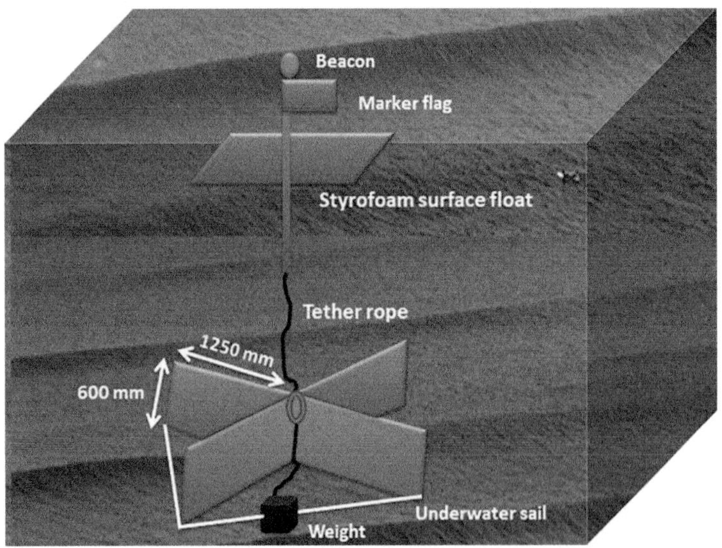

FIG. 12.1 Sketch of drogue used in this study.

The drogue continued traveling southward with irregular movement patterns. Eventually, it approached Tioman Island, the current starting to deviate to the coastal water on November 5, 2019, with a speed of $1.2\,\mathrm{m\,s^{-1}}$, which was associated with the flood tide. It is worth noting that the current slows down to the speed of approximately $0.4\,\mathrm{m\,s^{-1}}$ during ebb tide. From November 6 to November 9, 2019, the current tends to complete one large eddy along Tioman Island with an average speed of $0.97\,\mathrm{m\,s^{-1}}$. The wind direction throughout the observation periods was northeasterly, which drove the current in a southward direction. The fluctuation of the tide from ebb tide to flood tide with a combination of the wind fluctuation offshore in the coastal waters off Tioman Island contributes to generating eddy flows between latitude 30°30.0′N and longitude 103°54.0′E to latitude 3°18.0′N and longitude 104°00.0′E (Fig. 12.2).

12.4.3 Eulerian method

Eulerian measurements have been deployed by various different sorts of current meters attached to anchorages or ships. Anchorages are positioned by ships for periods of several months to more than a year. Since the anchorage must be positioned and maintained by deep-sea research ships, the technique is costly. Submerged anchorages are preferred for various reasons: they are not affected by high-frequency, strong surface currents; the anchorage is out of sight and does not draw the attention of fishermen; and the floatation is generally sufficiently deep to escape being trapped by fishing nets [6, 11]. However, Eulerian measurements involve the following uncertainties: Subsurface anchorages exchange tiniest. Surface anchorages in robust current, therefore, exchange greatest and are rarely depleted. Anchorages, consequently, incline not to remain longer sufficient to deliver precise approximations

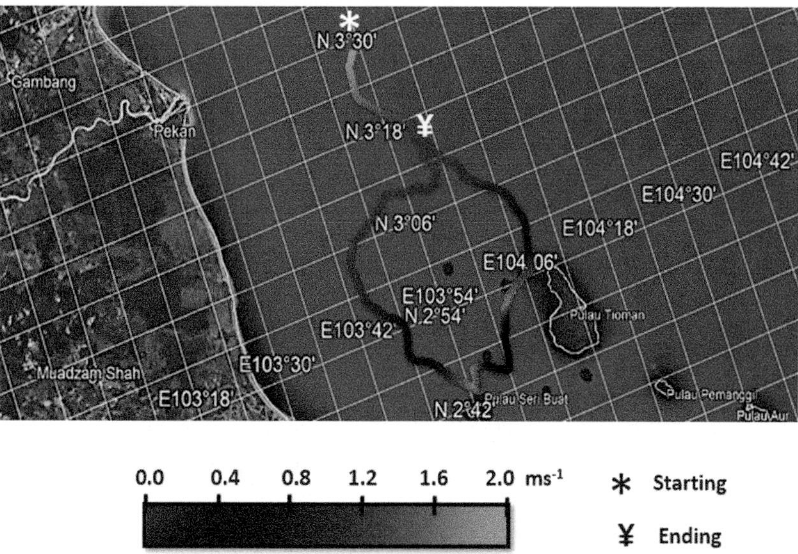

FIG. 12.2 Drogue's path from October 30, 2019, to November 9, 2019.

of average current velocity or interannual variability of the current velocity. Finally, entangling of the sensors by marine organisms can become a problem, especially for devices positioned for more than a few weeks closer to the surface.

12.4.3.1 *How to deploy a current meter to measure water column velocities*

Sensors deployed to gauge current velocity and direction at static positions are known as current meters. This technique uses sound pulses and takes advantage of the variation in the pitch or frequency of sound as it is reflected by particles suspended in the moving currents. When sound is reflected by particles moving toward the meter, the pitch rises; whereas when the sound is echoing off particles moving away from the meter, the pitch falls. In this way, the acoustic wave and current (AWAC) profiler uses the Doppler effect to measure current velocity by transmitting a short pulse of sound, listening to its echo, and measuring the change in pitch or frequency of the echo. The Doppler shift measured by each transducer is proportional to the velocity of the particles along its acoustic beam. Any particle motion perpendicular to the beam will not affect the Doppler shift. By merging the velocities from numerous transducers, and knowing the relative orientation of these transducers, the three-dimensional velocity can be calculated [12]. Acoustic wave and current (AWAC) is used in current measurement, and can measure current speed and direction in 1 m-thick layers from the bottom to the surface and can measure long waves, storm waves, short wind waves, or transient waves generated by local ship traffic [13].

Fig. 12.3 demonstrates the orientation of the three slanted transducers deployed by the AWAC. Each beam leans 25 degrees off the vertical axis and the three beams are correspondingly spread out at 120 degrees azimuth angles. Consequently, the strength of the return signal will decay as the pulse moves away from the transducer, owing to the symmetrical scattering of the sound waves and the absorptionof acoustic energy by the water.

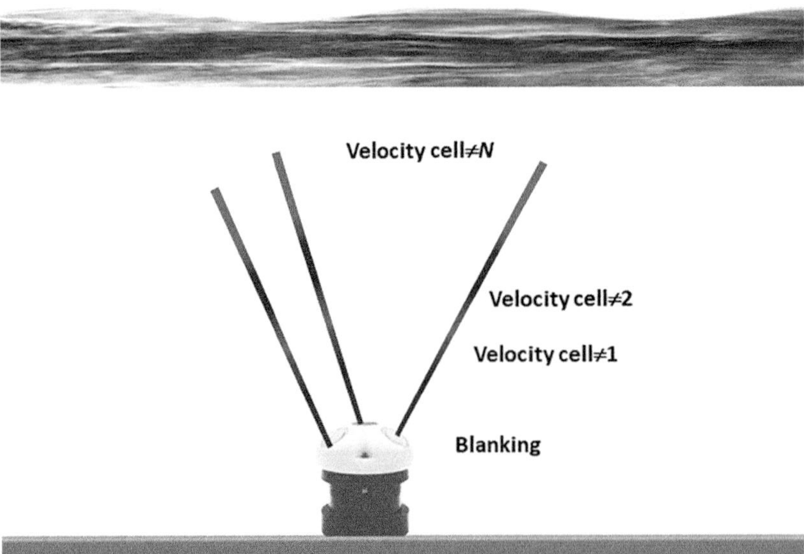

FIG. 12.3 Three slanted transducer orientations used by the AWAC.

In this chapter, AWAC is used to determine the current movement at 3°06′11″N and 103° 42′40″E (Fig. 12.4) from October 30, 2019, to November 9, 2019, during the Sentinel-1A pass over the east coast of Malaysia. The AWAC current data can be presented in a scatter plot in which there are two peaks describing current movement. The first peak indicates southeasterly current movement while the second one shows a northwesterly direction. Consequently, the main current flows indicate high concentrations of northeasterly scatter dots with a maximum velocity of 1.8 m s^{-1} (Fig. 12.5). However, the northwesterly current flows have

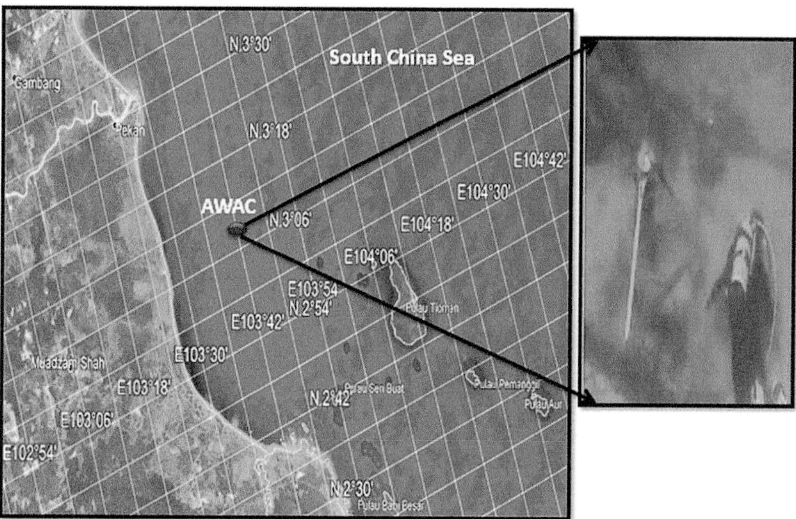

FIG. 12.4 The geographical location of AWAC deployment.

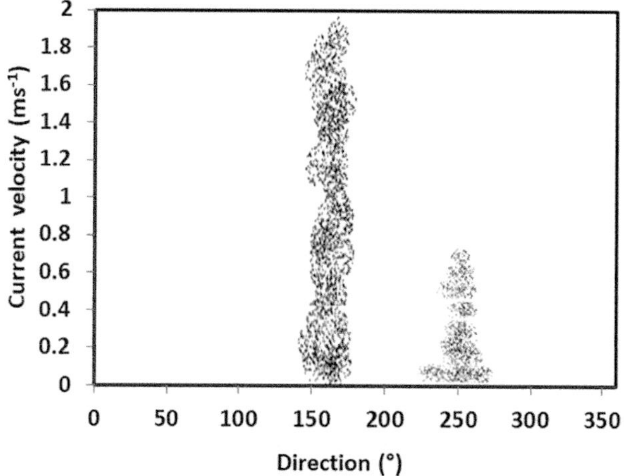

FIG. 12.5 Scatter plot retrieves from AWAC.

a maximum velocity of $0.75\,\mathrm{m\,s^{-1}}$. Consequently, the southeasterly peaks correspond to the rising tide of 2m, while the northwesterly peaks correspond to the falling tide of 0.8m. This example just demonstrates a simple method to represent the current data collected using AWAC, which can be used to validate the SAR data in the next sections.

Both Lagrangian and Eulerian techniques are considered as observation methods. However, these techniques are limited to observing small areas of the ocean, i.e., at a fixed point, which does not involve the complete current pattern of the entire sea. In this regard, the ocean current mathematical model is required to comprehend fully the scenario of the ocean current flow.

12.5 Governing equations of inviscid motion

Let us take the ocean as an inviscid fluid with density ρ. A water particle is spinning about the z-axis with constant angular velocity $\Omega \sin\varphi$, where φ is the latitude. Additionally, let us consider x and y are horizontal coordinate axes along the steady sea surface, and the z-axis is bound upwards. The location of the free surface is specified by$z=\zeta(x,y,t)$, where ζ denotes the surface height. The atmospheric pressure at the surface is signified by $P_s(x,y,t)$. The bottom topography does not diverge with time and is known by $z=-H(x,y)$ (Fig. 12.6).

The mathematical description of the water particle motion of unit mass can now be expressed as:

$$\frac{D\vec{v}}{dt}+f\vec{k}\times\vec{v}=-\frac{1}{\rho}\nabla p-g\vec{k}, \tag{12.1}$$

In Eq. (12.1), the term $\frac{D}{dt}\equiv\frac{\partial}{\partial t}+\vec{v}\cdot\nabla$ represents the total derivative following a fluid particle and $f=2\Omega\sin\phi$ is the Coriolis parameter, which causes water particle spinning. Let us consider that the mass of water particles within a geometrically stationary volume can only

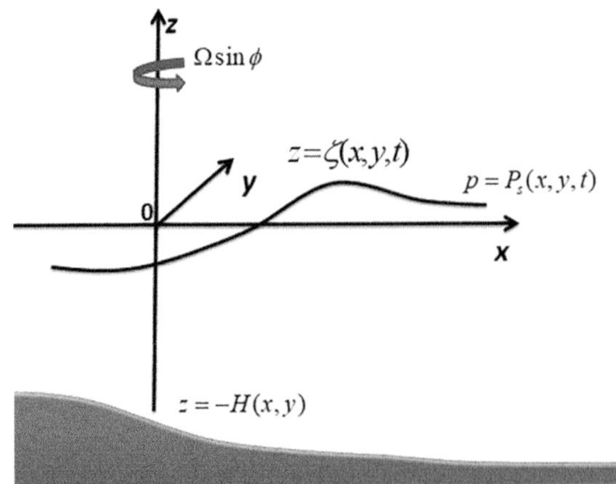

FIG. 12.6 Definition sketch.

modify as a consequence of the advection of water particles [14]. Therefore, the mathematical explanation of mass conservation is given by:

$$\frac{D\rho}{dt} = -\rho\nabla \cdot \vec{v}. \tag{12.2}$$

This is often referred to as the continuity equation (see Chapter 1). Therefore, let us presume that the density of a water particle is preserved $\frac{D\rho}{dt}=0$, which leads to the reduction of the continuity equation as:

$$\nabla \cdot \vec{v}=0. \tag{12.3}$$

In addition, let us consider a mixed water body with homogenous water particle densities. In this scenario, the coordinate system is orientated with the x-axis tangential to a latitudinal circle and the y-axis pointing northwards [14, 15] (Fig. 12.7).

On this understanding, the reference system f is merely a function of y, which is given by:

$$f = f_0 + \left(\frac{df}{dy}\right)_0 y = f_0 + \beta y, \tag{12.4}$$

where:

$$\left.\begin{aligned} f_0 &= 2\Omega \sin\varphi_0, \\ \beta &= \frac{1}{R}\frac{d}{d\varphi}(2\Omega\sin\varphi)_{\varphi_0} = \frac{2\Omega}{R}\cos\varphi_0. \end{aligned}\right\} \tag{12.5}$$

This is known as the beta-plane approximation. In this circumstance, f is almost constant in an (x, y)-area, the movement falls on f-plane. Consequently, the kinematic and dynamic boundary circumstances on the surface can be expressed, respectively, as:

$$\frac{D}{dt}(z-\zeta)=0,\ z=\zeta(x,y,t), \tag{12.6}$$

$$p=P_S,\ z=\zeta(x,y,t). \tag{12.7}$$

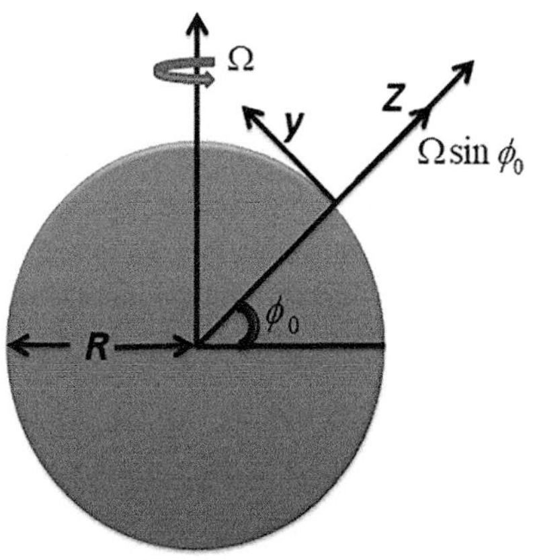

FIG. 12.7 The orientation of coordinate axes.

In this sense, the kinematic boundary condition at the bottom can be articulated as:

$$\vec{v} \cdot \nabla(z+H) = 0,\ z = -H(x,y). \tag{12.8}$$

Let us integrate the continuity equation, Eq. (12.3), from $z=-H$ to $z=\zeta(x, y, t)$:

$$\int_{-H}^{\zeta} w_z dz = -\int_{-H}^{\zeta} u_x dz - \int_{-H}^{\zeta} v_y dz. \tag{12.9}$$

Exploiting Eqs. (12.6) and (12.8):

$$\zeta_t = -\frac{\partial}{\partial x}\int_{-H}^{\zeta} u dz - \frac{\partial}{\partial y}\int_{-H}^{\zeta} v dz. \tag{12.10}$$

Therefore, the fundamental statement in shallow water speculation is that the pressure dissemination in the vertical direction is hydrostatic. By exploiting that fact that the density is constant, and employing Eq. (12.7) [14–17], this leads to:

$$p = \rho g(\zeta - z) + P_S(x,y,t). \tag{12.11}$$

This means the vertical component in Eq. (12.1) directs to the vertical acceleration $\frac{Dw}{dt}$, which must be extremely small. In this regard, it does not conspicuously modify the hydrostatic pressure variations [15, 18]. The horizontal components of Eq. (12.1), consequently, can thus be identified as:

$$\frac{Du}{dt} - fv = -g\zeta_x - \frac{1}{\rho}P_{Sx} \tag{12.12}$$

$$\frac{Dv}{dt} + fu = -g\zeta_y - \frac{1}{\rho}P_{Sy} \tag{12.13}$$

The right-hand sides of Eqs. (12.12) and (12.13), respectively, are independent of z. By exploiting that $v \equiv \frac{Dy}{dt}$ and $f=f_0+\beta y$, Eq. (12.12) can be modified as:

$$\frac{D}{dt}\left(u - f_0 y - \frac{1}{2}\beta y^2\right) = -g\zeta_x - \frac{1}{\rho}P_{Sx} \tag{12.14}$$

Eq. (12.14) reveals that $\frac{D}{dt}\left(u - f_0 y - \frac{1}{2}\beta y^2\right)$ is independent of z. Consequently, this is similarly real for $(u - f_0 y - \beta y^2/2)$, and thus correspondingly for u, if u and v are independent of z at time $t=0$. Correspondingly, Eq. (12.13) reveals that v is also independent of z, which is expressed as:

$$\left.\begin{aligned} u &= u(x,y,t), \\ v &= v(x,y,t). \end{aligned}\right\} \tag{12.15}$$

Additionally, by utilizing Eq. (12.3), then w_z is independent of z. Hence, by integrating into the vertical [16–18]:

$$w = -\left(u_x + v_y\right)z + C(x,y,t). \tag{12.16}$$

The function C is acquired by utilizing the boundary condition in Eq. (12.8) at the ocean bottom. Thus, the vertical velocity can thus be formulated as:

$$w = -(u_x + v_y)(z + H) - uH_x - vH_y. \tag{12.17}$$

Subsequently, u and v are independent of z, the integrations in Eq. (12.10) are straightforwardly accomplished. Consequently, the following nonlinear equations are obtained that involve the horizontal velocity components and the surface elevation:

$$u_t + uu_x + vu_y - fv = -g\zeta_x - \frac{1}{\rho}P_{Sx}, \tag{12.18}$$

$$v_t + uv_x + vv_y + fu = -g\zeta_y - \frac{1}{\rho}P_{Sy}, \tag{12.19}$$

$$\zeta_t + \{u(H+\zeta)\}_x + \{v(H+\zeta)\}_y = 0. \tag{12.20}$$

To solve this set of equations we require three initial conditions, e.g., the distribution of u, v, and ζ in space at time $t=0$, are required to solve these three sets of equations. Consequently, if the water flow is restricted by lateral boundaries, the solutions must satisfy the requirement of no flow through impermeable lateral boundaries. Fig. 12.8 shows the current pattern flows derived using Eqs. (12.18)–(12.20), the current velocities obtained from the AWAC, and sea-level elevation of the tide table of Malaysian coastal waters of October 2019 to November 2019. The current pattern indicates curvature eddy flows from the surface to a depth of 40 m with a maximum speed of 1.6 m s^{-1} from 30°30.0′N and longitude 103°54.0′E to latitude 3°18.0′N and longitude 104°00.0′E. This eddy flow coincides with the surface eddy flow observed by drogue (Fig. 12.2).

The current speed decreases with depth to approximately 0.4 m s^{-1}. This eddy is dominated from the surface to a depth of 40 m by anticlockwise rotation. It may be generated due to the strong monsoon wind stress and effective tidal-wind combination too.

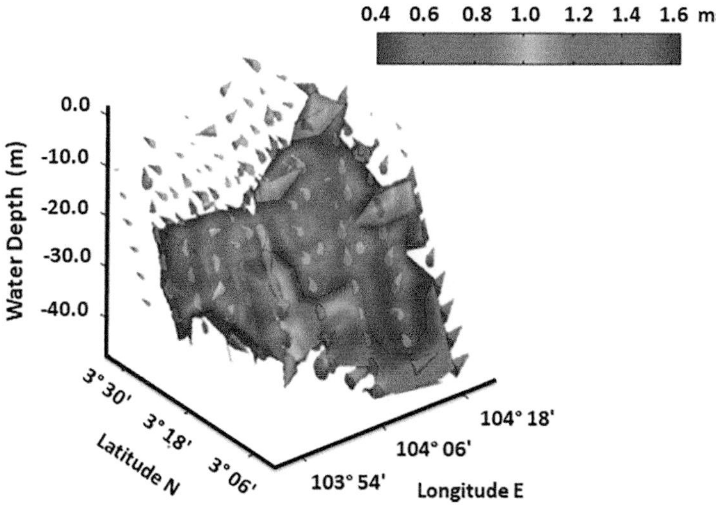

FIG. 12.8 Simulated current flow.

The above equations could represent the motion of a water particle in a given gravitational field that induces the tidal elevation fluctuations over the space-time coordinates. In this regard, the motion is principally free in curved space-time, the generalized gravitational potential being designated by the metric, $g_{\mu\upsilon}$, which can be identified as a function of the space-time coordinates. In the Newtonian approximation, tidal forces acting in a free-falling frame also show deviation from the Minkowskian geometry of special relativity. It is consequently noticeable that there must be a close relation between curvature and tidal forces that generate current movements. On this understanding, the geometry of space-time is the geometry of a four-dimensional space with local Minkowski metric; as will be discussed in the last chapter of this book. However, this geometry cannot be flat in the presence of a gravitational field. Space-time geometry must then be described by a metric, $g_{\mu\upsilon}$, which depends on the presence of a gravitational field, each entry being a function of the space-time coordinates The metric can be reduced to the special relativity form only locally, that is in an infinitely small neighborhood of an event, but not in a large region of space-time.

A free motion, therefore, in curved space-time is the analog of the inertial motion in flat space-time given by a straight line. Such a curve is called geodesic and is defined as the shortest line joining two points. In this context, geodesic describes the trajectory of a particle in a gravitational field, which must reduce to the usual Newtonian formula in the nonrelativistic state. From the point view of quantum mechanics, ocean current is the trajectory of water particles in a gravitational field, which obeys the curvature of space-time. In this understanding, all current movements must always be in a curved trajectory not in a straight line, for instance, the meandering of the Gulf Stream current. On this understanding, the presence of gravitation tells us the relativistic extension of the tidal forces and tidal current flow. In other words, the energy-momentum distribution, which is the content of tidal-energy and its current flow, determines the geometry of space-time.

12.6 Wind-driven current

Winds blowing across the surface of the ocean create the friction that directs the water in motion. That movement is a function of wind momentum variations and, consequently, of the energy conveyed to the ocean's surface. If there is no friction between wind and water, the motion does not occur. In the circumstance of strong winds, wind-drift currents are perhaps as deep as 100 m below the surface. Surface currents travel at approximately 2% of the wind speed that caused them. For instance, a wind blowing at $10\,\text{m}\,\text{s}^{-1}$ induces a surface current of approximately $20\,\text{cm}\,\text{s}^{-1}$. The progression of wind-driven currents occurs when winds blow across the ocean and drag on the surface. This induces a thin layer of motion, which, in turn, drags on the layer beneath, placing it also in motion. This progression continues downward, conveying momentum (product of mass and velocity) to consecutively deeper layers. Such transmissions of momentum between layers are disorganized, and energy dissipates through the water column. Accordingly, current velocity declines as the ocean becomes deeper [3, 16].

In general, the velocity introduced above is related to the displacement of individual oceanic particles. Nonetheless, numerous oceanic flow occurrences are *turbulent*, which means that it is challenging to keep track of individual particles. Nevertheless, the turbulent oceanic flow is not incapable of configuration. Even if the individual particles travel in a chaotic path, in an average sense they may delineate an ordered flow. In ideal circumstances, the mean oceanic flow has a considerably larger characteristic time scale than the rapid fluctuations of individual fluid particles [16, 19]. On this understanding, the mean velocities can be formulated by a time-averaging process as:

$$\overline{\vec{v}} = \frac{1}{\Delta T} \int_{t-\Delta T/2}^{t+\Delta T/2} \vec{v}\, dt. \tag{12.21}$$

Here ΔT is a period fluctuation that is large compared to the characteristic scale of the turbulence, but small compared to the typical time scale of the mean motion we intend to study. A similar sort of averaging can be performed for pressure and density. Accordingly, they can be expressed as:

$$\left.\begin{aligned} \vec{v} &= \overline{\vec{v}} + \vec{v}\,', \\ \rho &= \overline{\rho} + \rho', \\ p &= \overline{p} + p'. \end{aligned}\right\} \tag{12.22}$$

where $\vec{v}\,'$, ρ', and p' represent the fluctuating (turbulent) part of the motion.

The continuity equation—Eq. (12.3)—articulates the conservation of mass in a fluid and is a major law in nonrelativistic mechanics. Evidently, molecular processes such as heat and salt diffusion may alter the density of a fluid particle. This consequence is considered for the application of the *equation of state* for the fluid. The modification of density has an identical vital dynamical inference in that it modifies the buoyancy of a fluid particle, which in turn may lead to thermohaline circulation. Utilizing an averaging of the viscous Navier-Stokes equation in a reference system rotating with constant angular velocity $f/2$ about the vertical axis, then [16–20]:

$$\overline{\vec{v}}_t + \overline{\vec{v}} \cdot \nabla \overline{\vec{v}} + f\vec{k} \times \overline{\vec{v}} = -\frac{1}{\overline{\rho}} \nabla \overline{p} - g\vec{k} + \nabla \cdot \left[\mu \nabla \overline{\vec{v}} - \overline{\vec{v}\,'\vec{v}\,'} \right]. \tag{12.23}$$

Let neglect ρ' be compared to $\overline{\rho}$ in the pressure term, since usually in the ocean $|\rho'|/\overline{\rho} \sim 10^{-3}$. In addition, μ is the kinematic molecular viscosity coefficient ($\mu \approx 10^{-2}\ \mathrm{cm^2\ s^{-1}}$). The term $\mu \nabla \overline{\vec{v}}$ is usually much smaller than the turbulent momentum transport per unit mass, $-\overline{\vec{v}\,'\vec{v}\,'}$, and can, therefore, be neglected. The term $-\overline{\rho}\,\overline{\vec{v}\,'\vec{v}\,'}$ is usually referred to as the turbulent Reynolds stress tensor [14–18].

For large-scale ocean circulation, it is usual to presume that the turbulence is dependent on the direction. In this regard, let us assume A_H and A are turbulent eddy viscosity coefficients in the horizontal and the vertical directions, respectively. The surfaces of constant density are primarily horizontal in the ocean, and the stratification is steady. Therefore, the turbulent motion in the vertical direction is smaller than in the horizontal direction [14, 20]. Consequently, let us consider that $A_H \gg A$. For such motion, $\overline{w}_z$ is often negligible, i.e., $\overline{u}_x + \overline{v}_y \approx 0$. Furthermore, it must be typically equal: $u'^2 + v'^2 \gg w'^2$. Consequently, we can suppose for the regular stresses:

$$\left.\begin{aligned} -\overline{u'u'} &= 2A_H \overline{u}_x - K, \\ -\overline{v'v'} &= 2A_H \overline{v}_y - K, \\ -\overline{w'w'} &\approx 0, \end{aligned}\right\} \tag{12.24}$$

while the tangential stresses become:

$$\left.\begin{aligned} -\overline{u'v'} &= -\overline{v'u'} = A_H(\overline{v}_x + \overline{u}_y), \\ -\overline{u'w'} &= -\overline{w'u'} = A\overline{u}_z + A_H\overline{w}_x, \\ -\overline{v'w'} &= -\overline{w'v'} = A\overline{v}_z + A_H\overline{w}_y. \end{aligned}\right\} \tag{12.25}$$

Additionally, we suppose that A_H is constant, while we accept that $A = A(z)$. The horizontal component of Eq. (12.23) can then be written:

$$\overline{\vec{q}}_t + \overline{\vec{v}} \cdot \nabla\overline{\vec{q}} + f\vec{k} \times \overline{\vec{q}} = -\frac{1}{\rho}\nabla_H \overline{p}_* + A_H \nabla_H^2 \overline{\vec{q}} + \left(A\overline{\vec{q}}_z\right)_z', \tag{12.26}$$

where $\vec{q} \equiv u\vec{i} + v\vec{j}$, $\nabla_H^2 = \partial^2/\partial x^2 + \partial^2/\partial y^2$, and $\overline{p}_* = \overline{p} + \rho K$.

Fig. 12.9 shows the wind-driven current movement, as the wind is the main factor generating eddies in the coastal waters of Malaysia. This finding agrees with the results of the drogue, inviscid motion model, and AWAC current measurement. It is worth noting that the eddy flow from the surface to deeper water causes such a water slope. The wind-driven current also generates such turbulent flows from the surface to the bottom, especially in shallow coastal waters such as off the east coast of Malaysia. The water flows from surface to bottom appear to spin downward.

For large-scale ocean circulation, we can assume that the vertical pressure distribution is approximately hydrostatic. This means that we can replace the z-component in Eq. (12.23) [14, 16–21] by:

$$\overline{p} = g\int_z^{\eta} \overline{\rho}\, dz + P_S(x, y, t), \tag{12.27}$$

where P_S is the atmospheric pressure at the sea surface.

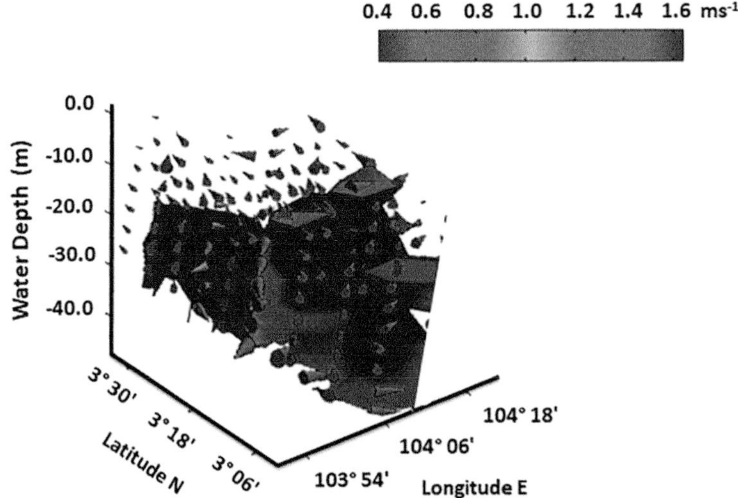

FIG. 12.9 Simulation of wind-driven current flow.

Currents also remain, since winds convey water, moving the sea surface (Fig. 12.9), and creating horizontal pressure gradients on which water is moved. Current speed and direction are revised by friction, the Coriolis effect, horizontal pressure gradients, and land and seafloor topography.

12.7 Ekman spiral

Tomczak and Godfrey [22] noticed that the generation of currents by wind requires the frictional transfer of momentum from the atmosphere to the ocean. The effect of wind stress on oceanic movement is best analyzed by excluding for the moment the effect of the pressure field on the balance of forces and assuming that the ocean is well-mixed, such as the uniform density, and its surface horizontal. The balance of forces is then between friction and the Coriolis force. Since wind blowing over water is always associated with turbulent mixing, the condition of uniform density is usually satisfied in the wind-affected surface layer, and the balance between friction and the Coriolis force prevails. This layer is often called the Ekman layer [3, 22].

Ekman assumed a steady, homogeneous, horizontal flow with friction on a rotating Earth. Thus, horizontal and temporal derivatives are zero [6]:

$$\frac{\partial}{\partial t}=\frac{\partial}{\partial x}=\frac{\partial}{\partial y}=0 \tag{12.28}$$

Ekman further assumed a constant vertical eddy viscosity:

$$T_{xz}=\rho\varpi A_z\frac{\partial u}{\partial z}\quad T_{yz}=\rho\varpi A_z\frac{\partial u}{\partial z} \tag{12.29}$$

where T_{xz} and T_{yz} are the components of the wind stress in the x and y directions, respectively, and $\rho\varpi$ is the density of seawater.

With these assumptions, the x and y components of the momentum equation have the simple form:

$$fv+A_z\frac{\partial^2 u}{\partial z^2}=0 \tag{12.30}$$

$$-fu+A_z\frac{\partial^2 v}{\partial z^2}=0 \tag{12.31}$$

where f is the Coriolis parameter.

It is easy to verify that the Eqs. (12.30) and (12.31) have solutions:

$$u=V_0\exp(az)\sin(\pi/4-az) \tag{12.32}$$

$$v=V_0\exp(az)\cos(\pi/4-az) \tag{12.33}$$

when the wind is blowing to the north ($T=T_{yz}$), the constants are:

$$V_0=\frac{T}{\sqrt{\rho^2\varpi f A_z}}\quad\text{and}\quad a=\sqrt{\frac{f}{2A_z}} \tag{12.34}$$

and V_0 is the velocity of the current at the sea surface.

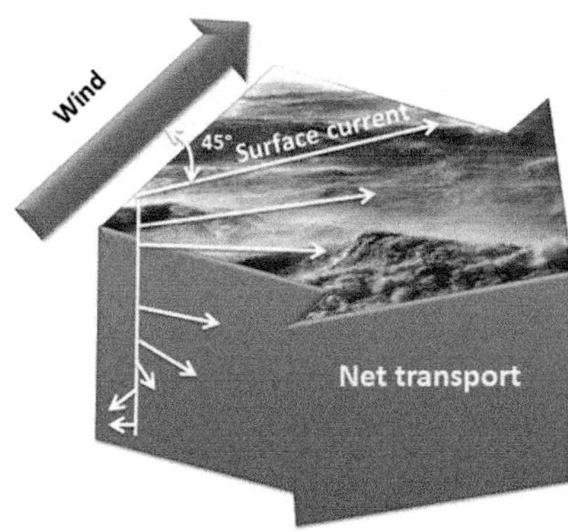

FIG. 12.10 Ekman spiral and net transport.

Now let us look at the form of the solutions. At the sea surface $z=0$, $\exp(z=0)=1$, and:

$$u(0)=V_0\cos(\pi/4) \tag{12.35}$$

$$v(0)=V_0\sin(\pi/4) \tag{12.36}$$

The current has a speed of V_0 to the northeast. In general, the surface current is 45 degrees to the right of the wind when looking downwind in the northern hemisphere. The current is 45 degrees to the left of the wind in the southern hemisphere. Beneath the surface, the velocity decays exponentially with depth [3–8] (Fig. 12.10).

$$\left[u^2(z)+v^2(z)\right]^{1/2}=V_0\exp(az) \tag{12.37}$$

The magnitude of the Ekman surface current, V_{EO}, is deliberated by:

$$V_{EO}=\frac{\sqrt{2\pi\tau}}{D_E\rho_0|2\Omega\sin\varphi|} \tag{12.38}$$

where τ is the magnitude of the surface wind stress, Ω the angular speed of rotation of the Earth, φ the latitude, and D_E the Ekman layer depth [3].

Tomczak and Godfrey [22] stated that the current speed and direction change with depth in an Ekman layer was generated by wind blowing over a deep ocean. The current speed is largest at the surface and decreases rapidly with depth (Figs. 12.8 and 12.9).

12.8 Quantum theory of the Ekman spiral

Wind kinetic energy transfers to the sea surface causing a friction force that induces surface motion. The fluctuation of this kinetic energy through the water column changes the quantum state from the surface to the ocean floor. The updating of the quantum state through the water column generates, for example, water particles spiraling through the water column beneath the sea surface. The differences in the phases of spinor components through the water column generate a spin spiral. Let us assume that there are a number of quantum states from the wind surface stress to the ocean depth. This number of quantum states is a function of the water depth. In other words, the quantum state numbers increase with water depth. Shallow water depth involves fewer quantum state numbers than a deeper ocean. Therefore, if $\Psi = [\psi_1, \psi_2, \psi_3, \ldots, \psi_N]$ as ψ_N is an N-component spinor from the sea surface to water depth, the scientific explanation of the water spiral pattern can mathematically be written as:

$$\Psi = \begin{pmatrix} \psi_1 \\ \psi_2 \\ \vdots \\ \psi_N \end{pmatrix} = \begin{pmatrix} e^{i(n\theta + \varphi_i(r) - \omega t)}\phi_1(r) \\ e^{i(n\theta + \varphi_i(r) - \omega t)}\phi_2(r) \\ \vdots \\ e^{i(n\theta + \varphi_i(r) - \omega t)}\phi_N(r) \end{pmatrix} \tag{12.39}$$

Eq. (12.39) reveals the number of spiral arms (n) through the water column. In an Ekman spiral pattern, the geometric shape is not well constructed due to the changing of the kinetic energy phase $\varphi_i(r)$ through the baroclinity of a stratified fluid owing to water density friction, which is a nonlinear form. The nonlinearity of the kinetic energy, phase-changeable through the water column, is formulated as [23]:

$$\phi_j'' + \left(r^{-1} + i\varphi_j'\right)\phi_j' = \left[\left(-\omega + n^2\left(r^{-2}\right) + \left(\varphi_j'\right)^2\right)\right]\phi_j \tag{12.40}$$

Eq. (12.40) reveals the radial modulation of amplitudes $\phi_j(r)$. The shape of the spiral is determined by the baroclinic term β_j, which mainly generates the changes $\varphi_i(r)$ through the water depth as:

$$\beta_j = \frac{\partial^2 \varphi}{\partial z} + r^{-1}\left(\frac{\partial \varphi_j}{\partial z}\right) \tag{12.41}$$

The spiral does not exist if β_j is zero. Consequently, this quantum spiral is a sort of "point vortex" that has zero vorticity excluding the central phase singularity. In addition, Eq. (12.41) demonstrates that the quantum baroclinic consequence, represented by β_j, is the non-Hermitian terms which is in dynamic fluctuation to create the spiral effective motion. Fig. 12.11 shows the Ekman quantum spiral in the waters off the east coast of Malaysia.

It is worth noting that the water density fluctuations from the surface to the water depth of 40 m reduce the speed of the water particle spinning due to friction. As the friction increases, so does the density. The maximum water density of 22 kg m^{-3} reduces the kinetic energy of water particle spinning, which leads to a spiral decaying near the water bottom of 40 m. From the point of view of the quantum mechanics, the novel explanation of Ekman's theory: it is just changing of the quantum state of the kinetic energy that transfers by the wind stress to the water body causing spinning of the water column particles. The spinning of the water particle has a different energy phase, which is a function of the quantum baroclinic of the water body.

Entanglement builds up between the state of the water spiral state, water density, and the state of the wind stress. In this scenario, water particle spirals reach equilibrium, or a state of uniform energy distribution, within an infinite amount of time by becoming quantum mechanically entangled with their surroundings.

In this understanding, wind stress entanglement occurs when wind particles and seawater particles remain connected so that actions performed on one effect the other, despite the

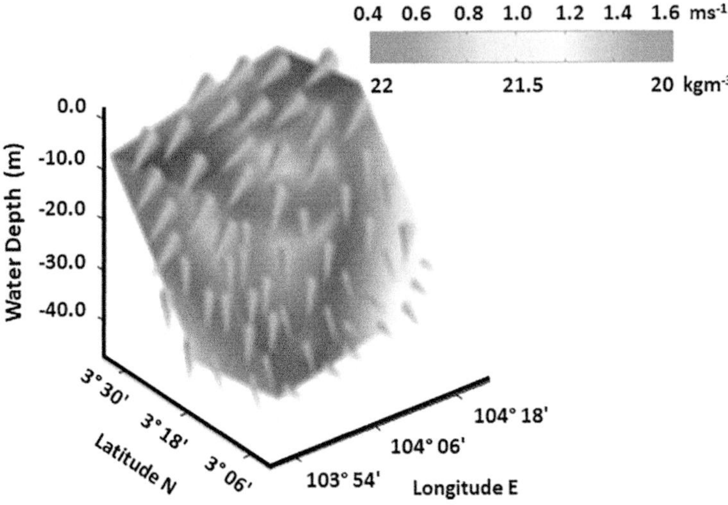

FIG. 12.11 Ekman quantum spiral through the water column.

distance between them (Fig. 12.12); as mentioned previously, this is known as "spooky action at a distance."

The Ekman spiral occurs due to wind particles transferring kinetic energy to seawater particles and producing entangled mixtures of wind kinetic energy and density fluctuation kinetic energy (Fig. 12.12). The length of the entanglement is proportional to water depth.

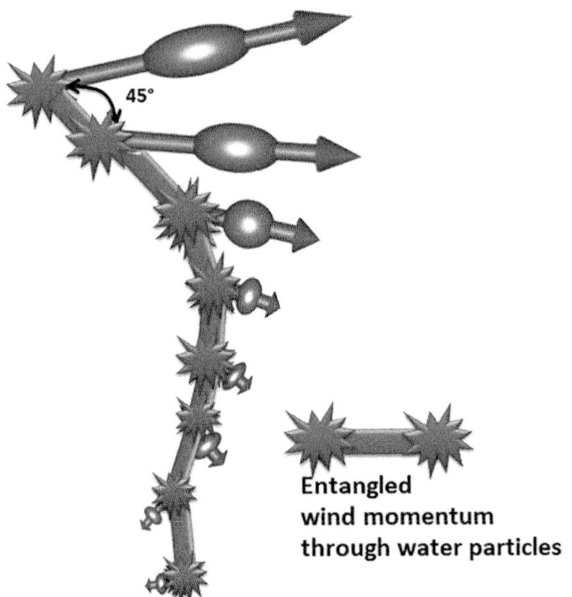

FIG. 12.12 Quantum view of the Ekman spiral.

Through the length of entanglement, the spiral water has space vibration and time vibration. Consequently, the Ekman spiral can be identified as a wave-particle duality as the water parcels or particles vibrate downward in space and time, traveling as correspondingly spiral waves until reaching a decaying state.

12.9 SAR Doppler shift frequency

In **quantum** mechanics, **Doppler** shifts swap the energy and momentum of the quanta and modify the wavenumber and frequency, but do not alter the ratio of the energy to the frequency or the momentum to the wavenumber.

Levita and Calderon [24] revealed that Doppler frequency shifts are a function of the relative velocities of the transmitter and reflecting object. The alteration in the frequency between the backscattered signal and the radiated signal is the Doppler frequency shift. The frequency shifts are due to the Doppler effect. Microwave signals radiated by the SAR antenna can be signified as a wave, if the wave theory of electromagnetic radiation is accepted. The frequency of the backscattered signal received by the SAR antenna can be exploited to reveal the relative location of an object based on the frequency shift of the backscattered signal.

Let us assume that the radar travels along an uncurving trajectory, at velocity V_S, and at a constant altitude, above a plane and a horizontal surface. The first object direction (O_1) creates an angle, θ_A, with the velocity vector (Fig. 12.13). A second object (O_2), however, is positioned in the identical range to the first one, creating angles equal to $\theta_A + \delta\theta_A$.

The dual objects are positioned in the antenna focal lobe. However, if the dual objects travel in an azimuth direction, the antenna beamwidth is not able to distinguish one from another. Relative to the SAR platform motion, each object has an imprecise velocity that is the ledge of the platform velocity along the azimuth direction X. This velocity is formulated as:

$$v_{r1} = V_S \cos\theta_A \tag{12.42}$$

$$v_{r2} = V_S \cos(\theta_A + \delta\theta_A) \tag{12.43}$$

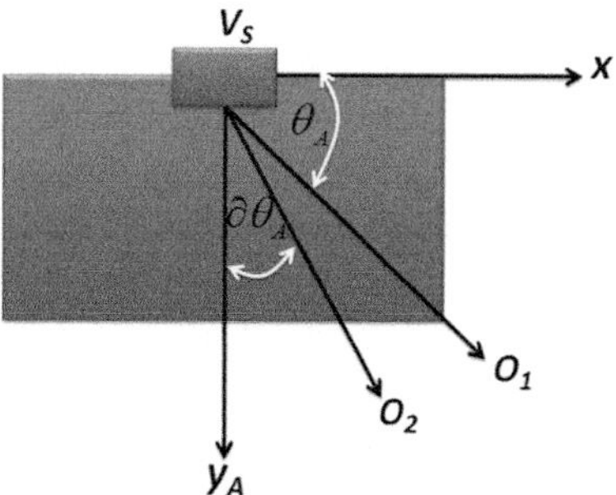

FIG. 12.13 Observation of object motion geometry.

The velocity of the SAR platform and changing of the imaging angles from O_1 to O_2 create a Doppler frequency for every object velocity. In this circumstance, if a target is shifting radially—that is, moving toward or away from the SAR antenna—the frequency of the reflected echo is shifted from the original frequency. In this view, the Doppler frequency shift occurs by quanta as reliant on the object or ocean surface radial velocity. In other words, if the object is traveling away, the backscattered signal would be less than the conveyed frequency by a similar quanta. If the backscattered signal is at a dissimilar quanta frequency to the radiated signal, the object velocity is radial. The Doppler frequency can be implemented to estimate the radial velocity of the ocean surface as the rate at which the ocean surface range is shifting, and it is generally denoted the range rate of the ocean surface motion [25].

SAR platform motions, however, cause Doppler frequency shifts. In this respect, numerous sorts of object motion can be designated that induce amplitude reduction of the backscatter signal. Consequently, an object that moves while it is being imaged would change the frequency of the SAR signal energy bounced off it. A frequency shift in the SAR system, therefore, alters the signal history of the object; as a consequence, the moving object might be identified as being in an image location other than the position on the ground where it was actually situated [26].

The effect of this is to swing the object position in the positive azimuth direction if the target is traveling toward the SAR, and in the negative azimuth direction for the object traveling away from the SAR. Consequently, the Doppler frequency shift can present in SAR as image smear or a blurred image. Amplitude reduction, which is return signal strength reduction owing to object motion, causes further degrading of the SAR image. For instance, azimuth defocusing is generated due to object motion in the azimuth direction, which causes SAR image smear. In general, the high azimuth resolution of SAR is accomplished by synthesizing over the finite integration of the known phase history of the backscattered signal that is Doppler-shifted owing to the relative motion between the radar antenna and the object being imaged [27].

There have been efforts to approximate the current patterns from microwave SAR platforms depleting RAR modulation, which were not entirely successful. In this view, SEASAT L-band data could not determine the Doppler shifts caused by sea surface motion. SEASAT had an L-band frequency SAR system, which provides small Doppler shifts caused by current velocity. However, C-band frequency data are sensitive to delivering larger Doppler shifts, which are used to detect large-scale ocean currents. Moreover, a larger incidence angle delivers a larger Doppler shift and hence further sensitivity for velocity. However, it increases the risk of Doppler centroid approximation problems owing to a low signal to noise ratio. In other words, a larger incidence angle delivers high sensitivity of larger Doppler shifts, but selective picking up of an extremely large incidence angle can cause an incredibly low signal to noise ratio, which results in feature smearing of the Doppler centroid calculation. Moreover, representative Doppler shifts for a $1\,\mathrm{m\,s^{-1}}$ current are in the order of 14 Hz for ERS SAR data and about 28 Hz for RADARSAT-1 at an incidence angle of 45 degrees [28–30].

The Doppler shift is caused by the relative motion between the satellite platform, the rotation of the Earth, and the velocity of the facets of the sea surface from which the backscattered SAR signal originates. The preliminary two former consequences are well-recognized, mainly for ENVISAT with its identical stable satellite orbit and attitude, and can be deducted to retrieve the sea surface velocity information. This surface velocity, in turn, is composed of contributions from the wind-wave induced motion and the background surface current, as discussed in previous chapters.

The Doppler centroid anomaly can be estimated as a mean, weighted generally scattering facet [31]:

$$\frac{\pi f_D}{k_R} = \frac{(u \sin\theta - \varpi \cos\theta)\sigma_0(\theta + \Delta\theta)}{\sigma_0(\theta + \Delta\theta)} \tag{12.44}$$

where u and ϖ include the horizontal and vertical velocity of the scattering facets and the surface current velocity, and $\Delta\theta$ is the local modification of the incidence angle θ.

An estimate of the surface current can be acquired, provided the contributions from wave- and wind-induced motion are first quantified and removed. This is a forward model that simulates radar cross-section signatures from specified fields of wind, surface current, and boundary layer stratification.

Since the challenging in measuring ocean surface motions, at least numerous contiguous resolution cells are prerequisites to retrieve a rate of the current variations in SAR images. With a coherent system, the surface current can be retrieved by the Doppler frequency shift of the backscattered signal.

12.10 SAR Doppler frequency shift model formulation

SAR utilizes the Doppler shift of the complex received field to detect scatterers in the flight direction. This complex field and its associated residual Doppler shift can be used to infer the velocity of these scatterers as advected by ocean currents [32]. This section provides the mathematical derivation of the SAR Doppler frequency shift.

The key to understanding of the Doppler frequency shift is the spectral density. The tiny scatterer points in SAR image involve the variation of Doppler spectral density as a function of the Doppler frequency shift. The relevant mathematical formula of Doppler spectral density is expressed as [30]:

$$F(x_0, \omega) = H(\omega)e^{i\omega' x_0/V_S} \int_{-\infty}^{\infty} e^{\left[-\frac{2}{T_s^2}\left(1+\left(\frac{T_s}{T}\right)^2 - \frac{ibT_s^2}{4}\right)\tau^2 - \frac{4}{T_s}\left(\frac{x_0}{V_S T_s} + i\frac{\omega' T_s}{4}\right)\tau - \frac{2x_0^2}{V_S^2 T_s^2}\right]} d\tau \tag{12.45}$$

here x_0 is the position of an object in the SAR data, T_s is the Gaussian width function, V_S is the SAR satellite velocity $= 6212\,\mathrm{m\,s^{-1}}$, τ is the delay time equals $\frac{t-x_0}{V_S}$, and $x_0 = vt$, and b is the chirp rate of $\frac{2kv^2}{R}$.

The Doppler spectral magnitude can be obtained by solving Eq. (12.45). The radial current velocity is obtained by correlating SAR image intensity to Doppler spectral magnitude. Therefore, Eq. (12.45) corresponds to the following equation:

$$F(x_0, \omega) = H(\omega)e^{i\omega' x_0/V_S} \int_{-\infty}^{\infty} e^{\left[-\frac{2}{T_s^2}\left(1+\left(\frac{T_s}{T}\right)^2 - \frac{ibT_s^2}{4}\right)\tau^2 - \frac{4}{T_s}\left(\frac{x_0}{V_S T_s} + i\frac{\omega' T_s}{4}\right)\tau - \frac{2x_0^2}{V_S^2 T_s^2}\right]} d\tau \times e^{\left[-\frac{2x_0^2}{V_S^2 T_s^2}\right] d\tau} \tag{12.46}$$

Eq. (12.46) multiplies the term $e^{\left[-\frac{2x_0^2}{V_S^2 T_s^2}\right] d\tau}$ by Eq. (12.45) to reduce the impact of the delay time, which can be simplified as [25,30]:

$$F(x_0,\omega)=H(\omega)e^{\left[\frac{i\omega' x_0}{V_S}-\frac{2x_0^2}{V_S^2T_s^2}\right]}\cdot\int_{-\infty}^{\infty}\left[-\frac{2}{T_s^2}\left(1+\left(\frac{T_s}{T}\right)^2-\frac{ibT_s^2}{4}\right)\tau-\frac{4}{T_s}\left(\frac{x_0}{V_ST_s}+i\frac{\omega' T_s}{4}\right)\tau\right]d\tau^2 \tag{12.47}$$

In this regard, Eq. (12.47) expresses the transfer modulus of the Fourier transform $F(x_0,\omega)$ and $H(\omega)$ is the Fourier transform of the matched frequency filter. Solving the integral in the right-hand side of Eq. (12.47) leads to:

$$F(x_0,\omega)=H(\omega)e^{\left[\frac{i\omega' x_0}{V_S}-\frac{2x_0^2}{V_S^2T_s^2}\right]}\cdot\frac{\sqrt{\pi}}{\sqrt{\frac{2}{T_s^2}\left(1+\left(\frac{T_s}{T}\right)^2-\frac{ibT_s^2}{4}\right)}}\cdot e^{\left[\frac{\left(\frac{4}{T_s}\left(\frac{x_0}{V_ST_s}+i\frac{\omega' T_s}{4}\right)\right)^2}{\frac{8}{T_s^2}\left(1+\left(\frac{T_s}{T}\right)^2-\frac{ibT_s^2}{4}\right)}\right]} \tag{12.48}$$

The mathematical simplification of Eq. (12.48) can be written as:

$$F(x_0,\omega)=H(\omega)e^{\left[\frac{i\omega' x_0}{V_S}-\frac{2x_0^2}{V_S^2T_s^2}\right]}\left(\frac{\sqrt{\pi}}{\sqrt{2}}\right)\cdot\frac{T_S}{\sqrt{\left[1+(T_s/T)^2-i(bT_s^2/4)\right]}}\cdot e^{\left[\frac{\frac{16}{T_s^2}\left[\left(\frac{x_0}{V_ST_s}\right)^2+2\left(\frac{x_0}{V_ST_s}\right)\left(\frac{i\omega' T_s}{4}\right)-\frac{\omega'^2T_s^2}{16}\right]}{\frac{8}{T_s^2}\left[1+(T_s/T)^2-i(bT_s^2/4)\right]}\right]} \tag{12.49}$$

The Doppler frequency shift spectral is obtained by the square modulus of the Fourier transform as:

$$|F(x_0,\omega)|^2\cong\frac{4\pi^2}{b^2}e^{\left[-\left(\frac{2}{bT_h}\right)^2\omega^2\right]}\cdot e^{\left[-\left(\frac{2}{bT}\right)^2(\omega-\omega_d)^2\right]}\cdot e^{\left[-\left(\frac{2}{bT}\right)^2\left(\omega-\omega_d-\frac{bx_0}{V_S}\right)^2\right]} \tag{12.50}$$

Eq. (12.50) involves large time-bandwidth products $(bT^2/4)$, $\left(bT_h^2/4\right)$, and $\left(bT_s^2/4\right)$ >, which are reasonable for most SAR data. On this understanding, $\omega_{\max}$ presents the peak of the Doppler spectral density, which is formulated as:

$$\omega_{\max}=\frac{\omega_d\left(1+\left(T_sT^{-1}\right)^2\right)+b\left(x_0V_S^{-1}\right)}{1+\left(T_sT^{-1}\right)^2+\left(T_sT_h^{-1}\right)^2} \tag{12.51}$$

Therefore, Eq. (12.51) can be expressed in terms of the bandwidths by:

$$\omega_{\max} = \frac{\omega_d\left(1+\left(\frac{\partial\omega_s}{\partial\omega_a}\right)^2\right)+b\left(x_0 V_S^{-1}\right)}{1+\left(\frac{\partial\omega_s}{\partial\omega_a}\right)^2+\left(\frac{\Delta\omega_s}{\Delta\omega_h}\right)^2} \tag{12.52}$$

here the $\Delta\omega_a$ represents the received signal with azimuthal bandwidth bT, $\Delta\omega_h$ represents the processor bandwidth bT_h, and $\Delta\omega_s$ is the laser weighting bandwidth. Eq. (12.52) can be reduced in the following form as:

$$\omega_{\max} = \frac{\omega_d + b\left(x_0 V_S^{-1}\right)}{1+\left[\frac{\partial\omega_s}{\partial\omega_a}\right]^2} \tag{12.53}$$

Eq. (12.55) reveals the conventional Doppler expression in terms of the temporal frequency. Consequently, according to the above perspective, the radial velocity of any object in the SAR image then can be computed as a function of the Doppler spectral density peak. Nonetheless, the radar cross-section of the sea surface must be considered rather than a single object location. Hence, the phase of the complex field at the frequency plane is uncorrelated such that [27]:

$$\left\langle dl(x_0)dl^*\left(x_0'\right)\right\rangle = \sigma(x_0)\delta\left(x_0 - x_0'\right)dx_0' \tag{12.54}$$

here $\sigma(x_0)$ is the radar scattering cross-section per unit surface area and the angular brackets denote averaging over subresolution point targets or equivalently infinitesimal scattering elements. The complex field at the frequency plane for a distributed surface is obtained by summing the spatial contributions from the individual infinitesimal scatterers:

$$F(\omega) = \int_{-\infty}^{\infty} F(x_0, \omega)dl(x_0) \tag{12.55}$$

Eq. (12.55) is depleted to determine the spectra frequency of the scattered surface such as land or ocean, which can be signified as a random collection of scatterers. Consequently, the mean Doppler spectral intensity is formulated as:

$$I(\omega) = \left\langle \left|F(\omega)^2\right| \right\rangle = \int_{-\infty}^{\infty} \sigma(x_0)|F(x_0, \omega)|^2 dx_0 \tag{12.56}$$

Eq. (12.56) clarifies numerous significant facts that affect the retrieving ocean current from the SAR image. In this view, the SAR cross-section varies with x_0 owing to the impact of wind gradients in the SAR image and tilt and hydrodynamical effects of the ocean waves, as discussed in previous chapters.

12.11 Radial current velocity based on Doppler spectral intensity

In the circumstance of short wave heights within an identical radar cross-section, let us approximate that $\sigma(x_0) \cong \sigma$. Thus, The Doppler spectral intensity is expressed as:

$$I(\omega) = \frac{2\sigma\pi^{5/2} V_S T_s}{b^2} e^{\left[-\left(\frac{2}{bt_h}\right)^2 \omega^2\right]} \cdot e^{\left[-\left(\frac{2}{bT}\right)^2 (\omega-\omega_d)^2\right]} \tag{12.57}$$

Eq. (12.59) demonstrates that the Doppler spectrum for an identically scattered scene is independent of T_s the laser weighting. Though for an object point or a nonuniform scattered scene, the laser weighting affects the spectral density locus [27, 31].

The Doppler spectral intensity can be formulated based on similar procedures to the Doppler spectral density peak as:

$$\omega_{\max} = \omega_d \left(1 + \left(\frac{T}{T_h}\right)^2\right)^{-1} = \omega_d \left(1 + \left(\frac{\Delta\omega_a}{\Delta\omega_h}\right)^2\right)^{-1} \tag{12.58}$$

Initially, if the processor bandwidth is much larger than the signal bandwidth, then $\Delta\omega_h \gg \Delta\omega_a$, then Eq. (12.58) reduces to $f_{\max} = 2V_r\lambda^{-1}$ or $\omega_{\max} = \omega_d = 2kV_r$, where λ is SAR wavelength and k is wave number. If the $\Delta\omega_h = \Delta\omega_a$ then $\omega_{\max} = 0.5\omega_d$, which denotes the Doppler shifts value. If the ration relating these bandwidths is $\Delta\omega_a = \alpha\Delta\omega_h$, where $0 < \alpha \leq 1.0$, then $\omega_{\max} = \omega_d/(1+\alpha^2)$, implying that $\omega_d < \omega_{\max} \leq \omega_d/2$.

Following Rufenach and Alpers [25], the SAR image intensity can be formulated as:

$$I(x) = \frac{\pi}{2} T^2 \int_{\infty}^{\infty} \sigma(x_0) \frac{\rho_a}{\rho_a'} e^{\left[-\frac{\pi^2}{\rho_a'^2}\left(x - x_0 - \frac{R}{V_S} V_r\right)^2\right]} \tag{12.59}$$

where:

$$\rho'_a = N'\rho_a \left[1 + \frac{1}{N'^2}\left(\frac{\pi T^2}{\lambda} a_r\right)^2\right]^{1/2} \tag{12.59.1}$$

where ρ_a is the stationary target resolution, and:

$$N' = \left[1 + \left(\frac{\Delta\omega_a}{\Delta\omega_s}\right)^2 + \left(\frac{\Delta\omega_a}{\Delta\omega_h}\right)^2\right]^{1/2} \tag{12.59.2}$$

The azimuthal resolution ρ_a' based on Eq. (12.59.1) is given by:

$$\rho_a' = 1.12\rho_a \sqrt{\left[1 + 0.8\left(\frac{\pi T^2}{\lambda} a_r\right)^2\right]} \tag{12.60}$$

which is the degraded resolution instigated by orbital acceleration a_2 and can be equated to the previous result grounded on an infinite processor bandwidth, $\Delta\omega_h > \Delta\omega_a$, implying $N' = 1$.

The radial component of ocean current deduced from SAR images is given in terms of the Doppler peak frequency shift, f_{max}, based on Eq. (12.59.1); consequently, the horizontal ocean current V_c is:

$$V_c = \frac{2}{N}\left[\frac{\lambda V(1+\Delta f_a/\Delta f_h)^2}{2\rho_a \sin\theta \sin\Phi}\right](\Delta f_a)^{-1} \tag{12.61}$$

In the circumstances of $\Delta f_a = (\Delta\omega_a/2\pi)$ and $\Delta f_h = (\Delta\omega_h/2\pi)$, the image frequency and the SAR image intensity spectral are simulated.

12.12 Robust model for simulating surface current in SAR imaging

This section addresses the question of Doppler centroid (f_{DC}) ambiguity impact on an accurate retrieving of sea surface current in synthetic aperture radar (SAR). Two hypotheses evaluated are: (i) accurate Doppler centroid can model using both Wavelength Diversity Ambiguity Resolving (WDAR) and Multilook Beat Frequency (MLBF) algorithms; and (ii) accurate sea surface current speed retrieved in SAR satellite data using the robust estimation [13, 33, 34].

The term *robust estimation* means estimation techniques that are robust concerning the presence of gross errors in the data. Gross errors are defined as observations which do not fit to the stochastic model of parameter estimation [35]. In this context, uncertainties in the estimation of Doppler centroid frequency can lead to completely false results of sea surface current and might even prevent convergence of adjustment. Robust estimators are estimators which are relatively insensitive to limited variations in the frequency distribution function of the Doppler centroid frequency f_{DC}. Chapron et al. [36], however, did not take into account the problems of estimating the Doppler centroid, which might begin from a range-compressed dataset acquired by conventional single pulse repetition frequency (PFR) [37,38]. De Stefano and Guarnieri [39] stated that for efficiency, the constraint of operating on range-compressed data is required. Following Stefano and Guarnieri [39], the ambiguous estimation and Wavelength Diversity Ambiguity Resolving algorithm (WDAR and Multilook Beat Frequency (MLBF have been implemented to correct f_{DC} ambiguity and to fit a fine polynomial estimate in SAR images. First, a SAR image is divided into several blocks. In each block, both a second-order statistic estimator (WDAR) and a higher-order technique (MLBF) have been exploited to resolve coarse unambiguous Doppler centroid. These techniques have been chosen due to the large variation of f_{DC} with range, as can be seen clearly in IW mode data. The polynomial inversion model as given by Stefano and Guarnieri [39] is used:

$$f_{DC}(a,r) = Xr^2 + Yr + Za + h \tag{12.62}$$

where a and r are range and azimuth indexes of the samples at the center of each block, respectively, and X, Y, Z, and h are the polynomial coefficients to be estimated.

Two steps are required to achieve the polynomial inversion technique: (i) the wrapped plane is regressed and (ii) the model is then inverted on the residuals (*res*). The selection between the two steps is mainly done utilizing a threshold on the contrast parameter, which is based on the pixel intensity of each block. For instance, unambiguous f_{DC} is computed with

WDAR in low contrast blocks as compared to MLBF. Taking into account the value of the ambiguity (p) and the polynomial parameters (X, Y, Z, h), the unambiguous f_{DC} polynomial can be given by this formula:

$$\widehat{f}_{DC}(a,r) = X_{res}r^2 + (Y_p + Y_{res})r + (Z_p + Z_{res})a + (h_p + h_{res}) \tag{12.63}$$

Finally, offset frequency is implemented by subtraction of the MLBF estimate from WDAR. This is done with an assumption that the ambiguity estimate based on the MLBF technique is correct. Following Rufench et al. [27], the SAR ocean current values must be converted from the radial component V_r to the horizontal ocean component V_c by a given equation:

$$V_c = \frac{c^* 0.5\lambda \widehat{f}_{DC}(a,r)}{\sin\theta \sin\varphi} \tag{12.64}$$

where θ is the incidence angle of SAR data, φ the azimuth angle, and c is a constant value, which is determined by using least square method between in situ measured ocean current and the Doppler centroid $\hat{f}_{DC}(a,r)$, which is a function of surface current velocity.

The crucial issue that can be raised due to the performing of the least square method is a lack of robustness. The least-squares error function to be minimized is as follows [30]:

$$e^2(V_c) = d^{-1}\sum_i w^{-1}\left[V_i - V_c\left(\hat{f}_{DC_i}(ar)\right)\right]^2 \tag{12.65}$$

where V_i is the real measurement of surface current by using AWAC equipment, i is the number of observations, w is a weight that is assigned to each respective observation, and d is the number of degrees of freedom.

The robust standard deviation $\hat{\sigma}$ is estimated by a combination of least median of squares (LMedS) method with a weighted least squares procedure, which can be expressed as [13,33,34]:

$$\hat{\sigma} = 1.5\{1 + 5/n - p\} med\sqrt{r_i^2} \tag{12.66}$$

where r_i is the residual value, med is median absolute deviation of residual value, the factor 1.4826 is for consistent estimation in the presence of Gaussian noise, and the term $5/(n-p)$ is recommended as a finite sample correction.

The parameters can then be estimated by solving the weighted least squares problem:

$$\min \sum_i w(r_i){r_i}^2 \tag{12.67}$$

Following Marghany and Mazlan [34], the quasilinear transforms of tidal current (V) can be given as:

$$V = H\left\{V_c;\ \min \sum_i w(r_i){r_i}^2;\ W\right\} \tag{12.68}$$

where H represents the linear operator, which is the tidal current-SAR transform, and W represents parameters of the tidal current-SAR map, which are readily based on the physical conditions of current pattern movements (i.e., velocities and direction) and SAR properties such as Doppler frequency shift.

12.13 Tidal current direction estimation

The main problem in simulating the ocean surface current direction is that the imaged current pattern presents in the range direction of the SAR data. In this study we introduce a new approach to simulate the ocean current direction, which was adopted from the study of Marghany [13]. According to Marghany and Mazlan [33], the tidal current has two components, which are in azimuth and range directions. In this study, the edge of the frontal zone area is chosen and then divided to sequence kernel windows with the frame size of $n \times n$. Since the frontal zone consists of several adjoining pixels that must have higher signal amplitude than the surrounding pixels, then the Doppler spectrum of range compressed SAR data is estimated by performing a fast Fourier transform (FFT) in the azimuth direction. Further details of this approach are in Marghany and Mazlan [33]. The current speed direction Θ can be given by:

$$\Theta = \tan^{-1}\left[\frac{\left(\lambda\hat{f}_{DC}(a,r)\right)(2\sin\theta)^{-1}}{v_s(1-(1-2\Delta x\partial x v_s))^{-0.5}\left(\Delta\hat{f}_{DC}(a,r)R\lambda\right)^{-1}}\right] \tag{12.69}$$

where v_s is satellite velocity, R is slant range, Δx is the displacement vector, and ∂x is the pixel spacing in the azimuth direction.

The robust model is examined with a Sentinel-1A SAR image, which was acquired on November 9, 2019.

12.14 Ocean current retrieving from SAR data, case study: East coast of Malaysia

The SAR data acquired in this study are derived from a Sentinel-1A SAR image, which involves C_{VV} band data (Fig. 12.14). The Sentinel-1 SAR satellite was launched by the European Space Agency in April 2014. It operates in C-band with diverse polarizations and beam modes. Dual IW mode Sentinel-1 SAR images are collected in this study, acquired in VV-polarization in eastern Malaysian coastal waters from October 2019 through November 2019. The radar incidence angle ranges from 20 to 47 degrees. The coverage of each image contains the AWAC deployment location and drogue deployment, as discussed in Sections 12.4.2 and 12.4.3.1. As an example, a SAR image was acquired at 22:47:34 UTC on November 9, 2019, in eastern Malaysia coastal waters. Sentinel-1 SAR has a high spatial resolution ranging from 10 to 20 m in data collection. Although a bit of distortion by low wind zone exists in the image, which may affect the homogeneity of the sea surface backscattering and the good-quality power spectra of NRCS. The normalized backscatter ranges between −30 dB and −10 dB. The maximum normalized radar cross-sections are found along the large eddy in IW mode data (Fig. 12.14).

The retrieved sea surface current velocities using IW mode have been considered across the location of the AWAC rider buoy (Fig. 12.4). The examined area extends from inshore to offshore and approximately covered 200 km. The offshore normalized backscatter seems to be instable, with a maximum value of −12 dB. As well as the normalized radar cross-section approaching the inshore zone, it tends to be irregular owning to low wind zone and a large

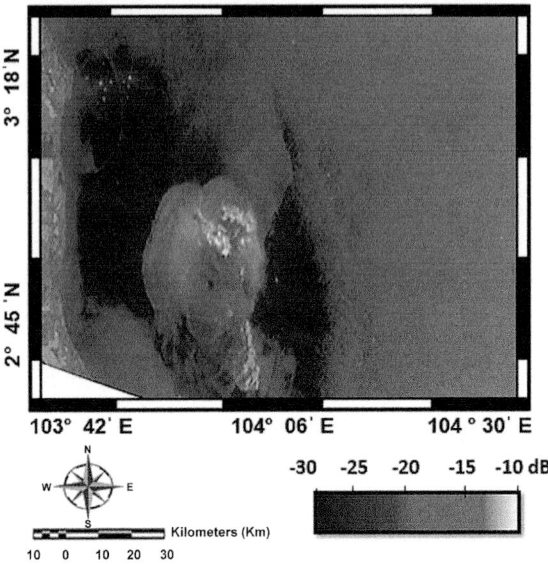

FIG. 12.14 Normalized radar-cross-section of IW mode data.

eddy. The low wind zone has a lower normalized radar cross-section of −28 dB than the surrounding environment (Figs. 12.14 and 12.15). On the other hand, the eddy has the largest normalized radar cross-section of −10 dB, compared to the surrounding sea. The diameter of this eddy is 40 km and its radius is 20 km and it is imaged by a shallow incidence angle between 34 and 35.8 degrees (Fig. 12.16). The influence of the incident angle of 42–45 degrees in the inshore surface current is characterized by changing of the normalized radar cross-section and frequency patterns. The mesoscale eddy dominates by a higher frequency of 320 Hz than the surrounding sea (Fig. 12.17) with the highest velocity of $1.8\,m\,s^{-1}$

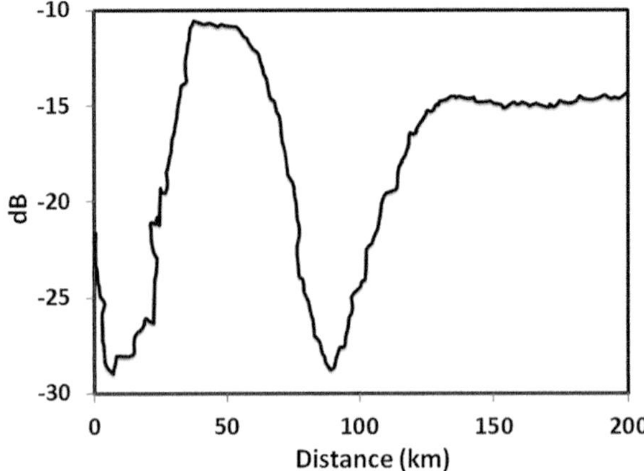

FIG. 12.15 Normalized radar cross-section along 200 km.

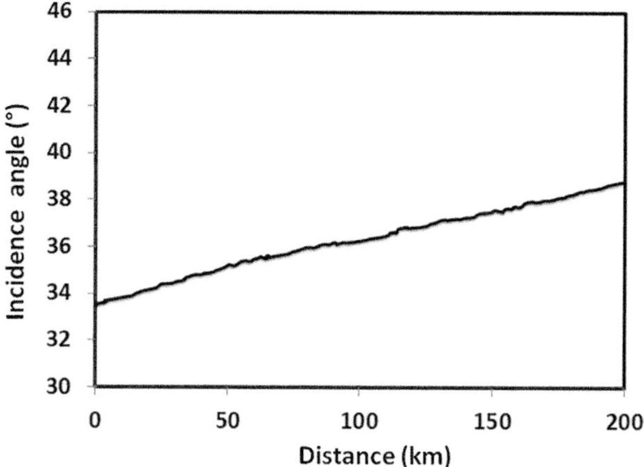

FIG. 12.16 Variation of incidence angle along 200 km.

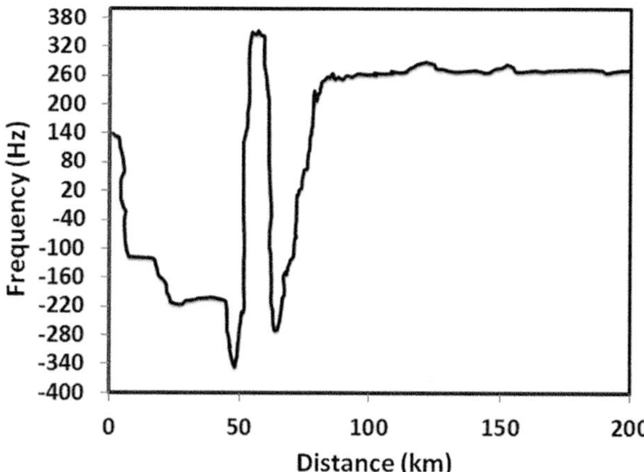

FIG. 12.17 Variations of Doppler spectral frequency.

(Fig. 12.18). On this understanding, the radar signature of the large-scale eddy (mesoscale) is very strong owing to the strong hydrodynamic modulation by horizontal current shear. The anticlockwise rotation indicates the downwelling of warm water. An eddy is then visualized on SAR images through a characteristic surface pattern consisting of bright (i.e., high radar backscatter), narrow, curvilinear, concentric bands that spiral inward into the pattern.

Eddies usually show up in SAR images as a result of wave/current interaction, which outlines the curved shape of the eddy, or are revealed indirectly through the presence of natural film trapped within spiraling lines associated with the eddy's orbital motion. In connection with eddies having strong thermal signals, such as warm or cold rings, the eddy can also be expressed by the change in wind stress across the temperature front [31]. In this regard, the offshore sea surface current starts with a maximum speed of $1.2\,\mathrm{m\,s^{-1}}$. In addition, the sea

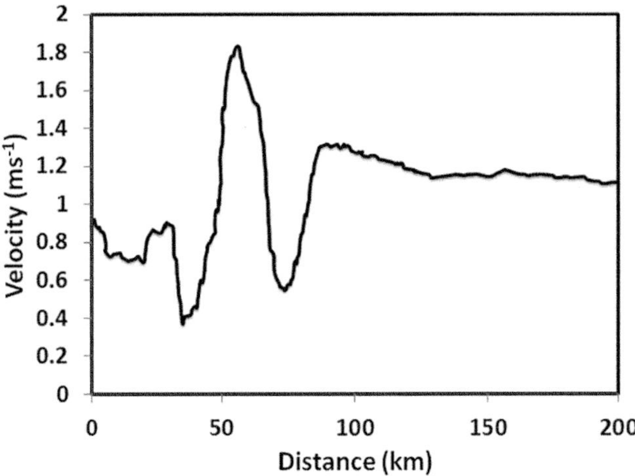

FIG. 12.18 The retrieved current velocity in Sentinel-1A IW mode.

surface current approaching the inshore zone tends to be a turbulent flow. The turbulent flows tend to range from 0.4 to 1.8 $\mathrm{m\,s^{-1}}$ (Fig. 12.18).

Fig. 12.19 reveals the Doppler velocity retrieved using an estimator (WDAR) and a higher-order technique (MLBF). The Doppler velocity ranges from 0 to 2 $\mathrm{m\,s^{-1}}$. Land and low wind zone have a zero Doppler rate. It is interesting to find the rate of Doppler spatial variation has a positive value both onshore and offshore. Indeed, WDAR and MLBF techniques exploited to resolve the coarse unambiguous of Doppler centroid effects in estimating ocean surface current.

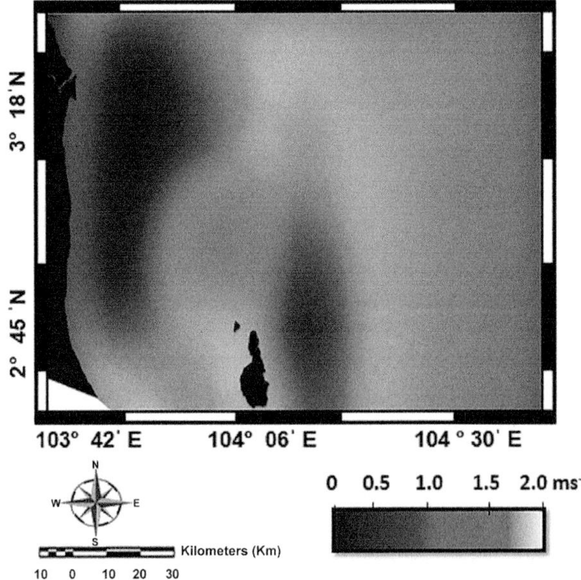

FIG. 12.19 Doppler velocity retrieved using WDAR and MLBF techniques.

Fig. 12.19 also demonstrates the large-eddy scale flows associated with boundary currents. The SAR frontal boundaries are represented by both dark and bright radar cross-sections of varying cross frontal width, in particular the bright front in the central part of the image. Spatial variations of the radar cross-section are usually induced by larger-scale features and processes such as long gravity waves, variable near-surface wind fields, air/sea temperature differences, and upper ocean circulation features, including mesoscale meandering fronts and eddies. Under moderate winds, between 3 and $10\,\mathrm{m\,s^{-1}}$, the SAR is seen to be capable of revealing current boundaries, including meanders and eddies. It is suggested that the corresponding imaging mechanisms alternate among the following:

(i) short gravity-wave/current interaction along with shear and/or convergence zones within the front;
(ii) changes in wind stress induced by strong gradients in the sea surface temperature; and
(iii) long-gravity-wave/current refraction.

Consistent with Johannessen et al. [31], the SAR image signatures are interpreted to be a manifestation of short-gravity wave/current interaction along the current fronts. As the short gravity waves propagate into and/or across the current front, they change steepness and propagation direction as long as the wind is moderate. Consequently, the SAR is capable of detecting the current fronts [40].

The Doppler spectra ambiguity is shown in Fig. 12.20A. It can be seen that the spectra intensity peaks are repeated along with azimuth and range directions. Fig. 12.20B, nevertheless, presents the Doppler spectrum frequency, which is positioned along neither the range direction nor the azimuth direction. The Doppler spectra are characterized by spectra intensity of 0.9 Hz and bandwidth of 50 Hz. In fact, the robust estimators (WDAR and MLBF) are estimators that are relatively insensitive to restricted variations in the frequency distribution function of the Doppler centroid frequency f_{DC}. Furthermore, both algorithms are capable of retrieving the correct Doppler Centroid ambiguity and fitting a fine polynomial estimate both on the uniform and contrasted scenes [39]. The sharp Doppler spectra peak is established by using WDAR and MLBF algorithms, as compared to one estimated directly by using a conventional algorithm. These results confirm the spatial variation of Doppler speed, which is illustrated in Fig. 12.9.

Fig. 12.21 depicts the sea surface current pattern retrieved using the robust technique. The current pattern movements are revealed clearly. The current velocity exceeds from offshore toward onshore as the onshore currents have highest velocity of $1.8\,\mathrm{m\,s^{-1}}$ than offshore one. The northeast current flow was the dominant feature along the coastal waters of Malaysia (Fig. 12.6) during November 2019. The retrieved eddy is well reconstructed, which matched accurately with the result of the drogue trajectory movement in Fig. 12.2. The drogue trajectory measurements, AWAC measurements, and retrieved surface current from IW are accurately correlated with r^2 of 0.87 and RMSE of $\pm 0.034\,\mathrm{m\,s^{-1}}$ (Fig. 12.22).

The computational efficiency of sea surface current from SAR data, consequential, is perfects and equips for real-time processing during SAR satellite overpasses. In general, SAR ocean current modeling, which is based on Doppler centroid analyses, through future research can perhaps deliver extra-precise and less ambiguous sea surface current flows in SAR data. This ratifies the study of Marghany and Mazlan [33]. Furthermore, the ambiguous estimated techniques are based on power spectrum estimation. Accordingly, the ambiguous estimator is the autocorrelation that comprises the expected phase of the first sample of the azimuth autocorrelation. Certainly, this estimator is implemented with an offset frequency.

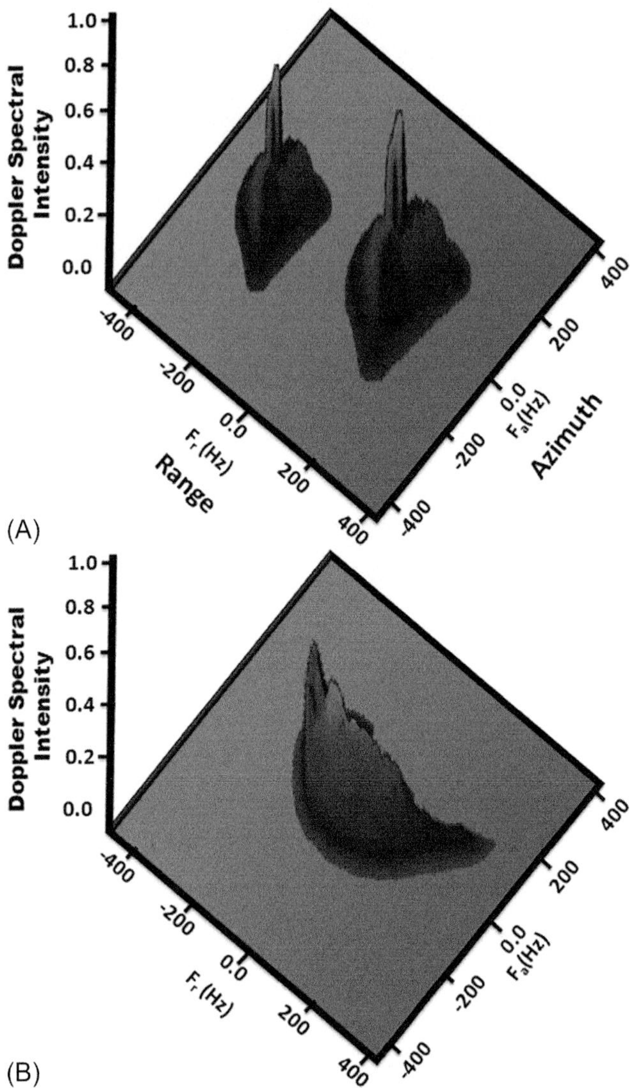

FIG. 12.20 Doppler spectra intensity (A) conventional algorithm and (B) robust estimator WDAR and MLBF.

Generally, the robust algorithm can retrieve the precise Doppler centroid uncertainty using a polynomial estimator. In separate blocks, subsequently, a coarse unambiguous estimator is delivered by exploiting both a second-order statistic estimator (WDAR) and a higher-order technique (MLBF). The weighted average of the block measures delivers the precise certainty of the robust estimator. This allows us to approximate and standardize the offset frequency constant. Incidentally, WDAR is precisely determined the offset frequency than MLBF. This confirms the previous work of De Stefano and Guarnieri [39]. Future work will aim to improve the accuracy of modeling surface current in SAR data by applying an appropriate algorithm and using a random variation of spatial in situ measurements.

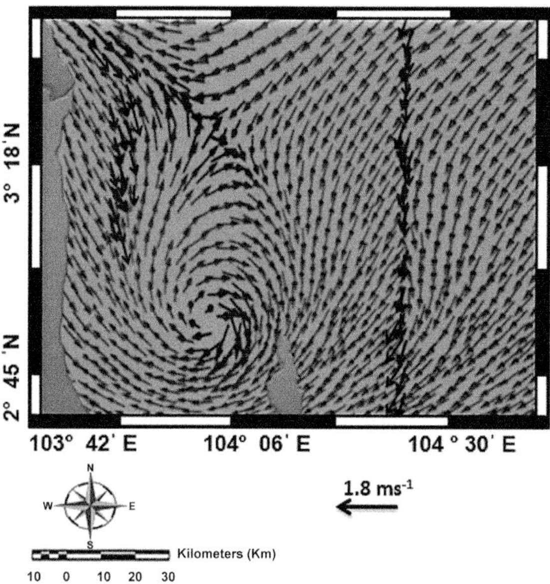

FIG. 12.21 Robust technique for retrieving sea surface current pattern.

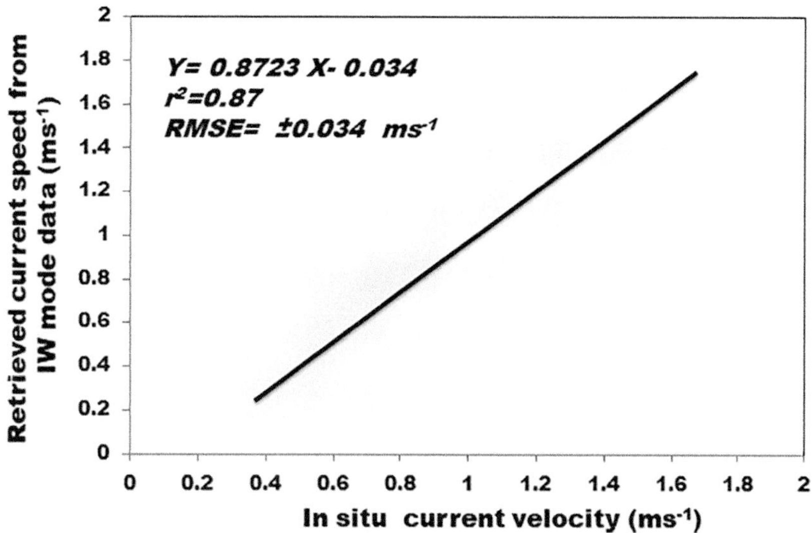

FIG. 12.22 Robust statistical analysis between in situ measurements and IW mode retrieved surface currents.

12.15 Quantization of large scale eddy in SAR image

The significant question is: can quantum mechanics explain the SAR imaged large-scale eddy? It is challenging to explain how the fluid begins to rotate and how rotating flow transfers angular momentum—a quality that speaks to how fast the liquids will spin. When rotating wind particles fall into a specific area on the sea, viscosity assists the wind particles to steer the surrounding water into the rotation, creating vortices or eddy currents in the process. This viscous drag diminishes the variance in motion between the two water density bodies. Shallow coastal water, conversely, tolerates this transformation. Reasonably throughout the convection of particles, it is further well-organized for wind particles, and wave-current interaction to transfer angular momentum through quantum mechanical state interfaces.

Moreover, entanglement occurs between the different quantum states of wind stress that transfer the momentum energy to the water body—as a stirring cup of coffee to build up water particle rotations through the water column. In this scenario, the wind stress momentum energy is acting as the force, or stirring force, that causes water particles to spin around its orbit to form eddies. The quantum entanglement builds up to steer this spinning movement downward through entanglement within their surroundings. However, viscosity friction of the surroundings can perturb this entanglement by decaying through bottom friction.

On the point of view of entanglement, ocean surface frequently shows a tendency for wind-driven ocean currents to be further in the direction of the wind than what realizes from an Ekman flow consideration with steady eddy viscosity. This can comparatively be owing to the circumstance that approximately of the transport is instigated by surface wind waves in a route close to the wind. Nonetheless, it also implies that the turbulent eddy viscosity must *grow* with depth in the upper part of the surface layer. Since the eddy coefficients are small in the interior of the ocean, the eddy viscosity must therefore increase to a maximum value at a certain depth, before it decreases to attain its deep ocean value.In this circumstance, the eddy flow is generated as entanglement to wind stress and it decays at a certain large depth as shown in an Ekman spiral owing to the collapsing of the entanglement wave function of wind-eddy generation through the deep water (Fig. 12.11).

References

[1] Gill AE. A discussion on ocean currents and their dynamics-ocean models. Philos Trans R Soc Lond A Math Phys Sci 1971;270(1206):391–413.

[2] Wahr JM, Jayne SR, Bryan FO. A method of inferring changes in deep ocean currents from satellite measurements of time-variable gravity. J Geophys Res Oceans 2002;107(C12):11.

[3] Officer CB. Physical oceanography of estuaries (and associated coastal waters). John Wiley & Sons, Inc.; 1976

[4] Talley LD. Descriptive physical oceanography: An introduction. Academic press; 2011.

[5] Bennett AF. Inverse methods in physical oceanography. Cambridge University Press; 1992.

[6] Stewart RH. Introduction to physical oceanography. College Station: Texas A & M University; 2008.

[7] Knauss JA, Garfield N. Introduction to physical oceanography. Waveland Press; 2016.

[8] Wyrtki K. Physical oceanography of the Indian Ocean. In: The biology of the Indian Ocean. Berlin, Heidelberg: Springer; 1973. p. 18–36.

[9] Kullenberg G. Physical oceanography. In: Elsevier oceanography series. vol. 30. Elsevier; 1981. p. 135–81.

[10] Coachman LK, Aagaard K. Physical oceanography of Arctic and subarctic seas. In: Marine geology and oceanography of the Arctic seas. Berlin, Heidelberg: Springer; 1974. p. 1–72.

[11] Maged Mahmoud M. Coastal circulation off Kuala Terengganu. Fakulti Pengajian Maritim dan Sains Marin; 1994 [Master dissertation].

[12] Marghany M. Robust model for retrieval sea surface current from different RADARSAT-1 SAR mode data, In: 2009 IEEE international conference on signal and image processing applications Nov 18, 2009IEEE; 2009. p. 492–5.

[13] Marghany M. Finite difference model for modeling sea surface current from RADARSAT-1 SAR data, In: 2009 IEEE international geoscience and remote sensing symposium Jul 12, 2009vol. 2. IEEE; 2009. p. II–487.

[14] Weber JE. Steady wind-and wave-induced currents in the open ocean. J Phys Oceanogr 1983;13(3):524–30.

[15] Grue J, Gjevik B, Weber JE, editors. Waves and nonlinear processes in hydrodynamics. Springer Science & Business Media; 2012.

[16] Weber JE. Ekman currents and mixing due to surface gravity waves. J Phys Oceanogr 1981;11(10):1431–5.

[17] Høydalsvik F, Weber JE. Mass transport velocity in free barotropic Poincaré waves. J Phys Oceanogr 2003;33 (9):2000–12.

[18] Weber JE. Heat, salt and mass transfer at a cold ice wall exposed to seawater near its freezing point. Geophys Astrophys Fluid Dyn 2003;97(3):213–24.

[19] Weber JE, Melsom A. Volume flux induced by wind and waves in a saturated sea. J Geophys Res Oceans 1993;98 (C3):4739–45.

[20] Weber JE, Broström G, Saetra Ø. Eulerian versus Lagrangian approaches to the wave-induced transport in the upper ocean. J Phys Oceanogr 2006;36(11):2106–18.

[21] Broström G, Christensen KH, Drivdal M, Weber JE. Note on Coriolis-Stokes force and energy. Ocean Dyn 2014;64 (7):1039–45.

[22] Tomczak M, Godfrey JS. Regional oceanography: An introduction. Elsevier; 2013.

[23] Yoshida Z, Mahajan SM. Quantum spirals. J Phys A Math Theor 2015;49(5):055501.

[24] Levita G, Calderon J. A new method of parameter estimation of oceanographic coastal zone by electromagnetic measurements, In: Proceedings of the 43rd IEEE Midwest symposium on circuits and systems (Cat. No. CH37144) Aug 8, 2000vol. 3. IEEE; 2000. p. 1434–7.

[25] Alpers WR, Ross DB, Rufenach CL. On the detectability of ocean surface waves by real and synthetic aperture radar. J Geophys Res Oceans 1981;86(C7):6481–98.

[26] Bahar E, Rufenach CL, Barrick DE, Fitzwater MA. Scattering cross section modulation for arbitrarily oriented composite rough surfaces: full wave approach. Radio Sci 1983;18(05):675–90.

[27] Rufenach CL, Shuchman RA, Lyzenga DR. Interpretation of synthetic aperture radar measurements of ocean currents. J Geophys Res Oceans 1983;88(C3):1867–76.

[28] Alpers W, Schröter J, Schlude F, Müller HJ, Koltermann KP. Ocean surface current measurements by an L band two-frequency microwave scatterometer. Radio Sci 1981;16(01):93–100.

[29] Barrick DE, Evans MW, Weber BL. Ocean surface currents mapped by radar. Science 1977;198(4313):138–44.

[30] Gonzalez FI, Rufenach CL, Shuchman RA. Ocean surface current detection by synthetic aperture radar. In: Oceanography from space. Boston, MA: Springer; 1981. p. 511–23.

[31] Johannessen JA, Chapron B, Collard F, Backeberg B. Use of SAR data to monitor the Greater Agulhas Current. In: Remote sensing of the African seas. Dordrecht: Springer; 2014. p. 251–62.

[32] Hansen MW, Kudryavtsev V, Chapron B, Johannessen JA, Collard F, Dagestad KF, Mouche AA. Simulation of radar backscatter and Doppler shifts of wave–current interaction in the presence of strong tidal current. Remote Sens Environ 2012;120:113–22.

[33] Marghany M, Hashim M. Robust model for sea surface current simulation from radarsat-1 SAT data. J Converg Inf Technol 2008;3(2):45–9.

[34] Marghany M. Developing robust model for retrieving sea surface current from RADARSAT-1 SAR satellite data. Int J Phys Sci 2011;6(29):6630–7.

[35] Messaoudi M, Sbita L, Abdelkrim MN. A robust nonlinear observer for states and parameters estimation and on-line adaptation of rotor time constant in sensorless induction motor drives. Int J Phys Sci 2007;2(8):217–25.

[36] Chapron B, Collard F, Ardhuin F. Direct measurements of ocean surface velocity from space: interpretation and validation. J Geophys Res Oceans 2005;110(C7):1–17.

[37] Hermansen S.C. The fate of the Atlantic water in the North Icelandic Irminger Current. The University of Bergen; 2012. [Master's thesis].

[38] Raney RK. Synthetic aperture imaging radar and moving targets. IEEE Trans Aerosp Electron Syst 1971;3:499–505.

[39] De Stefano M, Guarnieri AM. Robust Doppler Centroid estimate for ERS and ENVISAT, In: IGARSS 2003. 2003 IEEE international geoscience and remote sensing symposium. Proceedings (IEEE Cat. No. 03CH37477) Jul 21, 2003vol. 6. IEEE; 2003. p. 4062–4.

[40] Johannessen JA, Digranes G, Espedal H, Johannessen OM, Samuel P, Browne D, Vachon P. SAR ocean feature catalogue. vol. 1174. ESA Special Publication; 1994.

CHAPTER

13

Relativistic quantum of nonlinear three-dimensional front signature in synthetic aperture radar imagery

13.1 What is meant by quantum coastal front?

An ocean front is a frontier splitting two masses into waters of different densities and is the major cause of gradient variation of physical seawater properties (Fig. 13.1). For instance, different water masses are frequently separated by a front contrasting in temperature and salinity. Fronts, in this context, occur at a wide range of scales, commencing with those bent within an estuary between inflowing water and the estuary water. Other fronts consequently exist on the continental shelf, separating a zone of coastal water from oceanic water, or a stratified water mass from one that is vertically mixed. Fronts also occur on a large scale in the deep ocean, between water masses of different physical properties. The boundary between warm, salty subtropical waters and Antarctic waters (Fig. 13.2) is found in all three ocean basins. A commonly used criterion is that it is found at the latitude at which the salinity at a depth of 100m drops below 34.9 practical salinity units (PSU). The essential feature of a front bordering a plume is the density difference between the water on the two sides of it, but other features are often present, enabling it to be detected visually (Fig. 13.3). In this regard, the RADARSAT-1 SAR image captures the boundary of water plume out of the Columbia River (Fig. 13.2). There is often a color difference between water masses, arising from a greater concentration of phytoplankton or suspended particles in one than another. The front itself is frequently marked by a line of foam or floating debris. Bowman and Iverson [1] stated that the foam line (Fig. 13.4) is located at the surface convergence—the detritus line where buoyant objects are trapped by currents moving in opposite directions at the surface, and near the interface and the color front, where upwelled light undergoes a distinct spectral shift, approximating the steeply descending isopycnals [2]. Linear convergences are often known as fronts, especially when water properties (e.g., temperature, salinity, and productivity) are markedly different on either side of the convergence.

https://doi.org/10.1016/B978-0-12-820785-7.00017-4

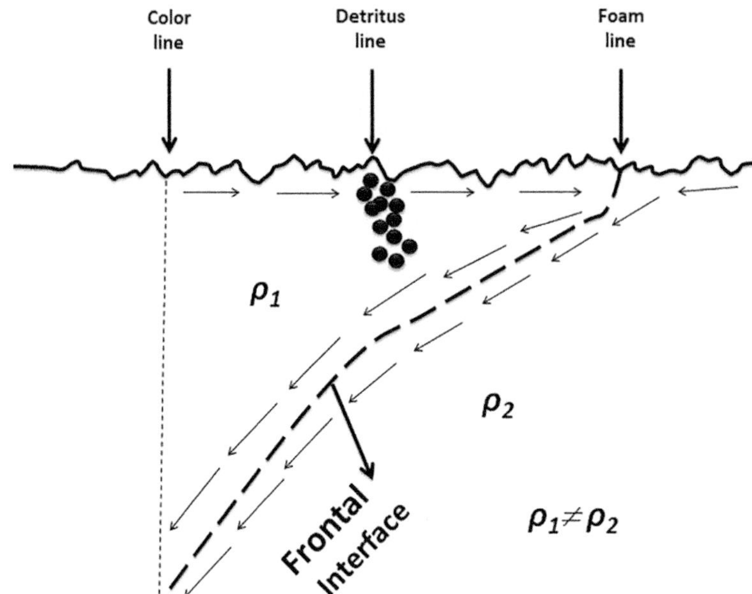

FIG. 13.1 Schematic cross-section of shallow water front with different densities ρ_1 and ρ_2.

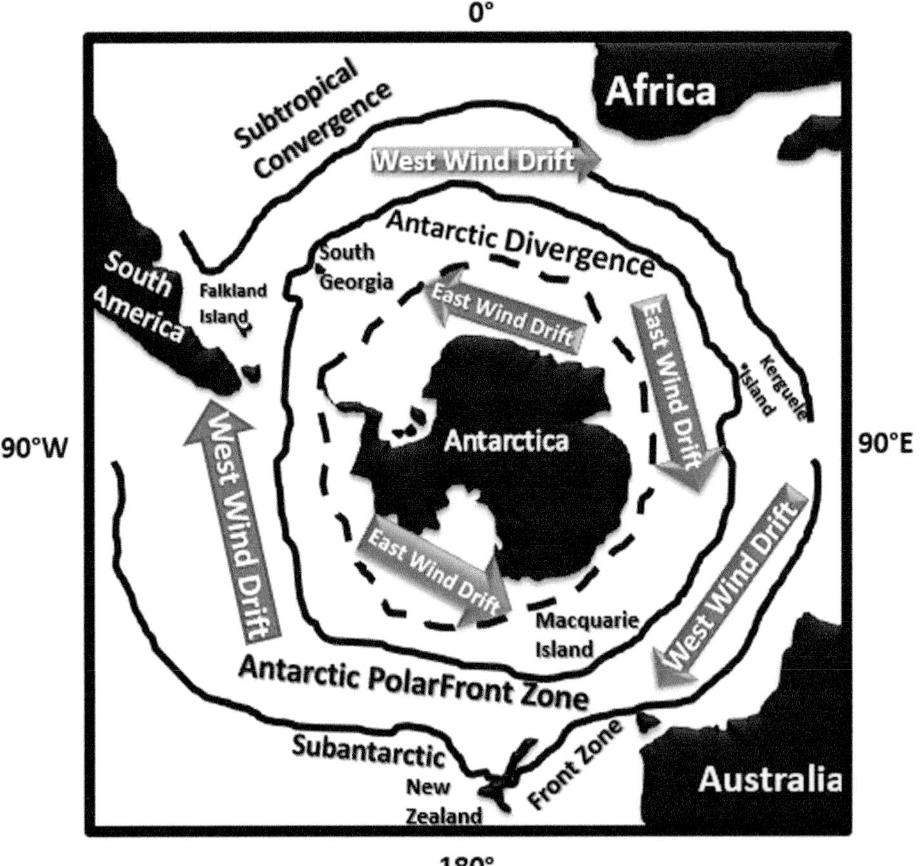

FIG. 13.2 The front between subtropical waters and Antarctic waters.

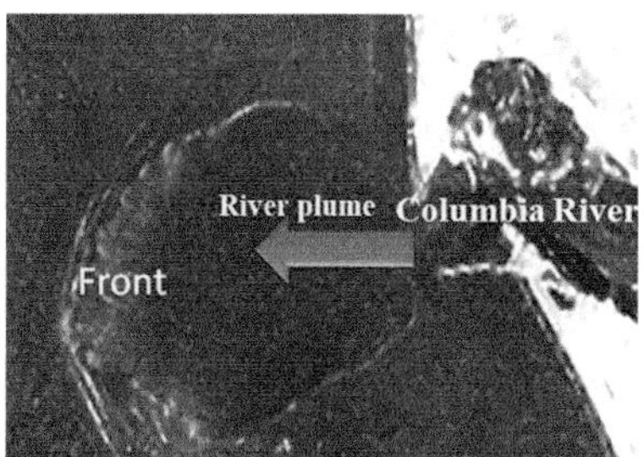

FIG. 13.3 Outflow river plume front in RADARSAT-1 SAR image.

FIG. 13.4 Foam lines as front indicators.

Flow is primarily parallel to temperature fronts, and strong currents can exist along fronts, even though the front may not move. Fronts therefore separate two different quantum states with different quantum entanglement. These two quantum states spin in opposite directions. One spins downward, due to high density, and one spins upward due to low density.

Let us assume that there is a duality of the quantum states from the density gradient to form two different water masses. This duality of quantum states is a function of the density gradient through water depth. In this view, let $\Psi=[\psi_1,\psi_2]$ as ψ_1 and ψ_N be two different quantum states of the water properties in the circumstance $\psi_1 \neq \psi_2$ and also $\psi_1 \notin \psi_2$. Therefore, let F be the boundary that separates between the quantum states in the circumstance of $F \in \psi_1$ and $F \in \psi_2$. On this understanding, the quantum mechanics of the different duality states can be mathematically expressed as:

$$\Psi=\begin{pmatrix}\psi_1\\ \psi_2\end{pmatrix}=\begin{pmatrix}F\left(u\dfrac{\partial\rho}{\partial x}\right)\\ F\left(v\dfrac{\partial\rho}{\partial y}\right)\end{pmatrix} \tag{13.1}$$

It is assumed in Eq. (13.1) that the front F separates two different quantum water masses: $\left(u\dfrac{\partial\rho}{\partial x}\right)$, which is the buoyancy flux along-shelf, and $\left(v\dfrac{\partial\rho}{\partial y}\right)$, which is cross-shelf buoyancy flux. The balance is achieved by decreasing the density of the time $\left(\dfrac{\partial\rho}{\partial t}<0\right)$, for instance. The front can be identified based on quantum mechanics as:

$$F = \{\psi_1 \in \Psi \neg \exists \psi_2 \in \Psi \ \ F|\psi_1\rangle \neq F|\psi_2\rangle\}. \tag{13.2}$$

On the other hand, the front F is mathematically expressed as:

$$F := \{F|\psi\rangle \,|\, |\psi\rangle \in \Psi\}. \tag{13.3}$$

Eq. (13.3) demonstrates that the front occurs in two different quantum states, which are spinning in different directions to sustain the occurrence of the separated boundary. On this understanding, F continues to belong to both quantum states as well, as they do not achieve steady-state conditions. Consequently, the spinning of the Earth around its axis significantly influences the dynamics of larger fronts, whereas small-scale fronts are believed to be influenced by nonlinear inertial and frictional effects alone.

In quantum mechanics, an ocean system can be specified by its wave function. The ocean is considered as a continuous quantum system; the wave function may be written as $\psi(\mathbf{r}, t)$, and the square of the absolute value of this wave function $|\psi(\mathbf{r}, t)|^2$ provides the probability of locating the water parcel at point $\mathbf{r}$ at time t. For discrete systems, such as the spin of the water parcel owing to density fluctuations, wind stress, and the Earth's rotation, the wave function can be presented as a ket vector: $|\psi\rangle = \alpha|+\rangle + \beta|-\rangle$. The vector $|+\rangle$ signifies spin $z+$, i.e., light density, and $|-\rangle$ embodies spin $z-$, i.e., high density. When the spin occurs, we obtain spin up with probability α^2 and spin down with probability β^2. In this regard, $|\psi\rangle$ is a superposition of the so-called basis states $|+\rangle$ and $|-\rangle$. On this understanding, two states of different water physical characteristics exist and spin in opposite directions without mixing owing to the existence of the front boundary. In other words, the front is considered as the spin boundary between two different quantum states and is represented as a wave function. In this scenario, it can be easy to determine the probability of this front existing in space and time owing to changes in the water's physical characteristics such as temperature, salinity, and density.

13.2 Signature of a front in a single SAR image

The critical question that arises is: what is the signature of a front in SAR images? The backscatter is a key to distinguish between features in SAR, so what is the functionality of radar backscatter regarding the image front?

The front feature appears brighter than the surrounding sea environment in SAR images. Local convergence of surface divergence creates current flow, which tends to modulate the energy of surface wind waves. Consequently, arrays steeper wave zones along with the convergences and lower amplitude waves at divergences zones are generated. In this context, SAR detects heterogeneities of the surface roughness as the strongest variations of backscatters, which can be identified clearly in radar backscatter cross-section. In this understanding, the surface features that outlook has the greatest strappingly in the SAR image; which is confined fronts where convergence or divergence is robust, steering to buoyant or obscure shapes.

For instance, RADARSAT-1 SAR F1 mode data acquired from the coastal waters of Kuala Terengganu, Malaysia, on March 26, 2004, reveals a curved front close to the mouth of the Terengganu River (Fig. 13.5). The RADARSAT-1 F1 mode backscatter cross-section of the front has a maximum value of −22 dB compared to the surrounding sea environment. The maximum backscatter value of −22 dB is found across a curved brightness front, and this

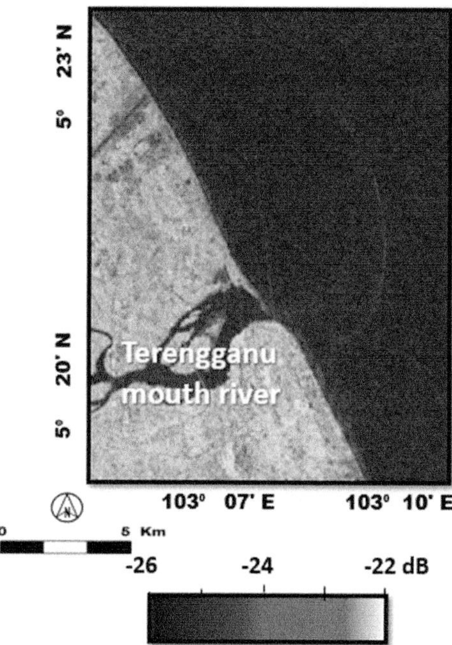

FIG. 13.5 Backscatter cross-section along the curved front in RADARSAT-1 SAR F1 mode data.

variation of radar backscatter cross-section is due to the current boundary gradient. Consistent with Vogelzang et al. [3], ocean current boundaries are often accompanied by changes in surface roughness, which can be detected by SAR. These surface roughness changes are due to the interaction of surface waves directly with surface current gradients. These interactions can cause an increase in surface roughness and radar backscatter [4, 5].

The single polarization HH in X-band Terra SAR strip map data identified a curved front along the coastal water of Terengganu, Malaysia, associated with turbulent flow on August 22, 2011 (Fig. 13.6). The curved front has a maximum backscatter cross-section of −10 dB compared to the surrounding sea environment. The dominant front occurrence along the Kuala Tereanggu River mouth is known as Marghany's front, as Marghany's studies are the primary studies that explained and demonstrated the occurrence of such fronts using SAR data [5–7]. Thus, the Terra SAR strip map image shows up fragments of the turbulence field where nonlinear manners of different current boundary interactions are close by driving the formation of fronts (frontogenesis) as the eddies grow. However, the hydrodynamic modulation for SAR imaging of the front and eddies is not only reliant on surface convergence or divergence velocity, but is also a function of the wind magnitude. In other words, variants in wind speed at the ocean surface cause local surface roughness and, in turn, may be distinguished utilizing imaging SAR. The dissimilarities in wind speed are perhaps allied with wave phenomena in the atmospheric boundary layer. In this regard, gravity waves are internal waves for which gravity is the reinstating power. They are perhaps allied with temperature inversions; these instigate steadily well-stratified layers in the lower troposphere, which work as waveguides.

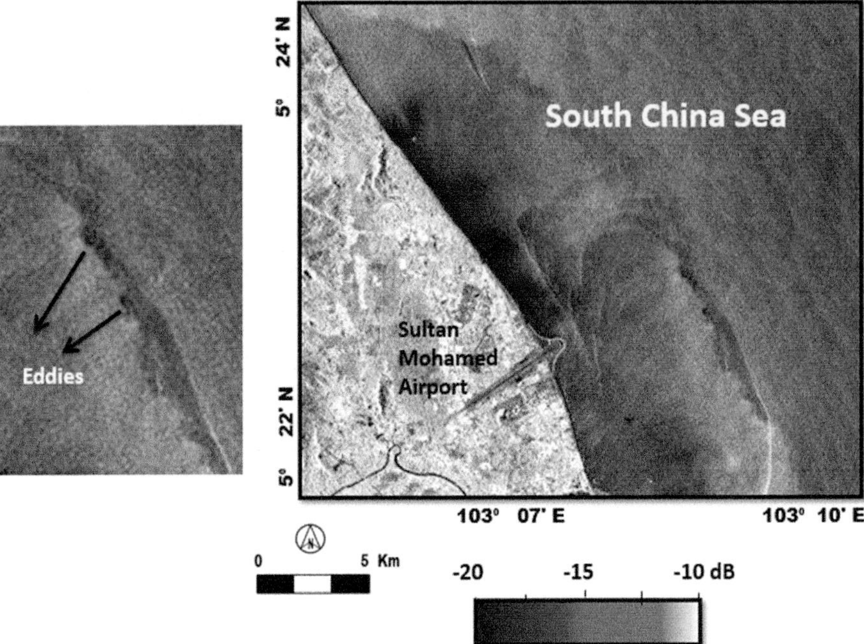

FIG. 13.6 Front occurrence in TerraSAR data.

Moreover, they are also perhaps combined with wind shear above and will commonly promulgate in the direction of the wind shear, which is often uncertain. Nonetheless, possible mechanisms are orographic interfaces, frontal disturbances, or convective instability, to specify a few.

In this regard, adequate strong wind stress must not cause spatially homogeneous sea roughness, which saturates the SAR backscatter, and prevents imaging of sea surface features such as fronts. The relative orientations of the convergence field, the wind, and the azimuth facing the SAR's direction influence the SAR front imaging mechanisms. In this circumstance, it requires to be finest the obvious imaging of an arched front. The front may swap its character from bright to dark along its length as a function of fluctuation of physical properties such as temperature, salinity, and density.

The SAR image signature, however, is not significantly modulated by wind variants caused by fluctuations in the boundary layer stratification, since the temperature fronts are reasonably weak. Moreover, salinity gradients are not reported to produce backscatter anomalies. Hence, the SAR image signatures are understood to be an indication of short gravity wave/current interaction along the current fronts. Since the short gravity waves are transmitted into and/or across the current front, they modify steepness and transmission route on condition that the wind is restrained. Accordingly, the SAR is able to distinguish the current fronts. The investigation of the front fluctuations in SAR images can be formulated mathematically based on the normalized radar cross-section (NRCS) as:

$$\frac{\Delta\sigma}{\sigma_0} \equiv \frac{\sigma_m - \sigma_0}{\sigma_0} = -A\frac{\partial u_r}{\partial x_r} \tag{13.4}$$

Eq. (13.4) shows that σ_0 is the undisturbed radar cross-section while σ_m is the maximum modulation of the NRCS within the frontal area. Consequently, u_r is the horizontal velocity along the radar look direction x_r, and A is a complex function relying, among other parameters, on radar wavelength, radar incidence angle, and wind speed and direction.

The detection of signatures of fronts in SAR images is relatively easy since fronts will usually become visible as distinct lines that are brighter or darker than the ambient regions or which separate regions of different mean image intensities. Nonetheless, the further interpretation of the signatures is difficult since the convergent currents at the ocean fronts can give rise to a variety of effects that have an impact on the backscattered radar signal, such as hydrodynamic wave-current interaction, wave breaking, or wave damping by accumulated surface films. Fig. 13.7 reveals **Sentinel-1A** satellite data acquired along the coastal water of the east coast of Johor to Pahang, Malaysia on May 19, 2020. The radar backscatter cross-section with the VV polarization band is −8 dB higher than the surrounding seawater environment. In this view, two different density quantum states of seawater, which are guided by the gravity force in the form of the gravity acceleration, cause strong entanglement that appears as an extremely sharp boundary (Fig. 13.7). In this scenario, the two different water masses are not coupled with them, but each of them individually interacts with the boundary, which validates Eq. (13.3).

It can also deliver a new definition of the front as a harmonic potential boundary placed by distance between different water masses. Consequently, these two different quantum states of water masses as a function of the gravity field are trapped by that harmonic potential boundary. In terms of SAR images of the front boundary, let us assume that the distance between the two water masses is d and the two oscillators and the gravitational interaction

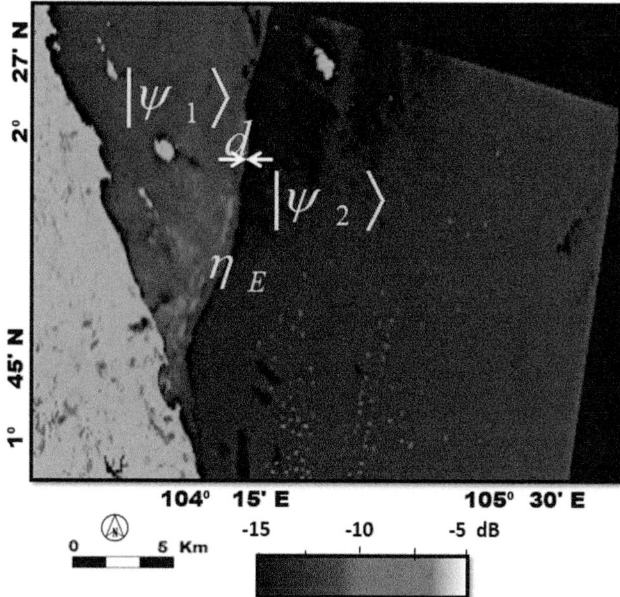

FIG. 13.7 Sharp Marghany's front boundary in Sentinel-1A satellite data.

between them give rise to the total Hamiltonian. The water quantum states of the water masses are then mathematically expressed as:

$$|\psi\rangle = \frac{\rho_1 u_1 + \rho_2 u_2}{\rho_1 + \rho_2} \tag{13.5}$$

where ρ_1 and ρ_2 are the two different water mass densities and u_1 and u_2 are their speed variations, respectively.

The distance d, from the two endpoints of any point on the line between water types, is inversely proportional to the ratio mixing and can calculate by:

$$\frac{\rho_1}{\rho_2} = \frac{d_1}{d_2} \tag{13.6}$$

In quantum mechanics, the two density modes undergo a squeezing effect, which entangles the water mass densities and provides sufficient strength to modulate the Bragg wave and induce a strong radar backscatter cross-section along with the distance d. To draw the frame mathematical expression of the quantum entanglement front that is imaging in SAR images, the study of Tanjung et al. [8] is adopted as the Hamiltonian Bragg wave should be comparable to the energy $\hbar\omega$ **of each** oscillator mass densities with them radar cross-section backscatter that $\frac{g(\partial\rho)^2(\sigma_1 - \sigma_2)^2}{d^3} \sim \hbar\omega$. In this regard, the dimensionless entanglement displacement identified in SAR images η_E is expressed as:

$$\eta_E \equiv \frac{2g(\partial\sigma)}{\omega^2 d^3} \tag{13.7}$$

In Eq. (13.7), the gravity acceleration induces the entanglement displacement of the radar cross-section along with the two different water mass densities, as they are trapped by the harmonic potential boundary η_E as inversely proportional to the angular frequency ω. Consequently, the maximum entanglement that occurs for the SAR imaging front can be given by:

$$\eta_{E_{max}} \equiv \eta_E(\ln \partial\sigma) \tag{13.8}$$

The time for maximum entanglement for the SAR imaging front can be estimated as:

$$t_{max} \equiv \pi(2\eta_E\omega)^{-1} \tag{13.9}$$

The angular frequency ω can be determined using the quantum fast Fourier transform (QFFT). Therefore, along the harmonic potential boundary, the consequence of gravity is the stronger attraction of dual different water densities of the spatial superposition that are closer, and hence generation of position and momentum correlations, leading to the growth of quantum entanglement.

13.3 Relativity of front signatures in polarimetric SAR

To date, few investigations have been dedicated to comprehending the polarimetric features of the SAR consequence to waves, currents, and other ocean surface aspects. The polarimetric SAR instantaneously assembles the backscattered consequences from horizontal transmitting and horizontal receiving (HH), horizontal transmitting and vertical receiving (HV), vertical transmitting and horizontal receiving (VH), and vertical transmitting and vertical receiving (VV) polarizations, from which a scattering matrix can be formed (Chapter 8). From the scattering matrix, the inherent scattering mechanism can be revealed by the polarization signature—for instance, the symmetry characteristics of polarimetric signatures of wind-roughened ocean surfaces can be investigated. Polarization signatures can be utilized to investigate sea spike and clutter echoes at low gazing angles. The polarimetric SAR response of an ambient ocean surface agrees with the tilted Bragg model (i.e., Bragg scattering on an azimuthal tilted surface) as based on the stochastic surface tilts. A higher degree of depolarization is present for returns originating within a current front. This depolarization is caused by the amalgamated nature of scattering mechanisms, such as tilted Bragg and specular, and the scatter instigated by breaking waves.

Generally, the AIRSAR imagery, for instance, shows that the signal-to-background ratio of radar cross-sections for the convergent front is higher than that of the shear front. The cross-polarization, HV, and VH responses are much higher than copolarization VV and HH responses owing to cross-polarization has the highest signal-to-clutter ratio than copolarization bands. Lee et al. [9] investigated a 45° flight path across a similar front, and found that the signal-to-background ratios are nearly equal for all three polarizations. Under the circumstance that the dominant wave direction is parallel to the SAR azimuthal direction, the dominant scattering mechanism causes an increase in HV and VH polarizations, which is due to the averaged azimuth surface tilts shaped by facets of long waves. In this regard, the resulting shift in the peak of the copolarization signature correlates with both the change in HV returns and with the direction of waves at the front.

The P-band and L-band HV images clearly show the convergent front as a vertical bright line with the highest backscatter radar cross-section −18 dB compared to the L_{HV} and C_{HV} bands (Fig. 13.8). The response of the convergent front is much weaker for VV polarization and for HH polarization. For the C-band images, conversely, the shear front is almost invisible, and the convergent front is detectable, but weaker than for P- and L-bands [9] (Fig. 13.9). In this respect, the P_{VV} bands sustain the highest backscatter radar cross-section of −15 dB compared to L_{VV} and C_{VV} bands, respectively. Consequently, the longer SAR integration time for the P-band causes wider fluctuations of normalized radar cross-section than L-band and C-band. In the circumstance that the integration time is longer than the ocean wave coherence time, it causes smear in the azimuthal direction.

The polarimetric SAR signatures correspondingly rely on the imaging angle between the aircraft pass and the front. This consequence, for instance, can be investigated by using the pass crossing a convergent front at 45°. As a result, signatures across the front are approximately equal at all three polarizations for all three bands except -band HV polarization. The transverse wavelike features imaged in the -band HV image are artifacts owing to radio interference. When the path crosses the front at 45°, the average surface tilt now has a component in the range direction, which reduces the local incidence angle and produces stronger responses for the HH and VV polarizations.

From the point of view of relativity theory, the size and length of the fronts have to be imagined differences among a variety of polarization, i.e., the cross-polarization, HV, and VH and copolarization VV and HH. Figs. 13.8 and 13.9 demonstrate the length contraction of a front in VV polarization and quadrature-polarimetric (quad-pol) synthetic aperture radar (SAR)

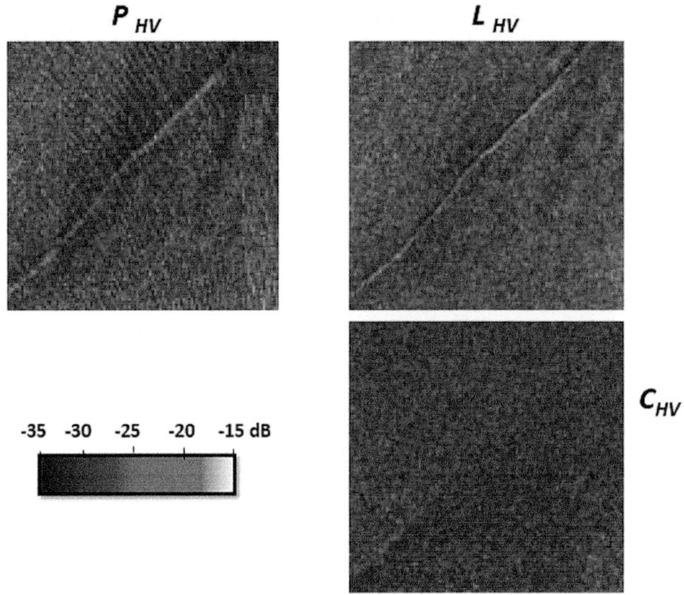

FIG. 13.8 Different backscatter radar cross-sections for different copolarizations.

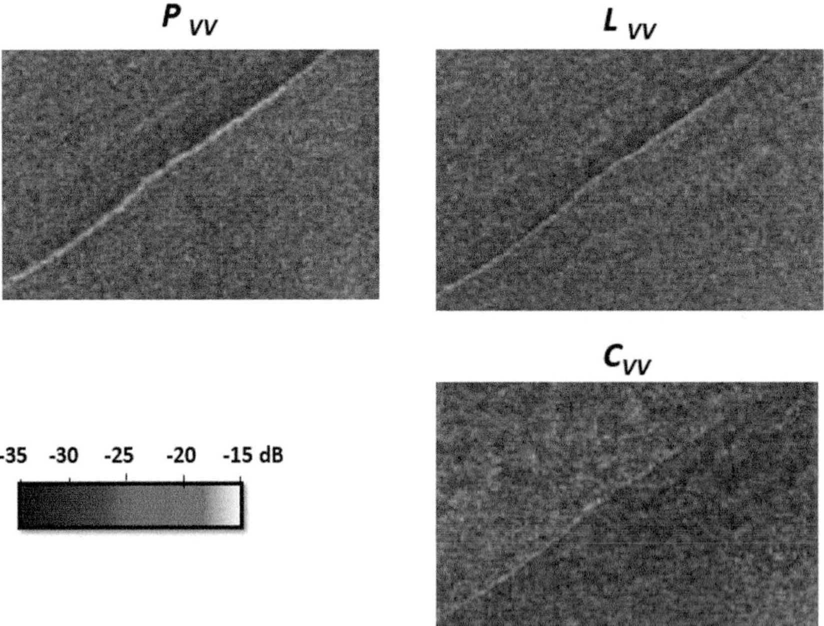

FIG. 13.9 Different backscatter radar cross-sections for different bands.

systems, which allow the measurement of the 2×2 complex scattering matrix S. The scattering matrix S is usually built up by operating the radar with the interleaved transmission of alternate H- and V-polarized pulses and simultaneous reception of both H and V polarizations (conventional quad-pol SAR). Length contraction takes place in the quad-pol SAR data as the front geometry and wavelength are constricted.

13.4 How does the tidal cycle effect front signature in SAR images?

Within the area enclosed by the frontal line, large zones characterized by a lower radar backscatter than in the outer area are often visible. The analysis of the SAR images suggests that the form and the location of the frontal features are mainly linked to the semidiurnal tidal phase in the outflow region, although their variability suggests also that they weakly depend on the river discharge, residual currents, and neap-spring tidal cycle. Hessner et al. [10] examined this hypothesis based on the comparison between SAR image interpretation and simulation results derived from the results obtained from the analysis of the ERS SAR images. Therefore, the validation was achieved by comparing with the results obtained from the nonlinear, hydrostatic shallow-water equations on an f plane. The model is forced by prescribing tidal and residual currents and river discharge at the open boundaries. Several simulations are performed by varying the values of these forcing parameters. The numerical results corroborate the observational conjecture: It is found that the form and the location of the simulated interface outcropping lines in the proximity of the river mouth are mainly determined by the semidiurnal tidal phase in the outflow region and that river discharge, residual currents, and neap-spring tidal cycle contribute only secondarily to their determination. Inserting the simulated surface velocity field into a simple radar-imaging model that relates the modulation of the backscattered radar power to the surface velocity convergence in radar look direction, narrow, elongated bands of enhanced radar backscatter emerge near the model frontal line while patches of low radar backscatter appear within the simulated river plume area. The consistency of the model results with the results obtained from the analysis of the SAR images enables one to infer a mean spatial and temporal evolution of the Rhine outflow plume over a semidiurnal tidal cycle from the analysis of spaceborne SAR images acquired during different tidal cycles over the river outflow area and suggests the possibility of using numerical modeling, in conjunction with the analysis of space-borne measurements, for monitoring the oceanic variability in any river outflow area.

13.5 Speckles impact on front signature in SAR images

The high speckle noise in SAR images has posed great difficulties in inverting SAR images for simulating coastal fronts. Speckle is a result of coherent interference effects among scatterers that are randomly scattered within each resolution cell. The speckle size, which is a function of the spatial resolution, induces errors in front signature detection. Reducing the speckle effect requires appropriate filters, i.e., Lee, Gaussian, etc., to be deployed in the preprocessing stage. The effectiveness of these speckle-reducing filters is, however, much influenced by causal factors and application. Since the SAR images the sea surface, all

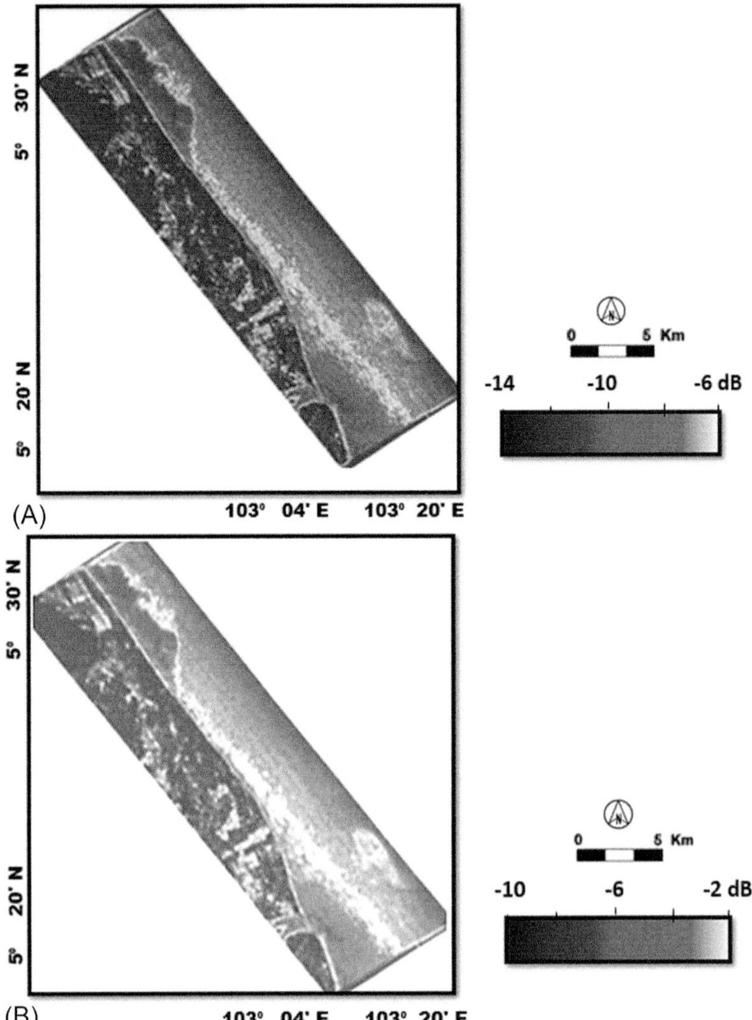

FIG. 13.10 Speckle in different TOPSAR bands: (A) L-HH band and (B) C-VV band, respectively.

speckles in SAR images are a function of local changes in the surface roughness, such as direct reduction of the wave height (because of slicks), wind impelled roughness changes (atmospheric effects), or wave-current interactions (fronts and bathymetry).

Fig. 13.10 shows the signature of the underwater topography. The signature of underwater topography is obvious as frontal lines parallel to the shoreline. The backscattered intensity is damped by −2 to −10dB compared to the surrounding water environment in L-band with HH polarization and −6 to −14dB in C-band with VV polarization data.

The SAR images are of numerous sorts: complex, intensity, amplitude, logarithm, single-look, and multilooks. On this basis, it is the complex image that experiences processes approximating pixels modulus calculation (amplitude image), the square of modulus (intensity image), the logarithm of intensity (logarithm image), or the more complex multilook processing consisting in pixels averaging and resampling. Therefore, in complex SAR images, the pixels have complex values. In this regard, the real part is the in-phase component of the measured SAR backscatter signal. However, the imaginary part is the quadrature component.

An SAR complex image involves fully developed speckle, in which both components have a Gaussian distribution, as given by:

$$P\left(R_e r^{-1}\right) = \frac{1}{\sqrt{\pi r}} e^{\left(-\frac{R_e^2}{r}\right)} \tag{13.10}$$

$$P\left(I_m R^{-1}\right) = \frac{1}{\sqrt{\pi R}} e^{\left(-\frac{I_m^2}{R}\right)} \tag{13.11}$$

Eqs. (13.5) and (13.6) reveal that in the front zone, due to constant reflectivity r, the SAR pixels are fluctuating because of speckles. In this regard, P represents the probability density function (PDF), R_e represents the real part, and I_m the imaginary part. By making $r=1$, a PDF of speckle in complex SAR images can be obtained.

The SAR intensity image I is obtained from the complex image by considering the square of the modulus:

$$I = R_e^2 + I_m^2 \tag{13.12}$$

In the SAR intensity image, the PDF of speckles can be expressed as an exponential distribution:

$$P_I\left(Ir^{-1}\right) = r^{-1} e^{\left(-\frac{I}{r}\right)} \quad I \geq 0 \tag{13.13}$$

However, the amplitude SAR image is considered as the square root of the intensity image, $A = \sqrt{I}$. In this circumstance, Rayleigh distribution can be used to express the SAR speckles as:

$$P_A\left(Ar^{-1}\right) = 2Ar^{-1} e\left(-A^2 r^{-1}\right) \quad A \geq 0 \tag{13.14}$$

Eqs. (13.8) and (13.9) reveal that both for the intensity and amplitude images, the variance of the pixel increases with r. This explains why the speckle, although stationary, appears as stronger in the light areas. It is, in fact, a consequence of the multiplicative nature of the speckle. On the other hand, the logarithm SAR image is the logarithm of the intensity ln(I). The logarithm SAR image has a Fisher-Tippet distribution, which is given by:

$$P_D\left(\frac{\ln(I)}{r}\right) = \frac{e^{\ln(I)}}{r} e^{\left(-\frac{e^{\ln(I)}}{r}\right)} \tag{13.15}$$

Eq. (13.15) shows that for various values of r, the distribution's mean fluctuates, but its variance does not. Since the logarithm alters any product into a sum, the speckle in the logarithm SAR image is an additive stationary noise, which clarifies the variance constancy.

The multilook SAR images are acquired by averaging, in columns, sets or three or four pixels. The subsequent SAR image must be intercalated and resampled to have a similar resolution on rows and columns. Owing to pixels averaging, the multilook image is less noisy. In this view, depending on the SAR image and the origin—intensity, amplitude, or logarithm—the multilook image can be of intensity, amplitude, or logarithm sort. The multilook treatment swaps the speckle scattering as a Gamma distribution in a multilook intensity SAR image:

$$P_I\left(Ir^{-1}\right) = \frac{1}{\Gamma(L)}\left(\frac{L}{r}\right)^L I^{L-1} e^{\left(-\frac{LI}{r}\right)} \quad I \geq 0 \tag{13.16}$$

where L is the number of looks, which is obtained by the number of averaged pixels.

Consequently, the mathematical expression of the generalized Gamma distribution in a multilook amplitude SAR image is given as:

$$P_A(A) = \frac{2}{\Gamma(L)}\left(\frac{L}{r}\right)^L A^{2L-1} e^{\left(\frac{-LA^2}{r}\right)} \quad A \geq 0 \tag{13.17}$$

Finally, Fisher-Tippet distribution in multilook logarithm SAR images is given by:

$$P_{\ln(I)}\left(\frac{\ln(I)}{r}\right) = \frac{L^L}{\Gamma(L)} e^{(-L(\ln(r) - \ln(I)))} e^{\left(-Le^{(-(\ln(r) - \ln(I)))}\right)} \tag{13.18}$$

In all types of multilook SAR image, the speckle variance decreases as the number of looks increases. However, as a consequence, delivered images have poor resolution [3].

13.6 Anisotropic diffusion algorithm for speckle reductions

Anisotropic diffusion is a technique that aims at reducing image noise while preserving edges, lines, and other details that are important for image interpretation [11]. Formally, let $\Omega \subset \mathbb{R}^2$ denote a subset of the SAR image plane $I_{SAR}(.,t): \Omega \to \mathbb{R}$, which is part of SAR intensity. Therefore, anisotropic diffusion is given by:

$$\begin{aligned} \frac{\partial I_{SAR}}{\partial t} &= \operatorname{div}\left(c(i,j,t)\Delta I_{SAR}\right) \\ &= \nabla c \cdot \nabla I_{SAR} + c(i,j,t)\Delta I_{SAR} \end{aligned} \tag{13.19}$$

where $c(i,j,t)$ is the diffusion coefficient and $c(i,j,t)$ controls the rate of diffusion that is usually chosen as a function of the SAR image intensity gradient ΔI_{SAR} to preserve bathymetry edge in SAR data, Δ denotes the Laplacian, ∇ denotes the gradient, and *div* (…) is the divergence operator.

Following Perona and Malik [11], two functions for the diffusion coefficient are considered:

$$c\left(\|\nabla I_{SAR}\|\right) = e^{-(\|\nabla I_{SAR}\|/K)^2} \tag{13.20}$$

and:

$$c\left(\|\nabla I_{SAR}\| = (1 + \|\nabla I_{SAR}\|)^{-1}\right. \tag{13.21}$$

The constant K controls the sensitivity to edges and is usually chosen as a function of the speckle variations in SAR data. In this regard, $c(\|\nabla I_{SAR}\| \to 0$ if $\|\nabla I_{SAR}\| \gg K$, which leads to all-pass filter. In addition to isotropic diffusion (Gaussian filtering achieved under boundary condition of $c(\|\nabla I_{SAR}\| \to 1$ when $\|\nabla I_{SAR}\| \ll K$), anisotropic diffusion resembles the process that creates a scale-space, where an image generates a parameterized family of successively more and more blurred images based on a diffusion process [12].

Fig. 13.11 shows that a sharper front signature is extracted by utilizing the anisotropic diffusion algorithm. L_{HH} band has more clear-cut features than C_{VV} band. The anisotropic

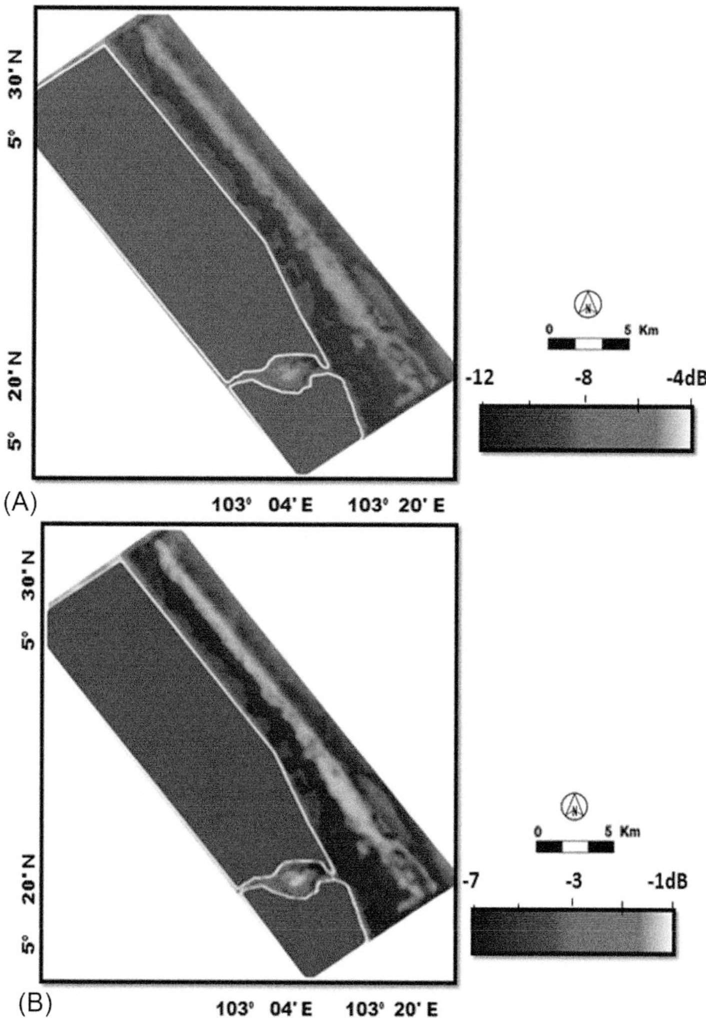

FIG. 13.11 Result of anisotropic diffusion algorithm for front signature from (A) C_{VV} and (B) L_{HH} bands.

diffusion algorithm can extract the boundary edge for both horizontal and vertical directions. In other words, it is a synthesized boundary edge. In this regard, anisotropic diffusion resembles the process that creates a scale-space, where an image generates a parameterized family of successively more and more blurred images based on a diffusion process. Each of the resulting images in this family is given as a convolution between the image and a 2-D isotropic Gaussian filter, where the width of the filter increases with the parameter (Fig. 13.11).

This diffusion process is a linear and space-invariant transformation of the original image. Anisotropic diffusion is a generalization of this diffusion process: it produces a family of parameterized images, but each resulting image is a combination of the original image and a filter that depends on the local content of the original image. As a consequence, anisotropic diffusion is a nonlinear and space-variant transformation of the original image. Finally, in C-band with VV polarization, this feature is weaker than in L-band with HH polarization. L_{HH} band has a backscatter value of −1 dB higher than the C_{VV} band. In this context, the character of the current gradient may be such that the L_{HH} band surface Bragg waves are more strongly modulated than for the C_{VV} band. This may explain weaker front signatures in the C_{VV} band [13].

13.7 3-D front model

There are three algorithms involved in 3-D front reconstruction: relativity velocity bunching, Volterra, and fuzzy B-spline. Significant wave heights are simulated from the SAR image by using the velocity bunching model; fuzzy B-spline uses significant wave height information to reconstruct the 3-D front; and the front flow pattern is modeled by the Volterra model.

13.7.1 Relativity velocity bunching model

In this study, two-dimensional quantum fast Fourier transform (2D-QFFT) has been applied to a single SAR image frame consisting of 512 × 512 image pixels, as demonstrated in Chapter 7. Consequently, the Gaussian algorithm is applied to remove the noise from the image and smooth the wave spectra into the normal distribution curve. The band used in this processing is C_{HH}-band for RADARSTA-1 SAR, Sentinel-1A, and TerraSAR strip map with X_{HH}-band. Each pixel represents a 12.5 m × 12.5 m area for SAR images. The entire image of SAR images corresponds to a 6.4 km × 6.4 km patches on the ocean surface. This frame size provides a sufficiently large area to include at least 10 cycles of very long surface waves, up to 640 m in length, which can be included in a single image frame. It is also small enough to show that the ocean can be reasonably assumed homogeneous within a frame [14].

The velocity bunching modulation transfer function (MTF) is the dominant component of the linear MTF for the ocean waves with an azimuth wave number (k_x). As stated by Alpers et al. [15] and Vachon et al. [16–19], velocity bunching can contribute to linear MTF based on the following equation:

$$T_{\eta_x} = \frac{R}{V_S}\left[\frac{\omega'}{(1-2u_r\cos\theta c^{-1}+u_r{}^2c^{-2})(1-u_r{}^2c^{-2})^{-1}}\right]\left[\frac{k_x}{k}\sin\theta + i\cos\theta\right] \tag{13.22}$$

where u_r is the dynamical sea surface moving at a radial velocity, $\frac{R}{V_S}$ is the scene range to the platform velocity ratio, which is 111 s in the case of RADARSAT-1SAR image data, and θ is, for instance, RADARSAT-1 SAR image incidence angel (35—49 degrees).

To estimate the velocity bunching spectra $S_{vb}(k)$, we modified the algorithm that was introduced by Krogstad and Schyberg [4] as a function of the relativity theory discussed in Chapter 6. The modification is to multiply the velocity bunching MTF T_{η_x} by the SAR image spectra variance of the azimuth shifts. This can be determined using the following formula:

$$S_{vb}(k) = \left[I_0 \sum_{n=1}^{\infty} \frac{\psi(k_x)^{2n}}{n!} S_{\zeta\zeta}{}^{*n}\psi(k)^2 e^{-K_x{}^2\rho_{\zeta\zeta}}\right]\left[T_{\eta_x}\right] \tag{13.23}$$

where $S_{\zeta\zeta}$ is the SAR spectra variance of the azimuth shifts owing to the surface motion, which is generated by the relativity velocity bunching impact in azimuth direction due to the high value of $\frac{R}{V_S}$.

Furthermore, SAR spectra of ocean wave images have a characteristic of azimuth cutoff. Moreover, they also have an intrinsic azimuth cutoff that in numerous circumstances fit precise bright with definite inspection and narrates to the cutoff; which is proportional directly to the standard deviation of the azimuth shift. In this view, standard deviation of the azimuth shift perhaps efficiently correlated to the essential sea state parameters such as significant wave height; wave period; and wind speed. In this regard, $\rho_{\zeta\zeta}$ is the variance of the derivative of displacements along the azimuth direction, I_0 is SAR image intensity, and "$*n$" means $(n-1)$-fold convolution, according to Krogsted and Schyberg [4]. Eq. (13.23) is used to draw the velocity bunching spectra energy contours.

In Chapter 6 we addressed the relativity of the velocity bunching theory due to the relativity in the Doppler frequency shift as demonstrated in the second term of Eq. (13.22). The consequence is the fluctuation of the wavelength changing in different polarization for the same object, as revealed in Section 13.3. From the point of view of relativistic theory, the achievement of the relative velocity bunching can cause fluctuation of the physical characteristics of the object in different incidence angles and different band and polarization frequencies.

Estimating significant wave height from velocity bunching spectra based on the azimuth cutoff arising from the velocity-bunching model, Eq. (13.23), the azimuth cutoff could be scaled by the standard deviation of the azimuth shift. Vachon et al. [18] suggested a relationship between the variance of the derivate of displacement along the azimuth direction $\rho_{\zeta\zeta}$ and the standard deviation of the azimuth shift η, which were estimated from the velocity bunching spectra as:

$$\eta = \sqrt{\rho_{\zeta\zeta}} \tag{13.24}$$

Significant wave height H_s can be computed using the relativity velocity bunching owing to the azimuth shift as:

$$\eta = \frac{R}{V_S}\sqrt{1-\frac{\sin^2(\theta)}{2}}\sqrt{\frac{\left(2\pi\left(T\sqrt{(gd)}\sqrt{1-\left(\frac{R}{V_S}\right)^2 c^{-2}}\right)^{-1}\right)g}{8}H_s} \tag{13.25}$$

The term $2\pi\left(T(\sqrt{gd})\sqrt{1-\left(\frac{R}{V_S}\right)^2 c^{-2}}\right)$ is the length contraction owing to the relativity of the azimuth shift, as discussed in Chapter 6. Moreover, θ is SAR image incident angle, $\frac{R}{V_S}$ is the scene range to platform velocity ratio, and g is the acceleration due to gravity. Note that the mean wave period T_0 is equal to $2\pi(\langle\langle k_x\rangle g)^{-0.5}$. Using Eqs. (13.24) and (13.25), the significant wave height H_s can be obtained:

$$H_s = \gamma\sqrt{\rho_{\zeta\zeta}}\frac{1+\theta^2}{4}\left[\frac{R}{V_S}\right]^{-1} \tag{13.26}$$

Here, γ is the fit parameter, which is determined using linear regression calculation between measured in situ significant wave height and the relativity of standard deviation fluctuation of the azimuth shift. Fig. 13.12, for instance, reveals the simulated significant wave height from different SAR sensors based on the relativity velocity bunching model in shallow water. The significant wave heights vary from one sensor to another due to different monsoon seasons. In this regard, the RADARSAT-1 SAR reveals significant wave heights ranging between 0.6m and 1.2m. The significant wave height represents the northeast monsoon as the SAR image was acquired in March. TerraSAR data, therefore, reveal significant wave heights that range between 0.4m and 1.3m, as TerraSAR data are acquired in August, which represents the southwest monsoon season. Sentinel-1A data demonstrates maximum significant wave height of 1.3m with different wave direction propagation. In fact, Sentinel-1A data

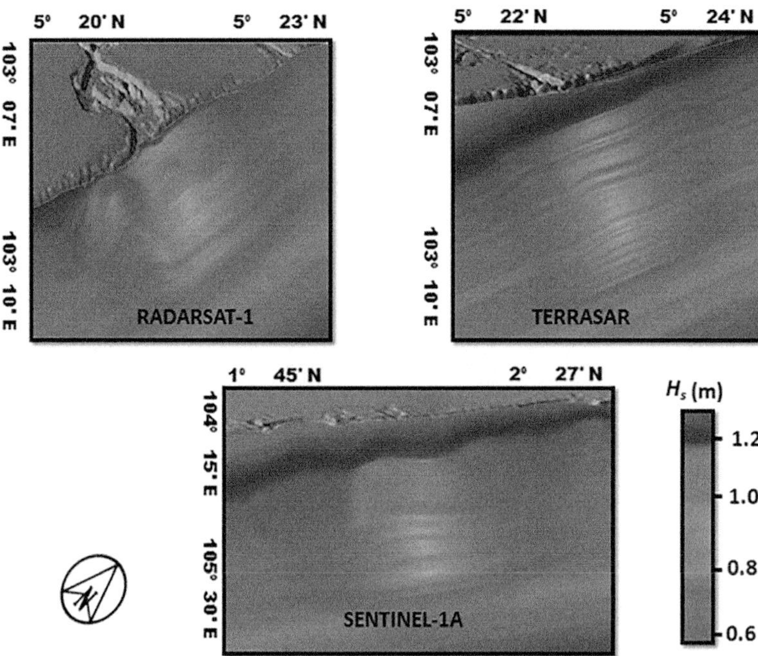

FIG. 13.12 Significant wave height is retrieved by relativity velocity bunching from different SAR sensors.

represent the transitional period between the northeast monsoon season and southwest monsoon season as the wave does not have a fixed direction. The relativity velocity bunching model delivers a wave refraction pattern as the wave tends to change its direction and pattern as it approaches shallow waters. The maximum significant wave height patterns are seen near the coastline at all SAR sensors, due to waves breaking.

13.7.2 Volterra model for front flow velocity

Referring to Inglada and Garello [20], the Volterra series can be used to model nonlinear imaging mechanisms of surface current gradients by RADARSAT-1 SAR image, for instance, as a result of the Volterra linear kernel containing most of the RADARSAT-1 SAR energy, which is used to simulate current flow along range direction. Following Inglada and Garello [20], the Volterra kernel filter has the following expression:

$$H_{1y}(v_x v_y) = k_y \vec{U} \cdot \frac{\partial x}{\partial u_x}\left[K^{\rightarrow} - 1\left[\frac{\partial}{\partial t} + \frac{\partial c_g^{\rightarrow}}{\partial x} + \frac{\partial \vec{u_x}}{\partial x} + 0.043\frac{(u_a^{\rightarrow} K^{\rightarrow})^2}{\omega_0}\right]\right] \cdot \left[\frac{\partial \psi}{\partial \omega}\right]$$
$$\cdot \frac{c_g^{\rightarrow}(K^{\rightarrow})U^{\rightarrow} + j \cdot 0.043(u_a^{\rightarrow} K^{\rightarrow})^2 \omega_0^{-1}}{\left[c_g^{\rightarrow}(K^{\rightarrow})U^{\rightarrow}\right]^2 + \left[0.043(u_a^{\rightarrow} K^{\rightarrow})^2 \omega_0^{-1}\right]^2} + j \cdot \left(0.6 \cdot 10^{-2} \cdot K^{\rightarrow} - 4\right)\left(\frac{R}{V_S}\right) u_x^{\rightarrow} \quad (13.27)$$

where $\vec{U}$ is the mean current velocity, $\vec{u}_x$ is the current flow and $\vec{u}_a$ is the current gradient along azimuth direction, respectively, k_y is the wavenumber along range direction, $\vec{K}$ is the spectra wave number, ω_0 is the angular wave frequency, $\vec{c}_g$ is the wave velocity group, ψ is the wave spectra energy, and R/V is the range to platform velocity ratio.

In reference to Inglada and Garello [20], the inverse filter $G(v_x, v_y)$ is used, since $H_{1y}(v_x, v_y)$ has a zero for (v_x, v_y), which indicates that the mean current velocity should have a constant offset [21]. The inverse filter $G(v_x, v_y)$ can be given as:

$$G(v_x, v_y) = \begin{cases} \left[H_{1y}(v_x, v_y)\right]^{-1} & \text{if } (v_x, v_y) \neq 0, \\ 0 & \text{otherwise.} \end{cases} \quad (13.28)$$

Substituting Eq. (13.26) into Eq. (13.27), range current velocity $U_y(0, y)$ can be predicted by:

$$U_y(0, y) = I_{SAR} \cdot G(v_x, v_y) \quad (13.29)$$

where I_{SAR} is the frequency domain of SAR images acquired by applying 2-D quantum Fourier transform to SAR images.

13.7.3 The fuzzy B-splines method

The fuzzy B-splines (FBS) are introduced to allow fuzzy numbers instead of intervals in the definition of the B-splines. Typically, in computer graphics, two objective quality definitions for fuzzy B-splines are used: triangle-based criteria and edge-based criteria. A fuzzy number is defined using interval analysis. There are two basic notions that we combine: confidence interval and assumption level. A confidence interval is a real value interval that provides the sharpest enclosing range for current gradient values. An assumption level, μ-level, is an estimated truth value in the [0,1] interval on our knowledge level of the gradient current [22]. The 0 values correspond to minimum knowledge of gradient current, and 1 to the maximum gradient current. A fuzzy number is then prearranged in the confidence interval set, each one related to an assumption level $\mu \in [0,1] \in [0, 1]$. Moreover, the following must hold for each pair of confidence intervals that define a number: $\mu \succ \mu' \Rightarrow H_s \succ H_s'$. Let us consider a function $f: H_s \rightarrow H_s'$, of N fuzzy variables $H_{s_1}, H_{s_2}, \ldots, H_{s_n}$, where H_{s_n} are the global minimum and maximum values of the significant wave height of the function on the space. Based on the spatial variation of the gradient current and significant wave height, the fuzzy B-spline algorithm is used to compute the function f.

Let $H(i,j)$ be the depth value at location i,j in the region D where i is the horizontal and j is the vertical coordinate of a grid of m times n rectangular cells. Let N be the set of eight neighboring cells. The fuzzy input variables are the amplitude differences of significant wave height defined by [23]:

$$\Delta H_{s_N} = H_{s_i} - H_{s_0}, \quad N=1,\ldots,8 \tag{13.30}$$

where the H_{s_i}, $N=1$, and 8 values are the neighboring cells of the processed cell H_{s_0} along the horizontal coordinate i.

To estimate the fuzzy number of significant wave height H_i, which is located along the vertical coordinate j, we estimate the membership function values μ and μ' of the fuzzy variables H_{s_i} and H_{s_j}, respectively, by the following equations, as described by Rövid et al. [24]:

$$\mu = \max\left\{\min\left\{m_{pl}(\Delta H_{s_i}) : H_{s_i} \in N_i\right\};\ \ N=1,\ldots,9\right\} \tag{13.31}$$

$$\mu' = \max\left\{\min\left\{m_{LNI}(\Delta H_{s_i}) : H_{s_i} \in N_i\right\};\ \ N=1,\ldots,9\right\} \tag{13.32}$$

where m_{pl} and m_{LNI} correspond to the membership functions of fuzzy sets. From Eqs. (13.31) and (13.32), one can estimate the fuzzy number of significant wave height H_{s_j}:

$$H_{s_j} = H_{s_i} + (L-1)\Delta\mu \tag{13.33}$$

where $\Delta\mu$ is $\mu - \mu'$ and $L = \{H_{s_1}, \ldots, H_{s_N}\}$. Eqs. (13.31) and (13.32) represent significant wave height in 2-D SAR images; to reconstruct fuzzy values of significant wave height in 3-D, then a fuzzy number of the significant wave in the z coordinate is estimated by the following equation proposed by Russo [25]:

$$H_{s_z} = \Delta\mu MAX\left\{m_{LA}\left|H_{s_{i-1,j}} - H_{s_{i,j}}\right|, m_{LA}H\left|H_{s_{i,j-1}} - H_{s_{i,j}}\right|\right\} \tag{13.34}$$

where H_{s_z} are fuzzy significant wave values in the z coordinate, which is a function of my and j coordinates, i.e., $H_{s_z} = F(H_{s_i}, H_{s_j})$.

Fuzzy number F_O for water depth in i,j and z coordinates can then be given by:

$$F_O = \left\{\min\left(H_{s_{z0}}, \ldots, d_{s_{z\Omega}}\right), \max\left(d_{s_{z0}}, \ldots, d_{s_{z\Omega}}\right)\right\} \tag{13.35}$$

where $\Omega = 1, 2, 3, 4$.

The fuzzy number of significant wave height F_O is then defined by B-spline to reconstruct 3-D of significant wave height. To this end, B-spline functions including the knot positions and set of control points are constructed. The requirements for B-spline surface are a set of control points, a set of weights, and three sets of knot vectors, and they are parameterized in the p and q directions.

A fuzzy B-spline surface $S(p,q)$ is described as a linear combination of basis functions in two topological parameters, p and q. Let $R=r_0;\ldots;r_m$ be a nondecreasing sequence of the real numbers. The r_i is called knots and R is the knot vector. The interval r_i and r_{i+1} is called knot span. According to Anile et al. [23], the Pth-degree *(order $P+1$)* piecewise polynomial function B-spline basis function, denoted by $\beta_{i,\,p}(r)$ is given by:

$$\beta_{i,1}(r)=\begin{cases}1 & \text{If } r_i\leq r\leq r_{i+1},\\ 0 & \text{Otherwise;}\end{cases} \tag{13.36}$$

$$\beta_{i,P}(r)=\frac{r-r_i}{r_{i+P-1}-r_i}\beta_{i,P-1}(r)+\frac{r_{i+P}-r}{r_{i+P}-r_{i+1}}\beta_{i+1,P-1}(r)\quad \text{for}\quad P>1 \tag{13.37}$$

To exercise more shape controllability over the surface, and invariance to perspective transformations, fuzzy B-spline is introduced. In addition to having the control point as in the B-spline, fuzzy B-spline also provides a set of weight parameters $w_{i,\,j}$ that exert more local shape controllability to achieve projective invariance. Following Russo [25], fuzzy B-spline surface that is composed of array grid ($O\times M$), i.e., O and M, are the element vectors that belonging to knot p and q, respectively, can be mathematically expressed as:

$$S(p,q)=\frac{\sum_{i=0}^{M}\sum_{j=0}^{O}F_O C_{ij}\beta_{i,4}(p)\beta_{j,4}(q)w_{i,j}}{\sum_{m=0}^{M}\sum_{l=0}^{O}\beta_{m,4}(p)\beta_{l,4}(q)w_{ml}}=\sum_{i=0}^{M}\sum_{j=0}^{O}F_o C_{ij}S_{ij}(p,q) \tag{13.38}$$

where $\beta_{i,\,4}(p)$ and $\beta_{j,\,4}(q)$ are two basis B-spline functions, $\{C_{ij}\}$ are the bidirectional controls net, and $\{w_{ij}\}$ are the weights. The curve points $S(p,q)$ are affected by $\{w_{ij}\}$ in the case of $p\in[r_i,r_{i+P+1}]$ and $q\in[r_j,r_{j+P'+1}]$, where P and P' are the degree of the two B-spline basis functions constituting the B-spline surface. Two sets of knot vectors are *knot* $p=[0,0,0,0,1,2,3,\ldots,O,O,O,O]$, and *knot* $q=[0,0,0,0,1,2,3,\ldots,M,M,M,M]$. Fourth-order B-spline bases are used $\beta_{j,\,4}(.)$ to ensure continuity of the tangents and curvatures on the whole surface topology including at the patch boundaries.

The construction begins with the same preprocessing aimed at the reduction of measured significant wave height and current values into a uniformly spaced grid of cells. As in the Volterra algorithm, data are derived from the SAR backscatter images by the application of a 2-dimensional quantum fast Fourier transform (2D-QFFT). First, each estimated current data value in a fixed kernel window size of 512×512 pixels and lines is considered as a triangular fuzzy number defined by a minimum, maximum, and measured value. Among all the fuzzy numbers falling within a kernel window size, a fuzzy number is defined whose range is given by the minimum and maximum values of significant wave height and current values along with each kernel window size. Furthermore, the identification of a fuzzy number is acquired to summarize the estimated significant wave height and current value data in a cell, and this is characterized by a suitable membership function. The choice of the most appropriate membership is based on triangular numbers, which are identified by minimum, maximum, and mean values of significant wave height and current values. Furthermore, membership support is the range of water depth data in the cell and whose vertex is the median value of significant wave height and current values data [22].

13.8 3-D front topology reconstruction in SAR data

Fig. 13.13 signifies the 3-D front topology reconstruction with significant wave heights, and current variations cross-front. Fig. 13.13 shows significant wave variation cross-front with a maximum significant wave height of 1.2 m and a gradient current of $0.9\,\mathrm{m\,s^{-1}}$. March represents the northeast monsoon period as coastal water currents in the South China Sea tend to move from the north [5]. Fig. 13.13 reveals a meandering current with a southward direction. This current is created because of the water inflow from the Terengganu River mouth. Furthermore, Marghany [6] stated that strong tidal current is a dominant feature in the South China Sea with a maximum velocity of $1.5\,\mathrm{m\,s^{-1}}$. The 3-D front (Fig. 13.14A) coincides with a water depth range between 10 and 20 m (Fig. 13.14B). This indicates shallow water, where the strong tidal stream (Fig. 13.13) causes vertical mixing.

The visualization of the 3-D front is sharp with the RADARSAT-1 SAR C_{HH} band because each operation on a fuzzy number becomes a sequence of corresponding operations on the respective μ and μ'-levels, and multiple occurrences of the same fuzzy parameters are evaluated as a result of the function on fuzzy variables [22]. Typically, in computer graphics, two objective quality definitions for fuzzy B-splines are used: triangle-based criteria and edge-based criteria. Triangle-based criteria follow the rule of maximization or minimization for the angles of each triangle. The so-called max-min angle criterion prefers short triangles with

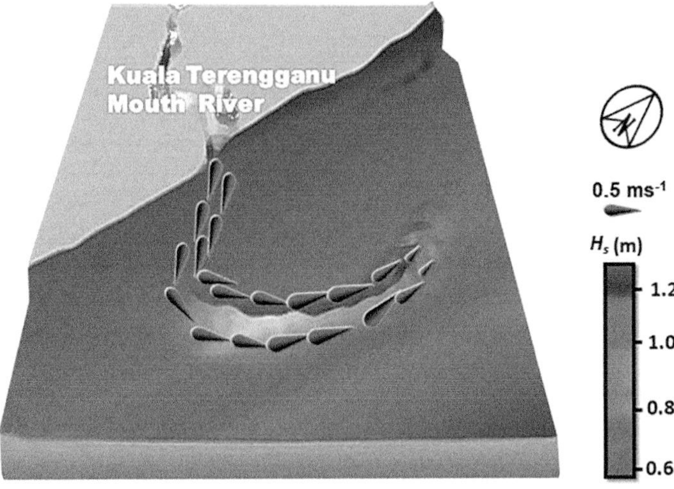

FIG. 13.13 3-D front reconstruction with significant wave height (H_s) and surface current variations (U_y).

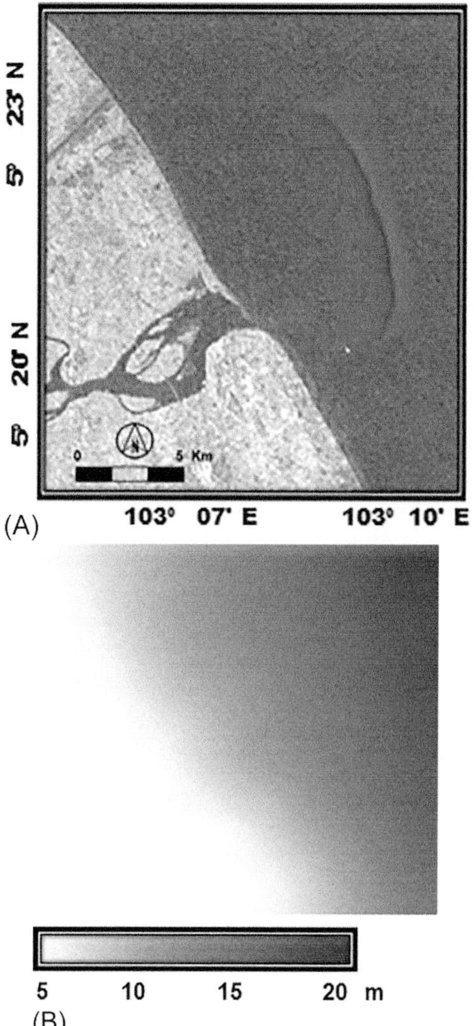

FIG. 13.14 RADARSAT-1 SAR F1 mode data for (A) 3-D front and (B) coastal bathymetry.

obtuse angles. Consequently, fuzzy B-splines construct a fine 3-D front topology from TerraSAR images (Fig. 13.15) and Sentinel-1A images (Fig. 13.16). In addition, the fuzzy B-spline depicts an optimized local triangulation between two different points [22]. This corresponds to the feature of deterministic strategies of finding only suboptimal solutions usually, which overcomes uncertainties. In this context, the spatial cluster of gradient flow at each triangulation point can be simulated (Figs. 13.14–13.16). Consequently, triangle-based criteria follow the rule of maximization or minimization of the angles of each triangle [24], preferring short triangles with obtuse angles. Furthermore, edge-based criteria prefer edges that are closely related.

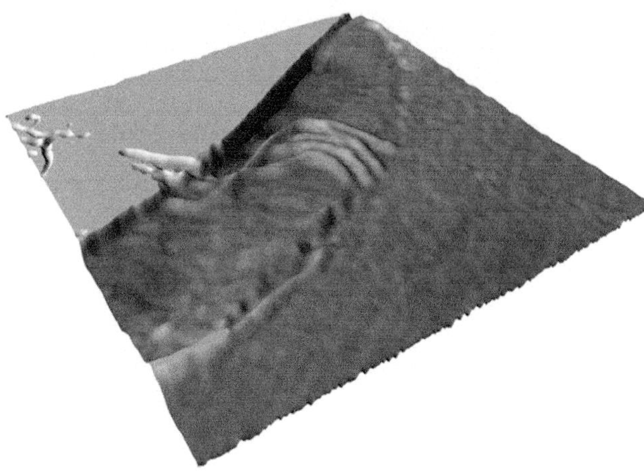

FIG. 13.15 3-D front reconstruction in TerraSAR image.

FIG. 13.16 3-D front reconstruction in Sentinel-1A image.

13.9 Quantized Marghany's front

The coastal waters of Malaysia are shallow, i.e., 50 m depth, and are dominated by mixing over different seasons owing to strong tidal current flow. Water mixing in the coastal waters of Malaysia hampers distinguishing variations in the coastal water's physical properties as the average temperature differences ranged between 2.8°C and 5.1°C during the southwest monsoon period (May to August). During the northeast monsoon period (October to March) the temperature differences were much smaller than the southwest monsoon season, at 1.3–2.2°C. In these circumstances, the possibility of thermal front occurrence is extremely small. However, a pioneering study discovered the sharp unique front occurrence along the coastal water of Malaysia using SAR images, which is Marghany's work [6, 7].

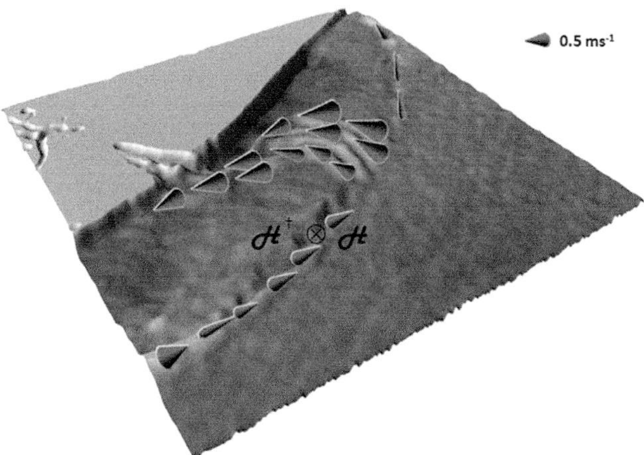

FIG. 13.17 Quantum front boundary flow in TerraSAR image.

Marghany's front is dominated by boundary current flows along the east coast of Malaysia with strong current flows of $1.2\,\mathrm{m\,s^{-1}}$, as demonstrated in TerraSAR images (Fig. 13.17). This curved front involves mesoscale eddies and sharp branches of the boundary current flow near the coastal waters. This perhaps occurs due to strong flood current flow from the mouth river of the Terengganu River and also due to strong diurnal tidal current flow along the east coast of Malaysia. However, this sharp boundary of current flow also occurs along the southern part of the east coast of Malaysia away from any river mouth (Fig. 13.18). In this case, the occurrence of such a front boundary is owing to tidal energy causing upwelling with a steady current flow of $1\,\mathrm{m\,s^{-1}}$. When the tidal current occurs, it experiences a centrifugal force, as well as a Coriolis force, which must be balanced by a horizontal gradient. If an average flow over the tidal period is considered, the total impact of the Coriolis force is balanced by the

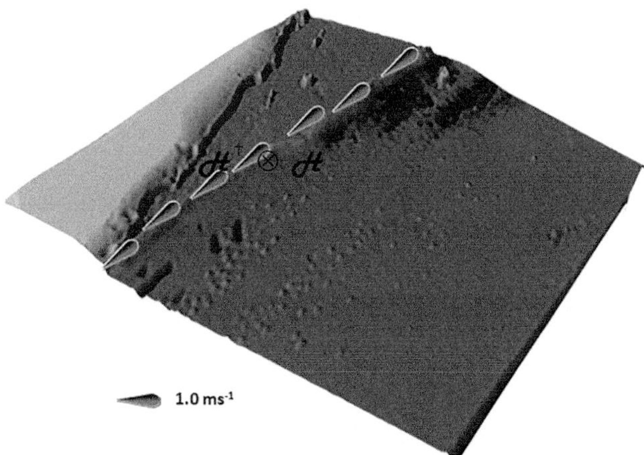

FIG. 13.18 Quantum front boundary flow in Sentinel-1A image.

density gradient. In the shallow boundary layer near the bottom, the tidal current is deformed due to friction of irregular bottom topography or the existence of sand waves. In this regard, the onshore density gradients drive an onshore drift in the boundary layer and then create upwelling.

Fig. 13.18 reveals a sharp quantum boundary between two different coastal water systems. The quantum boundary or quantum front encodes the dynamic interaction between the two different water systems through its boundary. Let us assume that boundary quantum space that separates the two quantized water systems in Fig. 13.18, for instance, is Q_1 and Q_2 on a common quantum Hilbert space $\mathcal{H}$. In this regard, the frontal boundary space can be expressed mathematically as:

$$\mathcal{H}_Q = \mathcal{H}^{\dagger} \otimes \mathcal{H} \tag{13.39}$$

The frontal boundary is considered as quantum space; which can be interpreted by the following approach: the vector of the boundary flow $|\vec{u}_i\rangle$ is assigned from $\mathcal{H}_Q$ to slightly couple of $\boldsymbol{u}$-dependent and i-dependent vectors $|\vec{u}\rangle$ and $|i\rangle$ by the following mathematical expression:

$$|\vec{u}_i\rangle = |u\rangle \otimes |i\rangle \tag{13.40}$$

Eq. (13.40) demonstrates that the quantum front is a quantum boundary that determines the quantum state in the two sides of the quantum front, which is identified as $Q \mid State \rangle = (Phys \mid State)\ \varepsilon\ \mathcal{H}_Q$. On this understanding, Marghany's coastal front is considered as an entangled quantum state boundary that separates different dynamical quantum states as the two different quantum states are only correlated to the quantum front boundary without any particular interaction between them. This definition can assist in determining the quantum front boundary in the SAR images as long it has sharp and steady strong peaks of backscatter that grow exponentially in $\mathcal{H}_Q$ as:

$$Q_i(\sigma_Q) = Q_i(\vec{u}\ (\sigma_Q)|\vec{u}\ (i)) = (\sigma_Q|i : \vec{u}\ ) = \langle \vec{u}\ : \sigma_Q|i : \sigma_Q\rangle \tag{13.41}$$

The quantum front boundary as encoded on the sharp radar quantum cross-section σ_Q can be employed to determine the exact location of the quantum front boundary Q_i in any SAR data. In this way, the changing in the current boundary conditions in the two different quantum states in SAR images can be used to determine transition amplitude, which involves the sharpest quantum radar-cross section along the quantum front boundary.

References

[1] Bowman MJ, Iverson RL. Estuarine and plume fronts. In: Oceanic fronts in coastal processes. Berlin, Heidelberg: Springer; 1978. p. 87–104.

[2] Bowden KF. Physical oceanography of coastal waters. E. Horwood; 1983.

[3] Vogelzang J, Wensink GJ, Calkoen CJ, Van der Kooij MW. Mapping submarine sand waves with multiband imaging radar: 2. Experimental results and model comparison. J Geophys Res Oceans 1997;102(C1):1183–92.

[4] Krogstad HE, Schyberg H. On Hasselmann's nonlinear ocean-SAR transformation. In: Remote sensing: Global monitoring for earth management. Espoo: IEEE; 1990 June 3–6, 1991.

[5] Marghany M, Mazlan H. Simulation of sea surface current velocity from synthetic aperture radar (SAR) data. Int J Phys Sci 2010;5:1915–25.
[6] Marghany M. Three-dimensional coastal water front reconstruction from RADARSAT-1 synthetic aperture radar (SAR) satellite data. Int J Phys Sci 2011;6(29):6653–9.
[7] Marghany M. Three-dimensional coastal front visualization from RADARSAT-1 SAR satellite data, In: International conference on computational science and its applications 2012 Jun 18Berlin, Heidelberg: Springer; 2012. p. 447–56.
[8] Tanjung K, Yao TG, Mauro P, Tomasz P. Observable quantum entanglement due to gravity. NPJ Quantum Inf 2020;6(1):1–10.
[9] Lee JS, Jansen RW, Schuler DL, Ainsworth TL, Marmorino GO, Chubb SR. Polarimetric analysis and modeling of multifrequency SAR signatures from Gulf Stream fronts. IEEE J Ocean Eng 1998;23(4):322–33.
[10] Hessner K, Rubino A, Brandt P, Alpers W. The Rhine outflow plume studied by the analysis of synthetic aperture radar data and numerical simulations. J Phys Oceanogr 2001;31(10):3030–44.
[11] Perona P, Malik J. Scale-space and edge detection using anisotropic diffusion. IEEE Trans Pattern Anal Mach Intell 1990;12(7):629–39.
[12] Sapiro G. Geometric partial differential equations and image analysis. Cambridge University Press; 2006.
[13] Romeiser R, Alpers W, Wismann V. An improved composite surface model for the radar backscattering cross section of the ocean surface: 1. Theory of the model and optimization/validation by scatterometer data. J Geophys Res Oceans 1997;102(C11):25237–50.
[14] Marghany M. Velocity bunching model for modelling wave spectra along east coast of Malaysia. J Indian Soc Remote Sens 2004;32(2):185–98.
[15] Alpers WR, Ross DB, Rufenach CL. On the detectability of ocean surface waves by real and synthetic aperture radar. J Geophys Res Oceans 1981;86(C7):6481–98.
[16] Vachon PW, Olsen RB, Krogstad HE, Liu AK. Airborne synthetic aperture radar observations and simulations for waves in ice. J Geophys Res Oceans 1993;98(C9):16411–25.
[17] Vachon PW, Krogstad HE, Paterson JS. Airborne and spaceborne synthetic aperture radar observations of ocean waves. Atmosphere-Ocean 1994;32(1):83–112.
[18] Vachon PW, Liu AK, Jackson FC. Near-shore wave evolution observed by airborne SAR during SWADE. Atmosphere-Ocean 1995;2:363–81.
[19] Vachon PW, Campbell JM, Dobson FW. Comparison of ERS and RADARSAT SARS for wind and wave measurement. ESA SP (Print); 19971367–72.
[20] Inglada J, Garello R. Depth estimation and 3D topography reconstruction from SAR images showing underwater bottom topography signatures, In: IEEE 1999 international geoscience and remote sensing symposium. IGARSS'99 (Cat. No. 99CH36293) Jun 28, 1999vol. 2. IEEE; 1999. p. 956–8.
[21] Khan M, Gondal MA. An efficient two step Laplace decomposition algorithm for singular Volterra integral equations. Int J Phys Sci 2011;6(20):4717–20.
[22] Anile AM, Deodato S, Privitera G. Implementing fuzzy arithmetic. Fuzzy Set Syst 1995;72(2):239–50.
[23] Anile AM, Gallo G, Perfilieva I. Determination of membership function for cluster of geographical data. Genova: Institute for Applied Mathematics, National Research Council, University of Catania; 1997 25 pp., Technical Report No.26/97.
[24] Rövid A, Varkonyi AR, Várlaki P. 3D Model estimation from multiple images, In: IEEE international conference on fuzzy systems (FUZZ-IEEE)vol. 1661. ; 2004.
[25] Russo F. Recent advances in fuzzy techniques for image enhancement. IEEE Trans Instrum Meas 1998;47 (6):1428–34.

CHAPTER

14

Automatic detection of nonlinear turbulent flow in synthetic aperture radar using quantum multiobjective algorithm

14.1 Introduction

A turbulent flow study requires standard mathematical procedures to understand its physical characteristics and mechanisms. In practice, there is a lot of emphasis on turbulence science due to its wide application in different fields of sciences, mechanical engineering, fluid dynamics, aerospace, physics, hydrodynamics, etc. Turbulent flow generally has a large number of spatial degrees of freedom and exhibits chaotic behavior in both time and space [1]. Fluid turbulence is sometimes referred to as spatial-temporal chaos. The term "turbulence" is used to describe the rapid flow of any fluid passing an obstacle or an airfoil; this creates turbulence in the boundary layers and develops a turbulent wake, which will generally increase the drag exerted by the flow on the obstacle. In this context, a turbulent flow is by nature unstable: a small perturbation will generally occur, due to the nonlinearities of the equations of motion.

Researchers are agreed that a turbulent flow is governed by the conversation of mass and momentum, as demonstrated earlier in Chapter 2. In this context, researchers have implemented the Navier-**Stokes** equations to describe the complexity of turbulent flow. The Navier-Stokes equations, however, cannot be solved analytically due to uncertainties, and the regularity of solutions of these equations is established rigorously. Consequently, remote sensing techniques have the potential to track the spatial variation of turbulent flow in an ocean or estuary.

https://doi.org/10.1016/B978-0-12-820785-7.00002-2

14.2 What is meant by quantum turbulence?

The term quantum turbulence denotes the turbulent motion of quantum fluids, systems such as superfluids, which are categorized by quantized vorticity, superfluidity, and, at finite temperatures, two-fluid behavior, as discussed in quantum front boundaries in Chapter 13. In this regard, ocean flow systems are considered superfluid.

Quantization of ocean flow involves three particulars: (i) it reveals two-fluid behavior at nonzero temperature or in the existence of layers; (ii) it can flow freely, without the dissipative consequence of viscous forces, especially in the circumstance of tsunami wave propagation; and (iii) its constrained rotation is limited to discrete vortex lines of known strength. Nevertheless, unlike the eddies in ordinary fluids, which are unceasing and can have random size, shape, and strength, superfluidity and quantized vorticity are extraordinary manifestations of quantum mechanics at macroscopic-length scales [1, 2].

On this understanding, the vortices have a constant size, and are indistinguishable. This is another surprising characteristic of superfluids, which is very different from the random vortices in a conventional fluid, and rises out of quantum physics, the consequences of which turn out to be observable on a large scale at low temperatures. Quantum turbulence, then, is a tangle of these quantized vortices, creating a pure structure of turbulence, which is far less complicated to model than conventional turbulence, in which countless feasible interactions of the eddies rapidly create a problem too complicated for it to be possible to predict what will occur [3].

Generally, quantum turbulence is the name given to the turbulent flow—the chaotic motion of a fluid at high flow rates—of quantum fluids. In this regard, ocean water particles spin up and down in an irregular pattern, creating chaotic movement on a large scale, as can be seen in the Gulf of Mexico loop current.

14.3 Turbulence imagined in SAR data

Turbulence is fairly pervasive in the layer of the atmosphere that is contiguous to the Earth's surface. It is powered by microscale gradients in momentum, temperature, and moisture, and its result is to extinguish similar gradients to those that contribute to its existence. This obliteration arises through turbulent transfer processes (fluxes), without which the transports of energy, moisture, and momentum between the Earth's surface and the atmosphere would be left to molecular diffusion. Subsequently, it has scales 3–6 orders of magnitude smaller than turbulent diffusion. Henceforth, turbulent flux energy contains much of the large-scale movements. The main question that arises is: how do SAR sensors image ocean turbulence?

When a SAR sensor illuminates the ocean surface at moderate incident angles between 20 and 60 degrees, the largest part of the backscattered energy is formed by ocean surface roughness on the scale of the radar wavelength projected on to the ocean surface—the "Bragg" wavelength, as demonstrated in previous chapters. Representative microwave radars function at wavelengths on the decameter and centimeter scales and consequently, the wind-generated roughness through the turbulent momentum flux is on these scales, which is responsible for the ocean surface radar signature. For instance, if the near-surface wind speed rises, subsequently the surface roughness, and consequently the backscattered energy, increases.

In the direction of the wind, the short waves are formed largely by wind propagation. In this regard, the backscattered energy from the turbulent surface is at maximum when the wind is blowing in the radar look direction. There is an analogous, although slightly smaller, local extreme in the backscattered power when the wind is driving away from the radar antenna. The minimum backscattered energy occurs when the radar look direction is perpendicular to the wind direction.

Accordingly, SAR senses the constraints that atmospheric and marine phenomena place on the centimeter-scale wave spectrum. At the same time, the prevailing atmosphere is very obvious to SAR and precipitation can at times disturb the radar signal. On the other hand, rain cells can appear as an irregular pattern in SAR images that indicate constrained cell flows.

The distinctive resolution of spaceborne SAR is of the order of 10–100 m with a swath width of the order of 100–1000 km. Consequently, spaceborne SAR is accomplished in delivering an exhaustive view of sea-surface stress-generated roughness patterns—the footprints of macroscale, mesoscale, and microscale meteorological phenomena and, one can suppose, the corresponding dynamic and thermodynamic environment allied with that phenomenon's presence (e.g., statically unstable versus statically stable; baroclinic versus barotropic). In the case of barotropic or weakly baroclinic convective marine atmospheric boundary layer quasitwo-dimensional signatures seen in SAR images, one can expect minimal directional shear and thus, the mottles tend to be elongated along, or to within a few degrees clockwise of, the near-surface mean wind direction.

In the case of the buoyancy-driven/shear-organized rolls, there are helical circulations generated through thermodynamic instability in an environment with sufficient vertical wind shear. For a specified quantity of marine atmospheric boundary layer buoyancy, as the amount of the wind shear strengthens, a field of arbitrarily prearranged lengthened mottle components grows into a field of linearly organized elongated mottle components. Rising and falling patterns of the circulation guide to the consistent amplified and declined sea surface roughness. The resulting SAR strength pattern shows the presence of irregular dark and bright mottled lines. The orientation of the surface footprint of this sort of roll, and thus its SAR signature, is forced in the same manner [4].

14.4 Can a quantum algorithm automatically detect turbulent flow in SAR images?

The segmentation problem of turbulent flow using partition clustering is viewed as a combinatorial optimization problem. But the prevailing optimal approaches, e.g., conventional genetic algorithms (CGAs), are often time-consuming [5], and their convergence speed is slow and easy to trap in local optimal values.

Conventional algorithms ensure a single phase at a time, since even a turning machine can only be in a single position at a time. Consistent with quantum physics, the elementary turbulence flow and ocean wave-particles of nature are not in one fixed state at any moment but can occupy several phases at ones, in what is called superposition. When these particles are scattered in the SAR images, they decohere into one state.

Quantum computing has its origins in quantum mechanics, for instance, uncertainty, superposition, interference, and implicit parallelism. These advantages of quantum computing make a superior variety and the superior steadiness between the investigation and the utilization of conventional evolutionary algorithms. Various scientists have introduced quantum codes into evolutionary computing. For instance, Han and Kim [6] suggested a quantum-inspired evolutionary algorithm for combinatorial optimization as a function of the perception and codes of quantum computing. In this regard, Jiao and Li et al. [7] developed a quantum-inspired immune clonal algorithm for global optimization.

Automatic detection of the turbulence boundary flow in SAR images is dominated by defecting of the single-objective clustering algorithms in addition to the uncertainty of brightness and darkness patch segmentations. Can a quantum multiobjective genetic algorithm automatically detect turbulence flow due to bottom features such as coral reefs?

14.5 Quantum computing

In the quantum computer, a qubit is used rather than the classical bit. Quantum systems are described by a wave function y that exists in a Hilbert space. The Hilbert space has a set of states, ψ, that form the basis, and the system is described by the quantum state, y, which is in a linear superposition of the basis states ψ, and in the general case, the coefficients $\mathbb{C}$ may be complex. A qubit is a vector in two-dimensional complex vector space [8]. In this space, a vector has two components and the projections of the vector based on the vector space are a complex number. In this regard, the quantum state can be described as:

$$|\psi\rangle = \alpha|0\rangle + \beta|1\rangle \tag{14.1}$$

$$|\dot{\alpha}^2| + |\beta^2| = 1 \tag{14.2}$$

where α and β are the complex numbers and $|0\rangle$ and $|1\rangle$ are the dual orthonormal bases for the dual dimensional vector space [6–9].

It can be shown that the qubit is a linear superposition of the bases state, by changing the values of the coefficients.

$$|\psi\rangle >= \cos\theta/2|0\rangle + \exp(i\varphi)\sin\theta/2|1\rangle \tag{14.3}$$

Eq. (14.3) demonstrates that the qubit can be represented as a vector $\boldsymbol{r}$ from the origin to a point of the three-dimensional sphere with a radius of one, which is known as the Bloch sphere (Fig. 14.1).

Qubits behave like bits, in that they can be on or off; however, in effect, they can be on and off at the same time, whereas the traditional bit can only be in either state 0 or state 1. Consequently, the qubit can take any value represented by the vector r on the Bloch sphere (Fig. 14.2).

The quantum system is said to be collapsed when the projection on one of the bases is achieved. That is also called decoherence or the measures. In other words, decoherence has been developed into a complete framework, but it does not solve the measurement problem. For instance, in the circumstance of the projection of $|\psi\rangle$ on the $|0\rangle$ basis, it should be $|\psi\rangle = \alpha|0\rangle$. Therefore, $|\dot{\alpha}^2|$ is the probability of the qubit to collapse on the state $|0\rangle$. Therefore, in quantum multiobjective evolutionary (QME), i.e. ,multiobjective evolutionary algorithm (MOEA) inspired by quantum computing, a novel Q-bit chromosome representation is adopted based on the concept and principles of quantum computing, such as Q-bits

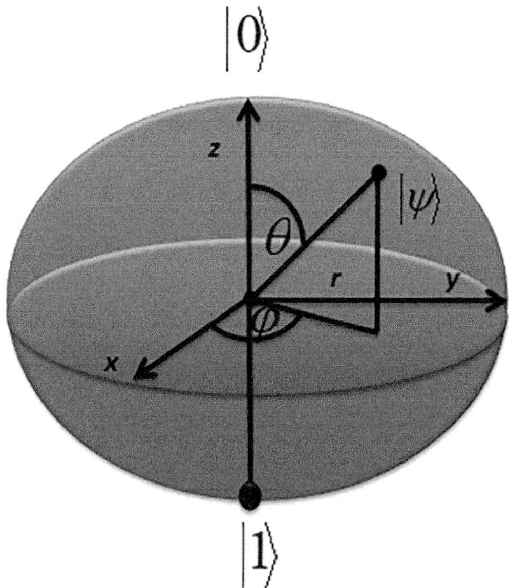

FIG. 14.1 Illustration of the Bloch sphere.

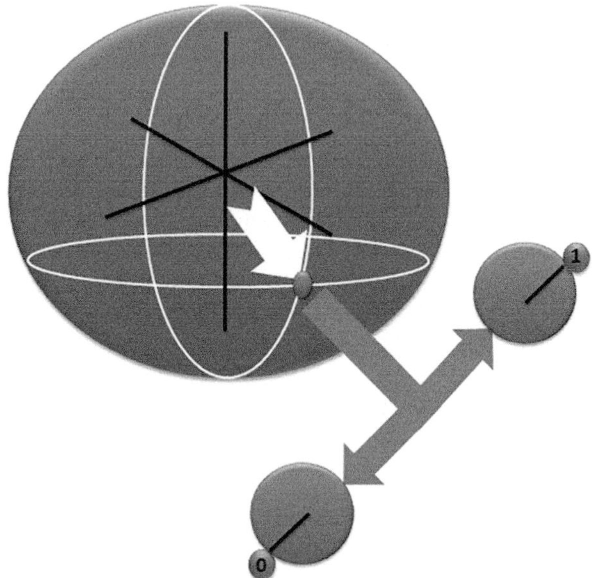

FIG. 14.2 Representation of qubit.

(a quantum bit which is a unit of quantum information) and linear superposition [6]. The characteristic of the representation is that any linear superposition of solutions can be represented. The smallest unit of information stored in a dual-state quantum computer is called a quantum bit (Q-bit), which may be in the "1" state, or in the "0" state, or in any superposition of the two. The state of a Q-bit can be represented by Eq. (14.3).

14.6 Quantum machine learning

In quantum machine learning, two phases are involved: learning and pattern recall.

14.6.1 Learning phase

Quantum machine learning is based on the Grover algorithm, which is based on constructing a coherent quantum system. In this regard, m represents the pattern in what is known as the pattern-sorting process. Let us assume a set $D=\{\psi(P)\}$ of SAR features, which has m patterns of the turbulence flow features as a training set. The aim is to produce $|\psi\rangle$ such that:

$$|\psi(P)\rangle = \sum\nolimits_{m\in P}\psi(P)|P\rangle \tag{14.4}$$

where P represents the different patterns in the training set $\boldsymbol{D}$.

P can be represented in a binary format, b_1, b_2, and b_3. Using Grover algorithms, n qubits are required to represent P patterns, in addition to $n+1$ qubits, required as control qubits in the learning algorithms. That gives the sum of $2n+1$ qubits, which are necessitated to characterize m patterns. It also delivers the sum of $2n+1$ registers in implanting the quantum network. A state $|\psi\rangle$ can be constructed using a polynomial number (n and m) of elementary operations on one, two, or three qubits. That means the $|\psi\rangle$ will be constructed recursively using three operations [9–13]. In this regard, the mathematical description of the coherence state of the SAR features $|\psi_{SAR}\rangle$ is given as:

$$|\psi_{PolSAR}\rangle = |x_1x_2\ldots, x_n, R_1R_2, \ldots, R_{n-1}, C_1C_2\rangle \tag{14.5}$$

Eq. (14.5) indicates the restoring P patterns as $x_1x_2\ldots,\ x_n$, with the total number n. Also, $R_1R_2, \ldots, R_{n-1}$ as well as C_1C_2 are controlled register patterns. The Grover algorithm, therefore, involves three operations, which deliver the coherent state of the SAR features $|\psi_{SAR}\rangle$. The first operation involves the matrix generation M, which is described by:

$$M = \begin{bmatrix} 1 & 0 & 0 & 0 \\ 0 & 1 & 0 & 0 \\ 0 & 0 & \sqrt{\frac{P-1}{P}} & \frac{\psi(m)}{\sqrt{P}} \\ 0 & 0 & \frac{\psi(m)}{\sqrt{P}} & \sqrt{\frac{P-1}{P}} \end{bmatrix} \tag{14.6}$$

M must be achieved under the circumstance of $1\le P\le m$ and $1 \le P \le m$. Then the second operation is presented by the flip transformation ψ^0 which is described by:

$$\Psi^0 = \begin{bmatrix} \psi & 0 \\ 0 & I \end{bmatrix} \tag{14.7}$$

This equation is used if there are dual qubits, i.e., one for the turbulent flow and another for the surrounding environment, one of which is required to control the flip transformation of the other one. In other words, let us assume the state of the turbulent flow in the SAR image is represented by Ψ^0, while the state of the surrounding environment is identified by $|\psi \geq \beta|0\rangle + \alpha|1\rangle$. On this understanding, the pattern state detection of turbulent flow using the M matrix is given by:

$$\Psi^0|\text{turbulent}\rangle\langle\text{non}-\text{turbulent}| = \begin{bmatrix} 0 & 1 & 0 & 0 & 1 & \beta \\ 1 & 0 & 0 & 0 & 1 & \alpha \\ 0 & 0 & 0 & 1 & 0 & \beta \\ 0 & 0 & 1 & 0 & 0 & \alpha \end{bmatrix} \tag{14.8}$$

The third operation is based on transformation Ψ^{00}, which is a 3-qubit operation. It flips the state of one qubit if and only if the other two are in the $|00\rangle$ states. In this regard, 3-qubits $|\psi_1 \geq 1|0\rangle$ and $|\psi_2 \geq 1|0\rangle$ $|\psi_3 \geq \beta|0\rangle + \alpha|1\rangle$ are assumed, and represent the levels of turbulent flow, i.e., strong, medium, and light, respectively. This can be mathematically described as:

$$\Psi^{00}|\psi_1\psi_2\psi_3\rangle = \begin{bmatrix} \alpha \\ \beta \\ 0 \\ 0 \\ 0 \\ 0 \\ 0 \\ 0 \end{bmatrix} \tag{14.9}$$

14.6.2 Pattern phase recall

A pattern phase examination is achieved in the post of the learning phase $\Psi^{00}|\psi_1\psi_2\psi_3\rangle$, which has all the patterns of the turbulent flow in the SAR data. The main hypotheses recall the pattern from the required pattern basis at the state $|\psi_{SAR}\rangle$. Recalling the turbulent flow pattern 01 from the 2-qubits. $|\psi_{\text{turbulent}}\rangle$ should collapse on the desired state after repeating the algorithm for $\left(\frac{3.14}{14}\right) \times 2$ one time iteration. For instance, $|\psi_{\text{turbulent}}\rangle$ can be achieved $-\left[\sqrt{3}\right]^{-1}|01\rangle + \left[\sqrt{3}\right]^{-1}|10\rangle - \left[\sqrt{3}\right]^{-1}|11\rangle$. Then the algorithm computes the average with $|01\rangle\left[\sqrt{3}\right]^{-1}(0, -1, 1, 1)$, which is equal to $0.25\sqrt{3}$. Conversely, the entire quantum set is rotated around the average, for instance, $|\psi > 0.5\sqrt{3}(1, 2, -1, -1)$. Finally, the probability of the algorithm to collapse on the turbulent flow state is, for instance, ¾=75%. In this circumstance, the higher the number of qubits, the higher the percentage of quick and accurate detection of the turbulent flow patterns.

14.7 Quantum multiobjective evolutionary algorithm (QMEA)

Consistent with Wang et al. [14], evolutionary computing with Q-bit representation has a better population diversity characteristic than other representations, since it can represent a linear superposition of state probabilities. The only single spinning upward Q-bit is enough to represent eight states, while in binary representation, at least eight strings (000), (001), (010), (011), (100), (101), (110), and (111) are needed (Fig. 14.3). Further, in QMEA crossover and

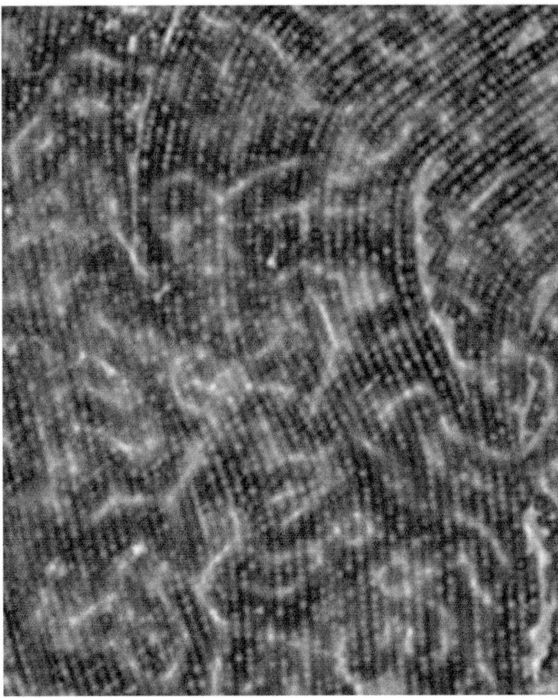

FIG. 14.3 The initial generation of qubits of SAR images.

mutation operations are used to maintain the diversity of the population, while the evolutionary operation that quantum gates operate on the probability amplitudes of basic quantum states is applied to maintain the diversity of the population in the quantum genetic algorithm (QGA) and quantum evolutionary algorithm (QEA).

14.8 Generation of qubit populations

Let us assume that $Q(t)=\{q_1^t, q_2^t, \dots, q_n^t\}$ (Fig. 14.4) is the individual qubit population, where $q_1^t=\left\{\begin{matrix}\alpha_{i1}^t, \alpha_{i2}^t, \dots, \alpha_{im}^t \\ \beta_{i1}^t, \beta_{i2}^t, \dots, \beta_{im}^t\end{matrix}\right\}$ $(i=1,2,\dots,n)$ is a qubit individual, n is the dimension of population and m is the span of the individual. In the preliminary stage, the qubit individuals have a similar probability of $\frac{1}{\sqrt{3}}$.

α_{i1}^t	....	....	....	α_m^t
β_1^t	....	....	....	β_m^t

FIG. 14.4 An array of one qubit q_1^t populations.

14.9 Generation of turbulent flow population pattern

Let us assume that the classical binary turbulent flow population $O(t)=\{o_1^t, o_2^t, \ldots, o_n^t\}$, where $(i=1,2,\ldots,n)$ with length m is a binary string, which is observed from q_i^t. The random number of turbulent population features is generated with the constraint of $N_O \in [0,1]$. In this circumstance, the random number N_O is validated as a turbulent flow when $N_O > |\alpha_i^t|^2$ in which the corresponding bit in o_n^t is equal to 1. On the contrary, if o_n^t equals 0, the SAR features do not represent a turbulence. Therefore, the clustering point center of the turbulence is represented by bit 1 in the binary individual.

14.9.1 Quantum fitness

The turbulent footprint boundaries must be coded into qubit chromosome form. In this case, the qubit chromosome q_m consists of several qubits, where every qubit corresponds to a coefficient in the north-order surface fitting polynomial, as given by:

$$f(i,j)_q = q_0 + q_1 i + q_2 j + q_3 i^2 + q_4 ij + q_5 j^2 + \cdots + q_m j^n \tag{14.10}$$

The fitness is then evaluated using the connectedness metric. In other words, the degree of the qubit chromosomes that are signed from surrounding points in the SAR image and clustered as the turbulent flow can be mathematically described as:

$$conn(q^t) = \sum_{i=1}^{m} \left(\sum_{j=1}^{L} \left(q_{i,j}^t \right) \right) \tag{14.11}$$

Eq. (14.11) shows the method of identification of turbulent clusters from the surrounding environment of several qubit chromosomes L with m size. In this circumstance, the turbulence is identified by every 10 qubit chromosomes for every point nearest point i and its neighboring j-th. Under the circumstance of $q_{i,j}^t = j^{-1}$ else, i.e., absence of turbulent occurrence in SAR image and feature belongs to the surrounding sea surface; the nonturbulent can be also identified.

14.9.2 Quantum mutation

In QEA, the qubit representation can be used as a mutation operator (Fig. 14.5). Directed by the current best individual, the quantum mutation is completed through the quantum rotation gate:

$$G(\theta) = \begin{bmatrix} \cos\theta & -\sin(\theta) \\ \sin\theta & \cos\theta \end{bmatrix} \tag{14.12}$$

where θ is the rotation angle of the quantum rotation gate (Fig. 14.6) and is described as:

$$\theta = k \times f(\alpha_i, \beta_i) \tag{14.13}$$

where k is a coefficient and the value of k has an impact on the speed of convergence.

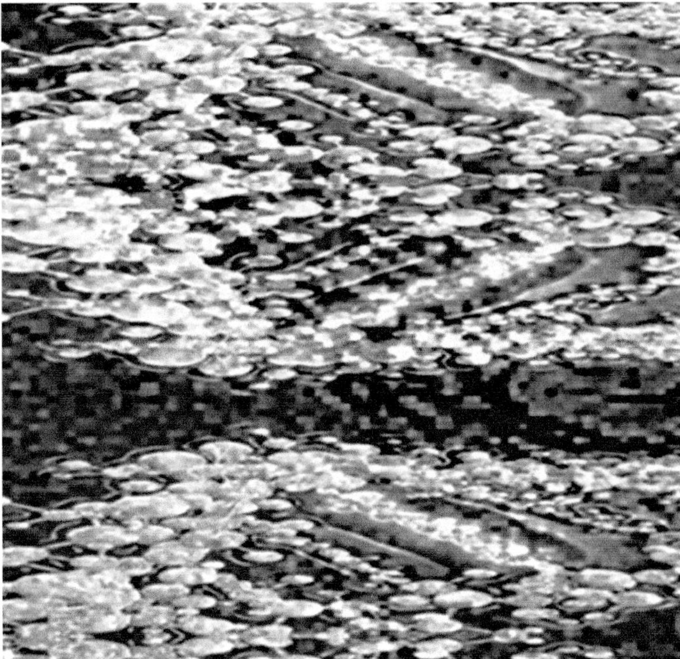

FIG. 14.5 Quantum mutation of SAR image for turbulent detection.

The value k must be chosen reasonably. If k is too big, the search grid of the algorithm is large and the solutions may diverge or have **a** premature convergence to a local optimum, and if it is too small, the search grid of the algorithm is also small and the algorithm may be in a stagnant state. Besides, $f(\alpha_i, \beta_i)$ determines the search direction of convergence to a global optimum [14]. The quantum gate is the key to determine the false alarm clustering of **turbulence detection** or the true clustering of **turbulence detection**. Fig. 14.6 demonstrates the initial recognition of **turbulent detection** as bright patches by using a quantum rotation gate. Table 14.1 is a lookup table to determine the best solution of the quantum chromosome of **turbulence detection**.

The updated procedures can be given by:

$$q_i^{t+1} = G(t) \times q_j^t \tag{14.14}$$

The probability amplitude of the qubit is denoted by q_i of **an** individual at t-th generation as t is evolutionary generation at the quantum gate $G(t)$.

14.10 Quantum nondominated sort and elitism (QNSGA-II)

This section presents a brief description of QNSGA-II relevant to this study. QNSGA-II is the second version of the famous "nondominated sorting genetic algorithm" based on the work of Prof. Kalyanmoy Deb for solving nonconvex and nonsmooth single and multiobjective optimization problems. Its main features are: (i) a sorting nondominated

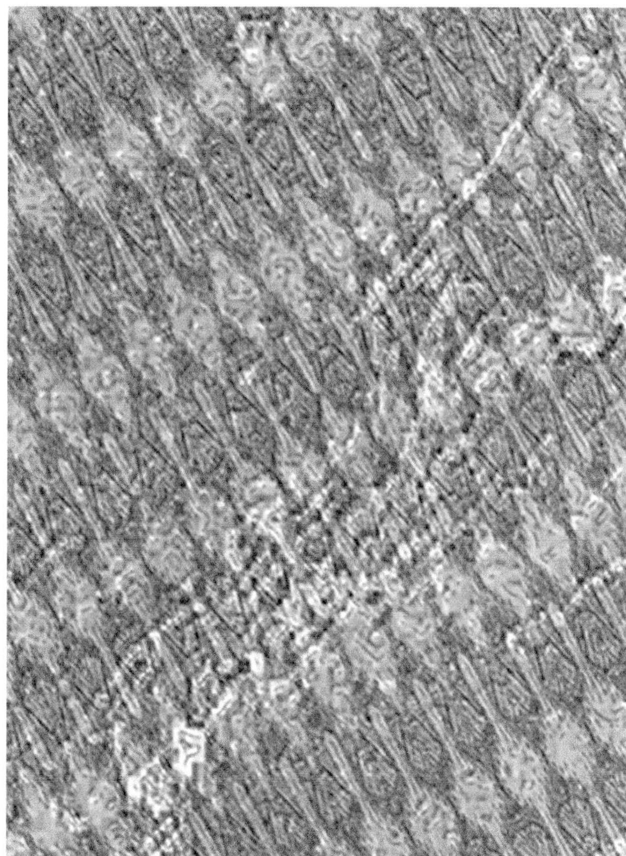

FIG. 14.6 Quantum rotation gate for turbulent detection in SAR data.

TABLE 14.1 Lookup table of $f(\alpha_i, \beta_i)$.

		$f(\alpha_i, \beta_i)$	
$\|\alpha_1\| * \|\beta_1\| > 0$	$\|\alpha_2\| * \|\beta_2\| > 0$	**arctan($\|\beta_1\|/\|\alpha_1\|$) (best solution)**	**arctan($\|\beta_2\|/\|\alpha_2\|$) (current solution)**
True	True	+1	−1
True	False	−1	+1
False	True	−1	+1
False	False	+1	−1

procedure of qubit chromosome $Q(q_N^t)$ where all the individual are sorted according to the level of nondomination; (ii) it implements elitism, which stores all nondominated solutions, hence enhancing convergence properties; (iii) it adopts suitable automatic mechanics based on the crowding distance to guarantee diversity and spread of solutions; and (iv) constraints are implemented using a modified definition of dominance without the use of penalty functions [14].

Perhaps one best solution does not exist in the case of multiple objectives. Therefore, there exists a set of solutions that are superior to the rest of the solutions in the search space when all objectives are considered but are inferior to other solutions in the space in one or more objectives. These solutions are known as Pareto-optimal solutions or nondominated solutions [15].

The efficiency of QNSGA lies in the way multiple objectives are reduced to dummy fitness functions using nondominated sorting procedures. Consequently, QNSGA can solve practically any number of objectives. In this regard, this algorithm can handle both minimization and maximization problems [16].

To sort a population of size N for the qubit chromosome $Q(q_1^t), \ldots, Q(q_N^t)$, according to the level of nondomination, each solution m must be compared with every other solution in the population to find if it is dominated. This requires comparisons $O(Q(q_m))_N$ for each solution, where m is the number of quantum qubit chromosomes of different pixels belonging to turbulence, sea roughness, and low wind zones. The initialized population $N\ Q(q_1^t), \ldots, Q(q_N^t)$ is sorted based on the level of nondomination. If $S_{Q(q1^{\prime})}$ is each quantum multiobjective solution that must be compared to other every solution to determine the level of domination, the fast sort algorithm, as given by Deb et al. [15], can be explored in turbulence automatic detection as follows:

For each individual $Q(q_1^t)$ in main population P, do the following:

Initialize $S_{Q(q_1^t)}=\Phi(Q(q_N^t))$. This set Φ would include all the individuals of $Q(q_N^t)$ which is being dominated by $Q(q_1)$.

Initialize $n_{Q(q_1)}=0$. This would be the number of individuals that dominate $Q(q_N^t)$ i.e., no individuals dominate $Q(q_1)$ then $Q(q_1)$ belongs to the first front; set rank for an individual $Q(q_1)$ to one i.e. $Q(q_1)_{rank}=1$.

for each individual m in $P(Q(q_1))$

if $Q(q_1)$dominated m then

. add m to the set $\Phi(Q(q_N^t))$ i.e. $\Phi(Q(q_N^t))=\Phi(Q(q_N^t))\cup\{m\}$

*else if m dominates $Q(q_1)$then

. the increment for domination counter for $Q(q_1)$ i.e. $n_{Q(q_1)}=n_{Q(q_1)}+1$

Let the first front set $F_1(Q(q_N^t))$ and then update by adding $Q(q_1)$ to front 1 i.e. $F_1(Q(q_N^t))=F_1(Q(q_N^t))\cup\{Q(q_1)\}$

Initialize the front counter to one. $i=1$

Then $F_i(Q(q_N^t))\neq\Phi(Q(q_N^t))$

Let $Q(q_1)\neq\Phi(Q(q_N^t))$. The set for sorting the individuals for $(i+1)^{th}$ the front

for each individual $Q(q_1)$ in front $F_i(Q(q_N^t))$

For every individual m in $S_{Q(q_1^t)}$ (is the set of individuals dominated by $Q(q_1)$)

. $n_{Q(q_1)}=n_{Q(q_1)}-1$, decrement the domination count for individual m.

. if $n_{Q(q_1)}=0$ then none of the individuals in the subsequent fronts would dominate m. Hence set $Q(q_1)_{rank}=i+1$. Update the set $Q(q_i^t)$ with individual m i.e. $Q(q_i^t)=Q(q_i^t)\cup m$.

-increment the front of one.

- For turbulence, now the set $Q(q_i^t)$ is the next front and hence $F_i(Q(q_i^t))=Q(q_i^t)$.

14.11 Quantum Pareto optimal solution

In this chapter, the multiobjective of SAR features for turbulence detection is considered. Let $P_{\max}(Q(q_i^t))$ be quantum qubit chromosomes of probability amplitude of turbulence footprint in SAR features, which should satisfy:

1. Turbulent flow pixels $P_{\max}(Q(q_i^t))$: the variation of maximum quantum qubit chromo somes of probability, which contain the turbulence quantum qubit chromosomes, i.e., $P_{\max}(Q(q_i^t))=\max\{P_1(Q(q_1^t),\ldots,P_k(Q(q_k^t)\}$, where $P_jQ(q_j^t)$ denotes probability occurrence of the turbulence in Pareto front j, $\forall j=1, 2, \ldots, k$.
2. Total probability of quantum qubit chromosomes of the turbulence $((\sum P_i(Q(q_i^t)_i)$: the sum of quantum qubit chromosomes in each row and column in SAR data.

Pareto optimal solutions are applied to retain the discrimination of turbulence diversity and its surrounding environment. The level of convergence to the Pareto front $P_f((P_i(Q(q_i^t))$ is calculated due to the Euclid distance, as follows:

$$P_f\left(P_i\left(Q\left(q_i^t\right)\right)\right)=\sum_{\omega=1}^{\omega}\sqrt{\sum_{m=1}^{N}\min_{u=1,U}\left(P_{um}\left(Q\left(q_{um}^t\right)\right)-a_{\omega m}\right)^2} \tag{14.15}$$

Eq. (14.15) determines the performance of the Pareto front for automatic detection of the turbulence flow using QEA. In this regard, N represents the number of optimized criteria for automatic detection of turbulent flow in SAR images and um is the number of Pareto front points (Fig. 14.7), which are delivered from the probability occurrence of quantum qubit chromosomes of the turbulence flow. The Pareto front algorithm generates the tested points $a_{\omega m}$.

14.12 Automatic detection of turbulent flow in SAR images

SAR images were acquired by the advanced land-observing satellite (ALOS) that was launched on January 24, 2006, by the Japan Aerospace and Exploration Agency (JAXA) and was operational until May 12, 2011. In this chapter, ALOS data was acquired on March

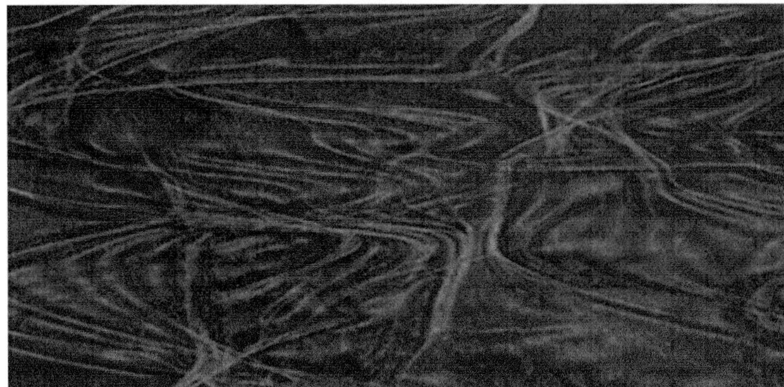

FIG. 14.7 Pareto front curves generated with quantum qubit chromosomes in SAR data.

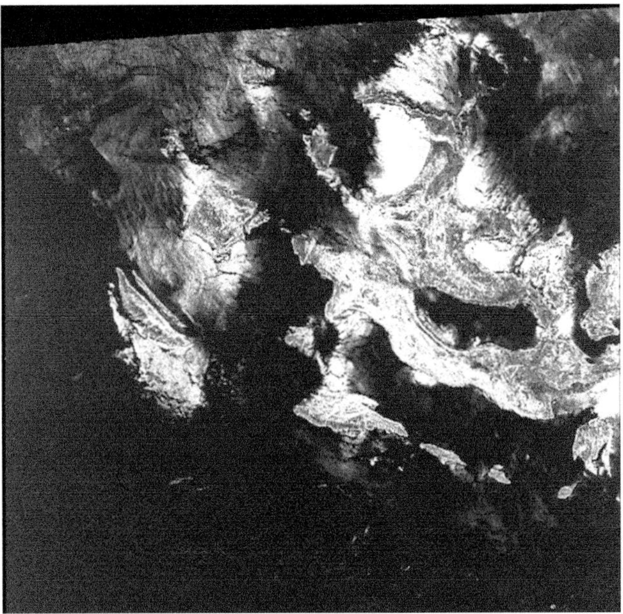

FIG. 14.8 ALOS data are acquired in this study.

22, 2007, covering the Farasan Islands (Fig. 14.8), Kingdom of Saudi Arabia, in the coastal waters of the Red Sea (Fig. 14.9).

ALOS (which carried PALSAR) was launched into a sun-synchronous orbit and revolved around the earth every 100 min or 14 times a day. ALOS PALSAR returned to the original path (repeat cycle) every 46 days. The interorbit distance was approximately 59.7 km at the equator. The PALSAR was an active microwave sensor that was not affected by weather conditions and operable both day and night. It was based on a synthetic aperture radar (SAR) carried

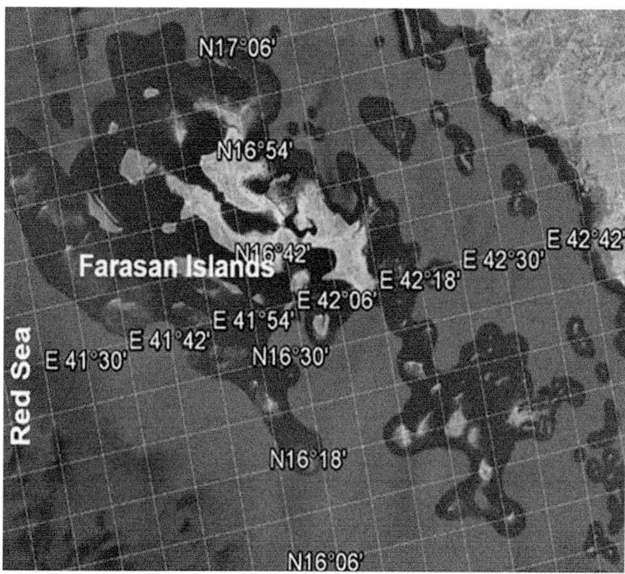

FIG. 14.9 Farasan Islands, in the Red Sea.

onboard Japan's first earth observation satellite (JERS-1). The PALSAR was a right-looking SAR using L-band frequency with a cross-track pointing capability of 8–60 degrees with HH polarization.

ALOS PALSAR data in fine mode has been used to investigate the coral reef-induced turbulent flow. Fine mode data has a maximum ground resolution of 7 m, which is one of the highest for a SAR (for comparison, SAR onboard JERS-1 was about 18 m resolution) loaded on a satellite. Fine resolution mode gives 10 m spatial resolution in both range and azimuth (70 km of swath width, −25 dB of noise equivalent backscattering coefficient, and 25 dB of signal-to-ambiguity (S/A) ratio at a look angle of 35 degrees).

Fig. 14.10 shows the normalized radar backscatter cross-section NRSC variation in an ALOS PALSAR image. The maximum NRSC of −5 dB occurs around the Farasan Islands. The maximum peak of NRSC occurs owing to the Farasan Islands having sandy shores, mangroves, and algal flats, which reach their maximum extent near the coastline, with reefs restricted to offshore locations [2,17,18]. The existence of the coral reefs in the Farasan Islands can generate an irregular pattern of water flow due to friction, which leads to sea surface current modulating the sea surface wave, causing an extremely rough sea surface and maximum radar backscatter.

ALOS PALSAR fine mode instrument has a lower noise floor range, of −21 dB, which allows for better sea surface imaging. In fact, the instrument noise floor is grabbed from turbulent flow-covered only at the far edge of the swath for the HH band backscatter. This can assist to quantify the radar cross-section of water with an L-band radar, even with look-alikes damping the surface waves. Specific quantities are operated for parameter settings, i.e., quantum antibody population, encoding length, cloning scale, and maximum iteration number. In these regards, we set N equals 100, encoding length L equals 20, under the circumstance of 100 iteration number to ensure an accurate result. Fig. 14.11 delivers the results, which are

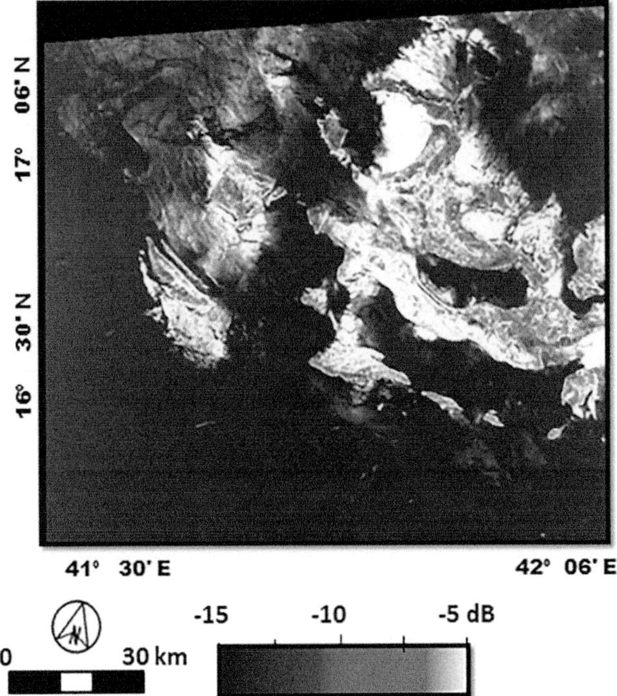

FIG. 14.10 Normalized radar backscatter cross-section NRSC variation.

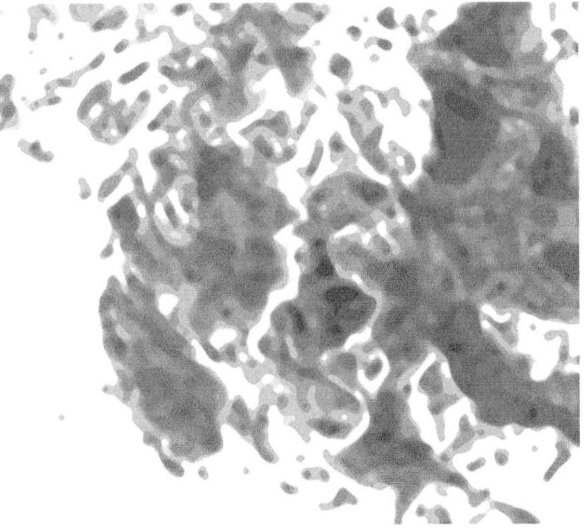

FIG. 14.11 Chaotic structure of the turbulent flow.

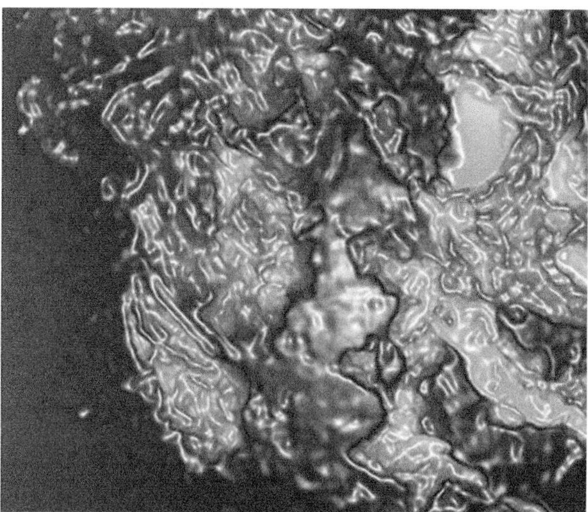

FIG. 14.12 Turbulent morphology flow discrimination in ALOS PALSAR data using QNSGA-II.

obtained by quantum NSGA-II (QNSGA-II). QNSGA-II can discriminate automatically between Bragg scattering and non-Bragg scattering [19–22]. The dynamic fluctuation between Bragg and non-Bragg scattering is accurately determined. Bragg scattering tends to trap the irregular turbulent flow between the Farasan Islands [22,23]. In this regard, the chaotic structure of the turbulent flow is well constructed by QNSGA-II (Fig. 14.11).

QNSGA-II can deliver several clusters in ALOS, for instance, surrounding sea surface and rotational mesoscale eddies (Fig. 14.12). Subsequently, by resetting to N equals 80, encoding length L equals 20, and increasing the iteration number to 100, we acquire a sharp chaotic turbulent morphology flow (Fig. 14.12).

The increment in the number of parameters leads to accurate clusters of the turbulent morphology flow. In other words, these turbulent morphology flow clusters have revealed the variety of the turbulent morphology flow characteristics, from strong, through moderate, to weak flow (Fig. 14.13). Using the Doppler frequency algorithm demonstrated in Chapter 12, the turbulent flows are simulated in ALOS PALSAR data. The water flow of $0.5\,\mathrm{m\,s^{-1}}$ cultivates an irregular pattern approaching the Farasan Islands and has a maximum magnitude of $1.5\,\mathrm{m\,s^{-1}}$. In fact, the coral-water interface generates a fluid viscosity that dampens turbulent motions and a molecular diffusion causes vertical transport [24]. The production or consumption of dissolved materials by the coral surface creates a concentration gradient, forming a diffusive boundary layer, as shown by Marghany and Hakami [25].

In fact, ALOS PALSAR characterizes a turbulent flow by detecting variations in the roughness of its surface and, for strong flow, changes in the electrical conductivity of its surface layer. In this regard, ALOS PALSAR "sees" strong turbulent flow as a rough (radar-bright) area against the smother (radar-dark) ocean surface because most of the radar energy that hits the smooth surface is deflected away from the radar antenna. Fine mode's high sensitivity and other capabilities facilitate separation of strong and weak flows for the first time using a radar system. It is preferable to implement HH polarization as it is sensitive to scattering from

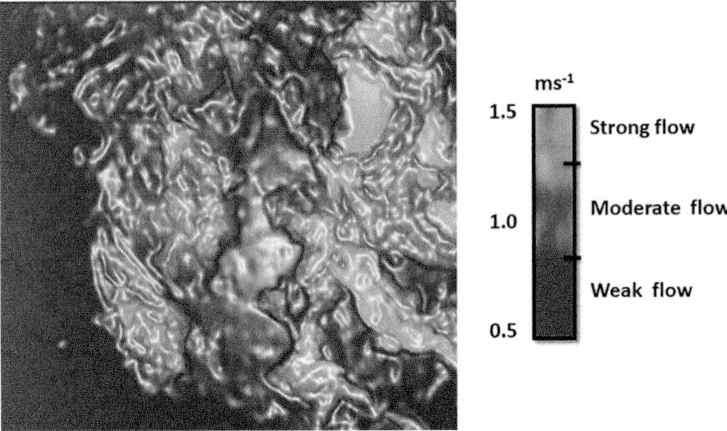

FIG. 14.13 Clustering of turbulent flows using QNSGA-II.

turbulent sea surfaces due to large tilt. QNSGAII delivers a clear detection of turbulent level flows, which involves strong, moderate, and weak flows (Fig. 14.13).

The advantages of global search in quantum immune and spectral embedding in ALOS PALSAR data are combined in QNSGA-II. QNSGA-II can find optimal solutions more frequently with maximum clustering accuracies in 100 runs. QNSGA-II is offered both optimal solution and optimal computing stability. It appears that the QNSGA-II output result is superior, which shows more of the turbulent flow clusters associated with coral reefs (Fig. 14.14).

It is easy to see the interaction between water flow and coral reefs in the QNSGA-II image. Moreover, QNSGA-II can distinguish between land, coral reefs, and turbulent flow. QNSGA-II

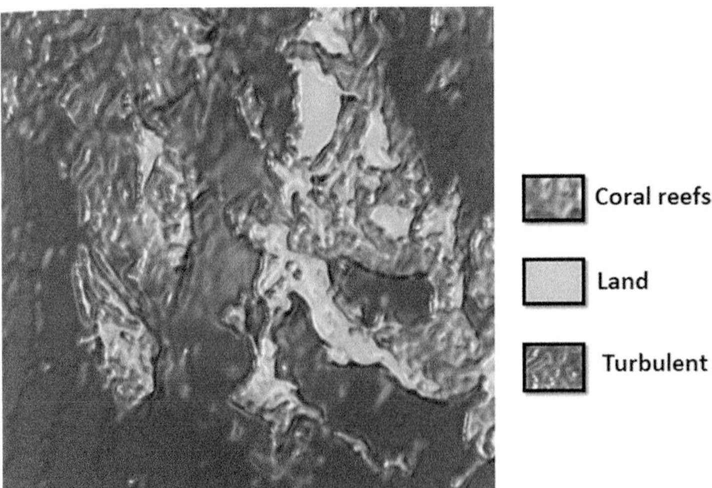

FIG. 14.14 Coral reefs cause turbulent flow using QNSGA-II.

can identify the boundary current impact generating turbulent level flow variation, i.e., thick and light (Fig. 14.14). The strong turbulent flow is dominated by a higher level of coherence than the surrounding sea rough surface.

14.13 Role of Pareto optimization in QNSGA-II

The QNSGA-II algorithm takes both local examination and global search into account, which enables it to handle the optimal solution with higher precision compared to conventional evolutionary algorithms. In this context, the QNSGA-II algorithm can be used for function optimization and multiuser detection of code division multiple access [19–22]. Quantum computing and the developed algorithm can be implemented with a genetic algorithm to improve its accuracy [19]. Finally, the quantum image processing is required for advanced SAR image segmentation and processing.

QNSGA-II can accurately identify the morphological boundary of turbulent flow magnitudes from strong to weak signified by different segmentation layers in SAR data. In fact, QNSGA-II provides a set of compromised solutions called Pareto optimal solution, since no single solution can optimize each of the objectives separately. The decision maker is provided with a set of Pareto optimal solutions in order to choose a solution based on the decision maker's criteria [26]. This sort of QNSGA-II solution technique is called an a posteriori method since the decision is taken after searching is finished [22,27]. In this context, the quantum Pareto-optimization approach does not require any a priori preference decision between the conflicting of the turbulent flow magnitudes, coral reefs, land, and surrounding sea footprint boundaries. Further, quantum Pareto optimal points form a Pareto front, as shown in Figs. 14.13 and 14.14 in the multiobjective function of the SAR features space.

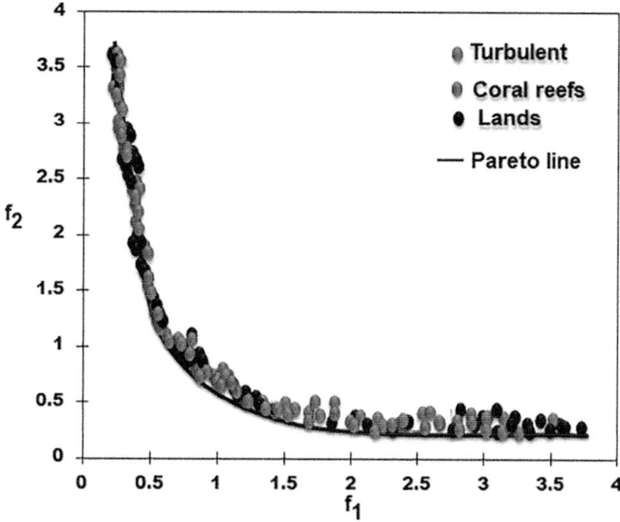

FIG. 14.15 Nondominated solutions by using QNSGA-II.

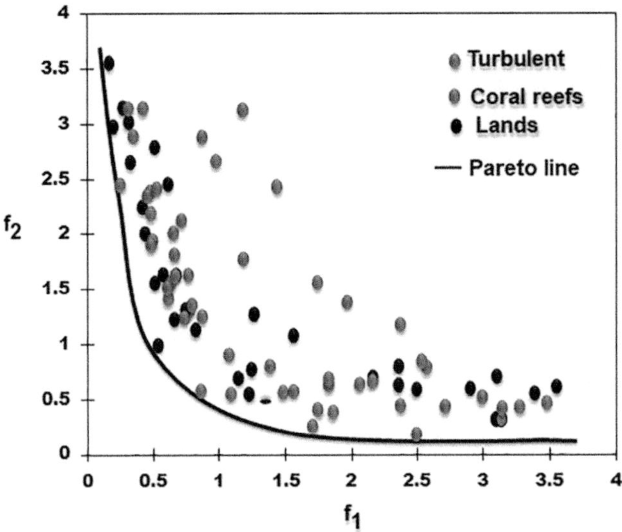

FIG. 14.16 Nondominated solutions by using Q-MEA.

The nondominated solution of different algorithms is presented in Figs. 14.15 and 14.16. It is clear that the solution of QNSGA-II (Fig. 14.15) is much better than Q-MEA (Fig. 14.16). Furthermore, the solution of Q-MEA is gathered around the center of the Pareto front. Under this circumstance, Q-MEA tends to concentrate on one part of the Pareto front. On the other hand, QNSGA-II maintained high degrees of diversity in their solutions during the search for the best optimal solution for turbulent flow magnitudes, coral reefs, land, and surrounding sea footprint boundaries levels in SAR data. In this regard, the QNSGA-II can better distribute its population along the obtained front than Q-MEA.

Moreover, QNSGA-II improves the quality of the nondominated set accompanied by the diversity of the population in multiobjective problems of polarimetric SAR features. QNSGA-II is designed to employ the concept and principles of quantum computing, for instance, uncertainty, superposition, and interference. Indeed, QNSGA-II maintains a better spread of nondominated sets and accurately locates the solutions of automatic detection of turbulent flow magnitudes, coral reefs, land, and surrounding sea footprint boundaries close to the Pareto optimal front (Fig. 14.15). In other words, QNSGA-II improves proximity to the Pareto optimal front, preserving integral diversity by retaining compensations of a quantum-inspired evolutionary algorithm (QEA). In this understanding, improving the accuracy of the Pareto optimal front to locate the superior solutions which are appraised as virtuous individuals by a fitness function.

14.14 Quantum coherence of turbulent flow magnitudes in SAR imaging and QNSGA-II

The turbulent flow magnitudes are considered a strong coherence zone. In quantum mechanics, quantum coherence is the gain in coherence or strength of the phase angles between the constituents of turbulent flow magnitudes and surrounding sea surface in a quantum superposition. One significance of this depressing is conventional or probabilistically addictive behavior. Quantum coherence offers the inclusion of wave function collapse, i.e., Bragg scattering and non-Bragg scattering. In other words, quantum coherence is considered the growth of the physical characteristics of sea roughness backscatter into a single possibility of turbulent flow magnitudes, which is imagined as coherence pixels in SAR images. On this understanding, coherence speculation of turbulent flow magnitudes in SAR data is the mechanism that determines the location of the quantum-classical boundary, i.e., Bragg scattering. Coherence occurs when turbulent flow interacts with its surrounding environment in a delectrical irreversible approach, i.e., a very poor conductor of electric current. In this circumstance, strong turbulent flow with higher delectrical than its surrounding prevents Bragg and non-Bragg scattering in the quantum superposition of the entire SAR scene's wavefunction from interfering with each other.

This is why a higher strength of backscatter occurs in the turbulent flow area than its surroundings. However, coherence spawns definite wave function deformation. It only delivers a description of the accomplishment of wave function construction, as the quantum nature of the turbulent flow in the SAR data "combines" from its surrounding environment. That is, components of the wavefunction are included from a coherent sea surface and attain phases from their immediate surroundings. In this circumstance, a total superposition of the global or universal wavefunction immobile exists (and remains coherent at the global level of the sea roughness). Consequently, it extremely provides accurate interpretations of the turbulent flow in SAR data. In this regard, QNSGA-II can play a vital role in determining the modulated hydrodynamics information in the turbulent flow pixels. Besides, QNSGA-II can solve the computational gridlock problems of conventional spectral clustering algorithms on large-scale SAR image segmentation. QNSGA-II denotes the turbulent representative pixels and boundaries through an operative encoding technique [28]. The QNSGA-II optimizes the selection of representative turbulent pixels, while the calculated affinity function for the clustering turbulent and its surrounding environments diminishes the computation periods to handle a huge number of SAR data. Subsequently, QNSGA-II can count the huge amount of eigenvalues with a value of one, which leads to the precise determination of cluster numbers across the SAR data. In other words, the presence of qubit distribution in this algorithm allows for excellent separation between different clusters. The boundaries between different turbulent clusters are well identified. Indeed, the largest eigenvalue is set as one, and the other eigenvalues diminish steadily, while the breach between the second and third eigenvalues is prevalent, which reveals that the three or two classes of the turbulence can be easily distinguished.

References

[1] Barevnghi CF, Skrbek L, Sreenivasan KR. Introduction to quantum turbulence. Proc Natl Acad Sci 2014;111 (Suppl. 1):4647–52.

[2] Andréfouët S, Mumby P, McField M, Hu C, Muller-Karger F. Revisiting coral reef connectivity. Coral Reefs 2002;21(1):43–8.

[3] Vinen WF, Niemela JJ. Quantum turbulence. J Low Temp Phys 2002;128(5–6):167–231.

[4] Sikora TD, Young GS, Beal RC, Monaldo FM, Vachon PW. Applications of synthetic aperture radar in marine meteorology. Atmos Ocean Interact 2006;2:83–105.

[5] Hole MK, Gulhane VS, Shellokar ND. Application of genetic algorithm for image enhancement and segmentation. Int J Adv Res Comput Eng Technol 2013;2(4):1342.
[6] Han KH, Kim JH. Quantum-inspired evolutionary algorithms with a new termination criterion, H/sub/spl epsi//gate, and two-phase scheme. IEEE Trans Evol Comput 2004;8(2):156–69.
[7] Jiao L, Li Y, Gong M, Zhang X. Quantum-inspired immune clonal algorithm for global optimization. IEEE Trans Syst Man Cybern B Cybern 2008;38(5):1234–53.
[8] Williams CP. Explorations in quantum computing. Springer Science & Business Media; 2010.
[9] Kaye P, Laflamme R, Mosca M. An introduction to quantum computing. Oxford University Press; 2007.
[10] Warren WS. The usefulness of NMR quantum computing. Science 1997;277(5332):1688–90.
[11] Braunstein SL, Caves CM, Jozsa R, Linden N, Popescu S, Schack R. Separability of very noisy mixed states and implications for NMR quantum computing. Phys Rev Lett 1999;83(5):1054.
[12] Platzman PM, Dykman MI. Quantum computing with electrons floating on liquid helium. Science 1999;284 (5422):1967–9.
[13] Beige A, Braun D, Tregenna B, Knight PL. Quantum computing using dissipation to remain in a decoherence-free subspace. Phys Rev Lett 2000;85(8):1762.
[14] Wang L, Tang F, Wu H. Hybrid genetic algorithm based on quantum computing for numerical optimization and parameter estimation. Appl Math Comput 2005;171(2):1141–56.
[15] Deb K. Nonlinear goal programming using multi-objective genetic algorithms. J Oper Res Soc 2001;52 (3):291–302.
[16] Deb K, Agrawal S, Pratap A, Meyarivan T, Fast Elitist A. Non-dominated sorting genetic algorithm for multi-objective optimization: NSGA-II; in parallel problem solving from nature-PPSN VI, In: Schoenauer M, Deb K, Rudolph G, Yao X, Lutton E, Merelo JJ, Schwefel HP, editors. Proceeding of 6th international conference, Paris, France, 18–20 Septembervol. 1917. Berlin: Springer; 2000. p. 849–58.
[17] Abu-Zied RH, Bantan RA, Basaham AS, El Mamoney MH, Al-Washmi HA. Composition, distribution, and taphonomy of nearshore benthic foraminifera of the Farasan Islands, southern Red Sea, Saudi Arabia. J Foraminifer Res 2011;41(4):349–62.
[18] Al-Anbaawy MI. Reefal facies and diagenesis of pleistocene sediments in Kamaran Island, Southern Red Sea, Republic of Yemen. Egypt J Geol 1993;37(2):137–51.
[19] Marghany M. Genetic algorithm for oil spill automatic detection from MultiSAR satellite data, In: Proc 34th Asian conference on remote sensing 2013, Bali, Indonesia, October 20-24, 2013; 2013. p. SC03671–7.
[20] Marghany M. Utilization of a genetic algorithm for the automatic detection of oil spill from RADARSAT-2 SAR satellite data. Mar Pollut Bull 2014;89:20–9.
[21] Marghany M. Multi-objective evolutionary algorithm for oil spill detection from COSMO-SkyMed satellite. In: Murgante B, Misra S, Carlini M, Torre CM, Nguyen H-Q, Taniar D, Apduhan BO, Gervasi O, editors. Computational science and its applications—ICCSA 2014. Springer; 2014. p. 355–71.
[22] Marghany M. Flock 1 data multi-objective evolutionary algorithm for turbulent flow detection, In: CD of 36th Asian conference on remote sensing: Fostering resilient growth in Asia, ACRS 2015, Quezon City, Manila, Philippines, 24 October 2015–28 October 2015; 2015. p. 1–6.
[23] Hedley JD, Roelfsema CM, Chollett I, Harborne AR, Heron SF, Weeks S, Skirving WJ, Strong AE, Eakin CM, Christensen TR, Ticzon V. Remote sensing of coral reefs for monitoring and management: a review. Remote Sens 2016;8(2):118.
[24] Stocking JB, Rippe JP, Reidenbach MA. Structure and dynamics of turbulent boundary layer flow over healthy and algae-covered corals. Coral Reefs 2016;35(3):1047–59.
[25] Marghany M, Hakami M. Automatic detection of coral reef induced turbulent boundary flow in the red sea from flock-1 satellite data. In: Oceanographic and biological aspects of the red sea. Cham: Springer; 2019. p. 105–22.
[26] Zhang Y, Wang S, Ji G, Dong Z. Genetic pattern search and its application to brain image classification. Math Probl Eng 2013;2013:1–9.
[27] Zhang Y, Gong DW, Ding Z. A bare-bones multi-objective particle swarm optimization algorithm for environmental/economic dispatch. Inform Sci 2012;192:213–27.
[28] Hey T. Quantum computing: an introduction. Comput Control Eng J 1999;10(3):105–12.

CHAPTER

15

Four-dimensional along-track interferometry for retrieving sea surface wave-current interaction

15.1 What is meant by four-dimensional and why?

The conventional well-recognized dimensions involve one-dimensional (1-D); two-dimensional (2-D); and three-dimensional (3-D). In this view, these dimensions are well-known as a space dimension that involves x for 1-D, x, y for 2-D, and x, y, z for 3-D. Needless to say, 2-D is encoded in the 1-D and the 3-D object is encoded in the 2-D (Fig. 15.1). In this understanding, 4-D would be encoded into 3-D. The significant question is: how can 4-D encode and reconstruct from 3-D? As well as the conventional dimensions being defined by space, any dynamic fluctuations involve space as a function of change over time. Therefore, time is considered as another dimension in addition to the 3-D, regarding time as being exclusive of the three dimensions of space. In previous chapters, it has been shown that time fluctuations as the inverse of the frequency lead to the Doppler frequency shift. This is clear in SAR object or sea surface imaging as a function of a delay time of the backscattered signal. As this delay time demonstrates relativity of velocity bunching due to length contraction, it can be useful to investigate the possibility of 4-D imaging in SAR.

Consequently, there are three spatial dimensions in physical space. The fourth dimension, time, interacts with the three spatial dimensions in a relationship determined by the theory of relativity. Utilization of 4-D in ocean dynamic studies must involve an understanding of the full scenario of sea surface dynamics, not just dynamic surface components, e.g., wave, current, or sea level, individually. The critical questions are: can 4-D deliver perfect dynamic sea surface fluctuations, and how is 4-D reconstructed in 3-D? This chapter delivers a transitional theory to rationalize the physics of 4-D ocean surface dynamic reconstruction based on SAR interferometry [1]. The 4-D theory may represent a novel model of ocean surface dynamic retrieving in SAR images.

Nonlinear Ocean Dynamics
https://doi.org/10.1016/B978-0-12-820785-7.00006-X

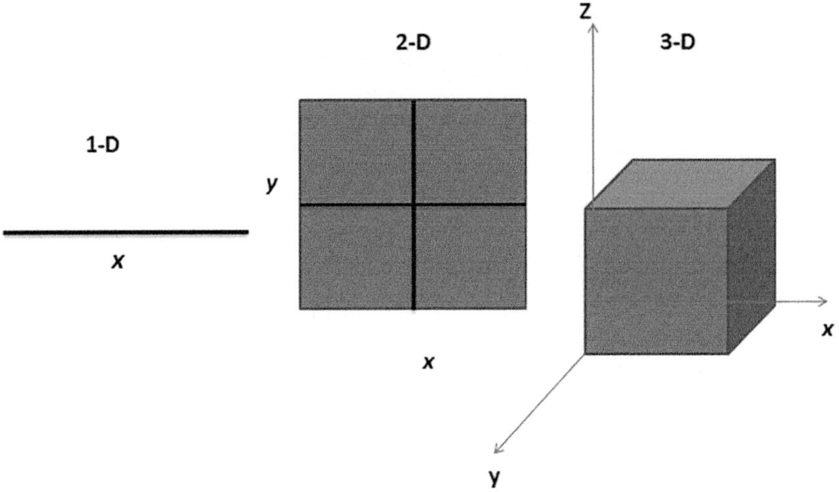

FIG. 15.1 Development from 1-D to 3-D.

15.2 Does *n*-dimensional exist?

A critical question is: do other dimensions exist—the so-called 5th, 6th, 7th, 8th, 9th, 10th, and 11th, and the hyperdimensional manifolds of string theory? In this view, these mentioned dimensions have assumed to exist in the space-time dimensions that can be occurred as innermost component particles, for instance, inner manifolds that fabricate the entirety of the universe. It develops as created entirely of the space-time dimensions, at present it is well-known as 11-D. The Calabi-Yau manifolds are the mathematical hyperdimensional structural entities composed of the extra dimensions that exist within all things, according to the superstring theory or theory of everything. In other words, in algebraic geometry, a Calabi-Yau manifold, otherwise known as a Calabi-Yau space, is a unique sort of manifold, which has properties, such as Ricci flatness, yielding functions in theoretical physics. Particularly in superstring theory, the greater dimensions of space-time are once in a while conjectured to take the structure of a 6-dimensional Calabi-Yau manifold, which leads to the concept of replicate symmetry (Fig. 15.2). They must also exist in space and time. The 11 dimensions have to occur together—the 4 that are easy to recognize, and the residual 7 that are extremely complicated. From the point of view of the Hilbert space and the Hilbert-Einstein collaborations, an infinite number of vectors can exist around the Hilbert space. If this concept is extended to dimensions, hypothetically, an infinite number of dimensions can be generated. This leads to mathematical abstractions that we do not have mathematical schemes to explain or describe.

Consequently, theoretically and abstractly interdimensional existences are perhaps possible. Moreover, they probably would have complexity intermingling with the recent shape of the universe. The growth in nanotechnology, and Femto-technology that are involved in quantum computing, perhaps can overcome the absence of the specific mathematical algorithms to reconstruct interdimensional complicated form.

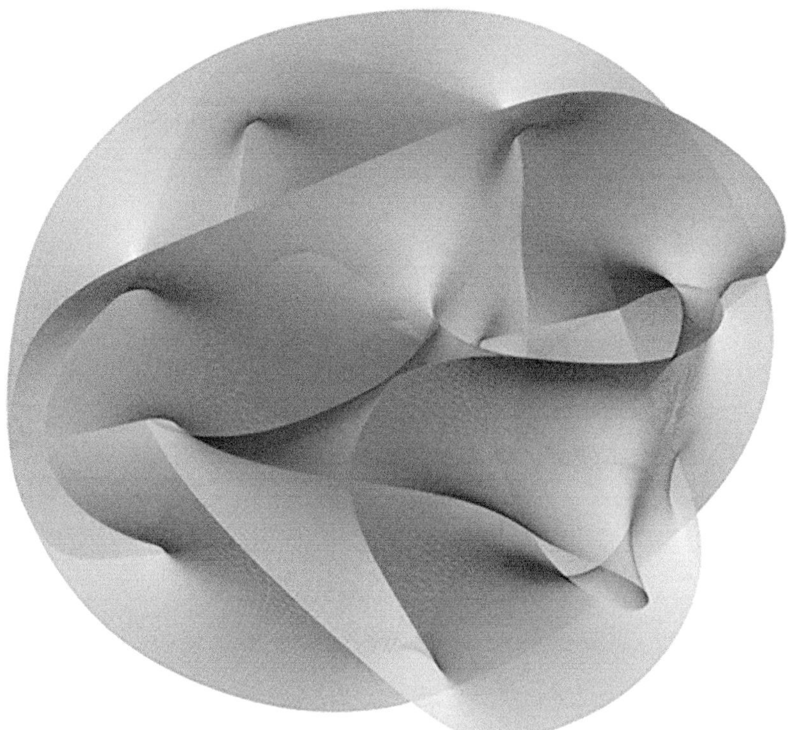

FIG. 15.2 6-dimensional Calabi-Yau manifold.

15.3 Physics of Interferometry

Interferometry is a technique in which electromagnetic waves are superimposed, instigating the phenomenon of interference, which is exploited to extract information [1]. In this regard, Young's slits experiment is the fundamental concept underlying interferometry (Fig. 15.3). Thomas Young exploited this experiment to prove that the light "beam" was a wave, at a time when light was thought to be a particle. Briefly, light is shone through dual slits, interferes, and creates a pattern that is straightforward to explain using a wave model, but which cannot be explained if light behaves like particles.

The phenomenon of interference between electromagnetic waves is primarily based on this idea. When two or more waves pass through the same space, the remaining amplitude at every point is the sum of the amplitudes of the individual waves. In some cases, such as in a line array, the summed variant will have a greater amplitude than any of the components individually; this is known as constructive interference (Fig. 15.4).

In other cases, such as in noise-canceling, the summed variant has a smaller amplitude than the aspect variations; this is referred to as destructive interference (Fig. 15.5).

This demonstrates wave-particle duality: if the light beams are just a wave, then the electrons would absorb some energy no matter what the frequency. If the light beams are just a

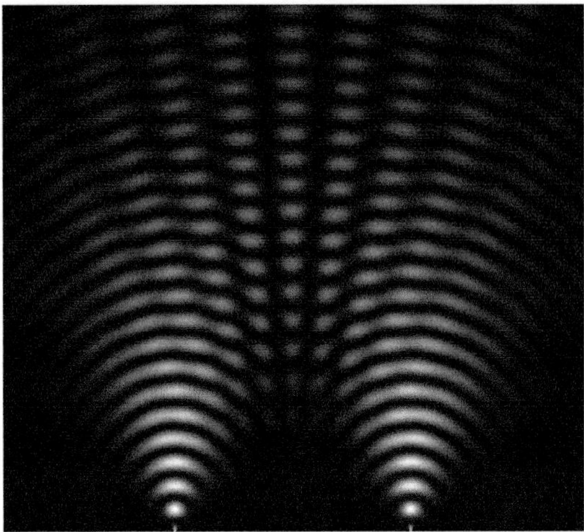

FIG. 15.3 Young's slits experiment.

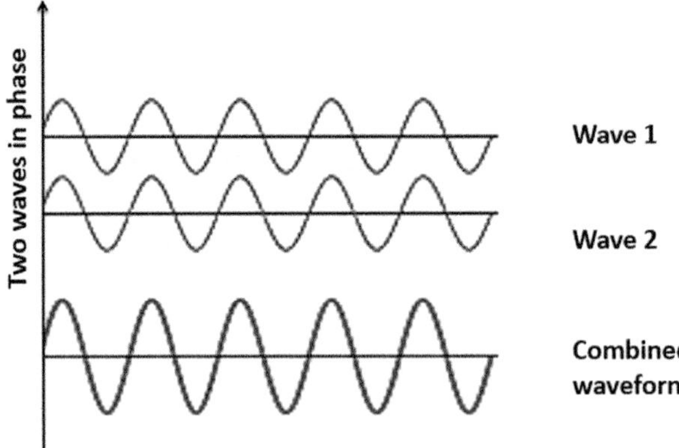

FIG. 15.4 Constructive interference.

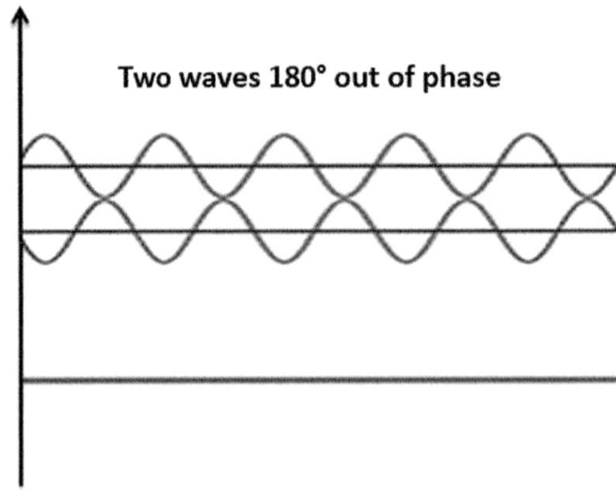

FIG. 15.5 Destructive interference.

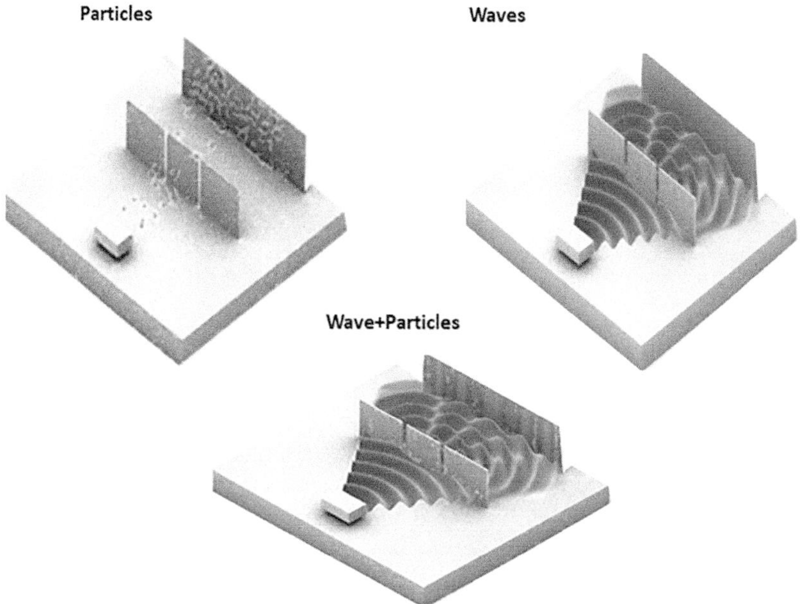

FIG. 15.6 Sketch of the wave-particle duality of microwave signals.

wave, then the emission of an electron would take longer when a lower intensity light was used, not instantaneously (Fig. 15.4).

Therefore, light is not just a wave. It can also behave as though it was made of tiny energy packets or particles. We call these particles, photons (Fig. 15.6). It is one of these photons that, if it hits one electron on a surface, the electron will absorb the energy and it will fly off the surface. So if the intensity is greater, i.e., there are more photons, then more electrons can be knocked off.

15.4 What is synthetic aperture interferometry?

InSAR (interferometric synthetic aperture radar) is a technique that maps millimeter-scale deformations of the Earth's surface with radar satellite measurements. Because of the continuous change of the Earth's surface, the ability to yield measurements at night and in any weather condition makes this technique extremely valuable [2]. In this regard, let us assume that two complex SAR images S_1 and $S_2$are separated by the baseline B (Fig. 15.7). The radar phase difference ϕ_1 for a common transmitter is mathematically given by:

$$\phi_1 = \frac{4\pi}{\lambda}(\Delta R + \zeta) \tag{15.1}$$

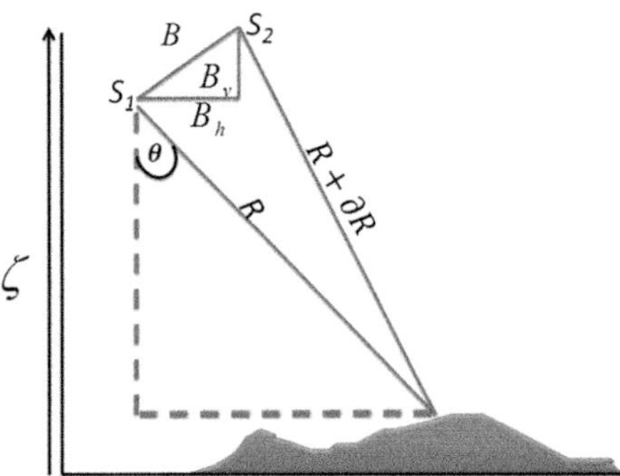

FIG. 15.7 Principle of InSAR geometry.

where ΔR is the slant range difference from satellite to target, respectively, at different times, and λ is the SAR fine mode wavelength. For the surface displacement measurement, the zero-baseline InSAR configuration is the ideal as $\Delta R = 0$, so that:

$$\phi = \phi_d = \frac{4\pi}{\lambda}(\zeta) \tag{15.2}$$

Eq. (15.2) demonstrates that the phase consists of the records of the distance between the antenna and the ground surface and is acquired as a fraction (remainder) of the distance (actually, twice the distance, due to the return-trip) when divided by the wavelength of the radio wave. Nonetheless, it is challenging to use the phase as significant statistics owing to the fact we are aware of solely a fractional portion, no longer the entire wave quantity inclusive of an integer portion.

In actual fact, zero-baseline, repeat-pass InSAR configuration is hardly achievable for either spaceborne or airborne SAR. Therefore, a method to remove the topographic phase as well as the system geometric phase in a nonzero baseline interferogram is needed. If the interferometric phase from the InSAR geometry and topography can be stripped off the interferogram, the remnant phase would be the phase from block surface movement, providing the surface maintains high coherence [3]. Zebker et al. [4] exploited the three-pass method to remove the topographic phase from the interferogram. This method requires a reference interferogram, which must contain the topographic phase only. The three-pass approach has the advantage that all data is kept within the SAR data geometry, while the DEM method can produce errors by misregistration between SAR data and cartographic DEM [4]. The three-pass approach is restricted by data availability. The three-pass DInSAR (differential interferometric synthetic aperture radar) technique uses another InSAR pair as a reference interferogram that does not contain any surface movement event, as in:

$$\phi' = \frac{4\pi}{\lambda}\Delta R' \tag{15.3}$$

Incorporating Eqs. (15.2) and (15.3) delivers the phase difference, from the surface displacement only, as:

$$\phi_d = \phi - \frac{\Delta R}{\Delta R'}\phi' = \frac{4\pi}{\lambda}\zeta \tag{15.4}$$

For an exceptional case where $\frac{\Delta R}{\Delta R'}$ in Eq. (15.4) there is a positive integer number, phase unwrapping may not be necessary [1]. However, this is not practical and it is difficult to achieve from the system design for a repeat-pass interferometer. From Eq. (15.4) the displacement sensitivity of DInSAR is given as:

$$\frac{\partial \phi_d}{\partial \zeta} = \frac{4\pi}{\lambda} \tag{15.5}$$

The important dimension in InSAR is the component of the surface topography that changes in the slant range direction. The resulting range displacement ΔR is given by:

$$\Delta R = |\mathrm{m}| \sin(\theta_i - \zeta) \tag{15.6}$$

where m is the surface topography deformation vectors that describe the direction and magnitude of the surface changes, θ_i is the SAR data incident angle, and ζ is the surface topography's elevation. Incorporating Eq. (15.4) into Eq. (15.6), the sea surface topography deformation is estimated by:

$$|\mathrm{m}| = \frac{\lambda \phi_d}{2\pi \sin(\theta_i - \zeta)} \tag{15.7}$$

15.5 Interferograms

Interferograms are the result of subtracting the phases of two SAR images through the D-InSAR process. The phase difference is depicted by the color of the pixels in an interferogram, which displays repeating color cycles. Because the same phase value is repeated in every cycle of the radar wave, the colors repeat. In this way, the colors of fringes in an interferogram show the difference in phase resulting from the difference in distance between two SAR data at the target point. For instance, an area where the difference in phase is 0 degrees is shown in light-blue, 60 degrees in blue, 180 degrees shown in red, and so on. If the difference in phase is 360 degrees, the distance that a radio wave travels back and forth between the radar and the ground changes by just one wavelength [5–8].

For example, at an SAR wavelength of 23.6 cm, the phase difference of 360 degrees is equivalent to a movement of 11.8 cm (one-way distance is half of round-trip distance). Similarly, the point whose difference in phase is 60 degrees shifts 2.0 cm: $23.6 \times \left(\frac{60}{360}\right) 0.5 \approx 2.0$ cm. In this regard, the difference in phase is equivalent to the quantity of deformation. The color of fringes represents the magnitude of deformation at the target point (Fig. 15.8).

In Fig. 15.8, it can be seen that the equivalent color cycle repeats three times, in the direction of increasing phase. This means that the western part has subsided three times a full wavelength with respect to the eastern part between the two image acquisitions that were used to produce the interferogram. Because we know the exact length of a full-wave cycle, the deformation can be derived [6, 9–11]. This process is called unwrapping. Nonetheless, if there is strong atmospheric activity during one or both of the image acquisitions used for producing the interferogram, the deformation signal can be polluted with atmospheric noise. To resolve deformation from the atmosphere, a time series of interferograms can be used to model the atmospheric signals [12, 13].

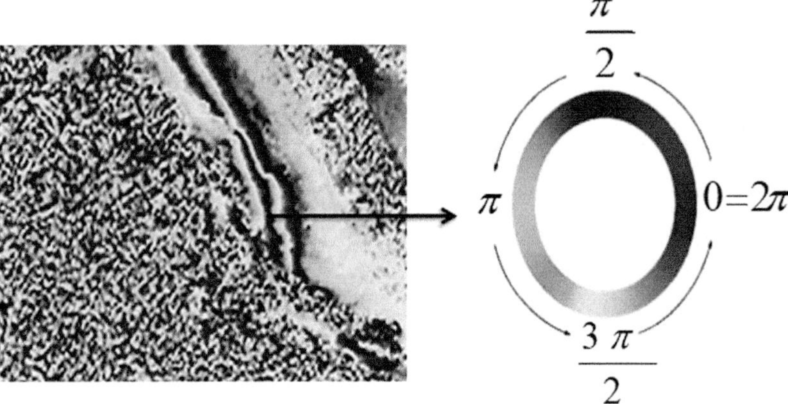

FIG. 15.8 Wave cycles in an interferogram.

15.6 Phase unwrapping

The phase acquired by InSAR is only measured in values between 0 and 360 degrees. Therefore, the magnitude of deformation cannot be determined from it without modification. The phase that originally takes on a wide range of value is folded (wrapped) in the range between 0 and 360 degrees. The work of unwrapping and turning the phase back to the magnitude of deformation is termed phase unwrapping. This technique is crucial to the 3-pass and 4-pass processes and can be applied to the automatization and precision enhancement of the baseline approximation, and perform quantitative analysis by applying it to the baseline estimation and conclusive results [14–18].

In the process mentioned above, if the atmosphere delay affects the whole of the image, the image often has a large margin of error. Therefore, it is essential to perform atmospheric compensation and calculate the time average of interferograms formed at different times. In addition to comprehending how precise the InSAR data is, it is important to distinguish how consistent it is. To conduct deformation measurements, the phase recorded is used by the satellite. This is the fraction of a complete wave cycle that is in addition to the number of full wavelengths the signal has traveled. Phase measurements can be thought of as the time on a clock. There are a finite number of times that can be demonstrated on a clock and every 12h, those times repeat. Similarly, there are a finite number of phase measurements that can be recorded within one wave cycle (0 to 2π). Once a wave cycle has completed, the phase measurements then repeat, in which case, the phase measurements are shown in a circle, like a clock (Fig. 15.9).

Nonetheless, the values have a new meaning as soon as a new revolution has taken place. For a clock, a new revolution indicates a new 12-h cycle has begun. To entirely comprehend what time it is, one needs to distinguish the date and/or which part of the day it currently is. Likewise, a new revolution of the phase cycle shows a full-wave cycle has been accomplished. To completely comprehend the meaning of the phase measurements, one needs to distinguish the value of full-wave cycles allied with phase measurements. Resolving this to determine the actual deformation is called "unwrapping" [18–22].

To generate the right deformation from sequential phase measurements in time, correlations in time and/or space are exploited to unwrap the measurements. For instance, the

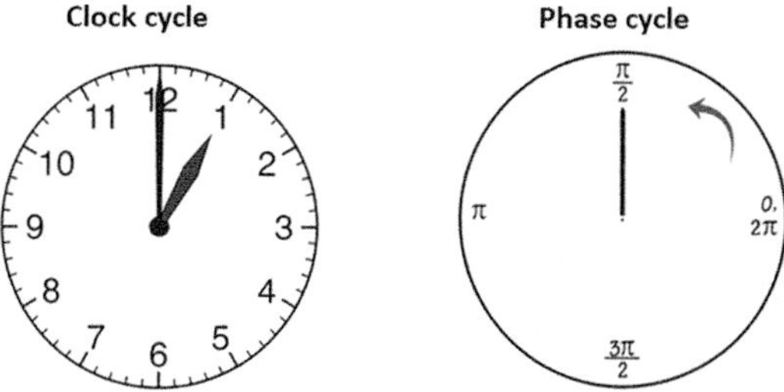

FIG. 15.9 Example of phase measurement as the time on a clock.

occurred deformation can be assumed having a linear relationship with phase measurements during the mission duration of TanDEM-X. If a wrong assumption is made, this may lead to supposed "phase unwrapping errors."

15.7 Understanding SAR interferograms

InSAR measures along a straight line between an SAR antenna and the ground in the direction of the satellite's line of sight (LOS). As mentioned in previous chapters, an SAR satellite transmits and receives a radio wave obliquely downward (obliquely upward from the ground). In addition, the track on which the satellite observes the ground from the east or the west is often used. The ground movement is three dimensional (east and west, north and south, up and down) but InSAR can measure it only in the LOS direction. Consequently, the direction the ground moved in is difficult to determine.

The "LOS distance becomes longer" indicates that the ground surface deformed in the direction away from the satellite. In other words, if the satellite observes the ground surface from the East, the ground surface shifted down or west. At this time, we cannot distinguish whether the ground surface relocated on the north-south direction. The ground surface might shift down, west, or downwest. Then, the ground surface might change in the direction away from the satellite since the ground surface shifted downeast but the downward component countered the east component. Conversely, these differences cannot be detected in InSAR technique [22–24].

The result of InSAR expresses ground deformation in the change of colors. A minor deformation of the ground causes a slight change of color; a major deformation of the ground causes a significant change of color (Fig. 15.10).

The difference in displacement between the different places is shown in the same color, which is one of the integral multiples of the half SAR wavelength (11.8cm). In this understanding, the existence displacement is just 11.8cm if the color turns back to the same color (Fig. 15.11).

Cross-track phase interferometry can be utilized to determine object movement on the ground. In this regard, the phase information is contained in the sea surface current velocity U_r as:

$$\phi=\frac{4\pi}{\lambda}(U_r\Delta t-B\sin(\theta-\alpha)) \tag{15.8}$$

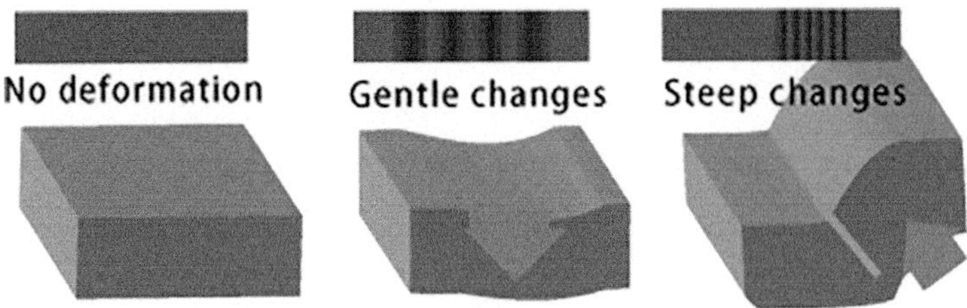

FIG. 15.10 Ground deformation shown by changes in color.

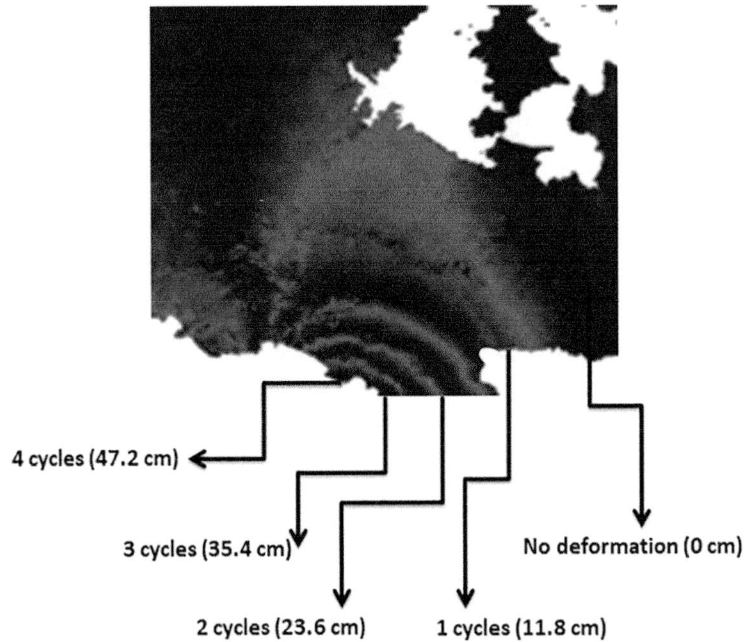

FIG. 15.11 Example of deformation magnitude from an interferogram pattern.

The interferometric phase involves the signature owing to topography φ_{topo}, displacement φ_{disp}, atmospheric effect φ_{atmo}, baseline error $\varphi_{baseline}$, and noise φ_{noise}, which is expressed mathematically as:

$$\phi = \varphi_{topo} + \varphi_{disp} + \varphi_{atmo} + \varphi_{baseline} + \varphi_{noise} \tag{15.9}$$

Eq. (15.9) shows that the phase concerning surface warp can be assimilated by eliminating other constituents. The topographic phase can be modeled by deploying a digital elevation model (DEM). The atmospheric effects, therefore, can contain spatially correlated artifacts of a few centimeters, and the baseline error can be simulated and eliminated. Consequently, the noise level of the interferometric phase relies on the coherence of the image pairs. In this regard, the displacement phase is grounded on the repeat-pass interferometric phase after eliminating the topographic effect, the atmospheric effect, and baseline error. In other words, phase displacement is considered as the incoherent absolute of changes in the surface backscattering phase, volume backscattering phase, and double-bounce backscattering phase. The mathematical description of phase displacement is given by:

$$\varphi_{disp} = +\varphi_{surface} + \varphi_{volume} + \varphi_{double-bounce} \tag{15.10}$$

Eq. (15.10) is the principle concept to determine water level changes from the interferometric phase value as a function of high coherent variation. In this context, the water level $\partial\zeta$ can be computed as a function of the displacement phase:

$$\partial\zeta = \lambda\phi_{disp}(4\pi\cos\theta)^{-1} + \aleph \tag{15.11}$$

Eq. (15.11) shows that the modeling of the water level is impacted by decorrelation noise $\aleph$. The sources of decorrelation are as follows [4, 6, 10, 11, 14–20]:

1. Temporal decorrelation (object fluctuation such as λ, vegetation, water, ice, dry soil, and urban developments).
2. Geometric decorrelation occurs due to variations of the interferometric phase within a resolution cell.
3. Increases with baseline causes total coherence loss.
4. Volume decorrelation occurs due to multiple targets with different interferometric phases at the same range. Consequently, it increases with baseline, and with penetration depth.
5. Doppler decorrelation occurs owing to acquisitions under different squint/Doppler Centroid. This implies that the resolution cell is being observed in a different direction. It can be mitigated by common band filtering at the expense of resolution. Complete decorrelation occurs when Doppler centroid difference equals the Doppler bandwidth.
6. Other sources of decorrelation occur due to co-registration errors, as well as ambiguities, quantization, and thermal/system noise.

15.8 Along-track interferometry

In along-track interferometric (ATI) data the interferometric phase at each pixel is correlated to the movement of the scatterers within the pixel. In this regard, the ATI technique is based on interferometric combination of dual complex SAR images of a similar scene, which are attained with a short time delay on the scale of milliseconds. Consequently, the phase alterations between pixels of the dual complex SAR images are proportionate for Doppler shifts of the reflected signal. In this regard, dual antenna separated by a consistent space in the flight direction is required to achieve dual interferometric SAR (InSAR) images in the circumstance of a short time-lag from a moving platform.

The technique is termed as along-track interferometry, not to be confused with cross-track interferometry (XTI) for topographic mapping. Unlike altimetry and cross-track interferometry, ATI delivers a straightforward measurement of the surface velocity in the circumstance that there is no geostrophic assumption is required.

The ATI technique is extremely useful for ocean dynamic studies. Ocean signatures of boundary layers are considerably superior in the phase than in the illumination of a radar backscattered signal, since the backscatter tends to be proportional to wave height while the phase is proportional to wave direction and speed. With an ATI technique, the coherence time is directly determined and is perhaps better for observing microwave-breaking events that manage the ocean-atmosphere gas transport/mixing process. Moreover, wave spectra measured by the ATI phase are less distorted than those obtained by single coherence SAR imaging.

Let us assume that dual complex SAR images S_1 and S_2 are implemented to form ATI by a conjugate-multiplication of S_1 by S_2:

$$\Delta\phi = \arg\left(S_1\, S_2^*\right). \tag{15.12}$$

Let us assume that U_r is the velocity of the scatterer movement with a line of sight (LOS) to the target, O. The phase changes owing to target and scatterer movement are given by:

$$\frac{\partial\phi}{\partial t} \approx U_r \cdot 4\pi O(\lambda)^{-1} \tag{15.13}$$

Eq. (15.13) reveals that the phase changes owing to the lag time changes can be used to retrieve the surface current movement. On this understanding, the signal is transmitted to the scattered and back at a similar location. It is vital that the spatial interferometric baseline for ATI data is zero, in which only the baseline is temporal. Preferably, one would exploit an array of large stationary antennas to retrieve the phase $\partial\phi$ of an image. This is unworkable. As an alternative, a pair of SAR antennas can be displaced in the direction of the travel of a moving platform. If the platform moves at a speed V_S and the phase centers of the antennas are moved a distance d, then the time interval between the dual SAR images formed using the two antennas is $\partial t = (V_S)^{-1} \times d$. In this circumstance, the radar transmits consecutively from the individual antenna, and obtains the signal with the same antenna used to transmit (known as "ping-pong" mode) [22].

For current measurements, time-lag needs to be adequately long to attain significant phase signatures from current differences of interest, but short enough to circumvent phase ambiguities and a decorrelation of the backscattered signal. The decorrelation time relies on radar frequency and wind/wave conditions [24].

Consequently, owing to the tiny delay of the SAR antenna in the flight direction, the realizable operative ATI baseline B_{ATI} is equivalently small, which accordingly picks up the sensitivity of any ground motion as:

$$U_r \approx \frac{\phi_{ATI}}{4\pi \cdot B_{ATI} \sin\theta(\lambda \cdot V_S)^{-1}} \tag{15.14}$$

Eq. (15.14) indicates the constraints of the low sensitivity of the revealing and the measurement of moving objects with comparatively high velocities in InSAR data. For instance, similar to vehicles in free-fluctuating traffic, the dynamic ocean surface fluctuations allow for strong spatial averaging (multilooking) of the interferogram phase [23].

15.9 Quantum of along-track interferometry

Let us assume dual antenna and a delay phase ϕ_{delay} that is acquainted with one of the lengths of the interferometer. Consequently, the quantity ϕ_{delay} is computed by evaluating the strength of the microwave signal of the dual output beams. In this regard, the value ϕ_{delay}

can be computed with a statistical error proportional to $\frac{1}{\sqrt{N}}$, as N is a nonentangled microwave signal—specifically, the direct simulation of the mean traveling time, which is determined for N, a nonentangled signal. In this instance, the miscalculation is expressed as:

$$\partial R \approx O\left[\Delta\omega\sqrt{N}\right]^{-1} \tag{15.15}$$

Eq. (15.15) confirms that the inverse of the frequency bandwidth $\Delta\omega$ causes the inaccuracy that is resulted from the oscillation of the range ∂R. Nonetheless, entangled signals can be utilized to overwhelm the conventional quantum constraint and achieve the Heisenberg limit. To handle the quantum mechanism of ATI, let us consider a dual-mode, path-entangled, signal-number state—a sort of Schrödinger's Cat state—which is recognized as the N00N state and mathematically is given by:

$$|\psi_{NOON}\rangle = \left[\sqrt{2}\right]^{-1}(|NO\rangle + |ON\rangle) \tag{15.16}$$

Eq. (15.16) shows the number of backscattered signals generated by ATI, which are either completely in the first complex SAR image S_1 or entirely in the second complex SAR image S_2. On the other hand, it is entirely removed into the entangled-signal source for generating superpositions of these dual states in the form:

$$|NOON\rangle = |S_1\rangle + |S_2\rangle = |N\rangle_{S_1}|O\rangle_{S_2} + |O\rangle_{S_1}|N\rangle_{S_2} \tag{15.17}$$

Consequently, each part of the entangled state tolerates diagonally a different dimension of the phase changing, which is accurately articulated as:

$$|\psi_{NOON}\rangle = \left(\sqrt{2}\right)^{-1}\left[|N\rangle_{S_1}|O\rangle_{S_2} + |O\rangle_{S_1}|N\rangle_{S_2}\right] \tag{15.18}$$

Nonetheless, the quantum states are in existing tremendously nonconventional states, to commence with their accomplishment in the phase-shifter, which is extremely unrelated. When a monochromatic beam of measured quantum states approves across a phase shifter, the phase shift is straight proportional to the number of photons N. Therefore, there is no n-dependence in the coherent state, where *induce* n is the steady signal intensity at ATI. In manifestations of a unitary development of the state, the growth of any interferometry state transient through a phase shifter ϕ is standardized by:

$$\widehat{U}(\phi) \equiv e^{(i\phi\hat{n})} \tag{15.19}$$

where $\hat{n}$ is the average interferometry fringes generated by a phase shift in the quantum along-track interferometry technique.

In this regard, the phase shift operator can be shown to have the consequent dual disparate influences on coherence in opposition to quantize the phase shift states as:

$$\hat{U}_\phi|O\rangle = e^{i\phi}|O\rangle, \tag{15.20}$$

$$\hat{U}_\phi|N\rangle = e^{i\phi}|N\rangle, \tag{15.21}$$

Eqs. (15.20) and (15.21) show that the phase shift for the coherent state is of known quantity; nonetheless, there is an N dependence on the exponential for the quantum phase shift state. The quantum phase shift state develops in phase N-times, which are further rapidly than the coherent state. In this instance, the $NOON$ state develops into:

$$\left|\psi_{NOON}^{\phi}\right\rangle = \frac{1}{\sqrt{2N!}}\left(|N\rangle_{S_1}|O\rangle_{S_2} + e^{iN\phi}|O\rangle_{S_1}|N\rangle_{S_2}\right) \tag{15.22}$$

Consequently, Eq. (15.22) can be articulated based on annihilation operators as:

$$\left|\psi_{NOON}^{(\phi)}\right\rangle = \frac{1}{\sqrt{2N!}}\left(\left(\hat{a}_{S_1}^{\dagger}\right)^{N} + e^{iN\phi}\left(\hat{a}_{S_2}^{\dagger}\right)^{N}|0\rangle_{S_1}|0\rangle_{S_2}\right) \tag{15.23}$$

where the bosonic creation operator $a_{\alpha}^{\dagger}$ is given by:

$$a_{\alpha}^{\dagger}|n_1, \ldots, n_{\alpha-1}, n_{\alpha}, n_{\alpha+1}, \ldots\rangle = \sqrt{n_{\alpha+1}}|n_1, \ldots, n_{\alpha-1}, n_{\alpha}, n_{\alpha+1}, \ldots\rangle, \tag{15.24}$$

Let us suppose that the ATI sea surface current U_s is delivered by:

$$\widehat{U}_s = \frac{1}{N!}\left(\left(\hat{a}_{S_1}^{\dagger}\right)^{N}|0\rangle\langle 0|\left(\hat{a}_{S_2}^{\dagger}\right)^{N} + \left(\hat{a}_{S_2}^{\dagger}\right)^{N}|0\rangle\langle 0|\left(\hat{a}_{S_2}^{\dagger}\right)^{N}\right) \tag{15.25}$$

The ATI observation measurements are vital to approximate the phase shift based on annihilation operators. In this regard, the fluctuation in sea surface current $\Delta\hat{U}_s$ can create an inaccuracy in the phase shift, which obeys:

$$\partial\phi = \frac{\Delta\hat{U}_s}{|-N\sin N\varphi|} \tag{15.26}$$

Eq. (15.26) reveals the Heisenberg constraint, as the ATI phase is deliberated as enormously entangled states. In other words, Eq. (15.26) can be conveyed as [25]:

$$\lim_{\alpha_1\to 1}\lim_{\alpha 2\to 1}\partial\phi = \frac{\Delta\hat{U}_s}{|-N\sin N\varphi|} \tag{15.27}$$

15.10 Quantum Hopfield algorithm for ATI phase unwrapping

To create quantum neural networks, let us consider the task of expending multiqubit quantum systems. In this regard, accurate correlation between neurons and qubits [26], unlocking access to quantum properties of entanglement and coherence of SAR data is instigated. In this view, let us instead encode the neural network into the energy of a quantum state. This is accomplished by accommodating a memory regulation between activation patterns of the neural network and pure states $|\phi\rangle$ of L-level of a quantum system. On this understanding, the determined state features of the current pattern can be expressed as $\phi \to |\phi\rangle|\phi|_2$ with $|\phi|_2 = \sqrt{\sum_{i=1}^{L}\phi_i^2}$ and $|\phi\rangle := |\phi|_2^{-1}\sum_{i=1}^{L}\phi_i|i\rangle$, which is encoded concerning the standard basis such that $\langle\phi|\phi\rangle = 1$. The L-level quantum system can be realized by a register of $N = [\log_2 L]$ qubits with the intention of the qubit directly above of signifying such a network scale logarithmically with the quantity of neurons [27].

The quantum Hebbian learning algorithm (qHob) is exploited for weighting matrix M of the Hopfield network. It relies on dual vital perceptions: (i) a direct alliance of the weighting matrix with a mixed state ρ, and (ii) one can efficiently perform quantum algorithms that harness the information contained in W. Consequently, ρ registers of N qubits is relative to:

$$\rho := \omega + \frac{\prod_L}{L} = W^{-1} \sum_{m=1}^{W} \left|\phi^{(m)}\right\rangle \left\langle \phi^{(m)}\right| \tag{15.28}$$

where $\prod_L$ is the L-dimensional identity matrix and W is a training set of activation patterns of phase ATI ϕ, with $m \in \{1, 2, \ldots, W\}$.

In this regard, the Hopfield neural network can be taught training set using the Hebbian learning rule, which sets the weighting matrix elements ω_{ij} along with the number of occasions in the training set that the neurons i and j mix together. In this sense, Hebbian learning is employed to set the weighting matrix M from the L-length training phase unwrapping data $\{\phi^{(m)}\}_{m=1}^{W}$. In this regard, the qHop algorithm continues to compute $|v\rangle = \frac{1}{\Delta\phi_{ATI}} |\phi\rangle$ where the pure state $|\phi\rangle$ is primarily arranged, which comprises user-identified neuron thresholds of the ATI data inverse matrix ϕ_{ATI}^{-1} of feature identifications and a partial memory pattern. In other words, the inverse matrix ϕ_{ATI}^{-1} containing information on the training data and regularization γ. To this end, the output of the pure state $|v\rangle$ contains information on the reconstructed state of phase $|\phi_{ATI}\rangle$ and Lagrange multiplier vector $\vec{L}_v$. The feature state matrix of ATI phase data can be expressed as a quantum state by:

$$\phi_{ATI} |v|_2 |v\rangle = |w|_2 |w\rangle \tag{15.29}$$

where:

$$|v\rangle := \frac{1}{|v|_2} \left(|\phi_{ATI}|_2 |0\rangle \otimes ||\phi_{ATI}|\rangle + \left|\vec{L}_v\right|_2 |1\rangle \otimes \left|\vec{L}_v\right\rangle \right) \tag{15.29.1}$$

$$|w\rangle := \frac{1}{|w|_2} \left(|\Xi|_2 |0\rangle \otimes |\Xi\rangle + \left|\varphi_{ATI}{}^{(inc)}\right|_2 |1\rangle \otimes \left|\varphi_{ATI}{}^{(inc)}\right\rangle \right) \tag{15.29.2}$$

where $\varphi_{ATI}{}^{(inc)}$ is the normalized quantum state corresponding to the incomplete activation pattern and $\left|\varphi_{ATI}{}^{(inc)}\right|_2 = l.$

The objective of Eqs. (15.29.1) and (15.29.2) is to optimize the energy function E in Eq. (15.29). In addition, Ξ represents each element of qHop, which should be set so that its magnitude is of the order of, at most, 1 qubit. In other words, $\Xi =: \{\Xi_i\}_{i=1}^{d} \in \mathbb{R}^d$ is a user-specified neuronal threshold vector that determines the switching threshold for each neuron. Let us identify the set of M unitary operators $\{U_k\}_{k=1}^{M}$ acting on an $N+1$ register of qubits consistent with:

$$U_k := |0\rangle\langle 0| \otimes \prod + |1\rangle\langle 1| \otimes e^{-i\left|\phi_{ATI}{}^{(k)}\right\rangle \left\langle \phi_{ATI}{}^{(k)}\right| \Delta t} \tag{15.30}$$

Eq. (15.30) reveals that the unitaries implement various phase pattern detections in ATI data for a small difference time Δt under the circumstance of $\left|\phi_{ATI}{}^{(k)}\right\rangle \left\langle \phi_{ATI}{}^{(k)}\right|$. Conversely, these unitaries can be simulated using the following mathematical equation:

$$\begin{aligned} &U_s := e^{-i|1\rangle\langle 1| \otimes s\Delta t} \\ &|0\rangle\langle 0| \otimes \prod + |1\rangle\langle 1| \otimes e^{-is\Delta t} \end{aligned} \tag{15.31}$$

where $|1\rangle\langle 1|\otimes s\Delta t$ is 1-sparse and efficiently simulatable.

Conversely, the trace tr_2 of U_s is over the second subsystem containing the state feature $|\phi_{ATI}\rangle$ in the ATI data, which is expressed by:

$$tr_2\left\{U_s\left(|q\rangle\langle q|\otimes\left|\phi_{ATI}{}^{(k)}\right\rangle\left\langle\phi_{ATI}{}^{(k)}\right|\otimes\sigma\right)U_s^\dagger\right\}=U_k(|q\rangle\langle q|\otimes\sigma)U_k^\dagger+\mathcal{O}\left(\Delta t^2\right). \tag{15.32}$$

In Eq. (15.32), the subsystem of ancilla qubit $|q\rangle\langle q|$ and σ efficiently experiences time evolution $\mathcal{O}\left(\Delta t^2\right)$ with U_k. The unitary $e^{i\phi_{ATI}t}$ is simulated to fix the error ϵ for arbitrary t. Indeed ϵ can be estimated by:

$$\epsilon:=\left\|\left(e^{-i\left|\phi_{ATI}{}^{(1)}\right\rangle\left\langle\phi_{ATI}{}^{(1)}\right|\frac{t}{nM}}\cdot e^{-i\left|\phi_{ATI}{}^{(M)}\right\rangle\left\langle\phi_{ATI}{}^{(M)}\right|\frac{t}{nM}}\right)^n-e^{-i\phi_{ATI}t}\right\|\in\mathcal{O}\left(\frac{t^2}{n}\right). \tag{15.33}$$

Eq. (15.33) indicates that the repetition of $n\in\mathcal{O}\left(\frac{t^2}{\epsilon}\right)$ is required along with M-sparse Hamiltonian simulations. The advantage of this approach is that the copies of the training states $\left|\phi_{ATI}{}^{(m)}\right\rangle$ can be used as quantum software states and, in addition, it does not require superpositions of the training states. Conversely, the ATI feature matrix ϕ_{ATI} can be identified as:

$$\arg\left(S_1\,S_2^*\right):=\begin{pmatrix}0 & P\\ P & 0\end{pmatrix}+\begin{pmatrix}-\left(\gamma+L_v{}^{-1}\right)\prod_{L_v} & 0\\ 0 & 0\end{pmatrix}+\begin{pmatrix}\rho & 0\\ 0 & 0\end{pmatrix} \tag{15.34}$$

In this regard, the simulation time is split into n small time steps $t=n\Delta t$. In this circumstance, the error ϵ can be extended into Taylor expansion as:

$$\epsilon_{\Delta t}:=\left\|e^{i\phi_{ATI}\Delta t}-U_B(\Delta t)U_C(\Delta t)U_D(\Delta t)\right\|\in\mathcal{O}\left(\Delta t^2\right) \tag{15.35}$$

where $U_B(\Delta t)$, $U_C(\Delta t)$, and $U_D(\Delta t)$ are the operators of $\begin{pmatrix}0 & P\\ P & 0\end{pmatrix}$, $\begin{pmatrix}-\left(\gamma+L_v{}^{-1}\right)\prod_{L_v} & 0\\ 0 & 0\end{pmatrix}$, and $\begin{pmatrix}\rho & 0\\ 0 & 0\end{pmatrix}$, respectively. On this understanding, at most $\mathcal{O}\left(\Delta t^2\right)$, the errors $e^{i\begin{pmatrix}0 & P\\ P & 0\end{pmatrix}\Delta t}$, $e^{i\begin{pmatrix}-\left(\gamma+L_v{}^{-1}\right)\prod_{L_v} & 0\\ 0 & 0\end{pmatrix}\Delta t}$, and $e^{i\begin{pmatrix}\rho & 0\\ 0 & 0\end{pmatrix}\Delta t}$ are simulated respectively, which allow simulation of the unitary $e^{i\phi_{ATI}t}$ to an error of $\epsilon\in\mathcal{O}\left(n\Delta t^2\right)$. Then the matrix of spectral energy of the qHop can be mathematically expressed as:

$$\phi_E=\sum_{j:\left|\lambda_j(\Delta\phi)\right|>\lambda}\lambda_j(\Delta\phi)\left|E_j(\Delta\phi)\right\rangle\left\langle E_j(\Delta\phi)\right|+\sum_{j:\left|\lambda_j(\Delta\phi)\right|<\lambda}\lambda_j(\Delta\phi)\left|E_j(\Delta\phi)\right\rangle\left\langle E_j(\Delta\phi)\right| \tag{15.36}$$

Eq. (15.36) demonstrates how to simulate the gradient energy of the various feature pattern variations in the ATI as a function of the size of the eigenvalues λ_j in comparison to a fixed user-defined number $\lambda>0$. In this circumstance, qHop maintains the polylogarithmic efficiency in run time whenever λ is such that $\lambda^{-1}\in O(poly(\log d))$. Therefore, the primary matrix inversion algorithm returns (up to normalization) as:

$$A_E^{-1}|E\rangle=\sum_{j:\left|\lambda_j(\Delta\phi)\right|\geq\lambda}\left\langle E_j(A_{SAR})|E\right\rangle\left(\lambda_j(\Delta\phi)\right)^{-1}\left|E_j(\Delta\phi)\right\rangle\left(\lambda_j(\Delta\phi)\right) \tag{15.37}$$

The input state of the energy gradient $|E\rangle$ is first achieved as it contains the threshold data and incomplete activation pattern. Therefore, it considers as the eigenbasis of the matrix of

ATI image energies A_E. Therefore, obtaining information from each energy pattern requires $O(K)$ operations of qHop. The qHop algorithm is then initialized along with sparse Hamiltonian simulation [28] to perform quantum phase estimation, allowing $\sum_j \langle E_j(\Delta\phi)|E\rangle|\widetilde{\lambda}_j(\Delta\phi)\rangle \otimes |E_j(\Delta\phi)\rangle$ to be obtained with $\widetilde{\lambda}_j(\Delta\phi)$ an approximation of the eigenvalue $(\lambda_j(\Delta\phi))$ to precision ϵ.

15.11 Quantum ATI Hopfield algorithm application to TanDEM-X satellite data

Two forms of knowledge are needed to retrieve sea surface current parameters: TanDEM-X of SAR and the real, unaltered sea surface current, measured through TanDEM-X satellite overpasses.

A pair of Terra-SAR-X satellite data was acquired by the TanDEM-X satellite on May 6, 2017. The first image was acquired at 7:27:17 am (Fig. 15.12A) and the second at 19:20:06 pm (Fig. 15.12B). The data are in spotlight mode with X-band and HH and VV polarization. These data are single look complex formatted data.

The TanDEM-X system involves the coordinated operation of dual satellites flying in an adjacent configuration. The alteration constraints for the formation are: (i) the orbits ascending nodes; (ii) the angle between the perigees; (iii) the orbital eccentricities; and (iv) the phasing between the satellites. The observance of ocean currents is a vital facet of assessing climate change, and spaceborne SAR along-track interferometry (ATI) promises to make a significant contribution to this field. It will offer large-area, worldwide surface current measurements.

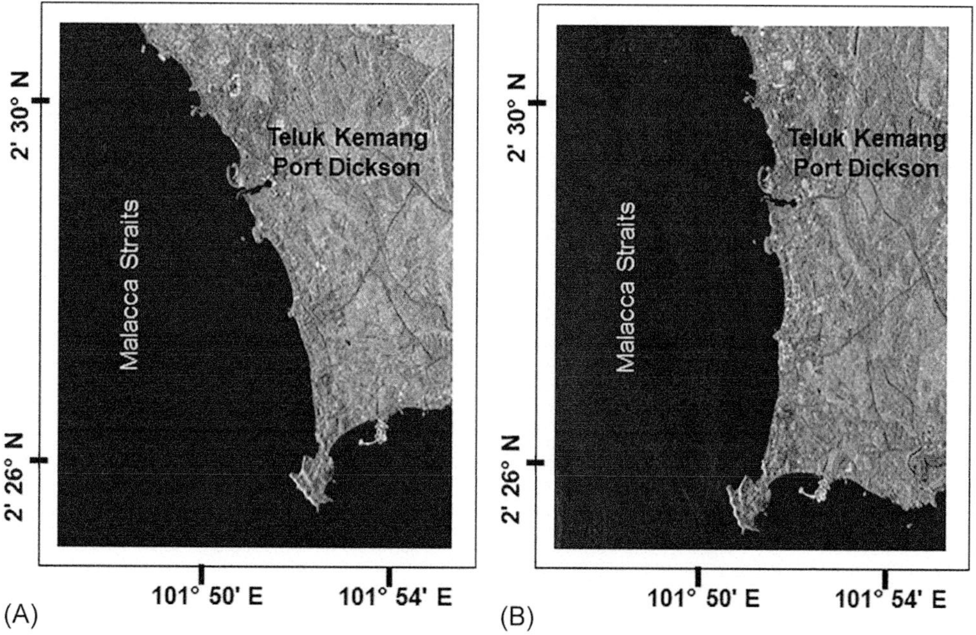

FIG. 15.12 TanDEM-X satellite (A) 7:27:17 am and (B) 19:20:06 pm.

According to Mittermayer and Runge [29], the velocity component of moving objects may be measured with ATI. The sensitivity of the instrument principally depends on the measuring device carrier frequency and consequently the effective time lag between the two measurements administered with two antennas and receiver chains. These parameters have to be tailored to the speed range of the objects of interest. High-speed objects like cars would require an extremely short time lag and would also need the two antennas to be separated by several meters.

The issue of mapping at relatively low velocity is often resolved by formations of SAR satellites that yield sufficiently sensitive ATI measurements [24]. In this study, the qHop algorithm relies on the ATI of TanDEM-X information. The TerraSAR-X and TanDEM-X satellites transmit from identical SAR instruments working at 9.65 GHz frequency (X-band). Throughout some devoted operations, both satellites are placed on exceeding link in a very special orbit configuration with a brief along-track baseline, which is providing a possibility for ocean current measurements. The data utilized in this study were acquired in the stripmap mode (SM), bistatic (TS-X active/TD-X passive) mode, and VV polarization. Consequently, these data have a swath width of 30 km as stripmap images are typically provided comprising an area of 30×50 km. The acquisition length in stripmap mode may be extended up to 1650 km (at 30 km width) [30, 31].

15.12 In situ measurement

The actual sea surface current speed and direction are collected by an Aquadopp 2 MHz current meter (Fig. 15.13). The surface current measurements, are acquired by using the Aquadopp 2 MHz current meter factory-made by Nortek AS (Fig. 15.13), the Scandinavian country. This sort of current meter is operated as a function of the Doppler based technology to measure surface currents at the deployment site (Fig. 15.14). The current meter device is intended with an electronic memory to store the sea surface information (velocity and direction) and an internal battery pack wherever it may be designed to record and store information internally for self-use.

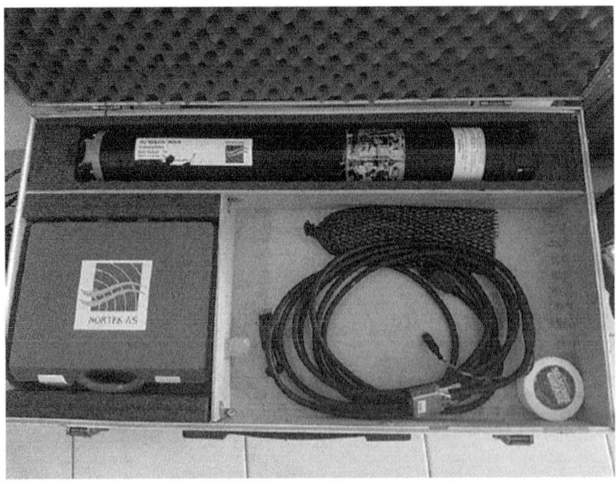

FIG. 15.13 Aquadopp 2 MHz current meter.

The Aquadopp 2MHz current meter was deployed on the coastal water of Teluk Kemang, Port Dickson, Malaysia on May 6, 2017 (Fig. 15.14). Two phases of data collection were carried out: (i) between 6:15am and 8:15am; and (ii) between 6:15pm and 8:15pm. The surface current data were measured in intervals of 2 h for both phases. The author then collected the current meter after the end of the TanDEM mission (Fig. 15.15).

FIG. 15.14 Aquadopp 2MHz current meter deployment.

FIG. 15.15 Current meter collected by the author after the end of the TanDEM-X satellite mission.

The in situ measurements are contacted along the coastal water of Teluk Kemang, Port Dickson, Malaysia, on May 6, 2017 (Fig. 15.16) using Aquadopp 2 MHz current meter.

15.13 Retrieving current from ATI TanDEM-X satellite data using qHop algorithm

The sea surface current velocities are simulated and modeled from the TanDEM-X data with VV polarization, with the simulation being performed along the range direction. The simulated velocity is taken across the location of the Aquadopp 2MHz current meter (Fig. 15.17). The test area is shown in Fig. 15.17, which is the inshore coastal water of the Malacca Straits, Malaysia. Fig. 15.17 shows that TanDEM-X data cross-section values increased with the increase of the incidence angle where the backscatter value is raised to −10 dB. The second curve is the result of the Doppler shift frequency. The curve shows that the Doppler shift frequency values fluctuated with the onshore value decreasing from 2 km to 5 km. The frequency value in the nearshore area was extremely low at 0.1 m/s. The spectral peak of the Doppler frequency is 0.04 with range frequency of −200 Hz. The average of the Doppler shift frequency in the onshore area was 0.01 Hz. This could be due to the low tide level of 0.3 m observed during in situ data collection.

The interferogram phase derived by direct ATI phase measurement involves a high level of noise and the fuzzy pattern of interferogram phase ranges between −0.9 and +0.9 degrees (Fig. 15.18). However, Fig. 15.19 shows an interferogram phase range between −0.8 and +0.8 degrees, derived from the qHop algorithm. A similar pattern is visible. This pattern signature represents current feature variations along the coastal waters. Demonstrably, the qHop algorithm provides interferogram clear phase without noise.

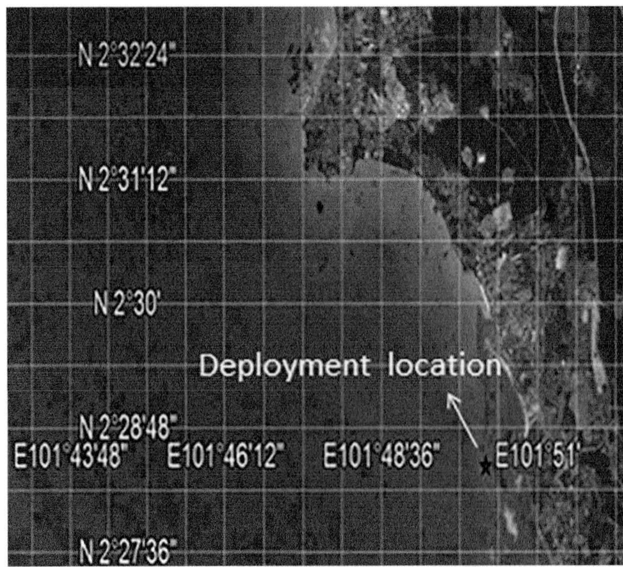

FIG. 15.16 Deployment site of current meter.

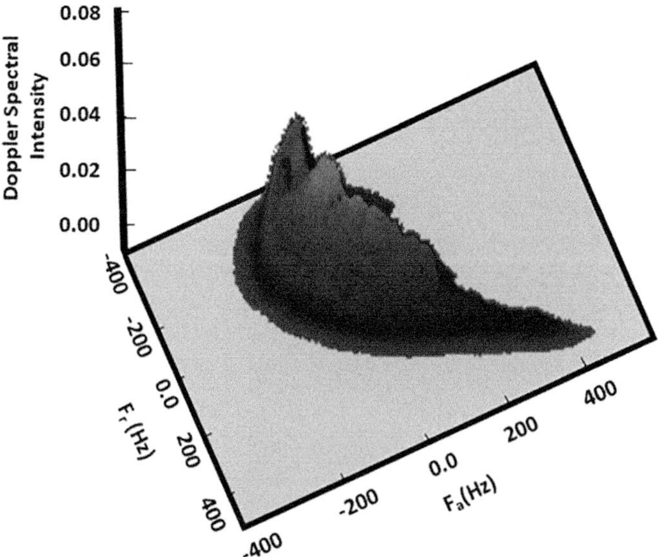

FIG. 15.17 Doppler spectra intensity of TanDEM-X data.

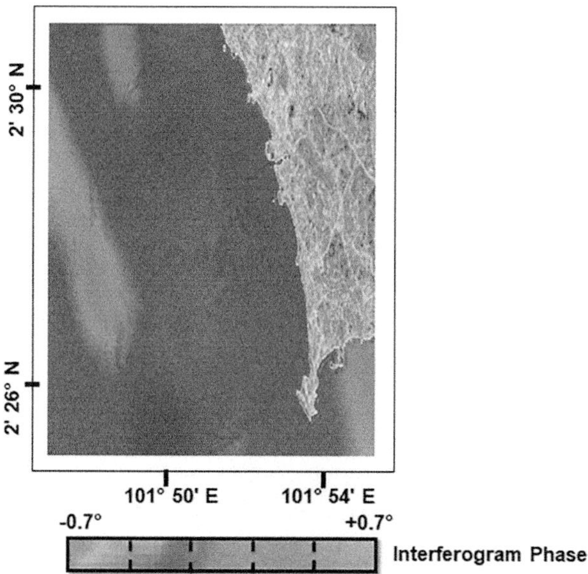

FIG. 15.18 Interferogram phase retrieved by direct ATI phase estimation.

The inverted interferogram phase can be used to compute the ATI Doppler sea surface current. The ATI Doppler shows a visible current pattern movement along the coastal waters with a minimum and a maximum speed of 0.05 m/s and 0.25 m/s, respectively (Fig. 15.20). In fact, an interferometric combination of the two images reveals phase alterations that are comparable to the backscatter variations of the Doppler frequency shift [24]. This speed conforms to the sea level differences of 0.4 m (Fig. 15.20).

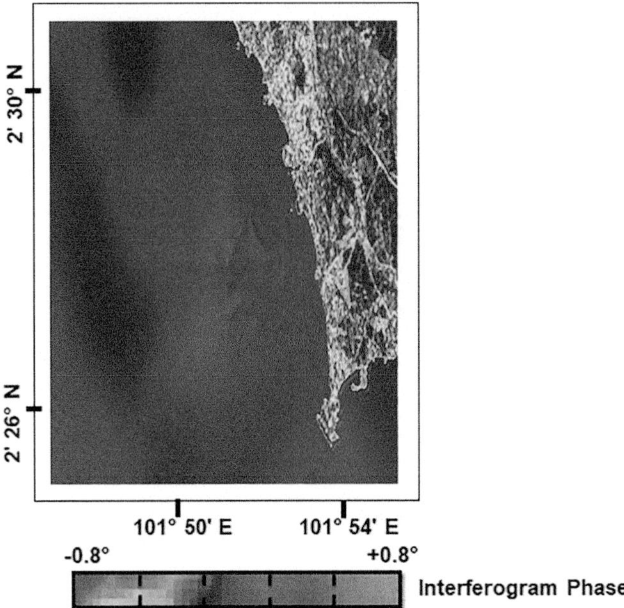

FIG. 15.19 Interferogram phase retrieved by qHop algorithm.

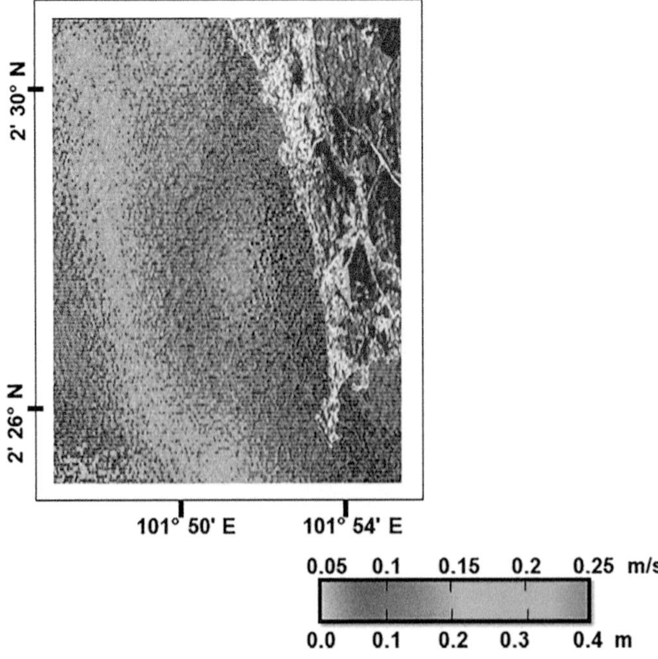

FIG. 15.20 Retrieving sea surface current and sea level variations from TanDEM-X data using the qHop algorithm.

Fig. 15.21 shows noteworthy correspondences between the consequence of sea surface flow speeds, calculated from TanDEM-X satellite data, and the consequence delivered in the in situ quantity. Fig. 15.21 demonstrates how the correlation coefficient alteration shows a direct correlation between the two different parameters that are modified. However, the failing of the emphasis is on forecasting of a single changing from the other, in correlation, the variation is on the grade to which a true model can designate the connection between two different measured parameters; for instance, in situ and TanDEM-X satellite measurements. There is a correlation between the retrieved sea surface flow and real in situ measured flow with r^2 of 0.83. Although this correlation is not faultless, it appears to have a reliable, direct association and resembles what one would guess when bearing in mind both sea surface flow simulations from satellite data and measurements in situ and then follow the hypothesis of normality.

Consequently, the qHop algorithm delivers actual sea surface current flow as well as it fits with real sea surface current measurements (Table 15.1) Clearly, the qHop algorithm gives a better performance than the direct ATI algorithm, with a lower P value of 0.00004 and RMSE of $\pm$0.00083 and maximumr^2 of 0.83 (Table 15.1). The rate of RMSE is generated owing to the Doppler frequency shift, which is caused by nonlinearity between ocean surface current and TanDEM-X data. The qHop solution has improved the accuracy of this study as it found the best optimal value for retrieving sea surface current from TanDEM-X SAR data.

In terms of accuracy, this study confirms the work done by Frasier and Camps [30], Myasoedov et al. [32], and Kudryavtsev et al. [33], showing that VV-polarization is an appropriate channel for retrieving sea surface current from SAR images. The accuracy of this work can be improved in the future based on the physical parameters of the quantum Hopfield algorithm, and also by using sequences of the TanDEM-X SAR data. On some occasions, the weak current

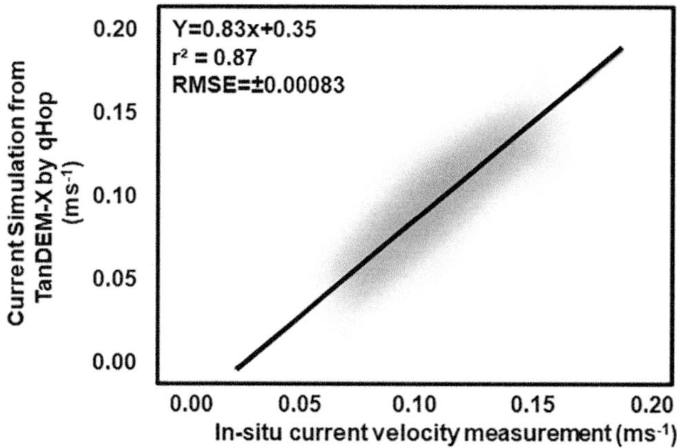

FIG. 15.21 Validation of qHop algorithm with in situ measurement.

TABLE 15.1 Statistical regression of current meter sea surface current and that retrieved by direct ATI technique and qHop algorithm.

Methods	r^2	RMSE ($\pm$m s^{-1})	P
Direct ATI technique	0.49	0.3	.0087
qHop algorithm	0.87	0.00083	.00004

can affect the accuracy of the algorithm. The fluctuation dynamic current flows are required to improve the accuracy and this can be achieved by acquiring sequences of TanDEM data.

15.14 Four-dimensional ATI quantum algorithm for wave-current interaction

The key innovation of this chapter is to develop a novel formula for the 4-D along-track interferometry phase unwrapping manipulating Hamming graph. The first idea is to reconstruct the fourth dimension of wave-current interaction by optimization of 4-D along-track interferometry. In this regard, this chapter postulates that 4-D can be inferred from the 3-D phase unwrapping of along-track interferometry. Along-track interferometry is technique used to measure the shifts of phase stability due to the Doppler frequency of sea surface dynamic fluctuations.

Following Marghany [34], the 4-D wrapped phase is calculated as:

$$\phi_w(x,y,z,t) = \text{mod}\left[\eta \sin\left(\frac{(x+y+z+t)2\pi}{\lambda}\right) + G(\sigma)\right] \tag{15.38}$$

Eq. (15.38) reveals the existence of a 4-D Cartesian coordinate space x, y, z, t of the predictable wrapped phase ϕ_{ATI}. In addition, "mod" demonstrates the modulo operator wrapping the data within the range of $0 \leq \phi_{ATI} \leq 2\pi$, which is governed by Gaussian additive $G(\sigma)$ for noise reduction through the procedures of 4-D wrapped phase encoding in 3-D as a function of standard deviation σ. Then the wrapped phase ϕ_{ATI} is associated with the accurate phase by depleting the Laplacian-based algorithm ∇^2. In this circumstance, the Laplacian of the real phase ϕ_{ATI} is stated in terms of the wrapped phase, which is gauged exploiting:

$$\nabla^2(\sin\Psi_{ATI}) = \frac{\nabla^2\phi_{ATI} + \sin\Psi_w \nabla^2(\cos\Psi_{ATI})}{\cos\Psi_{ATI}} \tag{15.39}$$

Using an inverse of ∇^{-2}, the real phase ϕ'_{ATI} can be approached in 4-D by manufacturing quantum Fourier-domain accelerative and ∇^{-2}, as expressed by [34]:

$$\phi'(i,j,k,t)_{ATI_Q} = \frac{1}{2^N}\sum_{J=0}^{2^n-1}\sum_{k=0}^{2^n-1} e^{\frac{-2\pi ijk}{2^n}}|j\rangle$$

$$\left[\frac{\sum_k \tilde{\alpha}_k|k\rangle\left[\cos\phi_{ATI}\left(\frac{1}{2^N}\sum_{J=0}^{2^n-1}\sum_{k=0}^{2^n-1} e^{\frac{-2\pi ijk}{2^n}}\right)|j\rangle\right]\left[|p^2+q^2+r^2+s^2\rangle\sum_k \tilde{\alpha}_k|k\rangle(\sin\phi_{ATI})\right]}{p^2+q^2+r^2+s^2}\right]$$

$$-\frac{1}{2^N}\sum_{J=0}^{2^n-1}\sum_{k=0}^{2^n-1} e^{\frac{-2\pi ijk}{2^n}}|j\rangle$$

$$\left[\frac{\sum_k \tilde{\alpha}_k|k\rangle\left[\sin\phi_{ATI}\left(\frac{1}{2^N}\sum_{J=0}^{2^n-1}\sum_{k=0}^{2^n-1} e^{\frac{-2\pi ijk}{2^n}}\right)|j\rangle\right]\left[(p^2+q^2+r^2+s^2)_Q\sum_k \tilde{\alpha}_k|k\rangle(\cos\phi_{ATI})\right]}{|p^2+q^2+r^2+s^2\rangle}\right] \tag{15.40}$$

Eq. (15.34) speculates the transfer of 2-D ATI phase wrapping into 4-D time-domain coordinates by the collaboration of sin and cosine transforms through new transform coordinates $|p^2+q^2+r^2+s^2\rangle$. Eq. (15.34) involves three sorts of operations: quantum forward and inverse cosine transforms, trigonometric operations, and the masking expression $|p^2+q^2+r^2+s^2\rangle$. Therefore, $\widetilde{\alpha}_k$ is mathematically expressed as:

$$\widetilde{\alpha}_k \equiv N^{-1}\sum_{j=1}^{N-1} e^{\frac{2\pi ijk}{N}}\alpha_j \tag{15.40.1}$$

Eq. (15.40.1) reveals that in quantum computing, loading the coefficient (α_j) for the computation bases is not very simple. In addition, there is no effective way to retrieve the coefficient (αk) of the computation bases after the transformation. The measurement returns the computation basis, not the coefficient.

Following Marghany [34], the relative ATI phase difference can be associated with sea surface elevation displacement, significant wave height, and radial sea surface current movements through the sensitivity vector found in the along-track interferometry in TanDEM; this can be expressed in 4-D as:

$$\begin{pmatrix}\partial\Phi_1\\ \partial\Phi_2\\ \partial\Phi_3\\ \partial\Phi_4\end{pmatrix}=\frac{2\pi}{\lambda}\begin{pmatrix}\left(\frac{\phi_{ATI}}{4\pi\cdot B_{ATI}\sin\theta(\lambda\cdot V_S)^{-1}}\right)_{1i} & \left(\frac{\phi_{ATI}}{4\pi\cdot B_{ATI}\sin\theta(\lambda\cdot V_S)^{-1}}\right)_{1j} & \left(\frac{\phi_{ATI}}{4\pi\cdot B_{ATI}\sin\theta(\lambda\cdot V_S)^{-1}}\right)_{1k} & \left(\frac{\phi_{ATI}}{4\pi\cdot B_{ATI}\sin\theta(\lambda\cdot V_S)^{-1}}\right)_{1p}\\ \left(\frac{\phi_{ATI}}{4\pi\cdot B_{ATI}\sin\theta(\lambda\cdot V_S)^{-1}}\right)_{2i} & \left(\frac{\phi_{ATI}}{4\pi\cdot B_{ATI}\sin\theta(\lambda\cdot V_S)^{-1}}\right)_{2j} & \left(\frac{\phi_{ATI}}{4\pi\cdot B_{ATI}\sin\theta(\lambda\cdot V_S)^{-1}}\right)_{2k} & \left(\frac{\phi_{ATI}}{4\pi\cdot B_{ATI}\sin\theta(\lambda\cdot V_S)^{-1}}\right)_{2t}\\ \left(\frac{\phi_{ATI}}{4\pi\cdot B_{ATI}\sin\theta(\lambda\cdot V_S)^{-1}}\right)_{3i} & \left(\frac{\phi_{ATI}}{4\pi\cdot B_{ATI}\sin\theta(\lambda\cdot V_S)^{-1}}\right)_{3j} & \left(\frac{\phi_{ATI}}{4\pi\cdot B_{ATI}\sin\theta(\lambda\cdot V_S)^{-1}}\right)_{3k} & \left(\frac{\phi_{ATI}}{4\pi\cdot B_{ATI}\sin\theta(\lambda\cdot V_S)^{-1}}\right)_{3t}\\ \left(\frac{\phi_{ATI}}{4\pi\cdot B_{ATI}\sin\theta(\lambda\cdot V_S)^{-1}}\right)_{4i} & \left(\frac{\phi_{ATI}}{4\pi\cdot B_{ATI}\sin\theta(\lambda\cdot V_S)^{-1}}\right)_{4j} & \left(\frac{\phi_{ATI}}{4\pi\cdot B_{ATI}\sin\theta(\lambda\cdot V_S)^{-1}}\right)_{4k} & \left(\frac{\phi_{ATI}}{4\pi\cdot B_{ATI}\sin\theta(\lambda\cdot V_S)^{-1}}\right)_{4t}\end{pmatrix}\begin{pmatrix}p\\ q\\ r\\ s\end{pmatrix}(t_n) \tag{15.41}$$

In Eq. (15.34), $\left(\frac{\phi_{ATI}}{4\pi\cdot B_{ATI}\sin\theta(\lambda\cdot V_S)^{-1}}\right)$ is the along-track sea current movements across orthogonal components of p, q, r, and s, in i, j, k, and t, respectively. Following Kojima [35], sea surface height distribution can be computed from ATI as the inverse of nonlinear sea surface velocity is given by:

$$H_S \approx \left[\sum_{m=1}^{\infty} U_r\left(x - U_r V_S^{-1}\sqrt{d^2+h^2}\right)\right]^{-1} \tag{15.42}$$

where d is the distance in the ground range direction and h is the flight altitude.

The phase 4-D unwrapping based on the 3-D wave height distribution and sea-level variation can be given by:

$$\begin{pmatrix}\partial\Phi_1\\ \partial\Phi_2\\ \partial\Phi_3\\ \partial\Phi_4\end{pmatrix}=\frac{2\pi}{\lambda}\begin{pmatrix}(H_s,\partial\zeta,U_r)_{1i} & (H_s,\partial\zeta,U_r)_{1j} & (H_s,\partial\zeta,U_r)_{1k} & (H_s,\partial\zeta,U_r)_{1p}\\ (H_s,\partial\zeta,U_r)_{2i} & (H_s,\partial\zeta,U_r)_{2j} & (H_s,\partial\zeta,U_r)_{2k} & (H_s,\partial\zeta,U_r)_{2t}\\ (H_s,\partial\zeta,U_r)_{3i} & (H_s,\partial\zeta,U_r)_{3j} & (H_s,\partial\zeta,U_r)_{3k} & (H_s,\partial\zeta,U_r)_{3t}\\ (H_s,\partial\zeta,U_r)_{4i} & (H_s,\partial\zeta,U_r)_{4j} & (H_s,\partial\zeta,U_r)_{4k} & (H_s,\partial\zeta,U_r)_{4t}\end{pmatrix}\begin{pmatrix}p\\ q\\ r\\ s\end{pmatrix}(t_n) \tag{15.43}$$

Consistent with Marghany [16, 34], the 4-D phase unwrapping can mathematically be expressed as:

$$\sum_{p,q,r,s} W^{p}_{p,q,r,s}\left|\left|\Delta\phi^{p}_{p,q,r,s}\right\rangle - \left|\Delta\psi^{p}_{p,q,r,s}\right\rangle\right|^{L} + \sum_{p,q,r,s} W^{q}_{p,q,r,s}\left|\left|\Delta\phi^{q}_{p,q,r,s}\right\rangle - \left|\Delta\psi^{q}_{p,q,r,s}\right\rangle\right|^{L}$$
$$+ \sum_{p,q,r,s} W^{r}_{p,q,r,s}\left|\left|\Delta\phi^{r}_{p,q,r,s}\right\rangle - \left|\Delta\psi^{r}_{p,q,r,s}\right\rangle\right|^{L} + \sum_{p,q,r,s} W^{s}_{p,q,r,s}\left|\left|\Delta\phi^{s}_{p,q,r,s}\right\rangle - \left|\Delta\psi^{s}_{p,q,r,s}\right\rangle\right|^{L} \quad (15.44)$$

where $|\Delta\phi\rangle$ and $|\Delta\psi\rangle$ are the quantum unwrapped and wrapped phase differences in p, q, r, and s, respectively, and W represents the user-defined weights.

The summations in Eq. (15.44) are carried out in both p, q, r, and s directions overall p, q, r, and ts, respectively [34]. The phase unwrapping problem can be solved by L^2 norm second differences reliability criterion. Consequently, L^2 can be extended into 4-D as:

$$\mathrm{Re}(|O\rangle) = \sum_{n=1}^{40} \sqrt{(|O\rangle \otimes |n_{+}\rangle)^2 + (|O\rangle \otimes |n_{-}\rangle)^2} \quad (15.45)$$

where $\mathrm{Re}(|O\rangle)$ is the reliability of the pixel $|O\rangle$, $|n_{+}\rangle$, a neighbor on the 4-D hypercube of neighbors to $|O\rangle$, and $|n_{-}\rangle$ the opposite neighbor.

In four dimensions, each voxel has 80 neighbors, resulting in a reliable criterion that is the sum of 40 measurements. To construct 4-D hyperspace or hypercube from physical sea surface parameters that are retrieved from 4-D phase unwrapping, a **Hamming graph formula is implemented as** a quantum walk as:

$$G_{H_C} \oplus H_m \left|\phi_{G_{H_C} \oplus H_m}(t)\right\rangle = \left|\phi_{G_{H_C}}(t)\right\rangle \otimes \left|\phi_{H_m}(t)\right\rangle \quad (15.46)$$

where $\phi_{G_{HC}}$ and ϕ_{H_m} are quantum phase unwrapping walks on G_{H_C} and H_m starting on vertices v_{m_1} and v_{m_2}, respectively. Eq. (15.46) is the key ingredient to construct hypercube ATI interferometry through the Hamming graphs. In general, the Hamming graph delivers tight characterization of quantum uniform mixing. In quantum mechanics, the natural approach to syndicate dual systems is across the tensor product $\otimes$. In this understanding, the Cartesian graph artifact provides a comparable function for quantum walks of the phase unwrapping [34].

The finishing form of the quantum walk for optimal exploration $2r$ can, consequently, be articulated by the blinking sequence of the unitary operators $U''^{(+)}_{m}$ as:

$$\left|\phi^{(e)}{}_{G_{H_C} \oplus H_m}(2r)\right\rangle = \left(U^{(+)}_{m} \sum_{\alpha,\vec{q}} \left|\left|\alpha, \vec{q} \oplus \vec{e}_{\alpha}\right\rangle\right|\alpha, \vec{q}\right\rangle\left|C''^{(+)}\right)^{r} \left|\phi^{e}{}_{G_{H_C}}(t)\right\rangle \otimes \left|\phi_{H_m}(t)\right\rangle \quad (15.47)$$

where $C''^{(+)}$ is the coin operator, which acts on the total Hilbert space $H^{C} \otimes H^{V}$; the $\sum_{\alpha,\vec{q}} \left|\left|\alpha, \vec{q} \oplus \vec{e}_{\alpha}\right\rangle\right|\alpha, \vec{q}\right\rangle\Big|$ is a propagator of a quantum walk across the hypercube vertices v_{m1} and v_{m2} with direction α through the vertex edge $\vec{e}$; $\oplus$ signifies the bitwise addition modulo 2 operator; and $\vec{q}$ is the Hamming weight of an integer which is the number of 1's in its binary string.

15.15 4-D visualization of wave-current sea level interactions

The phase unwrapping is the keystone for both cross-track and along-track interferometry. The 4-D phase unwrapping based on the Hamming weight graph is implemented, and Fig. 15.22 demonstrates the phase unwrapping development from 2-D to 4-D. 4-D phase unwrapping in the water body promises to deliver better sea surface pattern fluctuations than 2-D and 3-D phase unwrapping. Indeed, 4-D phase unwrapping exploiting Hamming weight and a quantum walk algorithm is beyond brilliant compared to 2-D and 3-D phase unwrapping. Mainly, 4-D phase unwrapping preserves the perfect phase cycle—the affinity of the entire cycle of phase along-track interferometry patterns. Moreover, the qHop and Hamming weight algorithms have minimized the error in the interferogram cycle because of the low coherence in water zones and along the coastline due to high decorrelation.

Fig. 15.23B shows the 4-D visualization derived from along-track interferometry. The 4-D visualization distinguishes between the current boundary, ocean wave height, and sea-level variation. Nonetheless, 3-D map cannot reconstruct a clear sea surface current boundary, but

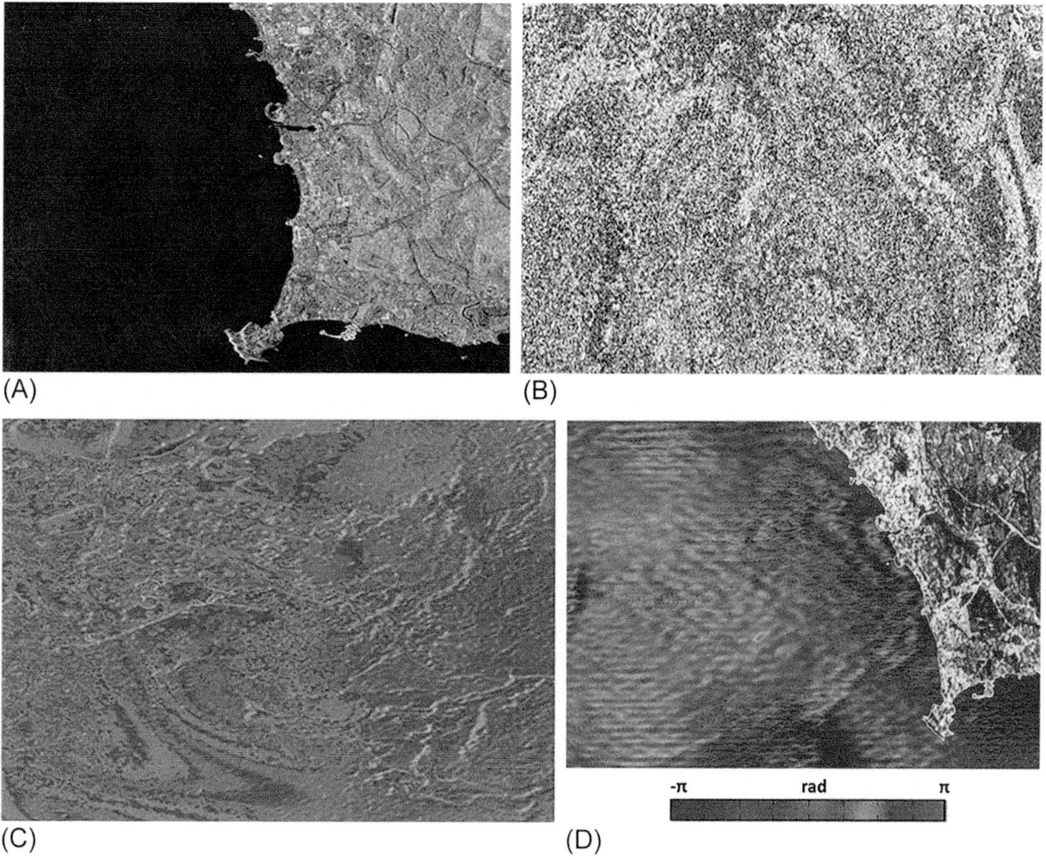

FIG. 15.22 Phase unwrapping (A) original data; (B) 2-D; (C) 3-D; and (D) 4-D.

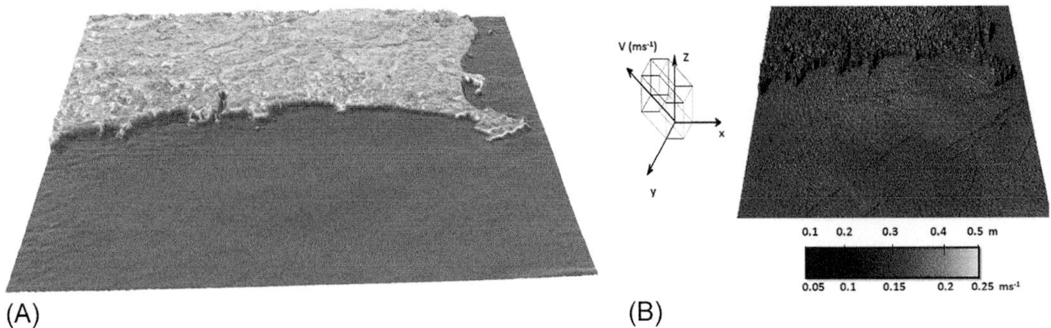

FIG. 15.23 Comparison between (A) 3-D and (B) 4-D visualization, created by quantum along-track interferometry.

it just can well-construct 3-D urban and land use features (Fig. 15.23A). In this regard, coding 3-D of TanDEM-X data into 4-D can visualize much information about coastal features, i.e., wave height peaks, clear sea surface current boundary, and ship peak signatures.

In fact, the automatic distinction of oil spill dark patches from the surrounding environment could ally with $|NOON\rangle$. In this circumstance, $|NOON\rangle$ is used to set dual state sea surface current boundary and wave height variation states in the 4-D image. The fourth dimension, consequently, demonstrates the trajectory movement of sea surface current flow along the coastal water. The maximum wave height delivered by ATI interferometry is 0.5 m, which corresponds to the trajectory movement of $0.3\,\text{m}\,\text{s}^{-1}$. Consequently, the quantum interferometry is considered as a deterministic algorithm, which is chosen to regulate a triangulation locally between twofold assorted pixels in TanDEM-X data. It could precisely be articulated as a complete three-dimensional high-resolution image (Fig. 15.23A). Quantum interferometry records every tiny detail of sea surface wave variation and sea surface current boundary flow. The Hamming graph delivers a wave-current interaction scenario, which can be viewed in neither 2-D nor 3-D in TanDEM-X data. This confirms that quantum interferometry based on the Hamming graph is a robust process for gauging movements of the order of a phase shift fluctuations owing to dynamic changes of the Doppler shift. Consequently, constructing a quantum interferometry based on the Hamming graph can contribute in recovering the matter of specular reflection owing to the Doppler frequency shift, in addition to permitting the nonlinearity of wave imaging in SAR and the nonlinearity of wave-current interaction.

The visualization of the wave-current interaction is extremely sharp by the reconstruction of the 4-D phase unwrapping by the Hamming graph algorithm as rooted in quantum processing. These results deliver an accurate geometry morphology of both wave height propagation and sea surface current boundary, which is indicated by the interference between current movement and wave height transformation in coastal waters.

15.16 Relativistic quantum 4-D of sea surface reconstruction in TanDEM data

In Einstein's theory of relativity, space-time is 4-D since a fourth dimension equivalent to time operates in 3-D space. In previous chapters, it was noted that, for fluid dynamic motion to have occurred, it must be constrained in space and time. On this understanding, the fluid flow is not only through space alone, but through 4-D space-time. The concept of space-time is recognized to be 4-D as a function of the Lorentz transformations. For instance, we have demonstrated previously the relativity impact on velocity bunching owing to the relativity of the Doppler frequency shift. In this circumstance, the dynamic sea surface imagined in SAR data are relocated from an inertial reference frame (x,y,z) for instance, to another (p,q,r,s) [34, 36–38]. Needless to say, the 4-D quantum interferometry could be a new approach retrieving a full scenario of ocean dynamic features in SAR images due to encompassing $|NOON\rangle$, Hamming graph, quantum walk search, and quantum hypercube algorithms. It is a promising technique to transfer or convert 2-D and 3-D images to a 4-D image. This work is a validation of Marghany [34].

Fig. 15.24 reveals that with different viewing angles, different features can be visualized in 4-D. The 4-D wave height, ship footprint, and the current boundary are perceived at 0-degree viewing angles with scattered short bright patches. At 90- and 360-degree viewing angles, wave-current interaction is recognized, and the turbulent movements along coastal water owing to wave-current interaction with extremely sharp crests are clearly distinguished at 180-degree viewing angles.

From the point of view of the hypercube, the edge waves are distinguished as shown in Fig. 15.25. This is confirmed with different viewing angles as every side of the hypercube shows different features of wave-current interaction. More specifically, as the current boundary and wave pattern are developed up and down, right and left in TanDEM-X data, the fourth dimension, which resembles tesseracts, lengthens can reconstruct the perfect pattern of wave-current interaction. Last but not least, as a consequence, 4-D develops a cubic mass of identical trivial tesseracts. In this view, when the tesseract is established in the global space, it achieves on the surfaces enfolding the upward and the right and left dimensions. From the point of view of duality, the current boundary movement and wave pattern turn into extending on one side of the hypercube and topological wave-current interaction feature

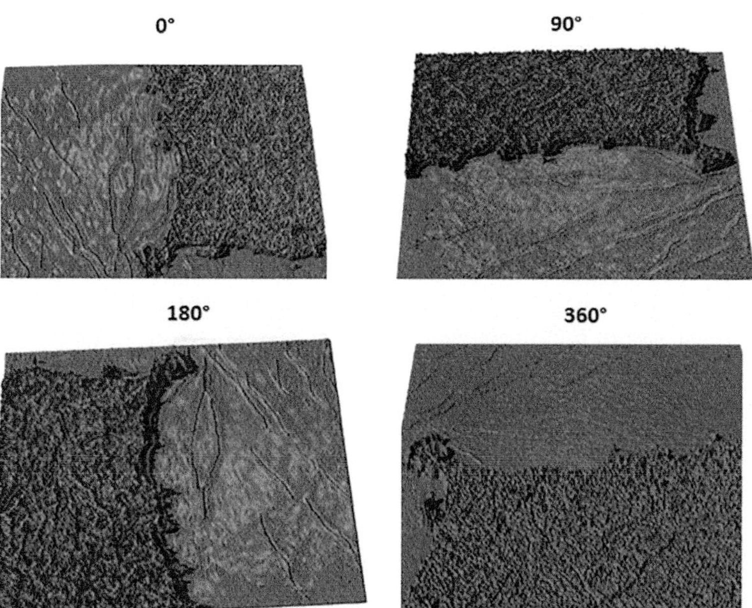

FIG. 15.24 Different view angels of 4-D quantum interferometry.

FIG. 15.25 Hypercube of 4-D wave-current interaction.

on the other side. In this view, the wave-current interaction features are deliberated as the releasing of dual compositions.

Moreover, the performance of a quantum random walk is rooted in the propagator across the n-dimensional of hypercube vertices. The base of the 4-D construction of wave-current interaction is achieved by a quantum random walk search self-loop, which determines the edges of hypercube vertices. The final version of the quantum walk for optimal search is based on a sequence of the unitary operators, which assist to reconstruct a 4-D wave-current interaction image. Consequently, for a quantum walk search with wave-current interaction, the querying operator inverts the phase of the bases at each step where boundary current patches are positioned [38]. This phase inversion creates the probability that amplitudes of the wave-current interaction vertices fluctuate more violently across the hypercube to form a 4-D wave-current boundary interaction.

The various viewing angles revolve a 4-D image into the diversity space of the perfect reconstruction scenario of wave-current duality interaction. In this regard, the meaning of the period or velocity as the presence of the fourth dimension is that period and space can spin into each other in a specific quantum interferometry algorithm. In other words, four-dimensional twists are precisely the revisions of space and time or speed that are dealt with by special relativity. In this understanding, space and time have advanced in an absolute ruling and lined by relativity.

Finally, such a study of synthetic aperture radar imaging of ocean dynamics cannot be the separated from or discount quantum mechanism. As we can see, quantum mechanics are involved from the beginning of the photon signal being radiated from antenna to the image processing that encompasses ocean dynamic retrieving parameters or reconstruction from 2-D to 4-D images.

References

[1] Hariharan P. Basics of interferometry. Elsevier; 2010.

[2] Massonnet D, Feigl KL. Radar interferometry and its application to changes in the Earth's surface. Rev Geophys 1998;36(4):441–500.

[3] Rosen PA, Hensley S, Joughin IR, Li FK, Madsen SN, Rodriguez E, Goldstein RM. Synthetic aperture radar interferometry. Proc IEEE 2000;88(3):333–82.
[4] Rao KS, Al-Jassar HK. Error analysis in the digital elevation model of Kuwait desert derived from repeat pass synthetic aperture radar interferometry. J Appl Remote Sens 2010;4(1):043546.
[5] Lu Z, Crane M, Kwoun OI, Wells C, Swarzenski C, Rykhus R. C-band radar observes water level change in swamp forests. EOS Trans Am Geophys Union 2005;86(14):141–4.
[6] Ramsey III E. Radar remote sensing of wetlands. Ann Arbor Press; 1999.
[7] Campbell BA. Radar remote sensing of planetary surfaces. Cambridge University Press; 2002.
[8] Kwoun OI, Lu Z. Multi-temporal RADARSAT-1 and ERS backscattering signatures of coastal wetlands in southeastern Louisiana. Photogramm Eng Remote Sens 2009;75(5):607–17.
[9] Marghany M. Hybrid genetic algorithm of interferometric synthetic aperture radar for three-dimensional coastal deformation. In: SoMeT. vol. 29. 2014. p. 116–31.
[10] Marghany M. Three-dimensional visualisation of coastal geomorphology using fuzzy B-spline of dinsar technique. Int J Phys Sci 2011;6(30):6967–71.
[11] Marghany M. Simulation of three-dimensional of coastal erosion using differential interferometric synthetic aperture radar. Global NEST J 2014;16(1):80–6.
[12] Zebker HA, Werner CL, Rosen PA, Hensley S. Accuracy of topographic maps derived from ERS-1 interferometric radar. IEEE Trans Geosci Remote Sens 1994;32(4):823–36.
[13] Marghany M. Three dimensional coastline deformation from Insar Envisat Satellite data, In: International conference on computational science and its applications 2013 Jun 24Berlin, Heidelberg: Springer; 2013. p. 599–610.
[14] Marghany M. Three-dimensional coastal geomorphology deformation modelling using differential synthetic aperture interferometry. Z Naturforsch A J Phys Sci 2012;67(6):419.
[15] Marghany M. DInSAR technique for three-dimensional coastal spit simulation from radarsat-1 fine mode data. Acta Geophys 2013;61(2):478–93.
[16] Marghany M. Four-dimensional earthquake deformation using ant colony based Pareto algorithm. Commun Appl Sci 2019;7(1).
[17] van Genderen J, Marghany M. A three-dimensional sorting reliability algorithm for coastline deformation monitoring, using interferometric data, In: IOP conference series: Earth and environmental sciencevol. 18, No. 1. IOP Publishing; 2014. p. 012116.
[18] Ferretti A, Monti-Guarnieri AV, Prati C, Rocca F, Massonnet D. INSAR principles B. ESA Publications; 2007.
[19] Xu W, Cumming I. A region-growing algorithm for InSAR phase unwrapping. IEEE Trans Geosci Remote Sens 1999;37(1):124–34.
[20] Bechor NB, Zebker HA. Measuring two-dimensional movements using a single InSAR pair. Geophys Res Lett 2006;33(16).
[21] Ferretti A, Prati C, Rocca F. Multibaseline InSAR DEM reconstruction: the wavelet approach. IEEE Trans Geosci Remote Sens 1999;37(2):705–15.
[22] Imel DD, Hensley S, Pollard B, Chapin E, Rodriguez E. AIRSAR along-track interferometry, In: Inchez 4th European conference on synthetic aperture radarvol. 117. 2002.
[23] Suchandt S, Runge H. Along-track interferometry using TanDEM-X: first results from marine and land applications, In: EUSAR 2012; 9th European conference on synthetic aperture radarVDE; 2012. p. 392–5.
[24] Romeiser R, Johannessen J, Chapron B, Collard F, Kudryavtsev V, Runge H, Suchandt S. Direct surface current field imaging from space by along-track InSAR and conventional SAR. In: Oceanography from space. Dordrecht: Springer; 2010. p. 73–91.
[25] Lanzagorta M. Quantum radar. Synth Lect Quantum Comput 2011;3(1):1–39.
[26] Behrman EC, Nash LR, Steck JE, Chandrashekar VG, Skinner SR. Simulations of quantum neural networks. Inform Sci 2000;128(3–4):257–69.
[27] Shukla A, Vedula P. Trajectory optimization using quantum computing. J Glob Optim 2019;1–27.
[28] Rucker R. Geometry, relativity and the fourth dimension. Courier Corporation; 2012.
[29] Mittermayer J, Runge H. Conceptual studies for exploiting the TerraSAR-X dual receive antenna, In: IGARSS 2003. 2003 IEEE international geoscience and remote sensing symposium. Proceedings (IEEE Cat. No. 03CH37477) 2003 Jul 21vol. 3. IEEE; 2003. p. 2140–2.
[30] Frasier SJ, Camps AJ. Dual-beam interferometry for ocean surface current vector mapping. IEEE Trans Geosci Remote Sens 2001;39(2):401–14.
[31] Gebhardt C, Pleskachevsky A, Rosenthal W, Lehner S, Hoffmann P, Kieser J, Bruns T. Comparing wavelengths simulated by the coastal wave model CWAM and TerraSAR-X satellite data. Ocean Model 2016;103:133–44.

[32] Myasoedov A, Kudryavtsev V, Chapron B. Dual co-polarized SAR imaging of the ocean surface phenomena, In: Conference proceedings of 2013 Asia-Pacific conference on synthetic aperture radar (APSAR)IEEE; 2013. p. 455–8.
[33] Kudryavtsev V, Kozlov I, Chapron B, Johannessen JA. Quad-polarization SAR features of ocean currents. J Geophys Res Oceans 2014;119(9):6046–65.
[34] Marghany M. Synthetic aperture radar imaging mechanism for oil spills. Gulf Professional Publishing; 2019.
[35] Ghiglia DC, Pritt MD, Unwrapping TD. Theory, algorithms, and software. New York: A Wiley-Interscience Publication; 1998.
[36] Kojima S. Numerical study for ocean wave measurement by high-resolution along-track interferometric Sar, In: IGARSS 2018–2018 IEEE international geoscience and remote sensing symposium 2018 Jul 22IEEE; 2018. p. 3208–11.
[37] Miller MA. Regge calculus as a fourth-order method in numerical relativity. Classical Quantum Gravity 1995;12 (12):3037.
[38] Moghaddam FF, Moghaddam RF, Cheriet M. Curved space optimization: A random search based on general relativity theory. arXiv; 2012 preprint arXiv:1208.2214.

Index

Note: Page numbers followed by *f* indicate figures and *t* indicate tables.

C

D

E

F

P

Q

R

U

V

W

Y

Printed in the United States
By Bookmasters